AF869529

Springer Optimization and Its Applications

Volume 177

Series Editors

Panos M. Pardalos, *University of Florida, Gainesville, FL, USA*

My T. Thai, *University of Florida, Gainesville, FL, USA*

Honorary Editor

Ding-Zhu Du, *University of Texas at Dallas, Dallas, TX, USA*

Advisory Editors

Roman V. Belavkin, *Middlesex University, London, UK*

John R. Birge, *University of Chicago, Chicago, IL, USA*

Sergiy Butenko, *Texas A&M University, College Station, TX, USA*

Vipin Kumar, *University of Minnesota, Minneapolis, MN, USA*

Anna Nagurney, *University of Massachusetts, Amherst, MA, USA*

Jun Pei, *Hefei University of Technology, Hefei, China*

Oleg Prokopyev, *University of Pittsburgh, Pittsburgh, PA, USA*

Steffen Rebennack, *Karlsruhe Institute of Technology, Karlsruhe, Germany*

Mauricio Resende, *Amazon, Seattle, WA, USA*

Tamás Terlaky, *Lehigh University, Bethlehem, PA, USA*

Van Vu, *Yale University, New Haven, CT, USA*

Michael N. Vrahatis, *University of Patras, Patras, Greece*

Guoliang Xue, *Arizona State University, Tempe, AZ, USA*

Yinyu Ye, *Stanford University, Stanford, CA, USA*

Aims and Scope

Optimization has continued to expand in all directions at an astonishing rate. New algorithmic and theoretical techniques are continually developing and the diffusion into other disciplines is proceeding at a rapid pace, with a spot light on machine learning, artificial intelligence, and quantum computing. Our knowledge of all aspects of the field has grown even more profound. At the same time, one of the most striking trends in optimization is the constantly increasing emphasis on the interdisciplinary nature of the field. Optimization has been a basic tool in areas not limited to applied mathematics, engineering, medicine, economics, computer science, operations research, and other sciences.

The series **Springer Optimization and Its Applications (SOIA)** aims to publish state-of-the-art expository works (monographs, contributed volumes, textbooks, handbooks) that focus on theory, methods, and applications of optimization. Topics covered include, but are not limited to, nonlinear optimization, combinatorial optimization, continuous optimization, stochastic optimization, Bayesian optimization, optimal control, discrete optimization, multi-objective optimization, and more. New to the series portfolio include Works at the intersection of optimization and machine learning, artificial intelligence, and quantum computing.

Volumes from this series are indexed by Web of Science, zbMATH, Mathematical Reviews, and SCOPUS.

More information about this series at http://www.springer.com/series/7393

Stefan M. Stefanov

Separable Optimization

Theory and Methods

Second Edition

Springer

Stefan M. Stefanov
Department of Mathematics
Faculty of Mathematics and Natural Sciences
South-West University Neofit Rilski
Blagoevgrad, Bulgaria

ISSN 1931-6828 ISSN 1931-6836 (electronic)
Springer Optimization and Its Applications
ISBN 978-3-030-78400-3 ISBN 978-3-030-78401-0 (eBook)
https://doi.org/10.1007/978-3-030-78401-0

Mathematics Subject Classification: 90-02, 90C08, 90C15, 90C25, 90C30

Originally published with the title: Separable Programming
1st edition: © Springer Science+Business Media Dordrecht 2001
2nd edition: © The Editor(s) (if applicable) and The Author(s), under exclusive license to Springer Nature Switzerland AG 2021
This work is subject to copyright. All rights are solely and exclusively licensed by the Publisher, whether the whole or part of the material is concerned, specifically the rights of translation, reprinting, reuse of illustrations, recitation, broadcasting, reproduction on microfilms or in any other physical way, and transmission or information storage and retrieval, electronic adaptation, computer software, or by similar or dissimilar methodology now known or hereafter developed.
The use of general descriptive names, registered names, trademarks, service marks, etc. in this publication does not imply, even in the absence of a specific statement, that such names are exempt from the relevant protective laws and regulations and therefore free for general use.
The publisher, the authors and the editors are safe to assume that the advice and information in this book are believed to be true and accurate at the date of publication. Neither the publisher nor the authors or the editors give a warranty, expressed or implied, with respect to the material contained herein or for any errors or omissions that may have been made. The publisher remains neutral with regard to jurisdictional claims in published maps and institutional affiliations.

This Springer imprint is published by the registered company Springer Nature Switzerland AG
The registered company address is: Gewerbestrasse 11, 6330 Cham, Switzerland

To my parents, in memoriam,
and to my sister Krassimira

Preface to the Second Edition

Twenty years have passed since the first edition of the book *Separable Programming: Theory and Methods*. During this period of time, many works in the area of separable optimization have appeared. This is the motivation for preparing a new edition of this book.

Some chapters of the original edition are revised and supplemented with new sections, and three new chapters and three new appendices are included in this edition.

New results of numerical experiments are included in Sect. 11.2 of Chap. 11 and at the end of Sect. 13.3 of Chap. 13.

In Chap. 12—*Applications of Convex Separable Unconstrained Nondifferentiable Optimization to Approximation Theory*—the approach, considered in Sects. 12.1–12.3, is extended to the case of numerical solution of some systems of nonlinear algebraic equations (new Sect. 12.4) and systems of nonlinear equations defined by convex functions (new Sect. 12.5). Section 12.4 of the original edition is renumbered as Sect. 12.6 and revised with some new computational results.

Chapter 14—*Valid Inequalities, Cutting Planes, and Integrality of the Knapsack Polytope*—is supplemented with results concerning valid inequalities generation (new Sects. 14.1 and 14.2), and the original text of this chapter concerning integrality of the knapsack polytope of the first edition constitutes now Sect. 14.3, complemented with the usage of totally unimodular matrices.

New Chap. 15—*Relaxation of the Equality Constrained Convex Continuous Knapsack Problem*—is devoted to further characterization of the optimal solution to the problem $(C^=)$ of Chap. 6, defined by a convex separable objective function subject to linear equality constraint and box constraints, through the optimal solution of a relaxed version of this problem.

In Chap. 16—*On the Solution of the Multidimensional Convex Separable Continuous Knapsack Problem with Bounded Variables*, a necessary and sufficient optimality condition (characterization theorem) for a feasible solution to be an optimal solution to the so-called multidimensional knapsack problem with bounded variables is formulated and proved, along with a primal-dual analysis.

In Chap. 17—*Characterization of the Optimal Solution of the Convex Generalized Nonlinear Transportation Problem*, the convex generalized nonlinear transportation problem and the convex generalized nonlinear transportation problem with box constraints are considered. A necessary and sufficient condition for a feasible solution to be an optimal solution for each of these two problems is stated and proved.

Optimization problems and methods for solving these problems, considered in this book, are interesting not only from the viewpoint of optimization theory, optimization methods, and their applications but also from the viewpoint of other fields of science, especially computer science. Artificial intelligence, machine learning, and other areas of computer science use various optimization methods, for example, convex optimization methods, large-scale optimization methods, discrete (including integer) optimization methods, stochastic gradient descent methods (including stochastic quasigradient methods), etc. In particular, machine learning algorithms can be formulated as optimization algorithms. These applications revived the optimization methods in the past few years.

In the new Appendix E, the solvability of a quadratic optimization problem with a feasible region defined as a Minkowski sum of a compact set and finitely generated convex closed cone is discussed.

In Appendix F, the Cauchy-Schwarz inequality, combined with properties of induced matrix norms, is used for solving a quadratic optimization problem.

In Appendix G, statements and proofs of some theorems of the alternative are presented.

Some minor revisions are made in the text of some chapters of the original edition.

The bibliography section is partially updated with new titles and new editions of some of the originally included reference titles. Bibliographical notes after Chap. 1, Parts I, II, and III are updated accordingly.

I am grateful to Series Editors of *Springer Optimization and Its Applications*, to the three anonymous reviewers, as well as to Elizabeth Loew of Springer for their help in preparing this edition for publication.

Blagoevgrad, Bulgaria Stefan M. Stefanov
March 2021

Preface to the First Edition

Mathematical optimization (mathematical programming) deals with the problem of optimizing (minimizing or maximizing) a function called the *objective function* subject to equality and/or inequality constraints that are defined by functions called the *constraint functions*.

In this book we consider a branch of mathematical programming – *separable programming*, where the objective function and the constraint functions can be expressed as the sum of single-variable functions. Such functions are said to be *separable*. Due to separability, separable programs have some interesting properties and can be solved by specific methods. Furthermore, many economic, industrial and other problems are described mathematically by separable programs. Thus, separable programming is significant from both the theoretical and practical points of view.

We also study one of the most important cases of separable programming – convex separable programming where the objective function and the constraint functions are convex. Convex functions have many special properties, for example, any local minimum of a convex function over a convex set is also a global minimum, and optimality criteria for convex programs are both necessary and sufficient conditions.

Some general results for separable programming are presented, techniques of approximating the original separable problem by linear program and the simplex method with the restricted basis entry rule for solving the resulting linear program are discussed. Due to convexity, linear programs that approximate *convex* separable problems can be solved by the *standard* simplex method, discarding the restricted basis entry rule. Because we solve the original separable problem by solving the approximate problem, then methods employing this approach are *approximate*. Some error estimations of approximating procedure for the convex separable problems are presented.

The dynamic programming approach to separable programs is also considered and some separable inventory and other models are reviewed.

The second part of this book is devoted to some special convex separable programs – minimization problems with convex and separable objective function over a feasible region defined by a separable convex inequality constraint of the form " $\leq$ " / linear equality constraint / linear inequality constraint of the form " $\geq$ ", and bounds on the variables. The three problems are denoted by (C), $(C^{=})$ and $(C^{\geq})$, respectively. A problem $(C_m^{=})$ which is a generalization of $(C^{=})$ with m equality constraints is also considered.

These problems have been the subject of intensive study in the last 35 – 40 years because they are interesting from both the theoretical and practical points of view.

Problems of these and related types arise in many cases, for example, in production planning and scheduling, in allocation of financial resources, in allocation of promotional resources among competing activities, in the theory of search, in subgradient optimization, in facility location, in the implementation of projection methods when the feasible region has the same form as the feasible sets under consideration, etc. That is why we need results and effective methods for solving such problems.

In this book, new iterative algorithms of polynomial complexity for problems (C), $(C^{=})$ and $(C^{\geq})$ are proposed.

Some applications of these algorithms are presented, in particular, application to stochastic quasigradient methods. Numerical approximations with respect to ℓ_1- and ℓ_∞-norms, as convex separable unconstrained nondifferentiable programs, are also considered.

This book consists of fourteen chapters, combined into three parts, and four appendices.

In Chapter One – *Preliminaries: Convex Analysis and Convex Programming*, we give some definitions and results connected with convex analysis (convex sets, projection of a point onto a set, separation of sets, convex functions, subgradients, subdifferentials, directional derivatives, etc.), convex programming and the Lagrangian duality. In Part Two, these concepts and results are utilized in developing suitable optimality conditions and numerical methods for solving convex separable problems. Chapter One can also be used as an introductory text in convex analysis and convex programming.

Chapters Two, Three and Four constitute *Part One – Separable Programming*. In Chapter Two – *Introduction. Approximating the Separable Problem*, a separable nonlinear program is defined, techniques of approximating the original separable problem by linear program are discussed, and the restricted basis entry rule of the simplex method for solving the linear approximating problem is considered.

In Chapter Three, some results for the *convex* separable case are presented.

Chapter Four – *Separable Programming: A Dynamic Programming Approach* is devoted to the dynamic programming approach to separable programming, the problem of dimensionality in dynamic programming (*the curse of dimensionality*, R. Bellman), and application of the Lagrange multiplier method for reducing the dimensionality of the problem. A dynamic programming approach to transportation problems with a nonlinear separable objective function as well as some separable inventory and other models are also considered.

Chapters Five to Eleven constitute *Part Two – Convex Separable Programming with Bounds on the Variables*. In Chapter Five – *Statement of the Main Problem. Basic Result* we state one of the basic problems under consideration, (C), and we prove a necessary and sufficient condition (Theorem 5.1) for a feasible solution to be an optimal solution to this problem.

Chapter Six – *Version One: Linear Equality Constraints* is devoted to a version $(C^=)$ of the basic problem (C), and a necessary and sufficient condition (Theorem 6.1), similar to Theorem 5.1, for a feasible solution to be an optimal solution to $(C^=)$ is proved. This chapter also considers a generalization $(C_m^=)$ of problem $(C^=)$ subject to m linear equality constraints, and Theorem 6.2 (parametric characterization of the solution to problem $(C_m^=)$) is proved.

Chapter Seven – *The Algorithms* contains statements of the iterative algorithms of polynomial complexity for both problems (C) and $(C^=)$, preceded by useful analyses of the solutions to these problems and followed by theorems (Theorem 7.1 and Theorem 7.2) which prove convergence of these algorithms.

In Chapter Eight – *Version Two: Linear Constraint of the Form* "$\geq$" we consider a version of the original problem for which the separable constraint is linear of the form "$\geq$" and positive coefficients. A sufficient optimality condition is proved (Theorem 8.1) and a convergent iterative algorithm of polynomial complexity is suggested for problems of this type.

Chapter Nine considers the concept of well-posedness of optimization problems in the sense of Tychonov, Hadamard, and in a generalized sense. Calmness in the sense of Clarke is also discussed. It is shown that the problems considered in Part Two are calm in the sense of Clarke. The concept of stability of the set of saddle points of the Lagrangian in the sense of Gol'shtein is also discussed and it is shown that this set is not stable for the "classical" Lagrangian. However, it is explained that despite this instability, due to the specificity of the approach suggested in this work, it is not necessary to use modified Lagrangians but the "classical" Lagrangians only.

In Chapter Ten we consider some extensions in order to overcome some theoretical and/or computational limitations both of the the approach suggested and the computer implementation.

Chapter Eleven – *Applications and Computational Experiments* consists of two sections. In the first one, ten examples of important convex objective functions and constraint functions (along with respective analytical expressions) for problems (C), $(C^=)$ and $(C^{\geq})$ are given. In the second section, some computational results, obtained by implementing the suggested algorithms to these examples, are presented.

Part Three – Selected Supplementary Topics and Applications consists of three chapters.

Chapter Twelve – *Applications of Convex Separable Unconstrained Nondifferentiable Optimization to Approximation Theory* deals with the data fitting problem and a problem for inconsistent systems of linear equations via minimizing the residual by using ℓ_1- and ℓ_∞-norms.

Chapter Thirteen – *About Projections in the Implementation of Stochastic Quasigradient Methods to Some Probabilistic Inventory Control Problems* deals with an application of algorithms suggested in Part Two of this book to the problems under consideration. The stochastic analogue of the nonsmooth ℓ_∞-problem is also formulated.

In Chapter Fourteen – *Integrality of the Knapsack Polytope*, a necessary and sufficient condition for integrality of the knapsack polytope is proved.

In Appendices A, B, C and D, we present some definitions and results of calculus, consider metric, Banach and Hilbert spaces, existence of solutions to optimization problems, and some theorems concerning best approximation theory, respectively.

I am grateful to series editors Panos M. Pardalos and Donald Hearn of the University of Florida. I am also grateful to John Martindale and Angela Quilici of Kluwer Academic Publishers for their help in preparing this book for publication.

Blagoevgrad – Sevlievo, 2000 Stefan M. Stefanov

Contents

Chapter 1
Preliminaries: Convex Analysis and Convex Programming

In this chapter, some definitions and results connected with convex analysis, convex programming, and Lagrangian duality are presented. In Part Two, these concepts and results are utilized in developing suitable optimality conditions and numerical methods for solving some convex problems.

1.1 Convex Sets. Sum of Sets and Product of a Set with a Real Number

Definition 1.1 Let $\mathbf{x}_1, \mathbf{x}_2 \in \mathbf{R}^n$. *Closed line segment* $[\mathbf{x}_1, \mathbf{x}_2]$ joining $\mathbf{x}_1$ and $\mathbf{x}_2$ is defined as

$$[\mathbf{x}_1, \mathbf{x}_2] = \{\mathbf{x} : \mathbf{x} = \lambda \mathbf{x}_1 + (1 - \lambda)\mathbf{x}_2, \quad \lambda \in [0, 1]\}.$$

Similarly, we can define *open line segment* $(\mathbf{x}_1, \mathbf{x}_2)$, *closed-open line segment* $[\mathbf{x}_1, \mathbf{x}_2)$, and *open-closed line segment* $(\mathbf{x}_1, \mathbf{x}_2]$.

Definition 1.2 A set $X \subseteq \mathbf{R}^n$ is said to be *convex* if the closed line segment joining every two points of X is in X, that is, X is *convex* if

$$\mathbf{x}_1, \mathbf{x}_2 \in X, \ \lambda \in \mathbf{R}, \ \lambda \in [0, 1] \quad \text{imply} \quad \lambda \mathbf{x}_1 + (1 - \lambda)\mathbf{x}_2 \in X.$$

Example 1.1 Each open ball and closed ball

$$B(\mathbf{x}_0; r) = \{\mathbf{x} \in \mathbf{R}^n : \|\mathbf{x} - \mathbf{x}_0\| < r\}, \quad \overline{B}(\mathbf{x}_0; r) = \{\mathbf{x} \in \mathbf{R}^n : \|\mathbf{x} - \mathbf{x}_0\| \le r\}$$

around a point $\mathbf{x}_0 \in \mathbf{R}^n$ is a convex set, where $\|\mathbf{x}\|$ denotes the Euclidean norm of $\mathbf{x}$ in $\mathbf{R}^n$.

© The Author(s), under exclusive license to Springer Nature Switzerland AG 2021
S. M. Stefanov, *Separable Optimization*, Springer Optimization and Its Applications 177,
https://doi.org/10.1007/978-3-030-78401-0_1

The proposition will be proved for $\overline{B}(\mathbf{x}_0; r)$; the open ball case is considered similarly. Let $\mathbf{x}_1, \mathbf{x}_2 \in \overline{B}(\mathbf{x}_0; r)$ (therefore $\|\mathbf{x}_1 - \mathbf{x}_0\| \leq r,\ \|\mathbf{x}_2 - \mathbf{x}_0\| \leq r$) and $\lambda \in [0, 1]$. Then

$$\begin{aligned}
&\|[\lambda \mathbf{x}_1 + (1-\lambda)\mathbf{x}_2] - \mathbf{x}_0\| \\
&\equiv \|\lambda \mathbf{x}_1 + (1-\lambda)\mathbf{x}_2 - \lambda \mathbf{x}_0 - (1-\lambda)\mathbf{x}_0\| \\
&= \|\lambda(\mathbf{x}_1 - \mathbf{x}_0) + (1-\lambda)(\mathbf{x}_2 - \mathbf{x}_0)\| \\
&\leq \|\lambda(\mathbf{x}_1 - \mathbf{x}_0)\| + \|(1-\lambda)(\mathbf{x}_2 - \mathbf{x}_0)\| \\
&= |\lambda|.\|\mathbf{x}_1 - \mathbf{x}_0\| + |1-\lambda|.\|\mathbf{x}_2 - \mathbf{x}_0\| \\
&= \lambda\|\mathbf{x}_1 - \mathbf{x}_0\| + (1-\lambda)\|\mathbf{x}_2 - \mathbf{x}_0\| \\
&\leq \lambda r + (1-\lambda) r \equiv r.
\end{aligned}$$

Therefore $\lambda \mathbf{x}_1 + (1-\lambda)\mathbf{x}_2 \in \overline{B}(\mathbf{x}_0; r)$, that is, $\overline{B}(\mathbf{x}_0; r)$ is a convex set by definition. ■

Theorem 1.1 (Intersection of convex sets)
Let $X_i,\ i \in I$, be a family of convex sets. Then the set $X \stackrel{\text{def}}{=} \bigcap_{i \in I} X_i$ is convex.

Proof. Let $\mathbf{x}_1, \mathbf{x}_2 \in X$. Therefore $\mathbf{x}_1, \mathbf{x}_2 \in X_i$ for each $i \in I$, and since X_i are convex sets, then for $\lambda \in [0, 1]$ we have that

$$\lambda \mathbf{x}_1 + (1-\lambda)\mathbf{x}_2 \in X_i \quad \text{for each} \quad i \in I.$$

Hence, $\lambda \mathbf{x}_1 + (1-\lambda)\mathbf{x}_2 \in \bigcap_{i \in I} X_i \stackrel{\text{def}}{=} X$.

Therefore X is a convex set by definition. ■

Definition 1.3 A point $\mathbf{x} \in \mathbf{R}^n$ is said to be a *convex combination* of $\mathbf{x}_1, \ldots, \mathbf{x}_m \in \mathbf{R}^n$ if there exist nonnegative numbers $\lambda_1, \ldots, \lambda_m$ such that $\sum_{i=1}^m \lambda_i = 1$ and

$$\mathbf{x} = \sum_{i=1}^{m} \lambda_i \mathbf{x}_i .$$

Definition 1.4 Let X be a set in $\mathbf{R}^n$. The *convex hull* (*convex envelope*, *convex closure*) of X is the set of all convex combinations of points of X,

$$\text{co } X = \left\{ \mathbf{z} \in \mathbf{R}^n : \mathbf{z} = \sum_{i=1}^{m} \lambda_i \mathbf{x}_i,\ \mathbf{x}_i \in X,\ \lambda_i \geq 0, i = 1, \ldots, m,\ \sum_{i=1}^{m} \lambda_i = 1 \right\}.$$

Convex combinations with arbitrary length are involved in co X.

Sometimes the convex hull of X is defined as the intersection of all convex sets in $\mathbf{R}^n$ containing X. Then the contents of Definition 1.4 is proved as a theorem. The two definitions are equivalent.

The following four theorems give some properties of convex combinations and convex hulls.

Theorem 1.2 (Convex sets and convex combinations)
Let $\mathbf{x}_1, \dots, \mathbf{x}_m \in X$ *where* X *is a set in* $\mathbf{R}^n$. *Then* X *is convex if and only if it contains the collection of all convex combinations of* $\mathbf{x}_1, \dots, \mathbf{x}_m$.

Proof. *Necessity.* (By induction) Let X be convex. If $m = 2$, then $\lambda\mathbf{x}_1 + (1-\lambda)\mathbf{x}_2 \in X$ for $\lambda \in [0, 1]$, $\mathbf{x}_1, \mathbf{x}_2 \in X$ because X is a convex set.

Induction hypothesis. Let the statement hold true for $m \leq k$. Verify it for $m = k+1$. Define the point

$$\mathbf{x} = \lambda_1\mathbf{x}_1 + \cdots + \lambda_k\mathbf{x}_k + \lambda_{k+1}\mathbf{x}_{k+1}$$

where $\lambda_i \geq 0$, $i = 1, \dots, k+1$, $\sum_{i=1}^{k+1} \lambda_i = 1$. (Without loss of generality assume $\lambda_i > 0$, because if some λ_i is zero we can reduce the length of convex combination under consideration and statement of Theorem 1.2 will be true by the induction hypothesis.) Then $1 - \lambda_{k+1} = \sum_{i=1}^{k} \lambda_i > 0$.

By the induction hypothesis (for $m = k$),

$$\mathbf{y} \stackrel{\text{def}}{=} \frac{\lambda_1}{1-\lambda_{k+1}}\mathbf{x}_1 + \cdots + \frac{\lambda_k}{1-\lambda_{k+1}}\mathbf{x}_k \in X$$

because

$$\frac{\lambda_i}{1-\lambda_{k+1}} > 0,\ i = 1, \dots, k \ \text{ and } \sum_{i=1}^{k} \frac{\lambda_i}{1-\lambda_{k+1}} = \frac{1}{1-\lambda_{k+1}} \sum_{i=1}^{k} \lambda_i = 1.$$

Then

$$\mathbf{x} \stackrel{\text{def}}{=} \lambda_1\mathbf{x}_1 + \cdots + \lambda_k\mathbf{x}_k + \lambda_{k+1}\mathbf{x}_{k+1} \equiv (1-\lambda_{k+1})\mathbf{y} + \lambda_{k+1}\mathbf{x}_{k+1} \in X$$

since $\mathbf{x}$ is expressed as a convex combination of the points $\mathbf{y}$ and $\mathbf{x}_{k+1}$ which belong to the convex set X.

Sufficiency. Let X contain all convex combinations of $\mathbf{x}_1, \dots, \mathbf{x}_m$. The case $m = 2$ implies that X is convex by definition. ∎

Theorem 1.3 (Convex sets and convex hulls)
The set X *in* $\mathbf{R}^n$ *is convex if and only if* $X = \text{co } X$.

Proof. (By double inclusion) Let X be convex. Then $\text{co } X \subseteq X$ according to Theorem 1.2. However, by convexity of X, $X \subseteq \text{co } X$ according to the alternative definition of convex hull. Therefore $X = \text{co } X$.

Conversely, let $X = \text{co } X$ and let $\mathbf{z}_1, \mathbf{z}_2 \in \text{co } X$. Then

$$\mathbf{z}_1 = \sum_{i=1}^{k} \alpha_i\mathbf{x}_i, \quad \mathbf{z}_2 = \sum_{j=1}^{l} \beta_j\mathbf{y}_j$$

where $\alpha_i \geq 0,\ \mathbf{x}_i \in X,\ i = 1, \ldots, k;\ \beta_j \geq 0,\ \mathbf{y}_j \in X,\ j = 1, \ldots, l$; and $\sum_{i=1}^{k} \alpha_i = 1;\ \sum_{j=1}^{l} \beta_j = 1$.

Consider $\mathbf{z} = \lambda \mathbf{z}_1 + (1 - \lambda)\mathbf{z}_2$ for $\lambda \in [0, 1]$. We have

$$\mathbf{z} = \lambda \mathbf{z}_1 + (1 - \lambda)\mathbf{z}_2 = \lambda \sum_{i=1}^{k} \alpha_i \mathbf{x}_i + (1 - \lambda) \sum_{j=1}^{l} \beta_j \mathbf{y}_j.$$

Since $\lambda\alpha_i \geq 0,\ i = 1, \ldots, k;\ (1 - \lambda)\beta_j \geq 0,\ j = 1, \ldots, l$; and

$$\lambda \sum_{i=1}^{k} \alpha_i + (1 - \lambda) \sum_{j=1}^{l} \beta_j = \lambda.1 + (1 - \lambda).1 = 1,$$

then $\mathbf{z}$ is a convex combination of the points $\mathbf{x}_1, \ldots, \mathbf{x}_k, \mathbf{y}_1, \ldots, \mathbf{y}_l$ of X, that is, $\mathbf{z} \in \operatorname{co} X$. Therefore $\operatorname{co} X$ is convex by definition. Then assumption $X = \operatorname{co} X$ implies that X is also convex. ■

Theorem 1.4 (Caratheodory's theorem)
Let X be a set in $\mathbf{R}^n$. Then each point of $\operatorname{co} X$ can be expressed as a convex combination of at most $n + 1$ points, not necessarily distinct, of X.

Proof. Let $\mathbf{z} \in \operatorname{co} X$. Then

$$\mathbf{z} = \sum_{i=1}^{k} \lambda_i \mathbf{x}_i, \quad \mathbf{x}_i \in X,\ \lambda_i \geq 0,\ i = 1, \ldots, k,\ \sum_{i=1}^{k} \lambda_i = 1,\ k = 1, 2, \ldots$$

by definition.

If $k \leq n + 1$, the proof is complete.

Let $k > n + 1$. Our purpose is to prove that $\mathbf{z}$ can be expressed as a convex combination of $k - 1$ points.

Consider the vectors $\mathbf{y}_i = (\mathbf{x}_i, 1) \in \mathbf{R}^{n+1},\ i = 1, \ldots, k;\ k > n + 1$. Since $k > n + 1$, then $\{\mathbf{y}_i\}_{i=1}^{k} \subset \mathbf{R}^{n+1}$ are linearly dependent in $\mathbf{R}^{n+1}$. Therefore there exist numbers $\alpha_i,\ i = 1, \ldots, k$ which are not equal to zero simultaneously such that $\sum_{i=1}^{k} \alpha_i \mathbf{y}_i = \mathbf{0}$. Taking into account that $\mathbf{y}_i = (\mathbf{x}_i, 1)$, we get

$$\sum_{i=1}^{k} \alpha_i \mathbf{x}_i = \mathbf{0}, \qquad \sum_{i=1}^{k} \alpha_i = 0.$$

Let $\alpha_{i_0} \neq 0$. Then

$$\mathbf{x}_{i_0} = -\frac{1}{\alpha_{i_0}} \sum_{\substack{i=1 \\ i \neq i_0}}^{k} \alpha_i \mathbf{x}_i.$$

Therefore we get

$$\mathbf{z} = \sum_{i=1}^{k} \lambda_i \mathbf{x}_i = \sum_{\substack{i=1 \\ i \neq i_0}}^{k} \lambda_i \mathbf{x}_i + \lambda_{i_0} \mathbf{x}_{i_0} = \sum_{\substack{i=1 \\ i \neq i_0}}^{k} \left(\lambda_i - \frac{\lambda_{i_0}}{\alpha_{i_0}} \alpha_i \right) \mathbf{x}_i .$$

In this expression,

$$\sum_{\substack{i=1 \\ i \neq i_0}}^{k} \left(\lambda_i - \frac{\lambda_{i_0}}{\alpha_{i_0}} \alpha_i \right) = \sum_{\substack{i=1 \\ i \neq i_0}}^{k} \lambda_i - \frac{\lambda_{i_0}}{\alpha_{i_0}} \sum_{\substack{i=1 \\ i \neq i_0}}^{k} \alpha_i = (1 - \lambda_{i_0}) - \frac{\lambda_{i_0}}{\alpha_{i_0}} (-\alpha_{i_0}) = 1.$$

We have to prove that $\lambda_i - \dfrac{\lambda_{i_0}}{\alpha_{i_0}} \alpha_i \geq 0,\ i = 1, \ldots, k,\ i \neq i_0$.

Since $\sum_{i=1}^{k} \alpha_i = 0$ and α_i's are not equal to zero simultaneously, there are positive and negative α_i's. Choose an index i_0 for which $\alpha_{i_0} \neq 0$ as follows. Let i_0 satisfy $\dfrac{\lambda_{i_0}}{\alpha_{i_0}} = \min \left\{ \dfrac{\lambda_i}{\alpha_i} \, : \, \alpha_i > 0 \right\}$. There are two cases to consider.

1. For indices i such that $\alpha_i > 0$ we have

$$\lambda_i - \frac{\lambda_{i_0}}{\alpha_{i_0}} \alpha_i = \alpha_i \left(\frac{\lambda_i}{\alpha_i} - \frac{\lambda_{i_0}}{\alpha_{i_0}} \right) \geq 0$$

   according to the assumption $\alpha_i > 0$ and the choice of i_0.
2. For indices i such that $\alpha_i \leq 0$, using that $\lambda_i \geq 0,\ \lambda_{i_0} \geq 0,\ \alpha_{i_0} \geq 0, \alpha_i \leq 0$ we have $\lambda_i - \frac{\lambda_{i_0}}{\alpha_{i_0}} \alpha_i \geq 0$.

Thus, $\mathbf{z}$ is expressed as a convex combination of $k - 1$ points in X.

Applying repeatedly this technique results in expressing $\mathbf{z}$ as a convex combination of at most $n + 1$ points of X. ∎

Theorem 1.5 (Convex hull of a compact set)
Let X be a compact set in $\mathbf{R}^n$. Then co X *is also a compact set in $\mathbf{R}^n$.*

Proof. Since X is bounded, there is some M with $\|\mathbf{x}\| < M$ for each $\mathbf{x} \in X$. Let $\mathbf{y} \in$ co X. Therefore

$$\mathbf{y} = \sum_{i=1}^{k} \lambda_i \mathbf{x}_i, \quad \lambda_i \geq 0,\ \mathbf{x}_i \in X,\ i = 1, \ldots, k,\ \sum_{i=1}^{k} \lambda_i = 1.$$

Then

$$\|\mathbf{y}\| = \left\| \sum_{i=1}^{k} \lambda_i \mathbf{x}_i \right\| \leq \sum_{i=1}^{k} \lambda_i \|\mathbf{x}_i\| \leq M \sum_{i=1}^{k} \lambda_i = M.1 = M.$$

Hence co X is bounded by definition.

Concerning closedness of co X, we will prove that each sequence $\{\mathbf{y}_k\}_{k=1}^{\infty} \subset$ co X has a subsequence which is convergent to a point of co X.

According to Caratheodory's theorem (Theorem 1.4),

$$\mathbf{y}_k = \sum_{i=1}^{n+1} \lambda_{ik}\mathbf{x}_{ik}, \quad \lambda_{ik} \geq 0, \ \mathbf{x}_{ik} \in X, \ i = 1, \ldots, n+1, \ \sum_{i=1}^{n+1} \lambda_{ik} = 1.$$

Sequences $\{\lambda_{ik}\}_{k=1}^{\infty}$ and $\{\mathbf{x}_{ik}\}_{k=1}^{\infty}$, $i = 1, \ldots, n+1$ are in the compact sets $\{0 \leq \lambda_{ik} \leq 1\}$ and X, respectively. Hence, convergent subsequences $\{\lambda_{ik_j}\}_{j=1}^{\infty}$ and $\{\mathbf{x}_{ik_j}\}_{j=1}^{\infty}$, $k_1 < k_2 < \ldots$ of these two sequences can be chosen, respectively, with $\lambda_{ik_j} \to \lambda_{i0}$, $\mathbf{x}_{ik_j} \to \mathbf{x}_{i0}$, $i = 1, \ldots, n+1$.

Consider the sequence $\{\mathbf{y}_{k_j}\}$:

$$\mathbf{y}_{k_j} = \sum_{i=1}^{n+1} \lambda_{ik_j}\mathbf{x}_{ik_j}, \quad \sum_{i=1}^{n+1} \lambda_{ik_j} = 1, \ \lambda_{ik_j} \geq 0, \ i = 1, \ldots, n+1.$$

Since $\lambda_{ik_j} \to \lambda_{i0}$, $\mathbf{x}_{ik_j} \to \mathbf{x}_{i0}$, $i = 1, \ldots, n+1$, then

$$\mathbf{y}_{k_j} \to \mathbf{y}_0 = \sum_{i=1}^{n+1} \lambda_{i0}\mathbf{x}_{i0}, \quad \sum_{i=1}^{n+1} \lambda_{i0} = 1, \ \lambda_{i0} \geq 0, \ i = 1, \ldots, n+1.$$

Therefore $\mathbf{y}_0$ is a convex combination of $\mathbf{x}_{i0} \in X$, $i = 1, \ldots, n+1$ (X is compact, therefore closed, set). Hence, $\mathbf{y}_0 \in$ co X, that is, co X is closed by definition.

Since co X is bounded and closed in $\mathbf{R}^n$ then it is compact by definition. ■

Definition 1.5 Let X and Y be sets in $\mathbf{R}^n$. The *Minkowski sum* of X and Y, denoted by $X + Y$, is defined as

$$X + Y = \{\mathbf{z} : \ \mathbf{z} = \mathbf{x} + \mathbf{y}, \ \mathbf{x} \in X, \ \mathbf{y} \in Y\}.$$

Definition 1.6 The *product γX of the set $X \subseteq \mathbf{R}^n$ with the real number γ* is defined as

$$\gamma X = \{\mathbf{z} : \ \mathbf{z} = \gamma \mathbf{x}, \ \mathbf{x} \in X\}.$$

Theorem 1.6 (Sum of convex sets and product of a convex set with a real number) *Let X and Y be convex sets in $\mathbf{R}^n$ and $\gamma \in \mathbf{R}$. Then the sets $X + Y$ and γX are convex.*

Proof. Let $\mathbf{z}_1, \mathbf{z}_2 \in X + Y$. Therefore $\mathbf{z}_1 = \mathbf{x}_1 + \mathbf{y}_1$ and $\mathbf{z}_2 = \mathbf{x}_2 + \mathbf{y}_2$ where $\mathbf{x}_1, \mathbf{x}_2 \in X$ and $\mathbf{y}_1, \mathbf{y}_2 \in Y$. Then

$$\lambda \mathbf{z}_1 + (1-\lambda)\mathbf{z}_2 = [\lambda \mathbf{x}_1 + (1-\lambda)\mathbf{x}_2] + [\lambda \mathbf{y}_1 + (1-\lambda)\mathbf{y}_2] \in X + Y$$

for $0 \leq \lambda \leq 1$. Therefore $X + Y$ is convex by definition.

Let $\mathbf{z}_1, \mathbf{z}_2 \in \gamma X$. Hence $\mathbf{z}_1 = \gamma \mathbf{x}_1,\ \mathbf{z}_2 = \gamma \mathbf{x}_2,\ \mathbf{x}_1, \mathbf{x}_2 \in X$. Then

$$\lambda \mathbf{z}_1 + (1 - \lambda)\mathbf{z}_2 = \gamma[\lambda \mathbf{x}_1 + (1 - \lambda)\mathbf{x}_2] \in \gamma X$$

for $0 \le \lambda \le 1$. Therefore γX is convex by definition. ■

Corollary 1.1 (Linear combinations of convex sets)
The linear combination $\sum_{i=1}^{m} \gamma_i X_i$ of convex sets $X_i,\ i = 1, \ldots, m$, is convex.

In particular, for $m = 2,\ \gamma_1 = 1$, and $\gamma_2 = -1$ we get $X_1 - X_2$ is a convex set.

Theorem 1.7 (Sum of closed sets, one of which is bounded)
Let X and Y be closed sets in $\mathbf{R}^n$ and at least one of $X,\ Y$ be bounded. Then the sum $Z = X + Y$ is a closed set.

Proof. Let X be bounded. Consider the convergent sequence $\{\mathbf{z}_k\}$, $\mathbf{z}_k \in Z$, $\mathbf{z}_k \to \mathbf{z}$. We have to prove that $\mathbf{z} \in Z$.

Since $\mathbf{z}_k \in Z$, then $\mathbf{z}_k = \mathbf{x}_k + \mathbf{y}_k,\ \mathbf{x}_k \in X,\ \mathbf{y}_k \in Y$ for $k = 1, 2, \ldots$. The sequence $\{\mathbf{x}_k\} \subset X$ is bounded. According to the Bolzano-Weierstrass theorem (Theorem A.1, Appendix A), a convergent subsequence $\{\mathbf{x}_{k_i}\}$ can be chosen with $\mathbf{x}_{k_i} \to \mathbf{x}$ where $\mathbf{x} \in X$ because X is closed. Then the sequence $\{\mathbf{y}_{k_i}\}$, where $\mathbf{y}_{k_i}$ is defined as $\mathbf{y}_{k_i} = \mathbf{z}_{k_i} - \mathbf{x}_{k_i}$, is convergent to some $\mathbf{y} = \mathbf{z} - \mathbf{x}$ and $\mathbf{y} \in Y$ because Y is closed.

Then $\mathbf{z} = \mathbf{x} + \mathbf{y},\ \mathbf{x} \in X,\ \mathbf{y} \in Y$, that is, $\mathbf{z} \in X + Y$ by definition. Therefore $Z = X + Y$ is closed in accordance with the definition of closed set. ■

1.2 Projection of a Point onto a Set. Separation of Sets and Support of Sets

Definition 1.7 A point $\mathbf{p} \in X$ is said to be a *projection of* $\mathbf{y}$ *onto* X if

$$\|\mathbf{y} - \mathbf{p}\| = \inf_{\mathbf{x} \in X} \|\mathbf{y} - \mathbf{x}\|.$$

Definition 1.8 The nonnegative number $\delta(\mathbf{y}, X) = \inf_{\mathbf{x} \in X} \|\mathbf{y} - \mathbf{x}\|$ is called the *distance from the point* $\mathbf{y}$ *to the set* X.

By definition, the projection of $\mathbf{y}$ onto X is an optimal solution to the problem

$$\min_{\mathbf{x} \in X} \{f(\mathbf{x}) = \|\mathbf{y} - \mathbf{x}\|\}.$$

Obviously, if $\mathbf{y} \in X$ then $\delta(\mathbf{y}, X) = 0$ and $\mathbf{y} = \mathbf{p}$.

Theorem 1.8 (Existence of projection)
Let X be a nonempty closed set in $\mathbf{R}^n$. Then every $\mathbf{y} \in \mathbf{R}^n$ has projection onto X.

The statement of Theorem 1.8 follows from a corollary of the Weierstrass theorem (Corollary C.1, Appendix C) because $\|\mathbf{y}-\mathbf{x}\|$ is a continuous function, $\lim_{\|\mathbf{x}\|\to\infty} \|\mathbf{y}-\mathbf{x}\| = +\infty$ for the fixed projected point $\mathbf{y}$, and X is a closed set by assumption.

Theorem 1.9 (Uniqueness of the projection onto a convex closed set)
Let X be a nonempty closed convex set in $\mathbf{R}^n$. Then every $\mathbf{y} \in \mathbf{R}^n$ has a unique projection onto X.

Proof. The existence of projection follows from Theorem 1.8 whose assumptions are also assumptions of Theorem 1.9.

Assume that $\mathbf{y} \notin X$ has two *distinct* projections $\mathbf{p}_1$ and $\mathbf{p}_2$ onto X. Hence $\|\mathbf{y}-\mathbf{p}_1\| = \|\mathbf{y}-\mathbf{p}_2\|$ by definition of projection.

Since X is convex, then $\mathbf{z} \stackrel{\text{def}}{=} \frac{1}{2}\mathbf{p}_1 + \frac{1}{2}\mathbf{p}_2 \in X$. Consider the triangle (in $\mathbf{R}^n$) whose vertices are $\mathbf{p}_1$, $\mathbf{p}_2$, and $\mathbf{y}$. Then

$$\|\mathbf{y}-\mathbf{z}\| = \left\|\mathbf{y} - \frac{1}{2}(\mathbf{p}_1+\mathbf{p}_2)\right\| \le \frac{1}{2}(\|\mathbf{y}-\mathbf{p}_1\| + \|\mathbf{y}-\mathbf{p}_2\|) = \|\mathbf{y}-\mathbf{p}_1\| = \|\mathbf{y}-\mathbf{p}_2\|.$$

On the other hand, $\|\mathbf{y}-\mathbf{z}\| \ge \|\mathbf{y}-\mathbf{p}_1\| = \|\mathbf{y}-\mathbf{p}_2\|$ according to definition of projection. Therefore the triangle inequality is satisfied as an equality for the considered triangle. Hence the vectors $\mathbf{y}-\mathbf{p}_1$ and $\mathbf{y}-\mathbf{p}_2$ are positively collinear: $\mathbf{y}-\mathbf{p}_1 = k(\mathbf{y}-\mathbf{p}_2)$, $k > 0$. From $\|\mathbf{y}-\mathbf{p}_1\| = \|\mathbf{y}-\mathbf{p}_2\|$ it follows that $k = 1$, that is, $\mathbf{y}-\mathbf{p}_1 = \mathbf{y}-\mathbf{p}_2$. Hence $\mathbf{p}_1 = \mathbf{p}_2$, a contradiction. Therefore the projection of a point onto a convex closed set is unique. ■

Theorem 1.10 (Characterization of the projection)
Let X be a convex closed set in $\mathbf{R}^n$ and $\mathbf{y} \in \mathbf{R}^n$. Then $\mathbf{p} \in X$ is a projection of $\mathbf{y}$ onto X if and only if

$$\langle \mathbf{x}-\mathbf{p},\ \mathbf{y}-\mathbf{p}\rangle \le 0 \quad \text{for every } \mathbf{x} \in X.$$

Proof. *Necessity.* Let $\mathbf{p} \in X$ be the projection of $\mathbf{y}$ onto X and let $\mathbf{x} \in X$. Since X is convex, $\mathbf{z} = \lambda\mathbf{x} + (1-\lambda)\mathbf{p} \in X$, $\lambda \in (0, 1]$. Then

$$\|\mathbf{z}-\mathbf{y}\|^2 = \|\lambda(\mathbf{x}-\mathbf{p}) + \mathbf{p}-\mathbf{y}\|^2 = \lambda^2\|\mathbf{x}-\mathbf{p}\|^2 + 2\lambda\langle \mathbf{x}-\mathbf{p}, \mathbf{p}-\mathbf{y}\rangle + \|\mathbf{p}-\mathbf{y}\|^2.$$

Since $\|\mathbf{z}-\mathbf{y}\| \ge \|\mathbf{p}-\mathbf{y}\|$ according to definition of projection, we obtain

$$\lambda^2\|\mathbf{x}-\mathbf{p}\|^2 + 2\lambda\langle \mathbf{x}-\mathbf{p}, \mathbf{p}-\mathbf{y}\rangle \ge 0,$$

that is,

$$\langle \mathbf{x}-\mathbf{p}, \mathbf{y}-\mathbf{p}\rangle \le \frac{1}{2}\lambda\|\mathbf{x}-\mathbf{p}\|^2$$

for every $\lambda \in (0, 1]$. Therefore $\langle \mathbf{x}-\mathbf{p},\ \mathbf{y}-\mathbf{p}\rangle \le 0 \ \ \forall \mathbf{x} \in X$.

Sufficiency. Let $\mathbf{p} \in X$ satisfy the hypothesis of Theorem 1.10. Equivalently, $\langle \mathbf{x}-\mathbf{p},\ \mathbf{p}-\mathbf{y}\rangle \ge 0 \ \ \forall \mathbf{x} \in X$. Therefore

$$\|\mathbf{x}-\mathbf{y}\|^2 \equiv \|\mathbf{x}-\mathbf{p}+\mathbf{p}-\mathbf{y}\|^2 = \|\mathbf{x}-\mathbf{p}\|^2 + 2\langle \mathbf{x}-\mathbf{p}, \mathbf{p}-\mathbf{y}\rangle + \|\mathbf{p}-\mathbf{y}\|^2$$

$$\geq \|\mathbf{p}-\mathbf{y}\|^2 \quad \forall \mathbf{x} \in X,$$

that is, $\mathbf{p}$ is a projection of $\mathbf{y}$ onto X by definition. ■

Geometrically, Theorem 1.10 states that the angle between the vectors $\mathbf{x}-\mathbf{p}$ and $\mathbf{y}-\mathbf{p}$ is nonacute, that is, right or obtuse, angle.

Definition 1.9 A *hyperplane* H in $\mathbf{R}^n$ is defined as

$$H = \{\mathbf{x} : \langle \mathbf{h}, \mathbf{x}\rangle = a\}$$

where $\mathbf{h} \in \mathbf{R}^n$, $\mathbf{h} \neq \mathbf{0}$, $a \in \mathbf{R}$, and $\langle \mathbf{h}, \mathbf{x}\rangle$ denotes the scalar (inner) product of $\mathbf{h}$ and $\mathbf{x}$ in $\mathbf{R}^n$.

A hyperplane H defines two *closed half-spaces* $H^+ = \{\mathbf{x} : \langle \mathbf{h}, \mathbf{x}\rangle \geq a\}$ and $H^- = \{\mathbf{x} : \langle \mathbf{h}, \mathbf{x}\rangle \leq a\}$ and two *open half-spaces* $\{\mathbf{x} : \langle \mathbf{h}, \mathbf{x}\rangle > a\}$ and $\{\mathbf{x} : \langle \mathbf{h}, \mathbf{x}\rangle < a\}$ in $\mathbf{R}^n$.

Definition 1.10 Let X and Y be nonempty sets in $\mathbf{R}^n$. A hyperplane $H = \{\mathbf{x} : \langle \mathbf{h}, \mathbf{x}\rangle = a\}$ is said to *separate* X and Y if $\langle \mathbf{h}, \mathbf{x}\rangle \geq a$ for each $\mathbf{x} \in X$ and $\langle \mathbf{h}, \mathbf{x}\rangle \leq a$ for each $\mathbf{x} \in Y$.

The hyperplane H is said to *properly separate* X and Y if, in addition, X and Y are not contained in H, that is, if $X \cup Y \not\subset H$.

The hyperplane H is said to *strictly separate* X and Y if $\langle \mathbf{h}, \mathbf{x}\rangle > a$ for each $\mathbf{x} \in X$ and $\langle \mathbf{h}, \mathbf{x}\rangle < a$ for each $\mathbf{x} \in Y$.

In these three cases, X and Y are called *separable*, *properly separable*, and *strictly separable*, respectively.

Definition 1.11 Let X be a nonempty set in $\mathbf{R}^n$ and $\overline{\mathbf{x}} \in \partial X$. A hyperplane $H = \{\mathbf{x} : \langle \mathbf{h}, \mathbf{x}-\overline{\mathbf{x}}\rangle = 0\}$ is called a *supporting hyperplane* of X at $\overline{\mathbf{x}}$ if $X \subseteq H^+$, that is, $\langle \mathbf{h}, \mathbf{x}-\overline{\mathbf{x}}\rangle \geq 0$ for each $\mathbf{x} \in X$, or $X \subseteq H^-$, that is, $\langle \mathbf{h}, \mathbf{x}-\overline{\mathbf{x}}\rangle \leq 0$ for each $\mathbf{x} \in X$.

Definition 1.12 A subset A of $\mathbf{R}^n$ is said to be an *affine set* (*affine manifold*, *affine variety*) if

$$\lambda\mathbf{x} + (1-\lambda)\mathbf{y} \in A \quad \text{for every} \quad \mathbf{x}, \mathbf{y} \in A,\ \lambda \in \mathbf{R}.$$

It can be proved that each affine subset of $\mathbf{R}^n$ is an intersection of a finite number of hyperplanes.

Since the intersection of an arbitrary collection of affine sets is also an affine set, then for any subset S of $\mathbf{R}^n$ there exists a smallest affine set containing S.

Definition 1.13 The *affine hull* aff S of S is the smallest affine set containing S.

Definition 1.14 The *relative interior* ri C of a convex set $C \subseteq \mathbf{R}^n$ consists of the points $\mathbf{x} \in$ aff C for which there exists an $\varepsilon > 0$ such that $\mathbf{y} \in C$ whenever $\mathbf{y} \in$ aff C

and *dist* $(\mathbf{x}, \mathbf{y}) \le \varepsilon$, where *dist* $(\mathbf{x}, \mathbf{y})$ is the Euclidean distance between $\mathbf{x}$ and $\mathbf{y}$ (see Appendix B), that is,

$$\text{ri}\, C = \{\mathbf{x} \in \text{ aff } C : \exists \varepsilon > 0,\ (\mathbf{x} + \varepsilon B) \cap (\text{aff } C) \subseteq C\},$$

where B is the unit ball in $\mathbf{R}^n$ around $\mathbf{0}$.

The set $r\partial C = (\text{cl } C) \setminus (\text{ri } C)$ is called the *relative boundary* of C, where cl C is the *closure* of C (Definition A.2, Appendix A).

C is said to be *relatively open* if ri $C = C$.

Obviously ri $C \subseteq C \subseteq$ cl C.

Theorem 1.11 (Separating a closed convex set and a point not belonging to it) *Let X be a nonempty closed convex set in $\mathbf{R}^n$ and $\mathbf{y} \notin X$. Then there exists a vector $\mathbf{h} \in \mathbf{R}^n$, $\mathbf{h} \neq \mathbf{0}$, and $a \in \mathbf{R}$ such that $\langle \mathbf{h}, \mathbf{y} \rangle < a$ and $\langle \mathbf{h}, \mathbf{x} \rangle \ge a$ for each $\mathbf{x} \in X$, that is, X and $\{\mathbf{y}\}$ can be strictly separated.*

Proof. By assumption, X is a nonempty closed convex set and $\mathbf{y} \notin X$. Then there exists a unique point $\mathbf{p}$ with $\langle \mathbf{x} - \mathbf{p}, \mathbf{y} - \mathbf{p} \rangle \le 0$ for all $\mathbf{x} \in X$ according to Theorem 1.9 and Theorem 1.10.

Set $\mathbf{h} = \mathbf{p} - \mathbf{y}$, $a = \langle \mathbf{h}, \mathbf{p} \rangle$. (Since $\mathbf{y} \notin X$ and $\mathbf{p} \in X$ then $\mathbf{h} \neq \mathbf{0}$.) Therefore $\langle \mathbf{x} - \mathbf{p}, -\mathbf{h} \rangle \le 0$, that is, $\langle \mathbf{h}, \mathbf{x} \rangle \ge \langle \mathbf{h}, \mathbf{p} \rangle = a$ for each $\mathbf{x} \in X$.

On the other hand,

$$a - \langle \mathbf{h}, \mathbf{y} \rangle = \langle \mathbf{h}, \mathbf{p} \rangle - \langle \mathbf{h}, \mathbf{y} \rangle = \langle \mathbf{h}, \mathbf{p} - \mathbf{y} \rangle = \langle \mathbf{h}, \mathbf{h} \rangle = \|\mathbf{h}\|^2 > 0,$$

that is, $\langle \mathbf{h}, \mathbf{y} \rangle < a$. ■

Theorem 1.12 (Supporting hyperplane at a boundary point) *Let X be a nonempty convex set in $\mathbf{R}^n$ and $\overline{\mathbf{x}} \in \partial X$. Then there exists a hyperplane that supports X at $\overline{\mathbf{x}}$, that is, there exists a vector $\mathbf{h} \in \mathbf{R}^n$, $\mathbf{h} \neq \mathbf{0}$ such that $\langle \mathbf{h}, \mathbf{x} - \overline{\mathbf{x}} \rangle \ge 0$ for each $\mathbf{x} \in$ cl X.*

If int $X \neq \emptyset$, *then $\langle \mathbf{h}, \mathbf{x} - \overline{\mathbf{x}} \rangle > 0$ for each $\mathbf{x} \in$ int X.*

Proof. By assumption, $\overline{\mathbf{x}} \in \partial X$. Since $\partial X = \partial(\text{cl } X)$, see Theorem A.5 of Appendix A, then $\overline{\mathbf{x}} \in \partial(\text{cl } X)$. Then there exists a sequence $\{\mathbf{y}_k\} \not\subset$ cl X such that $\mathbf{y}_k \to \overline{\mathbf{x}}$ when $k \to \infty$. From Theorem 1.11 it follows that for each $\mathbf{y}_k$ there exists an $\mathbf{h}_k$, $\|\mathbf{h}_k\| = 1$ such that $\langle \mathbf{h}_k, \mathbf{y}_k \rangle < \langle \mathbf{h}_k, \mathbf{x} \rangle$ for each $\mathbf{x} \in$ cl X. (If $\|\mathbf{h}_k\| \neq 1$, we can take the vector $\frac{\mathbf{h}_k}{\|\mathbf{h}_k\|}$ whose norm is equal to 1.) Since sequence $\{\mathbf{h}_k\}$ is bounded, according to the Bolzano-Weierstrass theorem (Appendix A, Theorem A.1) it has a convergent subsequence $\{\mathbf{h}_{k_l}\} \to \mathbf{h}$ with $\|\mathbf{h}\| = 1$. For each $\mathbf{x} \in$ cl X we have $\langle \mathbf{h}_{k_l}, \mathbf{y}_k \rangle < \langle \mathbf{h}_{k_l}, \mathbf{x} \rangle$. For a fixed $\mathbf{x} \in$ cl X and $l \to +\infty$ we have $\langle \mathbf{h}, \mathbf{x} - \overline{\mathbf{x}} \rangle \ge 0$. Since $\mathbf{x}$ is chosen arbitrarily in cl X, the proof of the first part of Theorem 1.12 is complete.

Let $\mathbf{u} \in$ int X and $\varepsilon > 0$ such that $\|\mathbf{x} - \mathbf{u}\| < \varepsilon$ implies $\mathbf{x} \in X$. Then $\mathbf{u} - \frac{\varepsilon}{2}\mathbf{h} \in X$ because $\|(\mathbf{u} - \frac{\varepsilon}{2}\mathbf{h}) - \mathbf{u}\| = \frac{\varepsilon}{2}\|\mathbf{h}\| = \frac{\varepsilon}{2} < \varepsilon$.

Assume that $\langle \mathbf{h}, \mathbf{u} - \overline{\mathbf{x}} \rangle = 0$. Then $\langle \mathbf{h}, \mathbf{u} \rangle = \langle \mathbf{h}, \overline{\mathbf{x}} \rangle$ and

$$\left\langle \mathbf{h}, \mathbf{u} - \frac{\varepsilon}{2}\mathbf{h} \right\rangle = \langle \mathbf{h}, \mathbf{u} \rangle - \frac{\varepsilon}{2}\langle \mathbf{h}, \mathbf{h} \rangle = \langle \mathbf{h}, \overline{\mathbf{x}} \rangle - \frac{\varepsilon}{2} < \langle \mathbf{h}, \overline{\mathbf{x}} \rangle$$

which contradicts the inequality $\langle \mathbf{h}, \mathbf{x} - \overline{\mathbf{x}} \rangle \geq 0$ for each $\mathbf{x} \in \text{cl } X$. ■

Corollary 1.2 (Separating a convex set and a noninterior point)
Let X be a nonempty convex set in $\mathbf{R}^n$ and $\overline{\mathbf{x}} \notin \text{int } X$. Then there is a vector $\mathbf{h} \in \mathbf{R}^n$, $\mathbf{h} \neq \mathbf{0}$ such that

$$\langle \mathbf{h}, \mathbf{x} - \overline{\mathbf{x}} \rangle \geq 0 \quad \text{for each } \mathbf{x} \in \text{cl } X.$$

Proof. If $\overline{\mathbf{x}} \in \partial X$, then Corollary 1.2 follows from Theorem 1.12. If $\overline{\mathbf{x}} \notin \text{cl } X$, then Corollary 1.2 follows from Theorem 1.11. ■

Theorem 1.13 (Separation and strict separation of sets)
X and Y are separable (strictly separable) if and only if $X - Y$ and $\{\mathbf{0}\}$ are separable (strictly separable).

Proof. *Necessity.* Let X and Y can be separated (strictly separated). Then there exists a nonzero vector $\mathbf{h} \in \mathbf{R}^n$ and a real number a such that the following inequalities are satisfied, respectively,

$$\langle \mathbf{h}, \mathbf{x} \rangle \geq a \quad \text{for each } \mathbf{x} \in X \quad \text{and} \quad \langle \mathbf{h}, \mathbf{y} \rangle \leq a \quad \text{for each } \mathbf{y} \in Y$$

$$\big(\langle \mathbf{h}, \mathbf{x} \rangle > a \quad \text{for each } \mathbf{x} \in X \quad \text{and} \quad \langle \mathbf{h}, \mathbf{y} \rangle < a \quad \text{for each } \mathbf{y} \in Y\big).$$

Hence, for each $\mathbf{z} = \mathbf{x} - \mathbf{y} \in X - Y$, $\mathbf{x} \in X$, $\mathbf{y} \in Y$, the following inequalities hold, respectively,

$$\langle \mathbf{h}, \mathbf{z} \rangle = \langle \mathbf{h}, \mathbf{x} - \mathbf{y} \rangle = \langle \mathbf{h}, \mathbf{x} \rangle - \langle \mathbf{h}, \mathbf{y} \rangle \geq 0 \equiv \langle \mathbf{h}, \mathbf{0} \rangle$$

$$(\langle \mathbf{h}, \mathbf{z} \rangle = \langle \mathbf{h}, \mathbf{x} - \mathbf{y} \rangle = \langle \mathbf{h}, \mathbf{x} \rangle - \langle \mathbf{h}, \mathbf{y} \rangle > 0 \equiv \langle \mathbf{h}, \mathbf{0} \rangle).$$

These inequalities mean that $X - Y$ and $\{\mathbf{0}\}$ can be separated (strictly separated, respectively).

Sufficiency. Let $X - Y$ and $\{\mathbf{0}\}$ can be separated (strictly separated). Therefore the inequalities established in the "necessity" part are satisfied. Moreover, the following inequalities hold true, respectively,

$$\sup_{\mathbf{x} \in X} \langle \mathbf{h}, \mathbf{x} \rangle - \inf_{\mathbf{y} \in Y} \langle \mathbf{h}, \mathbf{y} \rangle \geq 0 \qquad \Big(\sup_{\mathbf{x} \in X} \langle \mathbf{h}, \mathbf{x} \rangle - \inf_{\mathbf{y} \in Y} \langle \mathbf{h}, \mathbf{y} \rangle > 0\Big).$$

Therefore

$$\langle \mathbf{h}, \mathbf{x} \rangle \overset{\text{def}}{\leq} \sup_{\mathbf{x} \in X} \langle \mathbf{h}, \mathbf{x} \rangle \left\{ \begin{matrix} \leq \\ (<) \end{matrix} \right\} \inf_{\mathbf{y} \in Y} \langle \mathbf{h}, \mathbf{y} \rangle \overset{\text{def}}{\leq} \langle \mathbf{h}, \mathbf{y} \rangle \quad \forall \mathbf{x} \in X, \ \forall \mathbf{y} \in Y.$$

Hence, X and Y can be separated (strictly separated) by definition. ■

Corollary 1.3 (Strict separation of closed convex sets that do not have a point in common)
Let X and Y be nonempty closed convex sets that do not have a point in common and let Y be bounded. Then X and Y are strictly separable.

Proof. By Corollary 1.1 and Theorem 1.7, $X - Y$ is a closed and convex set. Clearly $\mathbf{0} \notin X - Y$, otherwise $\mathbf{0} = \mathbf{x} - \mathbf{y},\ \mathbf{x} \in X,\ \mathbf{y} \in Y$ and X and Y would have a point in common: $\mathbf{x} = \mathbf{y}$. Then Theorem 1.11 states that $X - Y$ and $\{\mathbf{0}\}$ can be strictly separated, and Theorem 1.13 states that X and Y can be strictly separated. ■

Theorem 1.14 (Separation of convex sets)
Let X and Y be nonempty convex sets and $X \cap Y = \emptyset$. Then X and Y can be separated by a hyperplane.

Proof. Denote $Z = X - Y$. Clearly $\mathbf{0} \notin Z$; otherwise X and Y would have a point in common.

We have to prove that Z and $\{\mathbf{0}\}$ are separable.

There are two cases to consider.

Case 1. Let $\mathbf{0} \notin \operatorname{cl} Z$. Since $\{\mathbf{0}\}$ is a bounded closed convex set, $\operatorname{cl} Z$ is a closed convex set and $\{\mathbf{0}\} \cap \operatorname{cl} Z = \emptyset$, then X and Y can be separated according to Corollary 1.3.

Case 2. Let $\mathbf{0} \in r\partial Z \equiv \operatorname{cl} Z \setminus \operatorname{ri} Z$.

We will prove as a subsidiary result the fact that if X is convex and $\mathbf{x}_0 \in \operatorname{cl} X \setminus \operatorname{ri} X$, then there is a supporting hyperplane H of X at $\mathbf{x}_0$ which does not contain X.

Without loss of generality assume that $\mathbf{0} \in \operatorname{aff} X$; if this is not the case, we will translate X to the set which contains $\mathbf{0}$. Then $\operatorname{aff} X = \mathbf{R}^k \subset \mathbf{R}^n$. The point $\mathbf{x}_0$ is a boundary point for X in $\operatorname{aff} X$ and X has an interior point (with respect to $\mathbf{R}^k$). By Theorem 1.12, there exists a supporting hyperplane H of X at $\mathbf{x}_0$. Since the inequalities are strict for the points of $\operatorname{int} X$, then H does not contain X.

Applying this result to $X = Z$ and $\mathbf{x}_0 = \mathbf{0}$ yields the statement of Theorem 1.14 in Case 2. ■

As an application of separation theorems, we will prove the following result.

Theorem 1.15 (Equality of compact convex sets from equality of supremums of a linear function on these sets)
Let X and Y be compact convex sets in $\mathbf{R}^n$ and

$$\sup_{\mathbf{z} \in X} \langle \mathbf{d}, \mathbf{z} \rangle = \sup_{\mathbf{z} \in Y} \langle \mathbf{d}, \mathbf{z} \rangle \quad \text{for all } \mathbf{d} \in \mathbf{R}^n.$$

Then $X = Y$.

Proof. Assume the contrary, that is, there is some $\mathbf{z}_0 \in X$ such that $\mathbf{z}_0 \notin Y$. Then there exists a nonzero vector $\mathbf{h}$ in $\mathbf{R}^n$ and a real number a with

$$\langle \mathbf{h}, \mathbf{y} \rangle \geq a > \langle \mathbf{h}, \mathbf{z}_0 \rangle \quad \text{for each } \mathbf{y} \in Y$$

according to Theorem 1.11. Denoting $\mathbf{d} = -\mathbf{h} \neq \mathbf{0}$, we get

$$\langle \mathbf{d}, \mathbf{y} \rangle < \langle \mathbf{d}, \mathbf{z}_0 \rangle \quad \text{for each } \mathbf{y} \in Y.$$

Since $\mathbf{z}_0 \in X$ we have

$$\langle \mathbf{d}, \mathbf{z}_0 \rangle \leq \sup_{\mathbf{z} \in X} \langle \mathbf{d}, \mathbf{z} \rangle.$$

Then

$$\sup_{\mathbf{y} \in Y} \langle \mathbf{d}, \mathbf{y} \rangle < \langle \mathbf{d}, \mathbf{z}_0 \rangle \leq \sup_{\mathbf{z} \in X} \langle \mathbf{d}, \mathbf{z} \rangle$$

which contradicts the hypothesis of Theorem 1.15. Therefore the assumption is incorrect, hence $X = Y$. ■

Theorem 1.16 (Proper separation of convex sets)
Let X_1 and X_2 be nonempty convex sets in $\mathbf{R}^n$. A necessary and sufficient condition for C_1 and C_2 to be properly separated by a hyperplane is that $\operatorname{ri} C_1 \cap \operatorname{ri} C_2 = \emptyset$.

The proof of this theorem is omitted.

1.3 Convex Functions

Definition 1.15 A function $f : X \to \mathbf{R}$ defined on a convex nonempty set $X \subseteq \mathbf{R}^n$ is said to be *convex* if for every $\mathbf{x}_1, \mathbf{x}_2 \in X,\ \lambda \in [0, 1]$ we have that

$$f(\lambda \mathbf{x}_1 + (1 - \lambda)\mathbf{x}_2) \leq \lambda f(\mathbf{x}_1) + (1 - \lambda) f(\mathbf{x}_2), \tag{1.1}$$

and f is said to be *strictly convex* if for every $\mathbf{x}_1, \mathbf{x}_2 \in X,\ \mathbf{x}_1 \neq \mathbf{x}_2,\ \lambda \in (0, 1)$ we have that

$$f(\lambda \mathbf{x}_1 + (1 - \lambda)\mathbf{x}_2) < \lambda f(\mathbf{x}_1) + (1 - \lambda) f(\mathbf{x}_2). \tag{1.2}$$

Taking "$\geq$" instead of "$\leq$" in (1.1) and "$>$" instead of "$<$" in (1.2), we obtain definitions of *concave* and *strictly concave* function, respectively.

f is concave (strictly concave) if and only if $-f$ is convex (strictly convex).

Note that convex, strictly convex, concave, and strictly concave functions are defined on a *convex* set because $\lambda \mathbf{x}_1 + (1 - \lambda)\mathbf{x}_2$ in (1.1) and (1.2) must be in X provided that $\mathbf{x}_1,\ \mathbf{x}_2 \in X$ and $\lambda \in [0, 1]$ ($\lambda \in (0, 1)$ for the strictly convex and strictly concave cases). That is why we will assume that the set on which a convex function is defined is convex even if this is not specified explicitly.

Example 1.2 The affine (linear) function $l(\mathbf{x}) = \langle \mathbf{c}, \mathbf{x} \rangle + a$, where $\mathbf{c}, \mathbf{x} \in \mathbf{R}^n, a \in \mathbf{R}$, is both convex and concave.

i) Let $l(\mathbf{x})$ be an affine function. For each $\mathbf{x}_1, \mathbf{x}_2 \in \mathbf{R}^n$ ($\mathbf{R}^n$ is a convex set by definition) and $\lambda \in [0, 1]$ we have that

$$\begin{aligned}
& l(\lambda \mathbf{x}_1 + (1-\lambda)\mathbf{x}_2) \\
&= \langle \mathbf{c}, \lambda \mathbf{x}_1 + (1-\lambda)\mathbf{x}_2 \rangle + a \\
&= \lambda \langle \mathbf{c}, \mathbf{x}_1 \rangle + (1-\lambda)\langle \mathbf{c}, \mathbf{x}_2 \rangle + \lambda a + (1-\lambda) a \\
&= \lambda [\langle \mathbf{c}, \mathbf{x}_1 \rangle + a] + (1-\lambda)[\langle \mathbf{c}, \mathbf{x}_2 \rangle + a] \\
&= \lambda l(\mathbf{x}_1) + (1-\lambda) l(\mathbf{x}_2).
\end{aligned}$$

The relation "=" (equals) can be considered as both "$\leq$" and "$\geq$".

Therefore $l(\mathbf{x})$ is both convex and concave (but not strictly) in $\mathbf{R}^n$ by definition.

ii) Let $l(\mathbf{x})$ be both convex and concave in $\mathbf{R}^n$. Hence

$$l(\lambda \mathbf{x}_1 + (1-\lambda)\mathbf{x}_2) \overset{\leq}{\underset{\geq}{}} \lambda l(\mathbf{x}_1) + (1-\lambda) l(\mathbf{x}_2),$$

that is,

$$l(\lambda \mathbf{x}_1 + (1-\lambda)\mathbf{x}_2) = \lambda l(\mathbf{x}_1) + (1-\lambda) l(\mathbf{x}_2)$$

for each $\mathbf{x}_1, \mathbf{x}_2 \in \mathrm{R}^n$ and $\lambda \in [0, 1]$. Therefore $l(\mathbf{x})$ is affine by definition. ■

Example 1.3 Quadratic function $f(x) = x^2,\ x \in \mathbf{R}$ is strictly convex.

First of all derive a subsidiary inequality. For each $x_1 \neq x_2$ we have that $(x_1 - x_2)^2 > 0$. Then $x_1^2 + x_2^2 > 2x_1x_2$ for $x_1 \neq x_2$.

Let $x_1, x_2 \in \mathrm{R},\ x_1 \neq x_2$, and $\lambda \in (0, 1)$. Then

$$\begin{aligned}
& f(\lambda x_1 + (1-\lambda)x_2) = (\lambda x_1 + (1-\lambda)x_2)^2 \\
&= \lambda^2 x_1^2 + 2\lambda(1-\lambda)x_1x_2 + (1-\lambda)^2 x_2^2 \\
&< \lambda^2 x_1^2 + \lambda(1-\lambda)(x_1^2 + x_2^2) + (1-\lambda)^2 x_2^2 \\
&\equiv \lambda x_1^2 + (1-\lambda)x_2^2 = \lambda f(x_1) + (1-\lambda) f(x_2).
\end{aligned}$$

Therefore $f(x) = x^2$ is strictly convex by definition. ■

Theorem 1.17 (Nonnegative linear combinations of convex and concave functions) *Let $f_i,\ i = 1, \ldots, m$ be numerical functions defined on $X \subseteq \mathbf{R}^n$. If f_i are convex (concave), then each linear combination with nonnegative coefficients of these functions*

$$f(\mathbf{x}) \overset{\text{def}}{=} \sum_{i=1}^{m} \alpha_i f_i(\mathbf{x}), \quad \alpha_i \geq 0,\ i = 1, \ldots, m$$

is convex (concave).

If f_i are convex (concave) and at least one of them is strictly convex (strictly concave) and corresponding α_i is positive, then the function $f(\mathbf{x})$ defined above is strictly convex (strictly concave).

Proof. Let $\mathbf{x}_1, \mathbf{x}_2 \in X$, $\lambda \in [0, 1]$, and f_i be convex. We have that

$$
\begin{aligned}
& f(\lambda \mathbf{x}_1 + (1-\lambda)\mathbf{x}_2) \\
&= \sum_{i=1}^{m} \alpha_i f_i(\lambda \mathbf{x}_1 + (1-\lambda)\mathbf{x}_2) \\
&\le \sum_{i=1}^{m} \alpha_i [\lambda f_i(\mathbf{x}_1) + (1-\lambda) f_i(\mathbf{x}_2)] \\
&= \lambda \sum_{i=1}^{m} \alpha_i f_i(\mathbf{x}_1) + (1-\lambda) \sum_{i=1}^{m} \alpha_i f_i(\mathbf{x}_2) \\
&\equiv \lambda f(\mathbf{x}_1) + (1-\lambda) f(\mathbf{x}_2).
\end{aligned}
$$

To obtain the inequality above we used that f_i are convex functions and $\alpha_i \ge 0,\ i = 1, \ldots, m$.

Therefore $f(\mathbf{x})$ is a convex function by definition.

The concave case is proved similarly.

In the strictly convex (strictly concave) case, we have strict inequality "$>$" ("$<$", respectively), and $\lambda \in (0, 1),\ \mathbf{x}_1, \mathbf{x}_2 \in X,\ \mathbf{x}_1 \neq \mathbf{x}_2$ in the proof above. ∎

Definition 1.16 The *effective domain* of a function f is the set

$$\operatorname{dom} f = \{\mathbf{x} \in \mathbf{R}^n : \ f(\mathbf{x}) < \infty\}.$$

Definition 1.17 Let X be a nonempty set in $\mathbf{R}^n$ and $f : X \to \mathbf{R}$. The set

$$\operatorname{epi} f = \{(\mathbf{x}, r) : \ f(\mathbf{x}) \le r,\ \mathbf{x} \in X,\ r \in \mathbf{R}\}$$

is called the *epigraph* of f.

Definition 1.18 f is said to be a *proper function* if $f(\mathbf{x}) < \infty$ for at least one $\mathbf{x} \in \mathbf{R}^n$ and $f(\mathbf{x}) > -\infty$ for all $\mathbf{x} \in \mathbf{R}^n$, or, in other words, if $\operatorname{dom} f$ is a nonempty set on which $f(\mathbf{x}) > -\infty$.

Otherwise f is called *improper*.

Theorem 1.18 (Necessary and sufficient condition for convexity of the epigraph) *Let X be a nonempty convex set in $\mathbf{R}^n$ and $f : X \to \mathbf{R}$. Then f is convex if and only if* $\operatorname{epi} f$ *is a convex set in* $\mathbf{R}^{n+1}$.

Proof. *Necessity.* Let f be a convex function and let $(\mathbf{x}_1, r_1),\ (\mathbf{x}_2, r_2) \in \operatorname{epi} f$. Therefore

$$f(\mathbf{x}_1) \le r_1, \quad f(\mathbf{x}_2) \le r_2, \quad \mathbf{x}_1,\ \mathbf{x}_2 \in X.$$

Let $\lambda \in [0, 1]$. Consider the convex combination of $(\mathbf{x}_1, r_1)$ and $(\mathbf{x}_2, r_2)$:

$$\lambda(\mathbf{x}_1, r_1) + (1-\lambda)(\mathbf{x}_2, r_2) = (\lambda \mathbf{x}_1 + (1-\lambda)\mathbf{x}_2, \lambda r_1 + (1-\lambda) r_2),$$

where $\lambda\mathbf{x}_1 + (1-\lambda)\mathbf{x}_2 \in X$ because X is convex. Then

$$\lambda r_1 + (1-\lambda)r_2 \geq \lambda f(\mathbf{x}_1) + (1-\lambda)f(\mathbf{x}_2) \geq f(\lambda\mathbf{x}_1 + (1-\lambda)\mathbf{x}_2),$$

where for the first inequality we have used the assumptions and that $\lambda \geq 0,\ 1-\lambda \geq 0$, and for the second inequality we have used that f is convex. Therefore

$$\lambda(\mathbf{x}_1, r_1) + (1-\lambda)(\mathbf{x}_2, r_2) = (\lambda\mathbf{x}_1 + (1-\lambda)\mathbf{x}_2, \lambda r_1 + (1-\lambda)r_2) \in \text{epi } f$$

by definition of epi f. Hence, epi f is a convex set by definition.

Sufficiency. Conversely, let epi f be convex and let $\mathbf{x}_1,\ \mathbf{x}_2 \in X$. Then $(\mathbf{x}_1, f(\mathbf{x}_1))$ and $(\mathbf{x}_2, f(\mathbf{x}_2))$ are in epi f with $r_i = f(\mathbf{x}_i),\ i = 1, 2$; that is, $(\mathbf{x}_1, f(\mathbf{x}_1))$ and $(\mathbf{x}_2, f(\mathbf{x}_2))$ satisfy the definition of epi f with equality. Since epi f is a convex set by assumption, then

$$\lambda(\mathbf{x}_1, f(\mathbf{x}_1)) + (1-\lambda)(\mathbf{x}_2, f(\mathbf{x}_2)) \in \text{epi } f \quad \text{for } \lambda \in [0, 1],$$

that is, $(\lambda\mathbf{x}_1 + (1-\lambda)\mathbf{x}_2,\ \lambda f(\mathbf{x}_1) + (1-\lambda)f(\mathbf{x}_2)) \in$ epi f, which means, by the definition of epi f, that

$$\lambda f(\mathbf{x}_1) + (1-\lambda)f(\mathbf{x}_2) \geq f(\lambda\mathbf{x}_1 + (1-\lambda)\mathbf{x}_2)$$

for $\lambda \in [0, 1]$. Therefore, f is convex by definition. ■

Sometimes a convex function is defined as a function f for which epi f is a convex set. Then (1.1) of Definition 1.15 is proved as a necessary and sufficient condition.

Theorem 1.19 (Convexity of the effective domain from convexity of the function) *If f is a convex function, then* dom f *is a convex set.*

Proof. Let $\mathbf{x}_1, \mathbf{x}_2 \in$ dom f. Then there exist $r_1, r_2 < \infty$ such that $f(\mathbf{x}_1) \leq r_1$, $f(\mathbf{x}_2) \leq r_2$, hence $(\mathbf{x}_1, r_1) \in$ epi f, $(\mathbf{x}_2, r_2) \in$ epi f. Therefore

$$\lambda(\mathbf{x}_1, r_1) + (1-\lambda)(\mathbf{x}_2, r_2) \in \text{epi } f \quad \text{for} \quad \lambda \in [0, 1]$$

because of the convexity of epi f (Theorem 1.18), that is,

$$f(\lambda\mathbf{x}_1 + (1-\lambda)\mathbf{x}_2) \leq \lambda r_1 + (1-\lambda)r_2 < \infty$$

by the assumption and definition of dom f. Hence $\lambda\mathbf{x}_1 + (1-\lambda)\mathbf{x}_2 \in$ dom f, therefore dom f is convex. ■

Theorem 1.20 (Convexity of the supremum of a family of convex functions) *Let $f_i : X \to \mathbf{R},\ i \in I$ be convex functions which are bounded from above on the convex set X in $\mathbf{R}^n$. Then the function*

$$f(\mathbf{x}) \stackrel{\text{def}}{=} \sup_{i \in I} f_i(\mathbf{x})$$

is convex on X.

f is strictly convex if each f_i is strictly convex and I is finite.

Proof. Let $\mathbf{x}_1,\ \mathbf{x}_2 \in X$ and $\lambda \in [0, 1]$. Since X is convex, then $\lambda\mathbf{x}_1 + (1-\lambda)\mathbf{x}_2 \in X$. Hence

$$\begin{aligned} f(\lambda\mathbf{x}_1 + (1-\lambda)\mathbf{x}_2) &\stackrel{\text{def}}{=} \sup_{i \in I} f_i(\lambda\mathbf{x}_1 + (1-\lambda)\mathbf{x}_2) \\ &\le \sup_{i \in I} [\lambda f_i(\mathbf{x}_1) + (1-\lambda) f_i(\mathbf{x}_2)] \\ &\le \lambda \sup_{i \in I} f_i(\mathbf{x}_1) + (1-\lambda) \sup_{i \in I} f_i(\mathbf{x}_2) \\ &\stackrel{\text{def}}{=} \lambda f(\mathbf{x}_1) + (1-\lambda) f(\mathbf{x}_2), \end{aligned}$$

where for the first inequality we have used that f_i are convex, and for the second inequality we have used that $\lambda \ge 0,\ 1-\lambda \ge 0$, and triangle inequality for the "supremum" function.

An alternative proof of Theorem 1.20 is the following.

According to definitions of $f(\mathbf{x})$ and epi f,

$$\text{epi } f = \bigcap_{i \in I} \text{epi } f_i,$$

where epi f_i are convex sets for $i \in I$ by Theorem 1.18. Then epi f is convex by Theorem 1.1 as an intersection of a family of convex sets. Therefore f is convex according to Theorem 1.18.

If each f_i is strictly convex and I is finite, then the inequalities in the proof above are strict and therefore $f(\mathbf{x})$ is strictly convex. ■

Theorem 1.21 (Convexity of a set defined by convex functions)
If $Y \subseteq \mathbf{R}^n$ is a convex set, $f_i : Y \to \mathbf{R},\ i = 1, \ldots, m$ are convex functions, then the set

$$X \stackrel{\text{def}}{=} \{\mathbf{x} \in Y : f_i(\mathbf{x}) \le 0,\ i = 1, \ldots, m,\ \mathbf{x} \ge \mathbf{0}\}$$

is convex.

Proof. Let $\mathbf{x}_1, \mathbf{x}_2 \in X$ (therefore, $\mathbf{x}_1, \mathbf{x}_2 \in Y,\ f_i(\mathbf{x}_1) \le 0,\ f_i(\mathbf{x}_2) \le 0, i = 1, \ldots, m,$ $\mathbf{x}_1 \ge \mathbf{0}, \mathbf{x}_2 \ge \mathbf{0}$) and $\lambda \in [0, 1]$.

Since Y is a convex set, $f_i(\mathbf{x})$ are convex functions for each $i = 1, \ldots, m$ and $\mathbf{x}_1, \mathbf{x}_2 \in X$, then

$$\lambda\mathbf{x}_1 + (1-\lambda)\mathbf{x}_2 \in Y,$$

$$f_i(\lambda\mathbf{x}_1 + (1-\lambda)\mathbf{x}_2) \le \lambda f_i(\mathbf{x}_1) + (1-\lambda) f_i(\mathbf{x}_2) \le 0$$

and

$$\lambda \mathbf{x}_1 + (1-\lambda)\mathbf{x}_2 \geq \mathbf{0},$$

where we have used that $\lambda \geq 0,\ 1-\lambda \geq 0;\ f_i(\mathbf{x}_1) \leq 0,\ f_i(\mathbf{x}_2) \leq 0, i = 1, \dots, m$ and $\mathbf{x}_1 \geq \mathbf{0}, \mathbf{x}_2 \geq \mathbf{0}$.

Therefore $\lambda \mathbf{x}_1 + (1-\lambda)\mathbf{x}_2 \in X$, that is, X is convex by definition.

An alternative way to prove this theorem is the following.

The set X can be represented as

$$X = X_1 \cap X_2 \cap \dots \cap X_m,$$

where

$$X_i \stackrel{\text{def}}{=} \{\mathbf{x} \in Y : f_i(\mathbf{x}) \leq 0,\ \mathbf{x} \geq \mathbf{0}\}, \quad i = 1, \dots, m.$$

As above, it is easy to prove that X_i is convex for each $i = 1, \dots, m$. Then $X = \bigcap_{i=1}^{m} X_i$ is convex according to Theorem 1.1. ■

Theorem 1.22 (Jensen's inequality)
Let $f : \mathbf{R}^n \to \overline{\mathbf{R}}$. *Then* f *is convex if and only if*

$$f\left(\sum_{i=1}^{m} \lambda_i \mathbf{x}_i\right) \leq \sum_{i=1}^{m} \lambda_i f(\mathbf{x}_i)$$

for $\lambda_1 \geq 0, \dots, \lambda_m \geq 0,\ \sum_{i=1}^{m} \lambda_i = 1$.

Proof. Without loss of generality let $\lambda_i > 0,\ i = 1, \dots, m$; otherwise the number of terms in the sum is less than m.

Necessity. (By induction) Let f be convex. If $m = 2$, the inequality is satisfied by definition of convex function.

Induction hypothesis. Let the inequality be satisfied for $m = k-1$.

Verify it for $m = k$. Consider the point

$$\mathbf{x} = \sum_{i=1}^{k} \lambda_i \mathbf{x}_i \quad \text{where} \quad \mathbf{x}_i \in X,\ \lambda_i > 0,\ i = 1, \dots, k,\ \sum_{i=1}^{k} \lambda_i = 1.$$

Then

$$f\left(\sum_{i=1}^{k} \lambda_i \mathbf{x}_i\right) \equiv f\left(\lambda_1 \mathbf{x}_1 + \sum_{i=2}^{k} \lambda_i \mathbf{x}_i\right) = f\left(\lambda_1 \mathbf{x}_1 + \lambda \sum_{i=2}^{k} \frac{\lambda_i}{\lambda} \mathbf{x}_i\right),$$

where $\lambda = \sum_{i=2}^{k} \lambda_i > 0$. Since $\frac{\lambda_i}{\lambda} > 0, \sum_{i=2}^{k} \frac{\lambda_i}{\lambda} = \frac{1}{\lambda} \sum_{i=2}^{k} \lambda_i = \frac{1}{\lambda}\lambda = 1, \mathbf{x}_i \in X$, and X is a convex set,

$$\mathbf{y} \stackrel{\text{def}}{=} \sum_{i=2}^{k} \frac{\lambda_i}{\lambda} \mathbf{x}_i \in X.$$

Because f is convex and $\mathbf{x}_1, \mathbf{y} \in X,\ \lambda_1 + \lambda = 1,\ \lambda_1, \lambda \in [0, 1]$,

$$f(\mathbf{x}) \equiv f(\lambda_1 \mathbf{x}_1 + \lambda \mathbf{y}) \leq \lambda_1 f(\mathbf{x}_1) + \lambda f(\mathbf{y}).$$

On the other hand,

$$f(\mathbf{y}) \equiv f\left(\sum_{i=2}^{k} \frac{\lambda_i}{\lambda} \mathbf{x}_i\right) \leq \sum_{i=2}^{k} \frac{\lambda_i}{\lambda} f(\mathbf{x}_i)$$

according to the induction hypothesis. From both inequalities it follows that

$$f\left(\sum_{i=1}^{k} \lambda_i \mathbf{x}_i\right) \equiv f\left(\lambda_1 \mathbf{x}_1 + \lambda \sum_{i=2}^{k} \frac{\lambda_i}{\lambda} \mathbf{x}_i\right) \equiv f(\lambda_1 \mathbf{x}_1 + \lambda \mathbf{y})$$

$$\leq \lambda_1 f(\mathbf{x}_1) + \lambda \sum_{i=2}^{k} \frac{\lambda_i}{\lambda} f(\mathbf{x}_i) = \lambda_1 f(\mathbf{x}_1) + \sum_{i=2}^{k} \lambda_i f(\mathbf{x}_i) \equiv \sum_{i=1}^{k} \lambda_i f(\mathbf{x}_i).$$

An alternative proof of "necessity" part of Theorem 1.22.

As above, without loss of generality, let $\lambda_i > 0,\ i = 1, \ldots, m$. If $\mathbf{x}_i \notin \operatorname{dom} f,\ i = 1, \ldots, m$, then $f(\mathbf{x}_i) = +\infty,\ \lambda_i f(\mathbf{x}_i) = +\infty$, and Jensen's inequality is trivially satisfied.

Otherwise, let $\mathbf{x}_i \in \operatorname{dom} f,\ i = 1, \ldots, m$. Since epi f is convex (Theorem 1.18) and $(\mathbf{x}_i, f(\mathbf{x}_i)) \in \operatorname{epi} f,\ i = 1, \ldots, m$ by Definition 1.17,

$$(\lambda_1 \mathbf{x}_1 + \cdots + \lambda_m \mathbf{x}_m,\ \lambda_1 f(\mathbf{x}_1) + \cdots + \lambda_m f(\mathbf{x}_m)) \in \operatorname{epi} f$$

according to Theorem 1.2. Therefore

$$f(\lambda_1 \mathbf{x}_1 + \cdots + \lambda_m \mathbf{x}_m) \leq \lambda_1 f(\mathbf{x}_1) + \cdots + \lambda_m f(\mathbf{x}_m)$$

by definition of epi f, that is, Jensen's inequality is satisfied.

Sufficiency. Let Jensen's inequality be satisfied for f with arbitrary m. Then this inequality with $m = 2$ implies convexity of f by definition of convex function. ■

Theorem 1.23 (Continuity of a convex function in the interior of a convex set) *Let X be a nonempty convex set in $\mathbf{R}^n$ and $f : X \to \mathbf{R}$ be a convex function. Then f is continuous on the interior* int X *of X.*

In particular, f is continuous in ri (dom f).

Proof. Let $\mathbf{x}_0 \in \operatorname{int} X$ and $\varepsilon > 0$ such that $f(\mathbf{x}) \leq r$ for every $\mathbf{x} \in \overline{B}(\mathbf{x}_0, \varepsilon)$, where $\overline{B}(\mathbf{x}_0, \varepsilon)$ is the closed ball with radius ε in $\mathbf{R}^n$ around $\mathbf{x}_0$. For each $\mathbf{y} \in B(\mathbf{x}_0, \varepsilon)$ and

$\mathbf{y} \neq \mathbf{x}_0$ denote

$$\mathbf{w} = \mathbf{x}_0 + \frac{\varepsilon(\mathbf{y} - \mathbf{x}_0)}{\|\mathbf{y} - \mathbf{x}_0\|}, \quad \mathbf{z} = \mathbf{x}_0 - \frac{\varepsilon(\mathbf{y} - \mathbf{x}_0)}{\|\mathbf{y} - \mathbf{x}_0\|}.$$

Clearly $\mathbf{w}, \mathbf{z} \in \overline{B}(\mathbf{x}_0, \varepsilon)$ because $\|\mathbf{w} - \mathbf{x}_0\| = \varepsilon,\ \|\mathbf{z} - \mathbf{x}_0\| = \varepsilon$.

The point $\mathbf{x}_0$ belongs to the line segment $[\mathbf{y}, \mathbf{z}]$ and it can be expressed as a convex combination of $\mathbf{y}$ and $\mathbf{z}$:

$$\mathbf{x}_0 = \frac{1}{1+\lambda}\mathbf{z} + \frac{\lambda}{1+\lambda}\mathbf{y},$$

where $\lambda = \dfrac{\varepsilon}{\|\mathbf{y} - \mathbf{x}_0\|} > 1$ because $\mathbf{y} \in B(\mathbf{x}_0, \varepsilon)$ and therefore $\|\mathbf{y} - \mathbf{x}_0\| < \varepsilon$. By convexity of f, we have

$$f(\mathbf{x}_0) \leq \frac{1}{1+\lambda} f(\mathbf{z}) + \frac{\lambda}{1+\lambda} f(\mathbf{y}).$$

Hence $\lambda(f(\mathbf{x}_0) - f(\mathbf{y})) \leq f(\mathbf{z}) - f(\mathbf{x}_0)$, and since $\mathbf{z} \in \overline{B}(\mathbf{x}_0, \varepsilon)$, then $\lambda(f(\mathbf{x}_0) - f(\mathbf{y})) \leq f(\mathbf{z}) - f(\mathbf{x}_0) \leq r - f(\mathbf{x}_0)$, that is,

$$f(\mathbf{x}_0) - f(\mathbf{y}) \leq \frac{1}{\lambda}(r - f(\mathbf{x}_0)) = \frac{\|\mathbf{y} - \mathbf{x}_0\|}{\varepsilon}(r - f(\mathbf{x}_0)).$$

On the other hand, since $\mathbf{y}$ belongs to the line segment joining $\mathbf{x}_0$ and $\mathbf{w}$, then $\mathbf{y}$ can be expressed as

$$\mathbf{y} = \frac{1}{\lambda}\mathbf{w} + \frac{\lambda - 1}{\lambda}\mathbf{x}_0,$$

where $\lambda = \dfrac{\varepsilon}{\|\mathbf{y} - \mathbf{x}_0\|} > 1$. Therefore $\frac{1}{\lambda}, \frac{\lambda-1}{\lambda} \in [0, 1]$ and $\frac{1}{\lambda} + \frac{\lambda-1}{\lambda} = 1$.

By convexity of f,

$$f(\mathbf{y}) \leq \frac{1}{\lambda} f(\mathbf{w}) + \frac{\lambda - 1}{\lambda} f(\mathbf{x}_0).$$

Since $\mathbf{w} \in \overline{B}(\mathbf{x}_0, \varepsilon)$, then $f(\mathbf{w}) \leq r$. Therefore

$$\lambda(f(\mathbf{y}) - f(\mathbf{x}_0)) \leq f(\mathbf{w}) - f(\mathbf{x}_0) \leq r - f(\mathbf{x}_0).$$

Hence

$$f(\mathbf{y}) - f(\mathbf{x}_0) \leq \frac{1}{\lambda}(r - f(\mathbf{x}_0)) = \frac{\|\mathbf{y} - \mathbf{x}_0\|}{\varepsilon}(r - f(\mathbf{x}_0)).$$

From both inequalities we have obtained it follows that

$$|f(\mathbf{y}) - f(\mathbf{x}_0)| \leq \frac{1}{\varepsilon}(r - f(\mathbf{x}_0))\|\mathbf{y} - \mathbf{x}_0\|,$$

that is, f is a continuous function by definition (see Appendix A, Definition A.7). ■

If $X = \mathbf{R}^n$, then $\operatorname{int} \mathbf{R}^n = \mathbf{R}^n$ and the convex function f is continuous in $\mathbf{R}^n$ according to Theorem 1.23.

If the set X is *open* in $\mathbf{R}^n$, then $\operatorname{int} X = X$ and the convex function f is continuous on X.

1.4 Directional Derivatives, Subgradients, and Subdifferentials. Differentiable Convex Functions

Definition 1.19 Let X be a nonempty set in $\mathbf{R}^n$, $f : X \to \mathbf{R}$, $\mathbf{x}_0 \in X$, and $\mathbf{d}$ be a nonzero vector in $\mathbf{R}^n$ such that $\mathbf{x}_0 + \lambda\mathbf{d} \in X$ for sufficiently small $\lambda > 0$. The *directional derivative* of f at $\mathbf{x}_0$ along the vector (direction) $\mathbf{d}$, denoted by $f'(\mathbf{x}_0; \mathbf{d})$ or $\frac{\partial f}{\partial \mathbf{d}}(\mathbf{x}_0)$, is the limit (if it exists)

$$f'(\mathbf{x}_0; \mathbf{d}) = \frac{\partial f}{\partial \mathbf{d}}(\mathbf{x}_0) = \lim_{\lambda \to +0} \frac{f(\mathbf{x}_0 + \lambda\mathbf{d}) - f(\mathbf{x}_0)}{\lambda}.$$

A function f for which the directional derivative at $\mathbf{x}_0$ exists is said to be *directionally differentiable* at $\mathbf{x}_0$.

Definition 1.20 A vector $\widehat{\mathbf{f}}(\mathbf{x}_0) \in \mathbf{R}^n$ is said to be a *subgradient* of a (convex) function $f : X \to \mathbf{R}$ at $\mathbf{x}_0$ if the inequality

$$f(\mathbf{x}) - f(\mathbf{x}_0) \geq \langle \widehat{\mathbf{f}}(\mathbf{x}_0), \mathbf{x} - \mathbf{x}_0 \rangle$$

holds for each $\mathbf{x} \in X \subseteq \mathbf{R}^n$, where $\langle \mathbf{x}, \mathbf{y} \rangle$ denotes the scalar (inner) product of $\mathbf{x}$ and $\mathbf{y}$ in $\mathbf{R}^n$.

This definition means that the graph of the affine function $h(\mathbf{x}) = f(\mathbf{x}_0) + \langle \widehat{\mathbf{f}}(\mathbf{x}_0), \mathbf{x} - \mathbf{x}_0 \rangle$ is a supporting hyperplane to the convex set epi f at the point $(\mathbf{x}_0, f(\mathbf{x}_0))$.

Definition 1.21 The set containing all subgradients of f at the point $\mathbf{x}_0$ is called the *subdifferential* of f at $\mathbf{x}_0$ and it is denoted by $\partial f(\mathbf{x}_0)$.

If $\partial f(\mathbf{x}_0) \neq \emptyset$, then f is said to be *subdifferentiable* at $\mathbf{x}_0$.

The subdifferential of f is defined as the multivalued mapping $\partial f : \mathbf{x} \to \partial f(\mathbf{x})$.

Example 1.4 Let $f(\mathbf{x}) = \|\mathbf{x}\|$, $\mathbf{x} \in \mathbf{R}^n$. Find $\partial f(\mathbf{0})$.

By definition, $\widehat{\mathbf{x}} \in \partial f(\mathbf{0})$ if and only if

$$f(\mathbf{x}) - f(\mathbf{0}) \geq \langle \widehat{\mathbf{x}}, \mathbf{x} - \mathbf{0} \rangle \quad \text{for all } \mathbf{x} \in \mathbf{R}^n,$$

that is,

$$\|\mathbf{x}\| - \|\mathbf{0}\| \equiv \|\mathbf{x}\| \geq \langle \widehat{\mathbf{x}}, \mathbf{x} \rangle \quad \text{for all } \mathbf{x} \in \mathbf{R}^n.$$

Therefore $\partial f(\mathbf{0}) = \{\widehat{\mathbf{x}} \in \mathbf{R}^n : \|\widehat{\mathbf{x}}\| \leq 1\}$.

When $n = 1$, $\partial f(0) = \{x \in \mathbf{R} : |x| \leq 1\}$.

Theorem 1.24 (Directional differentiability of a convex function) *Let X be a convex set in $\mathbf{R}^n$ and $f : X \to \mathbf{R}$ be a convex function. Then the directional derivative $\frac{\partial f}{\partial \mathbf{d}}(\mathbf{x}_0)$ of f at $\mathbf{x}_0 \in \operatorname{int} X$ in the direction $\mathbf{d} \in \mathbf{R}^n$, $\mathbf{d} \neq \mathbf{0}$ exists.*

Proof. Since $\mathbf{x}_0 \in \operatorname{int} X$, then there exists a $\lambda_0 > 0$ such that for $\lambda \in (0, \lambda_0)$ we have $\mathbf{x}_0 + \lambda \mathbf{d} \in X$.

Consider the difference quotient

$$q(\lambda) \stackrel{\text{def}}{=} \frac{f(\mathbf{x}_0 + \lambda \mathbf{d}) - f(\mathbf{x}_0)}{\lambda}.$$

Let $0 < \lambda_2 < \lambda_1 \leq \lambda_0$. Hence $0 < \frac{\lambda_2}{\lambda_1} < 1$. Since f is convex,

$$\begin{aligned} f(\mathbf{x}_0 + \lambda_2 \mathbf{d}) &\equiv f\left(\frac{\lambda_2}{\lambda_1}(\mathbf{x}_0 + \lambda_1 \mathbf{d}) + \left(1 - \frac{\lambda_2}{\lambda_1}\right)\mathbf{x}_0\right) \\ &\leq \frac{\lambda_2}{\lambda_1} f(\mathbf{x}_0 + \lambda_1 \mathbf{d}) + \left(1 - \frac{\lambda_2}{\lambda_1}\right) f(\mathbf{x}_0), \end{aligned}$$

whence

$$\frac{f(\mathbf{x}_0 + \lambda_2 \mathbf{d}) - f(\mathbf{x}_0)}{\lambda_2} \leq \frac{f(\mathbf{x}_0 + \lambda_1 \mathbf{d}) - f(\mathbf{x}_0)}{\lambda_1}$$

for $\lambda_1 > \lambda_2 > 0$. Therefore the difference quotient $q(\lambda)$ is a monotone nonincreasing function of λ, $\lambda \to +0$.

From convexity of f, for $\lambda > 0$ we have

$$\begin{aligned} f(\mathbf{x}_0) &\equiv f\left(\frac{\lambda}{1+\lambda}(\mathbf{x}_0 - \mathbf{d}) + \frac{1}{1+\lambda}(\mathbf{x}_0 + \lambda \mathbf{d})\right) \\ &\leq \frac{\lambda}{1+\lambda} f(\mathbf{x}_0 - \mathbf{d}) + \frac{1}{1+\lambda} f(\mathbf{x}_0 + \lambda \mathbf{d}), \end{aligned}$$

whence

$$\frac{\lambda}{1+\lambda}[f(\mathbf{x}_0) - f(\mathbf{x}_0 - \mathbf{d})] \leq \frac{1}{1+\lambda}[f(\mathbf{x}_0 + \lambda \mathbf{d}) - f(\mathbf{x}_0)],$$

$$f(\mathbf{x}_0) - f(\mathbf{x}_0 - \mathbf{d}) \leq \frac{f(\mathbf{x}_0 + \lambda \mathbf{d}) - f(\mathbf{x}_0)}{\lambda},$$

that is, $q(\lambda)$ is bounded from below by the constant $f(\mathbf{x}_0) - f(\mathbf{x}_0 - \mathbf{d})$.

We have proved that $q(\lambda)$ is a monotone nonincreasing function of $\lambda \to +0$ and it is bounded from below. Therefore the finite limit exists

$$\frac{\partial f}{\partial \mathbf{d}}(\mathbf{x}_0) = \lim_{\lambda \to +0} q(\lambda) = \inf_{\lambda > 0} \frac{f(\mathbf{x}_0 + \lambda \mathbf{d}) - f(\mathbf{x}_0)}{\lambda},$$

that is, $f(\mathbf{x})$ is directionally differentiable at $\mathbf{x}_0 \in \operatorname{int} X$ along each direction $\mathbf{d} \in \mathbf{R}^n$, $\mathbf{d} \neq \mathbf{0}$. ■

Theorem 1.25 (Properties of the subdifferential of a convex function)
Let f be a convex proper function on $\mathbf{R}^n$ and let $\mathbf{x}_0 \in \mathbf{R}^n$. Then the set $\partial f(\mathbf{x}_0)$ is nonempty, convex, closed, and bounded.

Proof. Consider the epigraph of f

$$\operatorname{epi} f = \{(\mathbf{x}, r) \in \mathbf{R}^{n+1} : \quad f(\mathbf{x}) \leq r, \ \ \mathbf{x} \in \mathbf{R}^n, \ \ r \in \mathbf{R}\}.$$

Since f is convex, then epi f is a convex set according to Theorem 1.18.

Point $(\mathbf{x}_0, f(\mathbf{x}_0))$ is a boundary point of epi f. From Theorem 1.12 it follows that there exist a $(\mathbf{h}, c) \in \mathbf{R}^{n+1}$, $(\mathbf{h}, c) \neq (\mathbf{0}, 0)$ and an $a \in \mathbf{R}$ such that

$$\langle(\mathbf{h}, c), (\mathbf{x}, r)\rangle \leq a \quad \text{for each} \ \ (\mathbf{x}, r) \in \ \operatorname{epi} f$$

and

$$\langle(\mathbf{h}, c), (\mathbf{x}_0, f(\mathbf{x}_0))\rangle > a,$$

that is,

$$\langle\mathbf{h}, \mathbf{x}\rangle + cr \leq a < \langle\mathbf{h}, \mathbf{x}_0\rangle + cf(\mathbf{x}_0).$$

When $\mathbf{x} = \mathbf{x}_0$, we get $cr \leq cf(\mathbf{x}_0)$, that is, $c(r - f(\mathbf{x}_0)) \leq 0$, and since $f(\mathbf{x}_0) \leq r$ then $c \leq 0$.

Assume that $c = 0$. Then $\langle\mathbf{h}, \mathbf{x} - \mathbf{x}_0\rangle \leq 0$ for each $\mathbf{x} \in \mathbf{R}^n$. Hence, $\mathbf{h} = \mathbf{0}$, that is, $(\mathbf{h}, c) = (\mathbf{0}, 0)$, a contradiction. Therefore the assumption is incorrect, and $c < 0$ strictly.

Then, setting $r = f(\mathbf{x})$, we obtain

$$\langle\mathbf{h}, \mathbf{x}\rangle + cf(\mathbf{x}) \leq \langle\mathbf{h}, \mathbf{x}_0\rangle + cf(\mathbf{x}_0),$$

$$\langle\mathbf{h}, \mathbf{x} - \mathbf{x}_0\rangle \leq c(f(\mathbf{x}_0) - f(\mathbf{x})).$$

Set $c_0 = -c > 0$. Then

$$\langle\mathbf{h}, \mathbf{x} - \mathbf{x}_0\rangle \leq c_0(f(\mathbf{x}) - f(\mathbf{x}_0)),$$

$$\langle c_0^{-1}\mathbf{h}, \mathbf{x} - \mathbf{x}_0\rangle \leq f(\mathbf{x}) - f(\mathbf{x}_0).$$

Hence $c_0^{-1}\mathbf{h} \in \partial f(\mathbf{x}_0)$ by Definition 1.20, that is, $\partial f(\mathbf{x}_0) \neq \emptyset$.

Let $\mathbf{d}_1, \mathbf{d}_2 \in \partial f(\mathbf{x}_0)$. Therefore

$$f(\mathbf{x}) - f(\mathbf{x}_0) \geq \langle\mathbf{d}_i, \mathbf{x} - \mathbf{x}_0\rangle, \quad i = 1, 2.$$

Then for $\lambda \in [0, 1]$ we have

$$\begin{aligned} f(\mathbf{x}) - f(\mathbf{x}_0) &\equiv \lambda[f(\mathbf{x}) - f(\mathbf{x}_0)] + (1-\lambda)[f(\mathbf{x}) - f(\mathbf{x}_0)] \\ &\geq \lambda\langle \mathbf{d}_1, \mathbf{x} - \mathbf{x}_0\rangle + (1-\lambda)\langle \mathbf{d}_2, \mathbf{x} - \mathbf{x}_0\rangle \\ &= \langle \lambda\mathbf{d}_1 + (1-\lambda)\mathbf{d}_2, \mathbf{x} - \mathbf{x}_0\rangle, \end{aligned}$$

that is, $\lambda\mathbf{d}_1 + (1-\lambda)\mathbf{d}_2 \in \partial f(\mathbf{x}_0)$. Therefore $\partial f(\mathbf{x}_0)$ is a convex set.

Let $\{\mathbf{d}_k\}_{k=1}^{\infty} \in \partial f(\mathbf{x}_0)$, $\mathbf{d}_k \to \overline{\mathbf{d}}$ when $k \to \infty$. Since $\{\mathbf{d}_k\} \in \partial f(\mathbf{x}_0)$,

$$f(\mathbf{x}) - f(\mathbf{x}_0) \geq \langle \mathbf{d}_k, \mathbf{x} - \mathbf{x}_0\rangle \quad \text{for any } \mathbf{x} \in \mathbf{R}^n.$$

Using the continuity of inner product, it follows that

$$f(\mathbf{x}) - f(\mathbf{x}_0) \geq \langle \overline{\mathbf{d}}, \mathbf{x} - \mathbf{x}_0\rangle \quad \text{for any } \mathbf{x} \in \mathbf{R}^n,$$

that is, $\overline{\mathbf{d}} \in \partial f(\mathbf{x}_0)$. Therefore $\partial f(\mathbf{x}_0)$ is a closed set by definition (Appendix A, Definition A.2).

Concerning boundedness of $\partial f(\mathbf{x}_0)$, assume the contrary, that there exists a sequence $\{\mathbf{d}_k\}_{k=1}^{\infty}$ such that $\{\mathbf{d}_k\} \in \partial f(\mathbf{x}_0)$ and $\|\mathbf{d}_k\| \to \infty$.

Set $\mathbf{x}_k = \mathbf{x}_0 + \mathbf{z}_k$, where $\mathbf{z}_k = \varepsilon_0 \frac{\mathbf{d}_k}{\|\mathbf{d}_k\|}$, $\varepsilon_0 > 0$. From the definition of $\partial f(\mathbf{x}_0)$ it follows that

$$f(\mathbf{x}_k) - f(\mathbf{x}_0) \geq \langle \mathbf{d}_k, \mathbf{x}_k - \mathbf{x}_0\rangle = \langle \mathbf{d}_k, \mathbf{z}_k\rangle = \varepsilon_0 \frac{\|\mathbf{d}_k\|^2}{\|\mathbf{d}_k\|} = \varepsilon_0\|\mathbf{d}_k\| \to \infty.$$

Since

$$\|\mathbf{x}_k - \mathbf{x}_0\| = \|\mathbf{z}_k\| = \left\| \varepsilon_0 \frac{\mathbf{d}_k}{\|\mathbf{d}_k\|} \right\| = \varepsilon_0$$

and since $f(\mathbf{x})$ is a continuous function because it is convex (Theorem 1.23), then $f(\mathbf{x})$ is bounded on the set

$$X_{\varepsilon_0}(\mathbf{x}_0) \stackrel{\text{def}}{=} \{\mathbf{x} \in \mathbf{R}^n : \|\mathbf{x} - \mathbf{x}_0\| \leq \varepsilon_0\}, \quad \varepsilon_0 > 0.$$

This contradicts the inequality

$$f(\mathbf{x}_k) - f(\mathbf{x}_0) \geq \varepsilon_0\|\mathbf{d}_k\| \to \infty$$

obtained above. Therefore the assumption is incorrect and $\partial f(\mathbf{x}_0)$ is bounded (Definition A.5, Appendix A). ■

Corollary 1.4 (Subdifferential of a convex function)
$\partial f(\mathbf{x}) \neq \emptyset$ if and only if f is a convex function.

Proof. By Theorem 1.25, if f is convex then $\partial f(\mathbf{x}) \neq \emptyset$.

Conversely, let $\partial f(\mathbf{x}) \neq \emptyset$ for any $\mathbf{x} \in \mathbf{R}^n$, let $\mathbf{x}_1, \mathbf{x}_2 \in \mathbf{R}^n$ and $0 < \lambda < 1$. Then there exists a subgradient $\widehat{\mathbf{x}} \in \partial f(\mathbf{x})$ for $\mathbf{x} = \lambda \mathbf{x}_1 + (1 - \lambda)\mathbf{x}_2$. Using the definition of subgradient, we get

$$f(\mathbf{x}_1) - f(\lambda \mathbf{x}_1 + (1 - \lambda)\mathbf{x}_2) \geq (1 - \lambda)\langle \widehat{\mathbf{x}}, \mathbf{x}_1 - \mathbf{x}_2 \rangle,$$
$$f(\mathbf{x}_2) - f(\lambda \mathbf{x}_1 + (1 - \lambda)\mathbf{x}_2) \geq \lambda \langle \widehat{\mathbf{x}}, \mathbf{x}_2 - \mathbf{x}_1 \rangle.$$

Multiplying the first inequality by $\lambda > 0$ and the second one by $1 - \lambda > 0$ and adding them yields

$$\lambda f(\mathbf{x}_1) + (1 - \lambda) f(\mathbf{x}_2) \geq f(\lambda \mathbf{x}_1 + (1 - \lambda)\mathbf{x}_2),$$

that is, f is convex by definition. ■

Theorem 1.26 (Relationship between directional derivatives and subgradients of a convex function)
Let X be an open convex set in $\mathbf{R}^n$, $f : X \to \mathbf{R}$ be convex, $\mathbf{x}_0 \in X$, $\mathbf{d} \in \mathbf{R}^n$. Then there exists a subgradient $\widehat{\mathbf{y}} \in \partial f(\mathbf{x}_0)$ such that

$$\frac{\partial f}{\partial \mathbf{d}}(\mathbf{x}_0) = \langle \widehat{\mathbf{y}}, \mathbf{d} \rangle = \max_{\widehat{\mathbf{x}} \in \partial f(\mathbf{x}_0)} \langle \widehat{\mathbf{x}}, \mathbf{d} \rangle.$$

Proof. By Theorem 1.24, f is differentiable along each direction $\mathbf{d}$.

Denote $q(\lambda) = \dfrac{f(\mathbf{x}_0 + \lambda \mathbf{d}) - f(\mathbf{x}_0)}{\lambda}$. Using the defining inequality for the subgradient

$$f(\mathbf{x}_0 + \lambda \mathbf{d}) - f(\mathbf{x}_0) \geq \langle \widehat{\mathbf{x}}, \lambda \mathbf{d} \rangle \quad \forall \widehat{\mathbf{x}} \in \partial f(\mathbf{x}_0)$$

we get

$$\frac{\partial f}{\partial \mathbf{d}}(\mathbf{x}_0) = \lim_{\lambda \to +0} q(\lambda) \geq \max_{\widehat{\mathbf{x}} \in \partial f(\mathbf{x}_0)} \langle \widehat{\mathbf{x}}, \mathbf{d} \rangle.$$

Consider the sets

$$X_1 = \{(\mathbf{x}, r) \in \mathbf{R}^{n+1} : \mathbf{x} \in X, \ f(\mathbf{x}) < r\}$$

and

$$X_2 = \left\{ \left(\mathbf{x}_0 + \lambda \mathbf{d}, \ f(\mathbf{x}_0) + \lambda \frac{\partial f}{\partial \mathbf{d}}(\mathbf{x}_0) \right) : \ \lambda > 0 \right\}.$$

Since $q(\lambda)$ is a monotone nonincreasing function for $\lambda \to +0$ (see the proof of Theorem 1.24) then

$$\frac{\partial f}{\partial \mathbf{d}}(\mathbf{x}_0) = \lim_{\lambda \to +0} q(\lambda) \leq q(\lambda) \quad \forall \, \lambda > 0,$$

that is,

$$f(\mathbf{x}_0 + \lambda \mathbf{d}) \geq f(\mathbf{x}_0) + \lambda \frac{\partial f}{\partial \mathbf{d}}(\mathbf{x}_0) \quad \forall \lambda > 0.$$

The sets X_1 and X_2 are convex (X_1 is the "open" epi f of the convex function f and X_2 is a ray) and $X_1 \cap X_2 = \emptyset$. If we assume that there exists a point $(\mathbf{x}_s, r_s) \in X_1 \cap X_2$, then $f(\mathbf{x}_s) < r_s$ where $r_s = f(\mathbf{x}_0) + \lambda_1 \frac{\partial f}{\partial \mathbf{d}}(\mathbf{x}_0)$, $\mathbf{x}_s = \mathbf{x}_0 + \lambda_1 \mathbf{d}$, $\lambda_1 > 0$, that is, $f(\mathbf{x}_0 + \lambda_1 \mathbf{d}) < f(\mathbf{x}_0) + \lambda_1 \frac{\partial f}{\partial \mathbf{d}}(\mathbf{x}_0)$, a contradiction.

Consider the set Y defined as

$$Y = X_1 - X_2 = \{\mathbf{y} : \mathbf{y} = \mathbf{x}_1 - \mathbf{x}_2,\ \mathbf{x}_1 \in X_1,\ \mathbf{x}_2 \in X_2\}.$$

Since X_1 and X_2 are convex sets, then Y is also convex (Corollary 1.1). Obviously $\mathbf{0}_{n+1} \notin Y$, otherwise $X_1 \cap X_2 \neq \emptyset$. From Corollary 1.2 it follows that there exists an $(\widehat{\mathbf{x}}, c) \neq (\mathbf{0}, 0)$ such that

$$\left\langle (\widehat{\mathbf{x}}, c),\ (\mathbf{x}, r) - \left(\mathbf{x}_0 + \lambda \mathbf{d},\ f(\mathbf{x}_0) + \lambda \frac{\partial f}{\partial \mathbf{d}}(\mathbf{x}_0)\right)\right\rangle \geq 0,$$

that is,

$$\langle \widehat{\mathbf{x}}, \mathbf{x}\rangle + cr \geq \langle \widehat{\mathbf{x}}, \mathbf{x}_0 + \lambda \mathbf{d}\rangle + c\left(f(\mathbf{x}_0) + \lambda \frac{\partial f}{\partial \mathbf{d}}(\mathbf{x}_0)\right).$$

As in the proof of Theorem 1.25 we get $c < 0$.

Letting $r = f(\mathbf{x}),\ \lambda = 0$ yields

$$f(\mathbf{x}) - f(\mathbf{x}_0) \geq -\frac{1}{c}\langle \widehat{\mathbf{x}}, \mathbf{x} - \mathbf{x}_0\rangle \quad \forall \mathbf{x} \in X,$$

that is, $\widehat{\mathbf{y}} = -\frac{\widehat{\mathbf{x}}}{c} \in \partial f(\mathbf{x}_0)$ by definition.

It follows that

$$f(\mathbf{x}) - f(\mathbf{x}_0) \geq \langle \widehat{\mathbf{y}}, \mathbf{x} - \mathbf{x}_0\rangle + \lambda\left(\frac{\partial f}{\partial \mathbf{d}}(\mathbf{x}_0) - \langle \widehat{\mathbf{y}}, \mathbf{d}\rangle\right).$$

For $\mathbf{x} = \mathbf{x}_0$ and $\lambda > 0$ we get

$$\frac{\partial f}{\partial \mathbf{d}}(\mathbf{x}_0) - \langle \widehat{\mathbf{y}}, \mathbf{d}\rangle \leq 0$$

whence, taking into account that $\widehat{\mathbf{y}} \in \partial f(\mathbf{x}_0)$, we have

$$\frac{\partial f}{\partial \mathbf{d}}(\mathbf{x}_0) \leq \langle \widehat{\mathbf{y}}, \mathbf{d}\rangle \leq \max_{\widehat{\mathbf{x}} \in \partial f(\mathbf{x}_0)} \langle \widehat{\mathbf{x}}, \mathbf{d}\rangle.$$

We have already obtained that

$$\frac{\partial f}{\partial \mathbf{d}}(\mathbf{x}_0) \geq \max_{\widehat{\mathbf{x}} \in \partial f(\mathbf{x}_0)} \langle \widehat{\mathbf{x}}, \mathbf{d} \rangle.$$

The two opposite inequalities imply the statement of Theorem 1.26. ■

Theorem 1.27 (Subdifferential of a product of convex function with a positive real number)
Let $f : \mathbf{R}^n \to \mathbf{R}$ *be a convex function. Then* $\partial(\alpha f)(\mathbf{x}_0) = \alpha\, \partial f(\mathbf{x}_0)$ *for each scalar* $\alpha > 0$.

Proof. By definition of subdifferential,

$$\begin{aligned}
\partial(\alpha f)(\mathbf{x}_0) &= \{\widehat{\mathbf{x}} \in \mathbf{R}^n : \alpha f(\mathbf{x}) - \alpha f(\mathbf{x}_0) \geq \langle \widehat{\mathbf{x}}, \mathbf{x} - \mathbf{x}_0 \rangle \quad \forall\, \mathbf{x} \in \mathbf{R}^n\} \\
&= \left\{\widehat{\mathbf{x}} \in \mathbf{R}^n : f(\mathbf{x}) - f(\mathbf{x}_0) \geq \left\langle \frac{1}{\alpha}\widehat{\mathbf{x}}, \mathbf{x} - \mathbf{x}_0 \right\rangle \quad \forall\, \mathbf{x} \in \mathbf{R}^n \right\} \\
&= \alpha \left\{\frac{1}{\alpha}\widehat{\mathbf{x}} \in \mathbf{R}^n : f(\mathbf{x}) - f(\mathbf{x}_0) \geq \left\langle \frac{1}{\alpha}\widehat{\mathbf{x}}, \mathbf{x} - \mathbf{x}_0 \right\rangle \quad \forall\, \mathbf{x} \in \mathbf{R}^n \right\} \\
&= \alpha \left\{\widehat{\mathbf{y}} \in \mathbf{R}^n : f(\mathbf{x}) - f(\mathbf{x}_0) \geq \left\langle \widehat{\mathbf{y}}, \mathbf{x} - \mathbf{x}_0 \right\rangle \quad \forall\, \mathbf{x} \in \mathbf{R}^n \right\} \\
&= \alpha\, \partial f(\mathbf{x}_0),
\end{aligned}$$

where $\widehat{\mathbf{y}} \overset{\text{def}}{=} \frac{1}{\alpha}\widehat{\mathbf{x}} \in \mathbf{R}^n$. ■

Theorem 1.28 (Subdifferential of a sum of convex functions)
Let $f(\mathbf{x}) = \sum_{i=1}^m f_i(\mathbf{x})$ *where* f_i *are proper convex functions on* $\mathbf{R}^n$ *and the convex sets* ri (dom f_i), $i = 1, \ldots, m$ *have a point in common. Then* $\partial f(\mathbf{x}) = \sum_{i=1}^m \partial f_i(\mathbf{x})$ *for each* $\mathbf{x} \in \mathbf{R}^n$.

Proof. The relation is proved by double inclusion.
Function f is convex according to Theorem 1.17.
Let $\widehat{\mathbf{x}}_i \in \partial f_i(\mathbf{x})$, $i = 1, \ldots, m$. Then

$$f_i(\mathbf{x}) - f_i(\mathbf{x}_0) \geq \langle \widehat{\mathbf{x}}_i, \mathbf{x} - \mathbf{x}_0 \rangle, \quad i = 1, \ldots, m.$$

Summing these inequalities for all $i = 1, \ldots, m$ yields

$$\sum_{i=1}^m f_i(\mathbf{x}) - \sum_{i=1}^m f_i(\mathbf{x}_0) \overset{\text{def}}{=} f(\mathbf{x}) - f(\mathbf{x}_0) \geq \left\langle \sum_{i=1}^m \widehat{\mathbf{x}}_i, \mathbf{x} - \mathbf{x}_0 \right\rangle,$$

that is, $\sum_{i=1}^m \widehat{\mathbf{x}}_i \in \partial f(\mathbf{x}_0)$. Therefore $\sum_{i=1}^m \partial f_i(\mathbf{x}) \subseteq \partial f(\mathbf{x})$.

The opposite inclusion "$\supseteq$" will be proved for $m = 2$; the general case follows from it by induction.

Let $\widehat{\mathbf{x}}_0 \in \partial(f_1 + f_2)(\mathbf{x}_0)$. We have to prove that $\widehat{\mathbf{x}}_0 \in \partial f_1(\mathbf{x}_0) + \partial f_2(\mathbf{x}_0)$.

Without loss of generality assume that $\widehat{\mathbf{x}}_0 = \mathbf{x}_0 = \mathbf{0},\ f_1(\mathbf{0}) = f_2(\mathbf{0}) = 0$ because we can reduce each case to this one by introducing

$$g_1(\mathbf{x}) = f_1(\mathbf{x}_0 + \mathbf{x}) - f_1(\mathbf{x}_0) - \langle \widehat{\mathbf{x}}_0, \mathbf{x} \rangle,$$

$$g_2(\mathbf{x}) = f_2(\mathbf{x}_0 + \mathbf{x}) - f_2(\mathbf{x}_0).$$

Consider the sets

$$C_1 = \{(\mathbf{x}, r) \in \mathbf{R}^{n+1} : f_1(\mathbf{x}) \le r\} \stackrel{\text{def}}{=} \text{epi}\, f_1,$$

$$C_2 = \{(\mathbf{x}, r) \in \mathbf{R}^{n+1} : f_2(\mathbf{x}) \le -r\}.$$

Under the assumptions of Theorem 1.28, C_1 and C_2 are convex sets.

It can be proved that for any convex function f, ri (epi f) consists of the pairs $(\mathbf{x}, r)$ for which $\mathbf{x} \in$ ri (dom f) and $f(\mathbf{x}) < r < \infty$.

Therefore

$$\text{ri}\, C_1 = \{(\mathbf{x}, r) \in \mathbf{R}^{n+1} : \mathbf{x} \in \text{ri}\,(\text{dom}\, f_1),\ f_1(\mathbf{x}) < r\},$$

$$\text{ri}\, C_2 = \{(\mathbf{x}, r) \in \mathbf{R}^{n+1} : \mathbf{x} \in \text{ri}\,(\text{dom}\, f_2),\ f_2(\mathbf{x}) < -r\}.$$

Since

$$\min_{\mathbf{x}} (f_1(\mathbf{x}) + f_2(\mathbf{x})) = (f_1 + f_2)(\mathbf{0}) = f_1(\mathbf{0}) + f_2(\mathbf{0}) = 0,$$

then $\text{ri}\, C_1 \cap \text{ri}\, C_2 = \emptyset$. If we assume that there exists an $(\mathbf{x}_1, r_1) \in \text{ri}\, C_1 \cap \text{ri}\, C_2$, then $f_1(\mathbf{x}_1) < r_1 < -f_2(\mathbf{x}_1)$, hence $(f_1 + f_2)(\mathbf{x}_1) < 0$, which contradicts the fact that $\min_{\mathbf{x}} (f_1 + f_2)(\mathbf{x}) = 0$.

Therefore C_1 and C_2 can be properly separated by a hyperplane H in $\mathbf{R}^n$ according to Theorem 1.16.

Then there exists an $\widehat{\mathbf{x}}_1 \in \mathbf{R}^n$ such that

$$\langle \widehat{\mathbf{x}}_1, \mathbf{x} \rangle \le r \quad \text{for all } (\mathbf{x}, r) \in C_1,$$

$$\langle \widehat{\mathbf{x}}_1, \mathbf{x} \rangle \ge r \quad \text{for all } (\mathbf{x}, r) \in C_2.$$

Equivalently,

$$f_1(\mathbf{x}) \ge f_1(\mathbf{0}) + \langle \widehat{\mathbf{x}}_1, \mathbf{x} - \mathbf{0} \rangle \quad \text{for all } \mathbf{x} \in \mathbf{R}^n,$$

$$f_2(\mathbf{x}) \ge f_2(\mathbf{0}) + \langle -\widehat{\mathbf{x}}_1, \mathbf{x} - \mathbf{0} \rangle \quad \text{for all } \mathbf{x} \in \mathbf{R}^n,$$

which means that $\widehat{\mathbf{x}}_1 \in \partial f_1(\mathbf{0}),\ -\widehat{\mathbf{x}}_1 \in \partial f_2(\mathbf{0})$. Hence, $\widehat{\mathbf{x}}_1 + (-\widehat{\mathbf{x}}_1) \in \partial f_1(\mathbf{0}) + \partial f_2(\mathbf{0})$, therefore, $\mathbf{0} \in \partial f_1(\mathbf{0}) + \partial f_2(\mathbf{0})$, that is, $\widehat{\mathbf{x}}_0 \in \partial f_1(\mathbf{x}_0) + \partial f_2(\mathbf{x}_0)$. ■

Theorem 1.29 (Subdifferential of a maximum of convex functions)
Let $f_i(\mathbf{x}),\ i = 1, \ldots, m$ be convex functions on $\mathbf{R}^n$ and

$$f(\mathbf{x}) = \max_{i=1,\ldots,m} f_i(\mathbf{x}).$$

Then

$$\partial f(\mathbf{x}) = \operatorname{co}\{\partial f_i(\mathbf{x}) : i \in I(\mathbf{x})\}$$

where

$$I(\mathbf{x}) = \{i \in \{1, \ldots, m\} : f_i(\mathbf{x}) = f(\mathbf{x})\}.$$

Proof. According to Theorem 1.20, f is a convex function.

Denote $D(\mathbf{x}) = \operatorname{co}\{\partial f_i(\mathbf{x}) : i \in I(\mathbf{x})\}$ where $I(\mathbf{x})$ is defined above.

For fixed $\mathbf{x}$ and $\mathbf{d}$ we have

$$f(\mathbf{x}+\lambda\mathbf{d}) = f(\mathbf{x}) + \lambda\frac{\partial f}{\partial \mathbf{d}}(\mathbf{x}) + o(\lambda) = \max_{i\in\{1,\ldots,m\}} \left\{ f_i(\mathbf{x}) + \lambda\frac{\partial f_i}{\partial \mathbf{d}}(\mathbf{x}) + o_i(\lambda) \right\},$$

where $\dfrac{o(\lambda)}{\lambda} \to 0,\ \dfrac{o_i(\lambda)}{\lambda} \to 0,\ i = 1, \ldots, m$ when $\lambda \to +0$. For sufficiently small $\lambda > 0$ we get

$$\begin{aligned} f(\mathbf{x}+\lambda\mathbf{d}) &= \max_{i\in I(\mathbf{x})} f_i(\mathbf{x}+\lambda\mathbf{d}) = \max_{i\in I(\mathbf{x})} \left\{ f_i(\mathbf{x}) + \lambda\frac{\partial f_i}{\partial \mathbf{d}}(\mathbf{x}) + o_i(\lambda) \right\} \\ &= f(\mathbf{x}) + \lambda \max_{i\in I(\mathbf{x})} \frac{\partial f_i}{\partial \mathbf{d}}(\mathbf{x}) + o(\lambda), \end{aligned}$$

where

$$\frac{o(\lambda)}{\lambda} \to 0,\ \lambda \to +0; \quad o(\lambda) \in \left[\min_{i\in I(\mathbf{x})} o_i(\lambda),\ \max_{i\in I(\mathbf{x})} o_i(\lambda) \right].$$

Then

$$\frac{\partial f}{\partial \mathbf{d}}(\mathbf{x}) = \max_{i\in I(\mathbf{x})} \frac{\partial f_i}{\partial \mathbf{d}}(\mathbf{x}) = \max_{i\in I(\mathbf{x})} \max_{\widehat{\mathbf{x}}_i\in\partial f_i(\mathbf{x})} \langle \widehat{\mathbf{x}}_i, \mathbf{d}\rangle = \max_{\widehat{\mathbf{x}}\in D(\mathbf{x})} \langle \widehat{\mathbf{x}}, \mathbf{d}\rangle.$$

However, by Theorem 1.26 there exists some $\widehat{\mathbf{y}} \in \partial f(\mathbf{x})$ such that

$$\frac{\partial f}{\partial \mathbf{d}}(\mathbf{x}) = \langle \widehat{\mathbf{y}}, \mathbf{d}\rangle = \max_{\widehat{\mathbf{x}}\in\partial f(\mathbf{x})} \langle \widehat{\mathbf{x}}, \mathbf{d}\rangle.$$

Therefore

$$\max_{\widehat{\mathbf{x}}\in\partial f(\mathbf{x})} \langle \widehat{\mathbf{x}}, \mathbf{d}\rangle = \max_{\widehat{\mathbf{x}}\in D(\mathbf{x})} \langle \widehat{\mathbf{x}}, \mathbf{d}\rangle$$

for all $\mathbf{d} \in \mathbf{R}^n$. Since $\partial f(\mathbf{x})$ is a compact convex set (Theorem 1.25) and $D(\mathbf{x})$ is also a compact convex set (Theorem 1.5 and Theorem 1.25) then Theorem 1.15 implies $\partial f(\mathbf{x}) = D(\mathbf{x})$. ■

Definition 1.22 Let f be a numerical function defined on an open set X in $\mathbf{R}^n$ and let $\overline{\mathbf{x}} \in X$. f is said to be *differentiable* at $\overline{\mathbf{x}}$ if for all $\mathbf{x} \in \mathbf{R}^n$ such that $\overline{\mathbf{x}} + \mathbf{x} \in X$ we have that

$$f(\overline{\mathbf{x}} + \mathbf{x}) = f(\overline{\mathbf{x}}) + \langle \mathbf{f}'(\overline{\mathbf{x}}), \mathbf{x} \rangle + \alpha(\overline{\mathbf{x}}, \mathbf{x})\|\mathbf{x}\|, \tag{1.3}$$

where $\lim\limits_{\mathbf{x} \to \mathbf{0}} \alpha(\overline{\mathbf{x}}, \mathbf{x}) = 0$.

Theorem 1.30 (Convexity, strict convexity, concavity, and strict concavity of differentiable multivariate functions)
Let f be a numerical differentiable function on an open convex set X in $\mathbf{R}^n$. f is convex on X if and only if

$$f(\mathbf{x}_1) - f(\mathbf{x}_2) \geq \langle \mathbf{f}'(\mathbf{x}_2), \mathbf{x}_1 - \mathbf{x}_2 \rangle$$

for each $\mathbf{x}_1, \mathbf{x}_2 \in X$.

f is concave on X if and only if

$$f(\mathbf{x}_1) - f(\mathbf{x}_2) \leq \langle \mathbf{f}'(\mathbf{x}_2), \mathbf{x}_1 - \mathbf{x}_2 \rangle$$

for each $\mathbf{x}_1, \mathbf{x}_2 \in X$.

f is strictly convex (strictly concave) on X if and only if these inequalities are strict, respectively, for each $\mathbf{x}_1, \mathbf{x}_2 \in X$, $\mathbf{x}_1 \neq \mathbf{x}_2$.

Proof. *Necessity.* Since f is convex on X, then f is convex at any point $\mathbf{x}_2 \in X$. Because X is an open set, there exists an open ball $B(\mathbf{x}_2, \varepsilon)$ with radius ε around $\mathbf{x}_2$, which is contained in X. Let $\mathbf{x}_1 \in X$ and $\mathbf{x}_1 \neq \mathbf{x}_2$. Then for some $\lambda \in (0, 1)$, $\lambda < \frac{\varepsilon}{\|\mathbf{x}_1 - \mathbf{x}_2\|}$ we have

$$\overline{\mathbf{x}} \stackrel{\text{def}}{=} \lambda\mathbf{x}_1 + (1 - \lambda)\mathbf{x}_2 = \mathbf{x}_2 + \lambda(\mathbf{x}_1 - \mathbf{x}_2) \in B(\mathbf{x}_2, \varepsilon) \subset X.$$

Since f is convex at $\mathbf{x}_2$, then from Definition 1.15, from convexity of $B(\mathbf{x}_2, \varepsilon)$ (Example 1.1) and $\overline{\mathbf{x}} \in B(\mathbf{x}_2, \varepsilon)$ it follows that for $\lambda \in (0, 1]$ we have

$$\lambda f(\overline{\mathbf{x}}) + (1 - \lambda) f(\mathbf{x}_2) \geq f(\lambda\overline{\mathbf{x}} + (1 - \lambda)\mathbf{x}_2)$$

or, equivalently,

$$f(\mathbf{x}_2) + \lambda(f(\overline{\mathbf{x}}) - f(\mathbf{x}_2)) \geq f(\mathbf{x}_2 + \lambda(\overline{\mathbf{x}} - \mathbf{x}_2)).$$

Then

$$\begin{aligned} f(\overline{\mathbf{x}}) - f(\mathbf{x}_2) &\geq \frac{f(\mathbf{x}_2 + \lambda(\overline{\mathbf{x}} - \mathbf{x}_2)) - f(\mathbf{x}_2)}{\lambda} \\ &= \frac{\lambda\langle \mathbf{f}'(\mathbf{x}_2), \overline{\mathbf{x}} - \mathbf{x}_2\rangle + \alpha(\mathbf{x}_2, \lambda(\overline{\mathbf{x}} - \mathbf{x}_2))\lambda\|\overline{\mathbf{x}} - \mathbf{x}_2\|}{\lambda} \\ &= \langle \mathbf{f}'(\mathbf{x}_2), \overline{\mathbf{x}} - \mathbf{x}_2\rangle + \alpha(\mathbf{x}_2, \lambda(\overline{\mathbf{x}} - \mathbf{x}_2))\|\overline{\mathbf{x}} - \mathbf{x}_2\|, \end{aligned}$$

where $\lim_{\lambda\to 0} \alpha(\mathbf{x}_2, \lambda(\overline{\mathbf{x}} - \mathbf{x}_2)) = 0$.

Taking the limit of the above expression as $\lambda \to 0$ yields

$$f(\overline{\mathbf{x}}) - f(\mathbf{x}_2) \geq \langle \mathbf{f}'(\mathbf{x}_2), \overline{\mathbf{x}} - \mathbf{x}_2\rangle.$$

Since f is convex at $\mathbf{x}_2$, since $\overline{\mathbf{x}} \in X$, and since $\overline{\mathbf{x}} = \lambda\mathbf{x}_1 + (1-\lambda)\mathbf{x}_2$, then

$$\begin{aligned} f(\overline{\mathbf{x}}) &\equiv f(\lambda\mathbf{x}_1 + (1-\lambda)\mathbf{x}_2) \leq \lambda f(\mathbf{x}_1) + (1-\lambda)f(\mathbf{x}_2) \\ &\equiv f(\mathbf{x}_2) + \lambda(f(\mathbf{x}_1) - f(\mathbf{x}_2)), \end{aligned}$$

that is,

$$\lambda(f(\mathbf{x}_1) - f(\mathbf{x}_2)) \geq f(\overline{\mathbf{x}}) - f(\mathbf{x}_2).$$

Using that $\overline{\mathbf{x}} - \mathbf{x}_2 = \lambda(\mathbf{x}_1 - \mathbf{x}_2)$, $\lambda > 0$ and previous inequalities, we get

$$\begin{aligned} f(\mathbf{x}_1) - f(\mathbf{x}_2) &\geq \frac{f(\overline{\mathbf{x}}) - f(\mathbf{x}_2)}{\lambda} \geq \frac{\langle \mathbf{f}'(\mathbf{x}_2), \overline{\mathbf{x}} - \mathbf{x}_2\rangle}{\lambda} \\ &= \frac{\langle \mathbf{f}'(\mathbf{x}_2), \lambda(\mathbf{x}_1 - \mathbf{x}_2)\rangle}{\lambda} = \langle \mathbf{f}'(\mathbf{x}_2), \mathbf{x}_1 - \mathbf{x}_2\rangle. \end{aligned}$$

The concave case is proved similarly.

Sufficiency. Let $\mathbf{x}_1, \mathbf{x}_2 \in X$, $\lambda \in [0, 1]$. Since X is convex then $\lambda\mathbf{x}_1 + (1-\lambda)\mathbf{x}_2 \in X$. Therefore by assumption we have

$$\begin{aligned} f(\mathbf{x}_1) - f(\lambda\mathbf{x}_1 + (1-\lambda)\mathbf{x}_2) &\geq (1-\lambda)\langle \mathbf{f}'(\lambda\mathbf{x}_1 + (1-\lambda)\mathbf{x}_2), \mathbf{x}_1 - \mathbf{x}_2\rangle, \\ f(\mathbf{x}_2) - f(\lambda\mathbf{x}_1 + (1-\lambda)\mathbf{x}_2) &\geq -\lambda\langle \mathbf{f}'(\lambda\mathbf{x}_1 + (1-\lambda)\mathbf{x}_2), \mathbf{x}_1 - \mathbf{x}_2\rangle. \end{aligned}$$

Multiplying the first inequality by $\lambda \geq 0$ and the second inequality by $1-\lambda \geq 0$ and adding them, we obtain

$$\lambda f(\mathbf{x}_1) + (1-\lambda)f(\mathbf{x}_2) \geq f(\lambda\mathbf{x}_1 + (1-\lambda)\mathbf{x}_2),$$

that is, f is convex by definition.

The concave case is considered similarly.

The strictly convex (strictly concave) case is proved in the same manner as the convex (concave) case, respectively. ■

Theorem 1.31 (Monotonicity of the gradient of a differentiable convex function) *Let f be a numerical differentiable function on an open convex set X in $\mathbf{R}^n$. A necessary and sufficient condition that f be convex (concave) on X is that for each $\mathbf{x}_1, \mathbf{x}_2 \in X$ we have*

$$\langle \mathbf{f}'(\mathbf{x}_2) - \mathbf{f}'(\mathbf{x}_1), \mathbf{x}_2 - \mathbf{x}_1 \rangle \geq 0 \qquad (\leq 0) \tag{1.4}$$

(*monotonicity of* $\mathbf{f}'(\mathbf{x})$).

A necessary and sufficient condition that f be strictly convex (strictly concave) on X is that for each $\mathbf{x}_1, \mathbf{x}_2 \in X$, $\mathbf{x}_1 \neq \mathbf{x}_2$ we have

$$\langle \mathbf{f}'(\mathbf{x}_2) - \mathbf{f}'(\mathbf{x}_1), \mathbf{x}_2 - \mathbf{x}_1 \rangle > 0 \qquad (< 0). \tag{1.4'}$$

Proof. *Necessity.* Let f be convex on X and let $\mathbf{x}_1, \mathbf{x}_2 \in X$. From Theorem 1.30 it follows that

$$f(\mathbf{x}_1) - f(\mathbf{x}_2) \geq \langle \mathbf{f}'(\mathbf{x}_2), \mathbf{x}_1 - \mathbf{x}_2 \rangle$$

and

$$f(\mathbf{x}_2) - f(\mathbf{x}_1) \geq \langle \mathbf{f}'(\mathbf{x}_1), \mathbf{x}_2 - \mathbf{x}_1 \rangle.$$

Summing these two inequalities, we get

$$\langle \mathbf{f}'(\mathbf{x}_2) - \mathbf{f}'(\mathbf{x}_1), \mathbf{x}_2 - \mathbf{x}_1 \rangle \geq 0.$$

Sufficiency. Let $\mathbf{x}_1, \mathbf{x}_2 \in X$, $\lambda \in [0, 1]$, and (1.4) be satisfied. By the Mean value theorem (Theorem A.2, Appendix A), there is some $\lambda_0 \in (0, 1)$ such that

$$f(\mathbf{x}_2) - f(\mathbf{x}_1) = \langle \mathbf{f}'(\mathbf{x}_1 + \lambda_0(\mathbf{x}_2 - \mathbf{x}_1)), \mathbf{x}_2 - \mathbf{x}_1 \rangle.$$

However,

$$\langle \mathbf{f}'(\mathbf{x}_1 + \lambda_0(\mathbf{x}_2 - \mathbf{x}_1)) - \mathbf{f}'(\mathbf{x}_1), \lambda_0(\mathbf{x}_2 - \mathbf{x}_1) \rangle \geq 0$$

by assumption (1.4), that is,

$$\langle \mathbf{f}'(\mathbf{x}_1 + \lambda_0(\mathbf{x}_2 - \mathbf{x}_1)), \mathbf{x}_2 - \mathbf{x}_1 \rangle \geq \langle \mathbf{f}'(\mathbf{x}_1), \mathbf{x}_2 - \mathbf{x}_1 \rangle.$$

Therefore,

$$f(\mathbf{x}_2) - f(\mathbf{x}_1) \geq \langle \mathbf{f}'(\mathbf{x}_1), \mathbf{x}_2 - \mathbf{x}_1 \rangle.$$

According to Theorem 1.30, f is convex on X.

The concave case is proved similarly.

The strictly convex (strictly concave) case is proved analogously to the convex (concave) case, respectively. ∎

Definition 1.23 Let f be a numerical function defined on an open set X in $\mathbf{R}^n$ and let $\bar{\mathbf{x}} \in X$. f is said to be *twice differentiable* at $\bar{\mathbf{x}}$ if for all $\mathbf{x} \in \mathbf{R}^n$ such that $\bar{\mathbf{x}} + \mathbf{x} \in X$ we have that

$$f(\bar{\mathbf{x}} + \mathbf{x}) = f(\bar{\mathbf{x}}) + \langle \mathbf{f}'(\bar{\mathbf{x}}), \mathbf{x} \rangle + \frac{1}{2}\langle \mathbf{x}, f''(\bar{\mathbf{x}})\mathbf{x} \rangle + \beta(\bar{\mathbf{x}}, \mathbf{x})\|\mathbf{x}\|^2, \tag{1.5}$$

where $\lim_{\mathbf{x} \to \mathbf{0}} \beta(\bar{\mathbf{x}}, \mathbf{x}) = 0$, and $f''(\bar{\mathbf{x}})$ is called the *Hessian* (*matrix*) of f at $\bar{\mathbf{x}}$.

Theorem 1.32 (Convexity, strict convexity, concavity, and strict concavity of twice differentiable multivariate functions)
Let f be a numerical twice differentiable function on an open convex set X in $\mathbf{R}^n$. f is convex (*concave*) *on X if and only if $f''(\mathbf{x})$ is positive* (*negative*) *semidefinite on X, that is, for each $\mathbf{x} \in X$,*

$$\langle \mathbf{y}, f''(\mathbf{x})\mathbf{y} \rangle \geq 0 \quad (\leq 0) \tag{1.6}$$

for all $\mathbf{y} \in \mathbf{R}^n$.

If $f''(\mathbf{x})$ is positive definite (*negative definite*) *on X, that is, for each $\mathbf{x} \in X$,*

$$\langle \mathbf{y}, f''(\mathbf{x})\mathbf{y} \rangle > 0 \quad (< 0) \tag{1.6$'$}$$

for all $\mathbf{y} \in \mathbf{R}^n$, $\mathbf{y} \neq \mathbf{0}$, then f is strictly convex (*strictly concave*) *on X.*

Conversely, if f is strictly convex (*strictly concave*) *on X then $f''(\mathbf{x})$ is positive semidefinite* (*negative semidefinite*), *but not necessarily positive definite* (*negative definite*), *respectively.*

Proof. *Necessity.* Let f be a convex twice differentiable function on X. Therefore it is convex at each $\mathbf{x} \in X$.

Let $\mathbf{y} \in \mathbf{R}^n$. Since X is an open set, there exists a $\lambda_0 > 0$ such that $\mathbf{x} \in X$ implies $\mathbf{x} + \lambda\mathbf{y} \in X$ for $\lambda \in (0, \lambda_0)$.

From Theorem 1.30 it follows that

$$f(\mathbf{x} + \lambda\mathbf{y}) - f(\mathbf{x}) - \lambda\langle \mathbf{f}'(\mathbf{x}), \mathbf{y} \rangle \geq 0 \quad \text{for } \lambda \in (0, \lambda_0).$$

However, f is twice differentiable at $\mathbf{x}$ by assumption. Therefore

$$f(\mathbf{x} + \lambda\mathbf{y}) - f(\mathbf{x}) - \lambda\langle \mathbf{f}'(\mathbf{x}), \mathbf{y} \rangle = \frac{1}{2}\lambda^2\langle \mathbf{y}, f''(\mathbf{x})\mathbf{y} \rangle + \lambda^2\beta(\mathbf{x}, \lambda\mathbf{y})\|\mathbf{y}\|^2,$$

where $\lim_{\lambda \to 0} \beta(\mathbf{x}, \lambda\mathbf{y}) = 0$. Therefore

$$\frac{1}{2}\langle \mathbf{y}, f''(\mathbf{x})\mathbf{y} \rangle + \beta(\mathbf{x}, \lambda\mathbf{y})\|\mathbf{y}\|^2 \geq 0 \quad \text{for } \lambda \in (0, \lambda_0).$$

If we take the limit as $\lambda \to 0$, we obtain

$$\langle \mathbf{y}, f''(\mathbf{x})\mathbf{y}\rangle \geq 0 \quad \text{for all } \mathbf{y} \in \mathbf{R}^n,$$

that is, $f''(\mathbf{x})$ is positive semidefinite by definition.

If f is strictly convex then f is convex and, as we have proved, $f''(\mathbf{x})$ is positive semidefinite on X. However, $f''(\mathbf{x})$ is *not* necessarily *positive* definite. This can be observed, for example, by the following counterexample:

$$f(x) = x^{2k}, \quad x \in \mathbf{R},\ k \text{ is a positive integer},\ k \geq 2.$$

It can be verified that f is strictly convex on $\mathbf{R}$. However $f''(x) = 2k(2k-1)x^{2k-2}$ is not positive definite because $f''(0) = 0$.

Sufficiency. Let f be twice differentiable on X and $f''(\mathbf{x})$ be positive semidefinite. For any $\mathbf{x}_1, \mathbf{x}_2 \in X$ we have by Taylor's second order theorem (Theorem A.3, Appendix A) that

$$f(\mathbf{x}_2) - f(\mathbf{x}_1) - \langle \mathbf{f}'(\mathbf{x}_1), \mathbf{x}_2 - \mathbf{x}_1\rangle = \frac{1}{2}\langle \mathbf{x}_2 - \mathbf{x}_1, f''(\mathbf{x}_1 + \gamma(\mathbf{x}_2 - \mathbf{x}_1))(\mathbf{x}_2 - \mathbf{x}_1)\rangle$$

for some $\gamma \in (0, 1)$.

Since $f''(\mathbf{x})$ is positive semidefinite on X by assumption, the right-hand side is nonnegative, and $\mathbf{x}_1 + \gamma(\mathbf{x}_2 - \mathbf{x}_1) \in X$ because X is convex. Therefore, the left-hand side is also nonnegative. From Theorem 1.30 it follows that f is convex on X.

If $f''(\mathbf{x})$ is positive definite, the right-hand side of the equality above is positive for each $\mathbf{x}_1, \mathbf{x}_2 \in X$, $\mathbf{x}_1 \neq \mathbf{x}_2$. Hence the same holds true for the left-hand side. According to Theorem 1.30, f is strictly convex on X.

The concave (strictly concave) case is proved similar to the convex (strictly convex) case, respectively. ∎

Example 1.5 Functions $f(x) = x^p$, $p \geq 1$, $x \in [0, \infty)$; $f(x) = -x^p$, $x \geq 0$, $0 \leq p \leq 1$; and $f(x) = x^p$, $x > 0$, $-\infty < p \leq 0$ are convex.

Functions $f(x) = x^p$, $p > 1$, $x \in (0, \infty)$; $f(x) = -x^p$, $x > 0$, $0 < p < 1$; and $f(x) = x^p$, $x > 0$, $-\infty < p < 0$ are strictly convex.

Since $(x^p)'' = p(p-1)x^{p-2} \geq 0$ for $p \geq 1$, $x \geq 0$, then x^p is convex according to Theorem 1.32, applied for a function of a single variable.

Moreover, x^p is convex for all $x \in \mathbf{R}$ if $p \geq 1$ is a positive even number.

When $p > 1$, $x \in (0, \infty)$, then $(x^p)'' > 0$ and x^p is strictly convex according to Theorem 1.32.

In particular, when $p = 2$, x^2 is strictly convex for all $x \in \mathbf{R}$. This is proved directly as Example 1.3.

An alternative proof of strict convexity. Let x_1, $x_2 \in (0, \infty)$, $\lambda \in (0, 1)$. Then, according to the binomial formula,

$$
\begin{aligned}
f(\lambda x_1 + (1-\lambda)x_2) &\stackrel{\text{def}}{=} (\lambda x_1 + (1-\lambda)x_2)^p \\
&= \sum_{k=0}^{p} \binom{p}{k} (\lambda x_1)^k [(1-\lambda)x_2]^{p-k} \\
&\equiv (1-\lambda)^p x_2^p + \sum_{k=1}^{p-1} \binom{p}{k} (\lambda x_1)^k [(1-\lambda)x_2]^{p-k} + \lambda^p x_1^p \\
&< (1-\lambda)^p x_2^p + \lambda^p x_1^p < (1-\lambda)x_2^p + \lambda x_1^p \\
&\stackrel{\text{def}}{=} (1-\lambda)f(x_2) + \lambda f(x_1).
\end{aligned}
$$

For the first inequality we have used that $\sum_{k=1}^{p-1} \binom{p}{k}(\lambda x_1)^k[(1-\lambda)x_2]^{p-k}$ is positive for $x_1,\ x_2 \in (0, \infty),\ \lambda \in (0, 1),\ p > 1$ and for the second inequality we have used that $\lambda^p < \lambda$ and $(1-\lambda)^p < 1-\lambda$ for $\lambda \in (0, 1)$ and $p > 1$.

Therefore x^p is strictly convex under the assumptions. Convexity (strict convexity) of the remaining functions of Example 1.5 is proved similarly.

Theorem 1.33 (Subgradient and subdifferential of a differentiable convex function) *Let X be a nonempty convex set in $\mathbf{R}^n$ and $f : X \to \mathbf{R}$ be a convex function which is differentiable at $\overline{\mathbf{x}} \in \operatorname{int} X$. Then $\widehat{\mathbf{f}}(\overline{\mathbf{x}}) = \mathbf{f}'(\overline{\mathbf{x}})$ and $\partial f(\overline{\mathbf{x}}) = \{\mathbf{f}'(\overline{\mathbf{x}})\}$ is a singleton.*

Proof. From Theorem 1.25 it follows that the set $\partial f(\overline{\mathbf{x}})$ is nonempty. Let $\mathbf{d} \in \partial f(\overline{\mathbf{x}})$. From definition of subgradient we have

$$f(\overline{\mathbf{x}} + \lambda \mathbf{s}) - f(\overline{\mathbf{x}}) \geq \lambda \langle \mathbf{d}, \mathbf{s} \rangle,$$

and from differentiability of f at $\overline{\mathbf{x}}$ we have

$$f(\overline{\mathbf{x}} + \lambda \mathbf{s}) = f(\overline{\mathbf{x}}) + \lambda \langle \mathbf{f}'(\overline{\mathbf{x}}), \mathbf{s} \rangle + \lambda \|\mathbf{s}\| \alpha(\overline{\mathbf{x}}, \lambda \mathbf{s})$$

for any vector $\mathbf{s} \in \mathbf{R}^n$ and sufficiently small λ, where $\lim_{\lambda \to +0} \alpha(\overline{\mathbf{x}}, \lambda \mathbf{s}) = 0$.

If we subtract this equation from the above inequality, we get

$$0 \geq \lambda \langle \mathbf{d} - \mathbf{f}'(\overline{\mathbf{x}}), \mathbf{s} \rangle - \lambda \|\mathbf{s}\| \alpha(\overline{\mathbf{x}}, \lambda \mathbf{s}).$$

Dividing by $\lambda > 0$, we obtain

$$0 \geq \langle \mathbf{d} - \mathbf{f}'(\overline{\mathbf{x}}), \mathbf{s} \rangle - \|\mathbf{s}\| \alpha(\overline{\mathbf{x}}, \lambda \mathbf{s}).$$

Setting $\lambda \to +0$, we get $\langle \mathbf{d} - \mathbf{f}'(\overline{\mathbf{x}}), \mathbf{s} \rangle \leq 0$.

Let $\mathbf{s} = \mathbf{d} - \mathbf{f}'(\overline{\mathbf{x}})$. Therefore $\langle \mathbf{s}, \mathbf{s} \rangle = \|\mathbf{s}\|^2 \leq 0$, which holds true if and only if $\mathbf{s} = \mathbf{0}$, that is, if and only if $\mathbf{d} = \mathbf{f}'(\overline{\mathbf{x}})$. Hence, $\mathbf{d} \in \partial f(\overline{\mathbf{x}})$ implies $\mathbf{d} = \mathbf{f}'(\overline{\mathbf{x}})$. Therefore, $\partial f(\overline{\mathbf{x}}) = \{\mathbf{f}'(\overline{\mathbf{x}})\}$ is a singleton. ■

The following two theorems give the rules for calculating subgradients of some types of functions.

Theorem 1.34 (Subgradient of a sum of univariate convex functions)
Let $f(x_1, \dots, x_n) = \sum_{j=1}^n f_j(x_j)$, $f_j(x_j)$ *be a convex function of* x_j *for each* $j = 1, \dots, n$.

Then $\widehat{\mathbf{f}}(\overline{\mathbf{x}}) = (\widehat{f_{x_1}}(\overline{x}_1), \dots, \widehat{f_{x_n}}(\overline{x}_n))$, *where*

$$\widehat{f_{x_j}}(\overline{x}_j) = \lambda_j f_j^+(\overline{x}_j) + (1 - \lambda_j) f_j^-(\overline{x}_j), \quad 0 \le \lambda_j \le 1, \ j = 1, \dots, n,$$

$f_j^+(\overline{x}_j) = \lim_{\varepsilon \to 0} f_j'(\overline{x}_j + \varepsilon)$, $f_j^-(\overline{x}_j) = \lim_{\varepsilon \to 0} f_j'(\overline{x}_j - \varepsilon)$ *are the derivatives of* f_j *on the right and on the left at* $\overline{x}_j$, *respectively.*

Proof. Since $f(\mathbf{x})$ is a convex differentiable function of $\mathbf{x}$ then according to Theorem 1.30 we have

$$f(\mathbf{y}) - f(\mathbf{x}) \ge \langle \mathbf{f}^{\pm}(\mathbf{x}), \mathbf{y} - \mathbf{x} \rangle,$$

that is, $\mathbf{f}^{\pm}(\mathbf{x}) \in \partial f(\mathbf{x})$ by the definition of subgradient (Definition 1.20).

Since the subdifferential $\partial f(\mathbf{x})$ of a convex function $f(\mathbf{x})$ is a nonempty, convex, and compact set (Theorem 1.25) and since $f_j^+(\overline{x}_j)$, $f_j^-(\overline{x}_j) \in \partial f(\overline{x}_j)$ according to the previous discussion, then

$$\lambda_j f_j^+(\overline{x}_j) + (1 - \lambda_j) f_j^-(\overline{x}_j) \in \partial f_j(\overline{x}_j), \quad j = 1, \dots, n.$$

Therefore

$$\begin{aligned} f(\mathbf{x}) - f(\overline{\mathbf{x}}) &\equiv \sum_{j=1}^n f_j(x_j) - \sum_{j=1}^n f_j(\overline{x}_j) \equiv \sum_{j=1}^n [f_j(x_j) - f_j(\overline{x}_j)] \\ &\ge \sum_{j=1}^n [\lambda_j f_j^+(\overline{x}_j) + (1 - \lambda_j) f_j^-(\overline{x}_j)](x_j - \overline{x}_j) \\ &\equiv \langle \widehat{\mathbf{f}}(\overline{\mathbf{x}}), \mathbf{x} - \overline{\mathbf{x}} \rangle, \end{aligned}$$

that is, $\widehat{\mathbf{f}}(\overline{\mathbf{x}}) = (\widehat{f_{x_1}}(\overline{x}_1), \dots, \widehat{f_{x_n}}(\overline{x}_n))$ with $\widehat{f_{x_j}}(\overline{x}_j)$, $j = 1, \dots, n$, defined above, is a subgradient of $f(x_1, \dots, x_n)$ at $\overline{\mathbf{x}}$ by definition. ■

Remark 1.1 *Since convex functions have derivatives on the right and on the left at each interior feasible point, then the assumption that* $f_j^+(\overline{x}_j)$ *and* $f_j^-(\overline{x}_j)$ *exist is reasonable.*

Theorem 1.35 (Subgradient of a function in two variables)
Let $f(x, y)$ *be a convex function of* x *for each* y, *let there exist a* $y(x)$ *such that*

$$f(x) \stackrel{\text{def}}{=} \max_{y \in Y} f(x, y) = f(x, y(x))$$

and let the subgradient $\widehat{f_x}(x, y)$ *of* $f(x, y)$ *with respect to* x *be known for each* y. *Then*

$$\widehat{f_x}(x) = \widehat{f_x}(x, y)_{|y=y(x)}.$$

Proof. Since $f(x, y)$ is a convex function of x for each y and since $y(x)$ is an optimal solution to problem $\max_{y \in Y} f(x, y)$, then

$$\begin{aligned} f(z) - f(x) &\equiv f(z, y(z)) - f(x, y(x)) \geq f(z, y(x)) - f(x, y(x)) \\ &\geq \langle \widehat{f_x}(x, y(x)), z - x \rangle. \end{aligned}$$

Therefore $\widehat{f_x}(x) = \widehat{f_x}(x, y)_{|y=y(x)}$ according to Definition 1.20 of subgradient. ■

1.5 Convex Programming

Let $f,\ g_i\ (i \in I_1),\ l_i\ (i \in I_2 \cup I_3)$ be numerical functions defined on a set $W \subseteq \mathbf{R}^n$. Consider the following problem.

Find $\mathbf{x}^*$ in X, if it exists, such that

$$f(\mathbf{x}^*) = \min_{\mathbf{x} \in X} f(\mathbf{x}) \tag{1.7}$$

where

$$X \stackrel{\text{def}}{=} \{\mathbf{x} \in W : g_i(\mathbf{x}) \leq 0,\ i \in I_1,\ \ l_i(\mathbf{x}) \leq 0,\ i \in I_2,\ \ l_i(\mathbf{x}) = 0,\ i \in I_3,$$

$$x_j \geq 0,\ j \in J_1\},$$

$I_1 \cup I_2 \cup I_3 = I \equiv \{1, \ldots, m\}$; $J_1 \equiv \{1, \ldots, n_1\} \subseteq J \equiv \{1, \ldots, n\}$.

As usual, the set X is called the *feasible region* or the *constraint set*, $\mathbf{x}^*$ the *minimum solution* or *solution*, and $f(\mathbf{x}^*)$ the *minimum*. All points $\mathbf{x}$ in the feasible region X are called *feasible points*.

Definition 1.24 The numerical function

$$L(\mathbf{x}, \boldsymbol{\lambda}) = f(\mathbf{x}) + \sum_{i \in I_1} \lambda_i g_i(\mathbf{x}) + \sum_{i \in I_2} \lambda_i l_i(\mathbf{x}) + \sum_{i \in I_3} \lambda_i l_i(\mathbf{x})$$

is called the *Lagrangian function* or simply the *Lagrangian* for problem (1.7), and components of the m-dimensional vector $\boldsymbol{\lambda}$, *Lagrange multipliers* or *dual variables*.

Here $(\mathbf{x}, \boldsymbol{\lambda}) \in X \times \Lambda \subset \mathbf{R}^{n+m}$, where

$$\Lambda = \{\boldsymbol{\lambda} = (\lambda_1, \ldots, \lambda_m) : \lambda_i \geq 0,\ i \in I_1 \cup I_2\}.$$

Definition 1.25 The point $(\mathbf{x}^*, \boldsymbol{\lambda}^*)$ is said to be a *saddle point* of $L(\mathbf{x}, \boldsymbol{\lambda})$ if the following inequalities hold true

$$L(\mathbf{x}^*, \boldsymbol{\lambda}) \leq L(\mathbf{x}^*, \boldsymbol{\lambda}^*) \leq L(\mathbf{x}, \boldsymbol{\lambda}^*) \quad \forall\, \mathbf{x} \in X, \ \ \forall\, \boldsymbol{\lambda} \in \Lambda,$$

that is, if the function $L(\mathbf{x}^*, \boldsymbol{\lambda})$ of $\boldsymbol{\lambda}$ attains its maximum at $(\mathbf{x}^*, \boldsymbol{\lambda}^*)$ and the function $L(\mathbf{x}, \boldsymbol{\lambda}^*)$ of $\mathbf{x}$ attains its minimum at $(\mathbf{x}^*, \boldsymbol{\lambda}^*)$.

Because there are inequality constraints in (1.7), the Lagrange multipliers associated with these constraints turn out to be nonnegative.

Let

(i) $f, g_i, \ i \in I_1$ be convex functions,
$l_i, \ i \in I_2 \cup I_3$ be affine (linear) functions;

(ii) (Slater's constraint qualification)
The convex functions $g_i, \ i \in I_1$, which are involved in definition of the feasible region X of problem (1.7), are said to satisfy *Slater's constraint qualification* if there exists a point $\overline{\mathbf{x}} \in X$ such that $g_i(\overline{\mathbf{x}}) < 0, \ i \in I_1$.

There are also other constraint qualifications (regularity conditions) except for the Slater's constraint qualification: Karlin's constraint qualification, the strict constraint qualification, reverse convex constraint qualification (Arrow-Hurwicz-Uzawa), modified Arrow-Hurwicz-Uzawa constraint qualification (Mangasarian-Fromovitz), Cottle's constraint qualification, Abadie's constraint qualification, etc.

In this book, only Slater's constraint qualification is considered. The importance of this concept will be explained later on.

As we will see below, Slater's constraint qualification is a prerequisite for the validity of the Karush-Kuhn-Tucker conditions.

If the condition (i) is satisfied, the minimization problem (1.7) is called a *convex programming problem* (*convex program*).

Theorem 1.36 (Local and global minimum of a convex function)
Let X be a convex set in $\mathbf{R}^n$ and f be convex on X. Then each local minimum of f on X is a global minimum as well.

Proof. Let $\mathbf{x}^*$ be a local minimum of f on X. Then there exists an $\varepsilon > 0$ such that $\mathbf{x} \in X$ and $\|\mathbf{x} - \mathbf{x}^*\| \leq \varepsilon$ imply $f(\mathbf{x}^*) \leq f(\mathbf{x})$.

Let $\mathbf{y} \in X$ and $\|\mathbf{y} - \mathbf{x}^*\| > \varepsilon$. Our purpose is to prove that $f(\mathbf{y}) \geq f(\mathbf{x}^*)$. It will then follow that $\mathbf{x}^*$ is a global minimum of f on X by definition.

Consider a point $\mathbf{z}$ such that $\|\mathbf{z} - \mathbf{x}^*\| = \varepsilon$ and

$$\mathbf{z} \in [\mathbf{x}^*, \mathbf{y}] \equiv \{\mathbf{x} : \mathbf{x} = \lambda \mathbf{x}^* + (1 - \lambda)\mathbf{y}, \ 0 \leq \lambda \leq 1\}.$$

Then $\mathbf{z}$ can be expressed as

$$\mathbf{z} = \frac{\varepsilon}{\|\mathbf{y} - \mathbf{x}^*\|}\mathbf{y} + \left(1 - \frac{\varepsilon}{\|\mathbf{y} - \mathbf{x}^*\|}\right)\mathbf{x}^* = \mathbf{x}^* + \frac{\varepsilon}{\|\mathbf{y} - \mathbf{x}^*\|}(\mathbf{y} - \mathbf{x}^*).$$

Since X is convex and $\mathbf{x}^*, \mathbf{y} \in X$, then $\mathbf{z} \in X$. Moreover, since $\|\mathbf{z} - \mathbf{x}^*\| = \varepsilon$ by definition of $\mathbf{z}$ and since $\mathbf{x}^* \stackrel{\text{def}}{=} \arg\min_{\|\mathbf{x}-\mathbf{x}^*\| \le \varepsilon} f(\mathbf{x})$, then $f(\mathbf{z}) \ge f(\mathbf{x}^*)$.

Because f is convex,

$$f(\mathbf{x}^*) \le f(\mathbf{z}) \equiv f\left(\frac{\varepsilon}{\|\mathbf{y} - \mathbf{x}^*\|}\mathbf{y} + \left(1 - \frac{\varepsilon}{\|\mathbf{y} - \mathbf{x}^*\|}\right)\mathbf{x}^*\right)$$

$$\le \frac{\varepsilon}{\|\mathbf{y} - \mathbf{x}^*\|} f(\mathbf{y}) + \left(1 - \frac{\varepsilon}{\|\mathbf{y} - \mathbf{x}^*\|}\right) f(\mathbf{x}^*),$$

that is,

$$0 \le \frac{\varepsilon}{\|\mathbf{y} - \mathbf{x}^*\|}(f(\mathbf{y}) - f(\mathbf{x}^*)),$$

and since $\varepsilon > 0$, $\|\mathbf{y} - \mathbf{x}^*\| > 0$, then $f(\mathbf{y}) \ge f(\mathbf{x}^*)$. ■

Theorem 1.37 (Convexity of the set of optimal solutions to a convex program) *Let X be a convex set in $\mathbf{R}^n$ and let the numerical function f be convex on X. Then the set of optimal solutions to the problem $\min_{\mathbf{x}\in X} f(\mathbf{x})$ is convex.*

Proof. Denote

$$X_{\min} = \left\{\mathbf{x}^* \in X : f(\mathbf{x}^*) = \min_{\mathbf{x}\in X} f(\mathbf{x})\right\}.$$

Let $\mathbf{x}_1, \mathbf{x}_2 \in X_{\min}$ $\left(\text{therefore } f(\mathbf{x}_1) = f(\mathbf{x}_2) = \min_{\mathbf{x}\in X} f(\mathbf{x}) \stackrel{\text{def}}{=} f_{\min}\right)$ and let $\lambda \in [0, 1]$. Since X is convex, then $\lambda\mathbf{x}_1 + (1 - \lambda)\mathbf{x}_2 \in X$, and since f is convex on X,

$$f(\lambda\mathbf{x}_1 + (1 - \lambda)\mathbf{x}_2) \le \lambda f(\mathbf{x}_1) + (1 - \lambda) f(\mathbf{x}_2) = \lambda f_{\min} + (1 - \lambda) f_{\min} = f_{\min}.$$

However, $f(\lambda\mathbf{x}_1 + (1 - \lambda)\mathbf{x}_2) \ge f_{\min}$ according to the definition of $f_{\min}$. From the two opposite inequalities it follows that $f(\lambda\mathbf{x}_1 + (1 - \lambda)\mathbf{x}_2) = f_{\min}$, that is, $\lambda\mathbf{x}_1 + (1 - \lambda)\mathbf{x}_2 \in X_{\min}$. Hence, $X_{\min}$ is a convex set by definition. ■

Theorem 1.38 (Uniqueness of the optimal solution to a strictly convex program) *Let X be a convex set in $\mathbf{R}^n$, f be a strictly convex numerical function on X, and $\mathbf{x}^*$ be a solution to the minimization problem $\min_{\mathbf{x}\in X} f(\mathbf{x})$. Then $\mathbf{x}^*$ is the unique solution to this problem.*

Proof. Assume that there exists an $\mathbf{x}_1 \in X$, $\mathbf{x}_1 \ne \mathbf{x}^*$ such that

$$f(\mathbf{x}_1) = f(\mathbf{x}^*) \stackrel{\text{def}}{=} f_{\min}.$$

Since X is a convex set, then $\lambda\mathbf{x}_1 + (1 - \lambda)\mathbf{x}^* \in X$ for $\lambda \in [0, 1]$, and according to strict convexity of f we have the strict inequality

$$f(\lambda \mathbf{x}_1 + (1-\lambda)\mathbf{x}^*) < \lambda f(\mathbf{x}_1) + (1-\lambda) f(\mathbf{x}^*) = \lambda f_{\min} + (1-\lambda) f_{\min} = f_{\min},$$

which contradicts the assumption that $f(\mathbf{x}^*) = \min_{\mathbf{x}\in X} f(\mathbf{x})$.

Therefore $\mathbf{x}^*$ is the only solution to the problem $\min_{\mathbf{x}\in X} f(\mathbf{x})$ with $f(\mathbf{x})$ strictly convex function. ■

Theorem 1.39 (Fermat's generalized rule)
Let $f : \mathbf{R}^n \to \mathbf{R}$ be convex function and X be a nonempty convex set in $\mathbf{R}^n$. The point $\mathbf{x}^*$ is an optimal solution to the minimization problem

$$\min_{\mathbf{x}\in X} f(\mathbf{x})$$

if and only if there exists a subgradient $\widehat{\mathbf{x}} \in \partial f(\mathbf{x}^*)$ such that for each $\mathbf{x} \in X$, the following inequality holds true

$$\langle \widehat{\mathbf{x}}, \mathbf{x} - \mathbf{x}^* \rangle \geq 0.$$

In particular, if $X = \mathbf{R}^n$, $\mathbf{x}^*$ is an optimal solution to the minimization problem $\min_{\mathbf{x}\in\mathbf{R}^n} f(\mathbf{x})$ if and only if $\mathbf{0} \in \partial f(\mathbf{x}^*)$.

Proof. Let $\widehat{\mathbf{x}}$ be a subgradient satisfying $\langle \widehat{\mathbf{x}}, \mathbf{x} - \mathbf{x}^* \rangle \geq 0$ for each $\mathbf{x} \in X$. Using this inequality and definition of subgradient, we obtain

$$f(\mathbf{x}) \geq f(\mathbf{x}^*) + \langle \widehat{\mathbf{x}}, \mathbf{x} - \mathbf{x}^* \rangle \geq f(\mathbf{x}^*) \text{ for each } \mathbf{x} \in X.$$

Hence $\mathbf{x}^*$ is an optimal solution to the minimization problem under consideration.

Conversely, let $\mathbf{x}^*$ be an optimal solution to the minimization problem $\min_{\mathbf{x}\in X} f(\mathbf{x})$. Consider the sets Y_1 and Y_2 in $\mathbf{R}^{n+1}$ defined through

$$Y_1 = \{(\mathbf{x} - \mathbf{x}^*, r) : \mathbf{x} \in \mathbf{R}^n,\ f(\mathbf{x}) - f(\mathbf{x}^*) < r\},$$
$$Y_2 = \{(\mathbf{x} - \mathbf{x}^*, r) : \mathbf{x} \in X,\ r \leq 0\}.$$

It can be verified that Y_1 and Y_2 are nonempty convex sets, and $Y_1 \cap Y_2 = \emptyset$. If we assume that there is some $(\mathbf{x}_1 - \mathbf{x}^*, r_1) \in Y_1 \cap Y_2$, then $\mathbf{x}_1 \in X$ and $f(\mathbf{x}_1) - f(\mathbf{x}^*) < r_1 \leq 0$, that is, $f(\mathbf{x}_1) < f(\mathbf{x}^*)$ for $\mathbf{x}_1 \in X$. This contradicts the assumption that $\mathbf{x}^*$ is an optimal solution to the minimization problem under consideration.

From Theorem 1.14 it follows that Y_1 and Y_2 can be separated, that is, there is a nonzero vector $(\mathbf{h}, \rho) \in \mathbf{R}^{n+1}$ and a real number a such that

$$\langle \mathbf{h}, \mathbf{x} - \mathbf{x}^* \rangle + \rho r \leq a \quad \forall\, (\mathbf{x}, r),\ \mathbf{x} \in \mathbf{R}^n,\ f(\mathbf{x}) - f(\mathbf{x}^*) < r,$$
$$\langle \mathbf{h}, \mathbf{x} - \mathbf{x}^* \rangle + \rho r \geq a \quad \forall\, (\mathbf{x}, r),\ \mathbf{x} \in X,\ r \leq 0.$$

Letting $\mathbf{x} = \mathbf{x}^*$ and $r = 0$ in the second inequality, we get $a \leq 0$.

Letting $\mathbf{x} = \mathbf{x}^*$ in the first inequality, we get $\rho r \leq a$ for each r such that $0 = f(\mathbf{x}^*) - f(\mathbf{x}^*) < r$. Therefore $\rho \leq 0$, and since $\rho r \leq a$ for all $r > 0$ and $\rho \leq 0$, then $a \geq 0$. From $a \leq 0$ and $a \geq 0$ it follows that $a = 0$.

Assuming that $\rho = 0$, from the first inequality it follows that $\langle \mathbf{h}, \mathbf{x} - \mathbf{x}^* \rangle \leq 0$ for each $\mathbf{x} \in \mathbf{R}^n$. Then $\mathbf{h} = \mathbf{0} \in \mathbf{R}^n$, a contradiction: vector $(\mathbf{h}, \rho)$ must be nonzero. Therefore ρ cannot be zero, and since $\rho \leq 0$ then $\rho < 0$.

Dividing both inequalities by $(-\rho) > 0$, denoting $-\frac{\mathbf{h}}{\rho} = \widehat{\mathbf{x}}$, and taking into account that $a = 0$ yields

$$r \geq \langle \widehat{\mathbf{x}}, \mathbf{x} - \mathbf{x}^* \rangle \quad \forall\, (\mathbf{x}, r),\ \mathbf{x} \in \mathbf{R}^n,\ f(\mathbf{x}) - f(\mathbf{x}^*) < r,$$

$$\langle \widehat{\mathbf{x}}, \mathbf{x} - \mathbf{x}^* \rangle \geq r \quad \forall\, (\mathbf{x}, r),\ \mathbf{x} \in X,\ r \leq 0.$$

Letting $r = 0$ in the last inequality, we get $\langle \widehat{\mathbf{x}}, \mathbf{x} - \mathbf{x}^* \rangle \geq 0$ for each $\mathbf{x} \in X$, that is, $\widehat{\mathbf{x}}$ satisfies the hypothesis of Theorem 1.39.

From the previous inequality it follows that

$$f(\mathbf{x}) - f(\mathbf{x}^*) \geq \langle \widehat{\mathbf{x}}, \mathbf{x} - \mathbf{x}^* \rangle \quad \forall\, \mathbf{x} \in \mathbf{R}^n,$$

that is, $\widehat{\mathbf{x}}$ is a subgradient of f at $\mathbf{x}^*$. We have already proved that for each $\mathbf{x} \in X$, this subgradient satisfies $\langle \widehat{\mathbf{x}}, \mathbf{x} - \mathbf{x}^* \rangle \geq 0$.

Consider the case $X = \mathbf{R}^n$.

Let f be convex and $\mathbf{x}^*$ be an optimal solution to the minimization problem $\min\limits_{\mathbf{x} \in \mathbf{R}^n} f(\mathbf{x})$. Hence $f(\mathbf{x}) \geq f(\mathbf{x}^*)$ for each $\mathbf{x} \in \mathbf{R}^n$. This inequality can be written as follows

$$f(\mathbf{x}) - f(\mathbf{x}^*) \geq 0 \equiv \langle \mathbf{0}, \mathbf{x} - \mathbf{x}^* \rangle \quad \forall\, \mathbf{x} \in \mathbf{R}^n,$$

that is, $\mathbf{0} \in \partial f(\mathbf{x}^*)$.

Conversely, if $\mathbf{0} \in \partial f(\mathbf{x}^*)$, then

$$f(\mathbf{x}) - f(\mathbf{x}^*) \geq \langle \mathbf{0}, \mathbf{x} - \mathbf{x}^* \rangle = 0 \quad \forall\, \mathbf{x} \in \mathbf{R}^n,$$

that is, $f(\mathbf{x}) \geq f(\mathbf{x}^*)\ \forall\, \mathbf{x} \in \mathbf{R}^n$, which means that $\mathbf{x}^*$ by definition is an optimal solution to the minimization problem under consideration.

The case $X = \mathbf{R}^n$ can also be obtained as a corollary of the inequality $\langle \widehat{\mathbf{x}}, \mathbf{x} - \mathbf{x}^* \rangle \geq 0$ proved above. Since this inequality must be satisfied for all $\mathbf{x}$ in $\mathbf{R}^n$, then $\widehat{\mathbf{x}} = \mathbf{0}$ and the necessary and sufficient condition becomes

$$\widehat{\mathbf{x}} = \mathbf{0} \in \partial f(\mathbf{x}^*).$$

■

If f is differentiable, then $\partial f(\mathbf{x}^*) = \{\mathbf{f}'(\mathbf{x}^*)\}$ according to Theorem 1.33, and the necessary and sufficient condition of Theorem 1.39 becomes $\mathbf{f}'(\mathbf{x}^*) = \mathbf{0}$, the well-known condition for multivariate differentiable functions.

Theorem 1.40 (Sufficient optimality theorem)
If $(\mathbf{x}^*, \boldsymbol{\lambda}^*)$ *is a saddle point of the Lagrangian* $L(\mathbf{x}, \boldsymbol{\lambda})$ *for problem* (1.7) *with* $x_j \geq 0,\ j \in J_1,\ \lambda_i \geq 0,\ i \in I_1 \cup I_2$, *then* $\mathbf{x}^*$ *is an optimal solution to problem* (1.7).

Proof. Let $(\mathbf{x}^*, \boldsymbol{\lambda}^*)$ be a saddle point of $L(\mathbf{x}, \boldsymbol{\lambda})$ for $x_j^* \geq 0, j \in J_1,\ \lambda_i^* \geq 0, i \in I_1 \cup I_2$. Then

$$
\begin{aligned}
& f(\mathbf{x}^*) + \sum_{i \in I_1} \lambda_i g_i(\mathbf{x}^*) + \sum_{i \in I_2} \lambda_i l_i(\mathbf{x}^*) + \sum_{i \in I_3} \lambda_i l_i(\mathbf{x}^*) \\
& \leq f(\mathbf{x}^*) + \sum_{i \in I_1} \lambda_i^* g_i(\mathbf{x}^*) + \sum_{i \in I_2} \lambda_i^* l_i(\mathbf{x}^*) + \sum_{i \in I_3} \lambda_i^* l_i(\mathbf{x}^*) \\
& \leq f(\mathbf{x}) + \sum_{i \in I_1} \lambda_i^* g_i(\mathbf{x}) + \sum_{i \in I_2} \lambda_i^* l_i(\mathbf{x}) + \sum_{i \in I_3} \lambda_i^* l_i(\mathbf{x})
\end{aligned}
$$

for all $\boldsymbol{\lambda} \in \Lambda = \{\boldsymbol{\lambda} = (\lambda_1, \ldots, \lambda_m) : \lambda_i \geq 0,\ i \in I_1 \cup I_2\}$ and $\mathbf{x} \in X$.

Note that $f(\mathbf{x}^*) + \sum_{i \in I_1} \lambda_i^* g_i(\mathbf{x}^*) + \sum_{i \in I_2 \cup I_3} \lambda_i^* l_i(\mathbf{x}^*)$ is a constant because components of $\mathbf{x}^*$ and $\boldsymbol{\lambda}^*$ are fixed.

Consider the first saddlepoint inequality. Since $\lambda_i \geq 0,\ i \in I_1 \cup I_2$, then $g_i(\mathbf{x}^*)$ must be nonpositive ($g_i(\mathbf{x}^*) \leq 0$) for each $i \in I_1$, and $l_i(\mathbf{x}^*) \leq 0$ for each $i \in I_2$; otherwise, if there is $g_i(\mathbf{x}^*) > 0$ for some $i \in I_1$ or $l_i(\mathbf{x}^*) > 0$ for some $i \in I_2$ and $\lambda_i > 0$ for this i, then the left-hand side would tend to $+\infty$ and the first saddlepoint inequality would not be satisfied. Similarly, $l_i(\mathbf{x}^*)$ must be equal to 0 for all $i \in I_3$; otherwise, if $l_i(\mathbf{x}^*) > 0$ for some $i \in I_3$ and $\lambda_i > 0,\ \lambda_i \to +\infty$ for this i, or if $l_i(\mathbf{x}^*) < 0$ for some $i \in I_3$ and $\lambda_i < 0,\ \lambda_i \to -\infty$ for this i, the first saddlepoint inequality would not be satisfied.

Therefore, $g_i(\mathbf{x}^*) \leq 0,\ i \in I_1,\ l_i(\mathbf{x}^*) \leq 0,\ i \in I_2,\ l_i(\mathbf{x}^*) = 0, i \in I_3$; and $x_j^* \geq 0,\ j \in J_1$ by assumption, that is, $\mathbf{x}^* \in X$ ($\mathbf{x}^*$ is a feasible point for problem (1.7)). According to these results,

$$
\sum_{i \in I_1} \lambda_i^* g_i(\mathbf{x}^*) + \sum_{i \in I_2} \lambda_i^* l_i(\mathbf{x}^*) + \sum_{i \in I_3} \lambda_i^* l_i(\mathbf{x}^*) \leq 0.
$$

However, from the first saddlepoint inequality with $\boldsymbol{\lambda} = \mathbf{0}$ we obtain

$$
f(\mathbf{x}^*) \leq f(\mathbf{x}^*) + \sum_{i \in I_1} \lambda_i^* g_i(\mathbf{x}^*) + \sum_{i \in I_2} \lambda_i^* l_i(\mathbf{x}^*) + \sum_{i \in I_3} \lambda_i^* l_i(\mathbf{x}^*),
$$

that is,

$$
\sum_{i \in I_1} \lambda_i^* g_i(\mathbf{x}^*) + \sum_{i \in I_2} \lambda_i^* l_i(\mathbf{x}^*) + \sum_{i \in I_3} \lambda_i^* l_i(\mathbf{x}^*) \geq 0.
$$

From both opposite inequalities,

$$\sum_{i\in I_1} \lambda_i^* g_i(\mathbf{x}^*) + \sum_{i\in I_2} \lambda_i^* l_i(\mathbf{x}^*) + \sum_{i\in I_3} \lambda_i^* l_i(\mathbf{x}^*) = 0.$$

Then the second saddlepoint inequality is satisfied as follows:

$$f(\mathbf{x}^*) \le f(\mathbf{x}) + \sum_{i\in I_1} \lambda_i^* g_i(\mathbf{x}) + \sum_{i\in I_2} \lambda_i^* l_i(\mathbf{x}) + \sum_{i\in I_3} \lambda_i^* l_i(\mathbf{x}).$$

For each $\mathbf{x} \in X$ we have $g_i(\mathbf{x}) \le 0,\ i \in I_1,\ l_i(\mathbf{x}) \le 0,\ i \in I_2,\ l_i(\mathbf{x}) = 0, i \in I_3$, and since $\lambda_i^* \ge 0,\ i \in I_1 \cup I_2$, then

$$\sum_{i\in I_1} \lambda_i^* g_i(\mathbf{x}) + \sum_{i\in I_2} \lambda_i^* l_i(\mathbf{x}) + \sum_{i\in I_3} \lambda_i^* l_i(\mathbf{x}) \le 0.$$

Hence, $f(\mathbf{x}^*) \le f(\mathbf{x})$ for any $\mathbf{x} \in X$, that is, $\mathbf{x}^* = \arg\min_{\mathbf{x}\in X} f(\mathbf{x})$. ■

Theorem 1.41 (Karush-Kuhn-Tucker saddlepoint necessary optimality theorem) *Let* (i) *and* (ii) *be satisfied. If* $\mathbf{x}^*$ *is an optimal solution to problem* (1.7) *then there exists some* $\boldsymbol{\lambda}^* \in \Lambda$ *such that* $(\mathbf{x}^*, \boldsymbol{\lambda}^*)$ *is a saddle point of* $L(\mathbf{x}, \boldsymbol{\lambda})$ *for* $x_j \ge 0,\ j \in J_1,\ \lambda_i \ge 0,\ i \in I_1 \cup I_2$, *and*

$$\sum_{i\in I_1} \lambda_i^* g_i(\mathbf{x}^*) + \sum_{i\in I_2} \lambda_i^* l_i(\mathbf{x}^*) + \sum_{i\in I_3} \lambda_i^* l_i(\mathbf{x}^*) = 0.$$

Proof. Let $\mathbf{x}^*$ be an optimal solution to problem (1.7) and conditions (i) and (ii) be satisfied. Consider the set

$$Y = \{\mathbf{y} = (y_0, y_1, \ldots, y_{m+n_1}) \mid \exists\, \mathbf{x} \in X : \ y_0 \ge f(\mathbf{x}),\ \ y_i \ge g_i(\mathbf{x}),\ i \in I_1,$$
$$y_i \ge l_i(\mathbf{x}),\ i \in I_2,\ \ y_i = l_i(\mathbf{x}),\ i \in I_3,\ \ y_{m+j} \le x_j,\ j \in J_1\} \subset \mathbf{R}^{m+n_1+1},$$

where $n_1 = |J_1|$. Y is a nonempty set, and Y is a convex set because $f,\ g_i,\ i \in I_1$ are convex functions and $l_i,\ i \in I_2 \cup I_3$ are affine functions (and therefore, convex functions, see Example 1.2, Theorem 1.21).

It is easy to verify that $\mathbf{z} \stackrel{\text{def}}{=} (f(\mathbf{x}^*), 0, \ldots, 0)_{m+n_1+1} \in Y$, $\mathbf{z}_\varepsilon \stackrel{\text{def}}{=} (f(\mathbf{x}^*) + \varepsilon, 0, \ldots, 0) \in Y$, $\mathbf{z}_{-\varepsilon} \stackrel{\text{def}}{=} (f(\mathbf{x}^*) - \varepsilon, 0, \ldots, 0) \notin Y$ for each $\varepsilon > 0$ because $\mathbf{x}^*$ is an optimal solution to problem (1.7) and according to definition of Y, the first inequality. Therefore $\mathbf{z} \in \partial Y$.

According to Theorem 1.12, there exists a $\mathbf{v} \in \mathbf{R}^{m+n_1+1}$, $\mathbf{v} \ne \mathbf{0}$ such that

$$\langle \mathbf{v}, \mathbf{y} \rangle \ge \langle \mathbf{v}, \mathbf{z} \rangle \quad \text{for all } \mathbf{y} \in Y. \tag{1.8}$$

Since components of elements of Y are unbounded above for $i \in \{0\} \cup I_1 \cup I_2$ and they are unbounded below for $m + j,\ j \in J_1$ by definition of Y, then $v_i \ge 0,\ i \in \{0\} \cup I_1 \cup I_2,\ \ v_{m+j} \le 0,\ j \in J_1$ according to (1.8).

Let $\mathbf{x} \in X$. Consider the point $\widehat{\mathbf{y}} = (\widehat{y}_0, \ldots, \widehat{y}_{m+n_1})$ whose components are defined as follows:

$$\widehat{y}_0 = f(\mathbf{x}), \quad \widehat{y}_i = g_i(\mathbf{x}),\ i \in I_1, \quad \widehat{y}_i = l_i(\mathbf{x}),\ i \in I_2 \cup I_3,$$
$$\widehat{y}_{m+j} = x_j,\ j \in J_1.$$

Obviously $\widehat{\mathbf{y}} \in Y$. Applying (1.8) to $\mathbf{y} = \widehat{\mathbf{y}} \in Y$, for all $\mathbf{x} \in X$ we get

$$v_0 f(\mathbf{x}) + \sum_{i \in I_1} v_i g_i(\mathbf{x}) + \sum_{i \in I_2} v_i l_i(\mathbf{x}) + \sum_{i \in I_3} v_i l_i(\mathbf{x}) + \sum_{j \in J_1} v_{m+j} x_j \\ \geq v_0 f(\mathbf{x}^*). \tag{1.9}$$

We want to prove that $v_0 \neq 0$.
Assume the contrary, that $v_0 = 0$. Then

$$\sum_{i \in I_1} v_i g_i(\mathbf{x}) + \sum_{i \in I_2} v_i l_i(\mathbf{x}) + \sum_{i \in I_3} v_i l_i(\mathbf{x}) + \sum_{j \in J_1} v_{m+j} x_j \geq 0 \tag{1.10}$$

for all $\mathbf{x} \in X$.

From $v_i \geq 0$ and $l_i(\mathbf{x}) \leq 0,\ i \in I_2;\ l_i(\mathbf{x}) = 0,\ i \in I_3;\ v_{m+j} \leq 0,\ x_j \geq 0,\ j \in J_1$ it follows that

$$\sum_{i \in I_2} v_i l_i(\mathbf{x}) + \sum_{i \in I_3} v_i l_i(\mathbf{x}) + \sum_{j \in J_1} v_{m+j} x_j \leq 0.$$

In order that (1.10) be satisfied, since $g_i(\mathbf{x}) < 0,\ i \in I_1$ for some $\mathbf{x} = \overline{\mathbf{x}}$ according to Slater's constraint qualification, v_i must be equal to 0 for each $i \in I_1$. Therefore

$$\sum_{i \in I_2} v_i l_i(\mathbf{x}) + \sum_{i \in I_3} v_i l_i(\mathbf{x}) + \sum_{j \in J_1} v_{m+j} x_j \geq 0. \tag{1.10$'$}$$

From both opposite inequalities it follows that

$$\sum_{i \in I_2} v_i l_i(\mathbf{x}) + \sum_{i \in I_3} v_i l_i(\mathbf{x}) + \sum_{j \in J_1} v_{m+j} x_j = 0.$$

The left-hand side of this expression is a linear function: $l_i(\mathbf{x}),\ i \in I_2 \cup I_3$ are linear by assumption, x_j's are also linear functions. Without loss of generality let $l_i(\mathbf{x}),\ i \in I_2 \cup I_3$ and x_j's be linearly independent (we can omit the linearly dependent functions and corresponding constraints, if there are any). We have obtained that a linear combination of linearly independent functions is equal to zero. Hence,

$$v_i = 0,\ i \in I_2 \cup I_3; \quad v_{m+j} = 0,\ j \in J_1,$$

that is, $v_0 = 0$ implies $\mathbf{v} = \mathbf{0}$, a contradiction with Theorem 1.12. Therefore, the assumption is incorrect. Hence $v_0 \neq 0$.

Dividing (1.9) by $v_0 \neq 0,\ v_0 > 0$ we get

$$f(\mathbf{x}) + \sum_{i\in I_1} \lambda_i^* g_i(\mathbf{x}) + \sum_{i\in I_2} \lambda_i^* l_i(\mathbf{x}) + \sum_{i\in I_3} \lambda_i^* l_i(\mathbf{x}) + \sum_{j\in J_1} \lambda_{m+j}^* x_j \geq f(\mathbf{x}^*)$$

for all $\mathbf{x} \in X$, where we have denoted $\lambda_i^* = \frac{v_i}{v_0}$ (≥ 0 for $i \in I_1 \cup I_2$ because $v_i \geq 0,\ i \in I_1 \cup I_2,\ v_0 > 0$), $\lambda_{m+j}^* = \frac{v_{m+j}}{v_0}$ (≤ 0 because $v_{m+j} \leq 0,\ v_0 > 0$).

Consider this inequality for $x_j \geq 0,\ j \in J_1$:

$$f(\mathbf{x}) + \sum_{i\in I_1} \lambda_i^* g_i(\mathbf{x}) + \sum_{i\in I_2} \lambda_i^* l_i(\mathbf{x}) + \sum_{i\in I_3} \lambda_i^* l_i(\mathbf{x}) \geq f(\mathbf{x}^*), \tag{1.11}$$

where the sum $\sum_{j\in J_1} \lambda_{m+j}^* x_j \leq 0$ is ignored from the left-hand side; this preserves (and even strengthens) the inequality. Letting $\mathbf{x} = \mathbf{x}^*$ in (1.11) yields

$$\sum_{i\in I_1} \lambda_i^* g_i(\mathbf{x}^*) + \sum_{i\in I_2} \lambda_i^* l_i(\mathbf{x}^*) + \sum_{i\in I_3} \lambda_i^* l_i(\mathbf{x}^*) \geq 0.$$

On the other hand, since $\lambda_i^* \geq 0,\ i \in I_1 \cup I_2,\ \ g_i(\mathbf{x}^*) \leq 0,\ i \in I_1,\ \ l_i(\mathbf{x}^*) \leq 0, i \in I_2,\ \ l_i(\mathbf{x}^*) = 0,\ i \in I_3$, then obviously

$$\sum_{i\in I_1} \lambda_i^* g_i(\mathbf{x}^*) + \sum_{i\in I_2} \lambda_i^* l_i(\mathbf{x}^*) + \sum_{i\in I_3} \lambda_i^* l_i(\mathbf{x}^*) \leq 0.$$

From both opposite inequalities it follows that

$$\sum_{i\in I_1} \lambda_i^* g_i(\mathbf{x}^*) + \sum_{i\in I_2} \lambda_i^* l_i(\mathbf{x}^*) + \sum_{i\in I_3} \lambda_i^* l_i(\mathbf{x}^*) = 0.$$

Adding this zero sum in the right-hand side of (1.11) and taking into account the definition of $L(\mathbf{x}, \boldsymbol{\lambda})$, we get $L(\mathbf{x}, \boldsymbol{\lambda}^*) \geq L(\mathbf{x}^*, \boldsymbol{\lambda}^*)$ for all $x_j \geq 0,\ j \in J_1$ and for $\boldsymbol{\lambda}^* \in \Lambda$.

Furthermore, for $\lambda_i \geq 0,\ i \in I_1 \cup I_2$ and $\mathbf{x}^* \in X$ we obviously have

$$f(\mathbf{x}^*) + \sum_{i\in I_1} \lambda_i g_i(\mathbf{x}^*) + \sum_{i\in I_2} \lambda_i l_i(\mathbf{x}^*) + \sum_{i\in I_3} \lambda_i l_i(\mathbf{x}^*) \leq f(\mathbf{x}^*)$$

$$\equiv f(\mathbf{x}^*) + \sum_{i\in I_1} \lambda_i^* g_i(\mathbf{x}^*) + \sum_{i\in I_2} \lambda_i^* l_i(\mathbf{x}^*) + \sum_{i\in I_3} \lambda_i^* l_i(\mathbf{x}^*),$$

that is, $L(\mathbf{x}^*, \boldsymbol{\lambda}) \leq L(\mathbf{x}^*, \boldsymbol{\lambda}^*)$ for all $\boldsymbol{\lambda} \in \Lambda$ and for $\mathbf{x}^* \in X$.

We have also proved that $L(\mathbf{x}, \boldsymbol{\lambda}^*) \geq L(\mathbf{x}^*, \boldsymbol{\lambda}^*)$ for all $x_j \geq 0,\ j \in J_1$ and for $\boldsymbol{\lambda}^* \in \Lambda$.

From both inequalities we obtain

$$L(\mathbf{x}^*, \boldsymbol{\lambda}) \le L(\mathbf{x}^*, \boldsymbol{\lambda}^*) \le L(\mathbf{x}, \boldsymbol{\lambda}^*) \quad \forall\, x_j \ge 0,\ j \in J_1,\ \ \forall\, \boldsymbol{\lambda} \in \Lambda.$$

Therefore, $(\mathbf{x}^*, \boldsymbol{\lambda}^*)$ is a saddle point of $L(\mathbf{x}, \boldsymbol{\lambda})$ by Definition 1.25.

Hence there exists a $\boldsymbol{\lambda}^*$ with the property required. Moreover, we have proved that

$$\sum_{i\in I_1} \lambda_i^* g_i(\mathbf{x}^*) + \sum_{i\in I_2} \lambda_i^* l_i(\mathbf{x}^*) + \sum_{i\in I_3} \lambda_i^* l_i(\mathbf{x}^*) = 0.$$

■

Theorem 1.42 (Optimality criterion in convex programming with differentiability: Karush-Kuhn-Tucker conditions)
Let f and g_i be numerical differentiable functions on the set X in $\mathbf{R}^n$.
For $\mathbf{x}^*$ to be an optimal solution to problem (1.7) *it is sufficient (and necessary, if conditions* (i) *and* (ii) *are satisfied) that there exists a $\boldsymbol{\lambda}^* \in \Lambda$ such that*

$$\frac{\partial L}{\partial x_j}(\mathbf{x}^*, \boldsymbol{\lambda}^*) \ge 0, \quad j \in J_1 \tag{1.12}$$

$$\frac{\partial L}{\partial x_j}(\mathbf{x}^*, \boldsymbol{\lambda}^*) = 0, \quad j \in J \setminus J_1 \tag{1.13}$$

$$x_j^* \frac{\partial L}{\partial x_j}(\mathbf{x}^*, \boldsymbol{\lambda}^*) = 0, \quad j \in J_1 \tag{1.14}$$

$$x_j^* \ge 0, \quad j \in J_1 \tag{1.15}$$

$$\frac{\partial L}{\partial \lambda_i}(\mathbf{x}^*, \boldsymbol{\lambda}^*) \le 0, \quad i \in I_1 \cup I_2 \tag{1.16}$$

$$\frac{\partial L}{\partial \lambda_i}(\mathbf{x}^*, \boldsymbol{\lambda}^*) = 0, \quad i \in I_3 \tag{1.17}$$

$$\lambda_i^* \frac{\partial L}{\partial \lambda_i}(\mathbf{x}^*, \boldsymbol{\lambda}^*) = 0, \quad i \in I_1 \cup I_2 \tag{1.18}$$

$$\lambda_i^* \ge 0, \quad i \in I_1 \cup I_2. \tag{1.19}$$

Proof. *Necessity.* Let $\mathbf{x}^*$ be an optimal solution to problem (1.7). According to Theorem 1.41, there exists a $\boldsymbol{\lambda}^* \in \Lambda$ such that $(\mathbf{x}^*, \boldsymbol{\lambda}^*)$ is a saddle point of $L(\mathbf{x}, \boldsymbol{\lambda})$ for $x_j \geq 0,\ j \in J_1,\ \lambda_i \geq 0,\ i \in I_1 \cup I_2$, that is,

$$L(\mathbf{x}^*, \boldsymbol{\lambda}) \leq L(\mathbf{x}^*, \boldsymbol{\lambda}^*) \leq L(\mathbf{x}, \boldsymbol{\lambda}^*).$$

The second saddlepoint inequality means that function $L(\mathbf{x}, \boldsymbol{\lambda}^*)$ attains its minimum at $(\mathbf{x}^*, \boldsymbol{\lambda}^*)$ in the region $\{\mathbf{x} \in \mathbf{R}^n : x_j \geq 0,\ j \in J_1\}$. Therefore conditions (1.12)–(1.15) are satisfied.

Similarly, the first saddlepoint inequality means that $L(\mathbf{x}^*, \boldsymbol{\lambda})$ attains its maximum at $(\mathbf{x}^*, \boldsymbol{\lambda}^*)$ in Λ. Therefore conditions (1.16)–(1.19) are satisfied.

Sufficiency. Let conditions (1.12)–(1.19) be satisfied. Since $L(\mathbf{x}, \boldsymbol{\lambda}^*)$ is a convex function of $\mathbf{x}$, then from Theorem 1.30 it follows that

$$\begin{aligned}
L(\mathbf{x}, \boldsymbol{\lambda}^*) - L(\mathbf{x}^*, \boldsymbol{\lambda}^*) &\geq \left\langle \frac{\partial L}{\partial \mathbf{x}}(\mathbf{x}^*, \boldsymbol{\lambda}^*), \mathbf{x} - \mathbf{x}^* \right\rangle \\
&= \left\langle \frac{\partial L}{\partial \mathbf{x}}(\mathbf{x}^*, \boldsymbol{\lambda}^*), \mathbf{x} \right\rangle - \left\langle \frac{\partial L}{\partial \mathbf{x}}(\mathbf{x}^*, \boldsymbol{\lambda}^*), \mathbf{x}^* \right\rangle \\
&= \sum_{j \in J} \frac{\partial L}{\partial x_j}(\mathbf{x}^*, \boldsymbol{\lambda}^*) x_j - \sum_{j \in J} \frac{\partial L}{\partial x_j}(\mathbf{x}^*, \boldsymbol{\lambda}^*) x_j^* \\
&= \sum_{j \in J_1} \frac{\partial L}{\partial x_j}(\mathbf{x}^*, \boldsymbol{\lambda}^*) x_j + \sum_{j \in J \setminus J_1} \frac{\partial L}{\partial x_j}(\mathbf{x}^*, \boldsymbol{\lambda}^*) x_j \\
&\quad - \sum_{j \in J_1} \frac{\partial L}{\partial x_j}(\mathbf{x}^*, \boldsymbol{\lambda}^*) x_j^* - \sum_{j \in J \setminus J_1} \frac{\partial L}{\partial x_j}(\mathbf{x}^*, \boldsymbol{\lambda}^*) x_j^* \geq 0
\end{aligned}$$

where the first sum (over J_1) in the last expression is nonnegative for each $x_j \geq 0,\ j \in J_1$ according to (1.12), the second sum is equal to zero according to (1.13), the third sum is equal to zero in accordance with (1.14), and the fourth sum is equal to zero according to (1.13).

Therefore $L(\mathbf{x}^*, \boldsymbol{\lambda}^*) \leq L(\mathbf{x}, \boldsymbol{\lambda}^*)$ for $x_j \geq 0,\ j \in J_1$.

Function $L(\mathbf{x}^*, \boldsymbol{\lambda})$ is a linear function of $\boldsymbol{\lambda}$. Then all second and higher-order derivatives of L with respect to $\boldsymbol{\lambda}$ are equal to 0 and according to Taylor's second order theorem (Theorem A.3, Appendix A) or by expression (1.5) of Definition 1.23 we get

$$\begin{aligned}
L(\mathbf{x}^*, \boldsymbol{\lambda}) - L(\mathbf{x}^*, \boldsymbol{\lambda}^*) &= \left\langle \frac{\partial L}{\partial \boldsymbol{\lambda}}(\mathbf{x}^*, \boldsymbol{\lambda}^*), \boldsymbol{\lambda} - \boldsymbol{\lambda}^* \right\rangle \\
&= \left\langle \frac{\partial L}{\partial \boldsymbol{\lambda}}(\mathbf{x}^*, \boldsymbol{\lambda}^*), \boldsymbol{\lambda} \right\rangle - \left\langle \frac{\partial L}{\partial \boldsymbol{\lambda}}(\mathbf{x}^*, \boldsymbol{\lambda}^*), \boldsymbol{\lambda}^* \right\rangle \\
&= \sum_{i \in I} \frac{\partial L}{\partial \lambda_i}(\mathbf{x}^*, \boldsymbol{\lambda}^*)\lambda_i - \sum_{i \in I} \frac{\partial L}{\partial \lambda_i}(\mathbf{x}^*, \boldsymbol{\lambda}^*)\lambda_i^* \\
&= \sum_{i \in I_1 \cup I_2} \frac{\partial L}{\partial \lambda_i}(\mathbf{x}^*, \boldsymbol{\lambda}^*)\lambda_i + \sum_{i \in I_3} \frac{\partial L}{\partial \lambda_i}(\mathbf{x}^*, \boldsymbol{\lambda}^*)\lambda_i \\
&\quad - \sum_{i \in I_1 \cup I_2} \frac{\partial L}{\partial \lambda_i}(\mathbf{x}^*, \boldsymbol{\lambda}^*)\lambda_i^* - \sum_{i \in I_3} \frac{\partial L}{\partial \lambda_i}(\mathbf{x}^*, \boldsymbol{\lambda}^*)\lambda_i^* \le 0
\end{aligned}$$

where the first sum (over $i \in I_1 \cup I_2$) in the last expression is nonpositive for each $\lambda_i \ge 0,\ i \in I_1 \cup I_2$ according to (1.16), the second sum is equal to zero according to (1.17), the third sum is equal to zero in accordance with (1.18), and the fourth sum is equal to zero according to (1.17).

Therefore $L(\mathbf{x}^*, \boldsymbol{\lambda}) \le L(\mathbf{x}^*, \boldsymbol{\lambda}^*)$ for $\lambda_i \ge 0,\ i \in I_1 \cup I_2$.

The two inequalities we have proved give

$$L(\mathbf{x}^*, \boldsymbol{\lambda}) \le L(\mathbf{x}^*, \boldsymbol{\lambda}^*) \le L(\mathbf{x}, \boldsymbol{\lambda}^*),$$

that is, $(\mathbf{x}^*, \boldsymbol{\lambda}^*)$ is a saddle point of $L(\mathbf{x}, \boldsymbol{\lambda})$ by Definition 1.25. According to Theorem 1.40, $\mathbf{x}^*$ is an optimal solution to problem (1.7). ■

Conditions (1.12)–(1.19) are called the *Karush-Kuhn-Tucker* (*KKT*) *conditions* for problem (1.7).

Conditions like (1.14) and (1.18) are said to be *complementary slackness conditions*, or simply *complementary conditions*.

1.6 Lagrangian Duality

Consider the following nonlinear minimization problem called the *primal problem*:
(P) Find

$$\min\ f(\mathbf{x}) \tag{1.20}$$

subject to

$$g_i(\mathbf{x}) \le 0,\ i \in I_1 \tag{1.21}$$

$$l_i(\mathbf{x}) = 0,\ i \in I_2 \tag{1.22}$$

$$\mathbf{x} \in X. \tag{1.23}$$

The maximization problem
(D) Find

$$\max \{\Psi(\mathbf{u}, \mathbf{v}) \stackrel{\text{def}}{=} \inf [L(\mathbf{x}, \mathbf{u}, \mathbf{v}) : \mathbf{x} \in X]\}$$

subject to

$$\mathbf{u} \geq \mathbf{0},$$

where

$$L(\mathbf{x}, \mathbf{u}, \mathbf{v}) = f(\mathbf{x}) + \sum_{i \in I_1} u_i g_i(\mathbf{x}) + \sum_{i \in I_2} v_i l_i(\mathbf{x}),$$

$\mathbf{u} = (u_i)_{i \in I_1}$, $\mathbf{v} = (v_i)_{i \in I_2}$, is said to be the *Lagrangian dual problem* to (P).

The Lagrangian dual function $\Psi(\mathbf{u}, \mathbf{v})$ may assume the value $-\infty$ for some vector $(\mathbf{u}, \mathbf{v})$, and the maximum of $\Psi(\mathbf{u}, \mathbf{v})$ may not exist.

Variables u_i, v_i of the dual problem (D) are the *Lagrange multipliers* of $L(\mathbf{x}, \mathbf{u}, \mathbf{v})$. The multipliers u_i associated with the inequality constraints $g_i(\mathbf{x}) \leq 0$, $i \in I_1$ are nonnegative whereas the multipliers v_i associated with the equality constraints $l_i(\mathbf{x}) = 0$, $i \in I_2$ are unrestricted in sign.

Since the dual problem (D) consists in maximizing the infimum of the function $L(\mathbf{x}, \mathbf{u}, \mathbf{v})$, it is sometimes called the *max-min problem*.

Except for the Lagrangian dual problem (D) written above, several problems have been proposed for dual problems: the conjugate dual problem, the surrogate dual problem (where the constraints are grouped into a single constraint by the use of Lagrange multipliers—Greenberg and Pierskalla), the mixed Lagrangian and surrogate (or composite) dual problem, etc.

In this book, only Lagrangian duality is considered.

Theorem 1.43 (Weak duality theorem)
Let $\mathbf{x}$ *be a feasible solution to problem* (P)*, that is,* $\mathbf{x} \in X$, $g_i(\mathbf{x}) \leq 0, i \in I_1$, $l_i(\mathbf{x}) = 0$, $i \in I_2$ *and let* $(\mathbf{u}, \mathbf{v})$ *be a feasible solution to problem* (D)*, that is,* $\mathbf{u} \geq \mathbf{0}$*. Then* $f(\mathbf{x}) \geq \Psi(\mathbf{u}, \mathbf{v})$.

Proof. By definition of $\Psi(\mathbf{u}, \mathbf{v})$ we have that

$$\Psi(\mathbf{u}, \mathbf{v}) = \inf \left\{ f(\mathbf{x}) + \sum_{i \in I_1} u_i g_i(\mathbf{x}) + \sum_{i \in I_2} v_i l_i(\mathbf{x}) : \mathbf{x} \in X \right\}$$

$$\leq f(\mathbf{x}) + \sum_{i \in I_1} u_i g_i(\mathbf{x}) + \sum_{i \in I_2} v_i l_i(\mathbf{x}) \leq f(\mathbf{x}),$$

where we have used that $\mathbf{u} \geq \mathbf{0}$, $g_i(\mathbf{x}) \leq 0, i \in I_1$, $l_i(\mathbf{x}) = 0$, $i \in I_2$, and $\mathbf{x} \in X$. ■

If the inequality $f(\mathbf{x}) \geq \Psi(\mathbf{u}, \mathbf{v})$ is strict, then a *duality gap* is said to exist.

Corollary 1.5 (Sufficient optimality condition from absence of duality gap)
If $f(\overline{\mathbf{x}}) = \Psi(\overline{\mathbf{u}}, \overline{\mathbf{v}}), \;\; \overline{\mathbf{u}} \geq \mathbf{0}$ *and* $\overline{\mathbf{x}} \in \{\mathbf{x} \in X : \; g_i(\mathbf{x}) \leq 0, \; i \in I_1, \;\; l_i(\mathbf{x}) = 0, \; i \in I_2\}$ *then* $\overline{\mathbf{x}}$ *and* $(\overline{\mathbf{u}}, \overline{\mathbf{v}})$ *are optimal solutions to problems* (P) *and* (D)*, respectively.*

Corollary 1.6 (Unboundedness from below of objective functions of primal and dual problems)
If $\inf \{f(\mathbf{x}) : \; g_i(\mathbf{x}) \leq 0, \; i \in I_1, \;\; l_i(\mathbf{x}) = 0, \; i \in I_2, \; \mathbf{x} \in X\} = -\infty$*, then* $\Psi(\mathbf{u}, \mathbf{v}) = -\infty$ *for each* $\mathbf{u} \geq \mathbf{0}$.

Proposition 1.1 (Theorem of the alternative)
Let X *be a nonempty convex set in* $\mathbf{R}^n$*,* $h : \mathbf{R}^n \to \mathbf{R}$ *and* $g_i : \mathbf{R}^n \to \mathbf{R}$*,* $i \in I_1$ *be convex functions, and* $l_i : \mathbf{R}^n \to \mathbf{R}$*,* $i \in I_2$ *be affine functions.*

If system

I. $h(\mathbf{x}) < 0, \;\; g_i(\mathbf{x}) \leq 0, \; i \in I_1, \;\; l_i(\mathbf{x}) = 0, \; i \in I_2,$ for some $\mathbf{x} \in X$

has no solution $\mathbf{x}$*, then system*

II. $u_0 h(\mathbf{x}) + \sum_{i \in I_1} u_i g_i(\mathbf{x}) + \sum_{i \in I_2} v_i l_i(\mathbf{x}) \geq 0$ for all $\mathbf{x} \in X$
$(u_0, \mathbf{u}) \geq \mathbf{0}, \quad (u_0, \mathbf{u}, \mathbf{v}) \neq (0, \mathbf{0}, \mathbf{0})$

has a solution $(u_0, \mathbf{u}, \mathbf{v})$.

If $u_0 > 0$*, the converse statement holds true.*

Proof. Let System I have no solution. Consider the set

$$\Lambda = \{(p, \mathbf{q}, \mathbf{r}) : \;\; p > h(\mathbf{x}), \;\; q_i \geq g_i(\mathbf{x}), \; i \in I_1, \;\; r_i = l_i(\mathbf{x}), \; i \in I_2 \quad \text{for some } \mathbf{x} \in X\}.$$

Under the assumptions of Proposition 1.1, Λ is a convex set.

Since System I has no solution by assumption, then $(0, \mathbf{0}, \mathbf{0}) \notin \Lambda$.

From Corollary 1.2 it follows that there exists a vector $(u_0, \mathbf{u}, \mathbf{v}) \neq (0, \mathbf{0}, \mathbf{0})$ such that $\langle (u_0, \mathbf{u}, \mathbf{v}), (p, \mathbf{q}, \mathbf{r}) - (0, \mathbf{0}, \mathbf{0}) \rangle \geq 0$ for each $(p, \mathbf{q}, \mathbf{r}) \in \operatorname{cl} \Lambda$, that is,

$$u_0 p + \langle \mathbf{u}, \mathbf{q} \rangle + \langle \mathbf{v}, \mathbf{r} \rangle \geq 0 \quad \text{for each} \;\; (p, \mathbf{q}, \mathbf{r}) \in \operatorname{cl} \Lambda.$$

Since p and components of $\mathbf{q}$ may be arbitrarily large in magnitude, this inequality holds true only if $u_0 \geq 0, \;\; \mathbf{u} \geq \mathbf{0}$. Moreover,

$$(p, \mathbf{q}, \mathbf{r}) \stackrel{\text{def}}{=} (h(\mathbf{x}), \mathbf{g}(\mathbf{x}), \mathbf{l}(\mathbf{x})) \in \operatorname{cl} \Lambda$$

where $\mathbf{g}(\mathbf{x}) = [g_i(\mathbf{x})]_{i \in I_1}, \;\; \mathbf{l}(\mathbf{x}) = [l_i(\mathbf{x})]_{i \in I_2}$. Therefore

$$u_0 h(\mathbf{x}) + \langle \mathbf{u}, \mathbf{g}(\mathbf{x}) \rangle + \langle \mathbf{v}, \mathbf{l}(\mathbf{x}) \rangle \geq 0.$$

This inequality is true for each $\mathbf{x} \in X$. Therefore, System II has a solution $(u_0, \mathbf{u}, \mathbf{v})$.

Assume now that System II has a solution $(u_0, \mathbf{u}, \mathbf{v}) \neq (0, \mathbf{0}, \mathbf{0})$ such that $u_0 > 0, \;\; \mathbf{u} \geq \mathbf{0}$, that is,

$$u_0 h(\mathbf{x}) + \langle \mathbf{u}, \mathbf{g}(\mathbf{x})\rangle + \langle \mathbf{v}, \mathbf{l}(\mathbf{x})\rangle \geq 0 \quad \text{for each } \mathbf{x} \in X.$$

Let $\mathbf{x} \in X$ and $\mathbf{g}(\mathbf{x}) \leq \mathbf{0}$, $\mathbf{l}(\mathbf{x}) = \mathbf{0}$. Since $\mathbf{u} \geq \mathbf{0}$, then $u_0 h(\mathbf{x}) \geq 0$ according to the above inequality. However, $u_0 > 0$ by assumption, therefore $h(\mathbf{x}) \geq 0$. Hence, System I has no solution. ■

Theorem 1.44 (Strong duality theorem)
Let X be a nonempty convex set in $\mathbf{R}^n$, let $f : \mathbf{R}^n \to \mathbf{R}$ and $g_i : \mathbf{R}^n \to \mathbf{R}$, $i \in I_1$ be convex functions, $l_i : \mathbf{R}^n \to \mathbf{R}$, $i \in I_2$ be affine functions, let Slater's constraint qualification hold true for $\overline{\mathbf{x}}$, and $\mathbf{0} \in \operatorname{int} \mathbf{l}(X)$ where $\mathbf{l}(\mathbf{x}) = [l_i(\mathbf{x})]_{i \in I_2}$ and $\mathbf{l}(X) = \{\mathbf{l}(\mathbf{x}) : \mathbf{x} \in X\}$. Then

$$\begin{aligned} &\inf \{f(\mathbf{x}) : \mathbf{x} \in X,\ g_i(\mathbf{x}) \leq 0,\ i \in I_1,\ l_i(\mathbf{x}) = 0,\ i \in I_2\} \\ &= \sup \{\Psi(\mathbf{u}, \mathbf{v}) : \mathbf{u} \geq \mathbf{0}\}. \end{aligned}$$

If $\inf f(\mathbf{x})$ *is finite, then* $\sup \{\Psi(\mathbf{u}, \mathbf{v}) : \mathbf{u} \geq \mathbf{0}\}$ *is achieved at* $(\overline{\mathbf{u}}, \overline{\mathbf{v}})$ *where* $\overline{\mathbf{u}} \geq \mathbf{0}$. *If the infimum is achieved at* $\overline{\mathbf{x}}$, *then* $\sum_{i \in I_1} \overline{u}_i g_i(\overline{\mathbf{x}}) = 0$.

Proof. Let

$$f_{\text{inf}} = \inf \{f(\mathbf{x}) : g_i(\mathbf{x}) \leq 0,\ i \in I_1,\ \ l_i(\mathbf{x}) = 0,\ i \in I_2,\ \ \mathbf{x} \in X\}.$$

According to this notation and the assumptions, $f_{\text{inf}} < +\infty$.

If $f_{\text{inf}} = -\infty$, then $\sup \{\Psi(\mathbf{u}, \mathbf{v}) : \mathbf{u} \geq \mathbf{0}\} = -\infty$ according to Corollary 1.6. Hence, $f_{\text{inf}} = \sup \{\Psi(\mathbf{u}, \mathbf{v}) : \mathbf{u} \geq \mathbf{0}\}$, that is, the statement of Theorem 1.44 holds true.

Let f_{inf} be finite, that is, $-\infty < f_{\text{inf}}$.

Consider the system

$$f(\mathbf{x}) < f_{\text{inf}}, \quad g_i(\mathbf{x}) \leq 0,\ i \in I_1, \quad l_i(\mathbf{x}) = 0,\ i \in I_2, \quad \mathbf{x} \in X.$$

This system has no solution according to definition of f_{inf}.

From Proposition 1.1 it follows that there exists a vector $(u_0, \mathbf{u}, \mathbf{v}) \neq (0, \mathbf{0}, \mathbf{0})$ with $(u_0, \mathbf{u}) \geq \mathbf{0}$ such that

$$u_0(f(\mathbf{x}) - f_{\text{inf}}) + \langle \mathbf{u}, \mathbf{g}(\mathbf{x})\rangle + \langle \mathbf{v}, \mathbf{l}(\mathbf{x})\rangle \geq 0 \quad \text{for all } \mathbf{x} \in X \tag{1.24}$$

where $\mathbf{g}(\mathbf{x}) = [g_i(\mathbf{x})]_{i \in I_1}$, $\mathbf{l}(\mathbf{x}) = [l_i(\mathbf{x})]_{i \in I_2}$.

We want to prove that $u_0 > 0$ (in general, $u_0 \geq 0$). Assume the contrary that $u_0 = 0$. By hypothesis of Theorem 1.44 (Slater's constraint qualification), there exists an $\overline{\mathbf{x}} \in X$ such that $\mathbf{g}(\overline{\mathbf{x}}) < \mathbf{0}$ and $\mathbf{l}(\overline{\mathbf{x}}) = \mathbf{0}$. From (1.24) it follows that $\langle \mathbf{u}, \mathbf{g}(\overline{\mathbf{x}})\rangle \geq 0$ for $u_0 = 0$. However, since $\mathbf{u} \geq \mathbf{0}$ and $\mathbf{g}(\overline{\mathbf{x}}) < \mathbf{0}$, then $\langle \mathbf{u}, \mathbf{g}(\overline{\mathbf{x}})\rangle \geq 0$ is possible only if $\mathbf{u} = \mathbf{0}$.

From (1.24), $u_0 = 0$ and $\mathbf{u} = \mathbf{0}$ it follows that $\langle \mathbf{v}, \mathbf{l}(\mathbf{x})\rangle \geq 0$ for all $\mathbf{x} \in X$. Since $\mathbf{0} \in \operatorname{int} \mathbf{l}(X)$, we can choose $\mathbf{l}(\mathbf{x}) = -\mu\mathbf{v}$, $\mu > 0$. Therefore $0 \leq \langle \mathbf{v}, \mathbf{l}(\mathbf{x})\rangle = \langle \mathbf{v}, -\mu\mathbf{v}\rangle =$

$-\mu\|\mathbf{v}\|^2$, hence $\mathbf{v} = \mathbf{0}$. Thus, $u_0 = 0$ implies $(u_0, \mathbf{u}, \mathbf{v}) = (0, \mathbf{0}, \mathbf{0})$, a contradiction. Therefore $u_0 \neq 0$, and since $u_0 \geq 0$, then $u_0 > 0$.

Dividing (1.24) by $u_0 > 0$ and denoting $\frac{\mathbf{u}}{u_0} = \overline{\mathbf{u}}$ (≥ 0 because $\mathbf{u} \geq \mathbf{0}$, $u_0 \geq 0$), $\frac{\mathbf{v}}{u_0} = \overline{\mathbf{v}}$, we get

$$f(\mathbf{x}) + \langle \overline{\mathbf{u}}, \mathbf{g}(\mathbf{x}) \rangle + \langle \overline{\mathbf{v}}, \mathbf{l}(\mathbf{x}) \rangle \geq f_{\inf} \quad \text{for all } \mathbf{x} \in X, \tag{1.25}$$

that is,

$$\Psi(\overline{\mathbf{u}}, \overline{\mathbf{v}}) \stackrel{\text{def}}{=} \inf \{ f(\mathbf{x}) + \langle \overline{\mathbf{u}}, \mathbf{g}(\mathbf{x}) \rangle + \langle \overline{\mathbf{v}}, \mathbf{l}(\mathbf{x}) \rangle : \ \mathbf{x} \in X \} \geq f_{\inf}.$$

From this result and Theorem 1.43 it follows that $\Psi(\overline{\mathbf{u}}, \overline{\mathbf{v}}) = f_{\inf}$, and $(\overline{\mathbf{u}}, \overline{\mathbf{v}})$ is an optimal solution to the dual problem (D) according to Corollary 1.5.

Suppose that $\overline{\mathbf{x}}$ is an optimal solution to the primal problem (P). Hence

$$\mathbf{g}(\overline{\mathbf{x}}) \leq \mathbf{0}, \quad \mathbf{l}(\overline{\mathbf{x}}) = \mathbf{0}, \quad \mathbf{x} \in X, \quad f(\overline{\mathbf{x}}) = f_{\inf}.$$

Setting $\mathbf{x} = \overline{\mathbf{x}}$ in (1.25), we obtain $\langle \overline{\mathbf{u}}, \mathbf{g}(\overline{\mathbf{x}}) \rangle \geq 0$. Since $\overline{\mathbf{u}} \geq \mathbf{0}$ and $\mathbf{g}(\overline{\mathbf{x}}) \leq \mathbf{0}$, then $\langle \overline{\mathbf{u}}, \mathbf{g}(\overline{\mathbf{u}}) \rangle \leq 0$, and from both opposite inequalities it follows that $\langle \overline{\mathbf{u}}, \mathbf{g}(\overline{\mathbf{u}}) \rangle = 0$. ■

Theorem 1.45 below is a saddlepoint optimality necessary and sufficient condition, and a necessary and sufficient condition in the absence of duality gap.

Theorem 1.45 (Saddlepoint optimality theorem)
A point $(\overline{\mathbf{x}}, \overline{\mathbf{u}}, \overline{\mathbf{v}})$ with $\overline{\mathbf{x}} \in X$, $\overline{\mathbf{u}} \geq \mathbf{0}$ is a saddle point of the Lagrangian $L(\mathbf{x}, \mathbf{u}, \mathbf{v})$ if and only if

1. $L(\overline{\mathbf{x}}, \overline{\mathbf{u}}, \overline{\mathbf{v}}) = \min \{ L(\mathbf{x}, \overline{\mathbf{u}}, \overline{\mathbf{v}}) : \mathbf{x} \in X \}$,
2. $g_i(\overline{\mathbf{x}}) \leq 0, \ i \in I_1, \ l_i(\overline{\mathbf{x}}) = 0, \ i \in I_2$,
3. $\sum_{i \in I_1} \overline{u}_i g_i(\overline{\mathbf{x}}) = 0$.

Moreover, $(\overline{\mathbf{x}}, \overline{\mathbf{u}}, \overline{\mathbf{v}})$ is a saddle point of $L(\mathbf{x}, \mathbf{u}, \mathbf{v})$ if and only if $\overline{\mathbf{x}}$ and $(\overline{\mathbf{u}}, \overline{\mathbf{v}})$ are optimal solutions to problems (P) and (D), respectively, with $f(\overline{\mathbf{x}}) = \Psi(\overline{\mathbf{u}}, \overline{\mathbf{v}})$, that is, with no duality gap.

Proof. Let $(\overline{\mathbf{x}}, \overline{\mathbf{u}}, \overline{\mathbf{v}})$ be a saddle point for the Lagrangian function $L(\mathbf{x}, \mathbf{u}, \mathbf{v})$. Condition 1. follows from the second saddlepoint inequality (Definition 1.25). Moreover, by the first saddlepoint inequality,

$$f(\overline{\mathbf{x}}) + \sum_{i \in I_1} \overline{u}_i g_i(\overline{\mathbf{x}}) + \sum_{i \in I_2} \overline{v}_i l_i(\overline{\mathbf{x}}) \geq f(\overline{\mathbf{x}}) + \sum_{i \in I_1} u_i g_i(\overline{\mathbf{x}}) + \sum_{i \in I_2} v_i l_i(\overline{\mathbf{x}}) \tag{1.26}$$

for all $(\mathbf{u}, \mathbf{v})$ such that $u_i \geq 0, \ i \in I_1$.

Therefore, we have $g_i(\overline{\mathbf{x}}) \leq 0, \ i \in I_1$ and $l_i(\overline{\mathbf{x}}) = 0, \ i \in I_2$; otherwise (1.26) would be violated (cf. proof of Theorem 1.40). This means that condition 2. holds true.

Setting $u_i = 0,\ i \in I_1$ in (1.26), we get $\sum_{i\in I_1} \overline{u}_i g_i(\overline{\mathbf{x}}) \geq 0$. However, since $u_i \geq 0,\ i \in I_1$ and $g_i(\overline{\mathbf{x}}) \leq 0,\ i \in I_1$ then obviously $\sum_{i\in I_1} \overline{u}_i g_i(\overline{\mathbf{x}}) \leq 0$. From both inequalities it follows that $\sum_{i\in I_1} \overline{u}_i g_i(\overline{\mathbf{x}}) = 0$, that is, condition 3. holds true.

Conversely, let $(\overline{\mathbf{x}}, \overline{\mathbf{u}}, \overline{\mathbf{v}})$ with $\overline{\mathbf{x}} \in X$ and $\overline{\mathbf{u}} \geq \mathbf{0}$ satisfy conditions 1., 2., and 3. Hence, $L(\overline{\mathbf{x}}, \overline{\mathbf{u}}, \overline{\mathbf{v}}) \leq L(\mathbf{x}, \overline{\mathbf{u}}, \overline{\mathbf{v}})$ according to condition 1. for all $\mathbf{x} \in X$, and

$$\begin{aligned} L(\overline{\mathbf{x}}, \overline{\mathbf{u}}, \overline{\mathbf{v}}) &= f(\overline{\mathbf{x}}) + \sum_{i\in I_1} \overline{u}_i g_i(\overline{\mathbf{x}}) + \sum_{i\in I_2} \overline{v}_i l_i(\overline{\mathbf{x}}) \overset{2.,3.}{=} f(\overline{\mathbf{x}}) \\ &\overset{2.}{\geq} f(\overline{\mathbf{x}}) + \sum_{i\in I_1} u_i g_i(\overline{\mathbf{x}}) + \sum_{i\in I_2} v_i l_i(\overline{\mathbf{x}}) = L(\overline{\mathbf{x}}, \mathbf{u}, \mathbf{v}) \end{aligned}$$

for all $(\mathbf{u}, \mathbf{v})$ such that $\mathbf{u} \geq \mathbf{0}$.

Therefore $(\overline{\mathbf{x}}, \overline{\mathbf{u}}, \overline{\mathbf{v}})$ is a saddle point of $L(\mathbf{x}, \mathbf{u}, \mathbf{v})$ by definition.

In order to prove the second part of Theorem 1.45, suppose that $(\overline{\mathbf{x}}, \overline{\mathbf{u}}, \overline{\mathbf{v}})$ is a saddle point for $L(\mathbf{x}, \mathbf{u}, \mathbf{v})$ with $\overline{\mathbf{x}} \in X$ and $\overline{\mathbf{u}} \geq 0$. According to the first part of this theorem, conditions 1., 2., and 3. are satisfied. From condition 2. it follows that $\overline{\mathbf{x}}$ is feasible to the primal problem (P), and since $\overline{\mathbf{u}} \geq \mathbf{0}$ then $(\overline{\mathbf{u}}, \overline{\mathbf{v}})$ is feasible to the dual problem (D).

From conditions 1., 2., and 3. it follows that

$$\begin{aligned} \Psi(\overline{\mathbf{u}}, \overline{\mathbf{v}}) &\overset{\text{def}}{=} \inf_{\mathbf{x}\in X} L(\mathbf{x}, \overline{\mathbf{u}}, \overline{\mathbf{v}}) \overset{1.}{=} L(\overline{\mathbf{x}}, \overline{\mathbf{u}}, \overline{\mathbf{v}}) \\ &\overset{\text{def}}{=} f(\overline{\mathbf{x}}) + \sum_{i\in I_1} \overline{u}_i g_i(\overline{\mathbf{x}}) + \sum_{i\in I_2} \overline{v}_i l_i(\overline{\mathbf{x}}) \overset{2.,3.}{=} f(\overline{\mathbf{x}}), \end{aligned}$$

that is, there is no duality gap. Corollary 1.5 implies that $\overline{\mathbf{x}}$ solves problem (P) and $(\overline{\mathbf{u}}, \overline{\mathbf{v}})$ solves problem (D), respectively.

Conversely, let $\overline{\mathbf{x}}$ and $(\overline{\mathbf{u}}, \overline{\mathbf{v}})$ be optimal solutions to problems (P) and (D), respectively, with no duality gap. Therefore $g_i(\overline{\mathbf{x}}) \leq 0,\ i \in I_1,\ l_i(\overline{\mathbf{x}}) = 0,\ i \in I_2,\ \overline{\mathbf{x}} \in X$, and $\overline{\mathbf{u}} \geq \mathbf{0}$.

By definition of $\Psi(\mathbf{u}, \mathbf{v})$,

$$\begin{aligned} \Psi(\overline{\mathbf{u}}, \overline{\mathbf{v}}) &= \min \left\{ f(\mathbf{x}) + \sum_{i\in I_1} \overline{u}_i g_i(\mathbf{x}) + \sum_{i\in I_2} \overline{v}_i l_i(\mathbf{x}),\ \mathbf{x} \in X \right\} \\ &\leq f(\overline{\mathbf{x}}) + \sum_{i\in I_1} \overline{u}_i g_i(\overline{\mathbf{x}}) + \sum_{i\in I_2} \overline{v}_i l_i(\overline{\mathbf{x}}) \\ &= f(\overline{\mathbf{x}}) + \sum_{i\in I_1} \overline{u}_i g_i(\overline{\mathbf{x}}) \leq f(\overline{\mathbf{x}}). \end{aligned}$$

(We can write "min" instead of "inf" because $(\overline{\mathbf{u}}, \overline{\mathbf{v}})$ is an optimal solution to problem (D).) However, $f(\overline{\mathbf{x}}) = \Psi(\overline{\mathbf{u}}, \overline{\mathbf{v}})$ by assumption (there is no duality gap). Therefore all inequalities above are satisfied as equalities. It follows that $\sum_{i\in I_1} \overline{u}_i g_i(\overline{\mathbf{x}}) = 0$

and

$$L(\overline{\mathbf{x}}, \overline{\mathbf{u}}, \overline{\mathbf{v}}) = f(\overline{\mathbf{x}}) = \Psi(\overline{\mathbf{u}}, \overline{\mathbf{v}}) = \min\{L(\mathbf{x}, \overline{\mathbf{u}}, \overline{\mathbf{v}}) : \mathbf{x} \in X\}.$$

Therefore conditions 1., 2., and 3. are satisfied and $\overline{\mathbf{x}} \in X$, $\overline{\mathbf{u}} \geq \mathbf{0}$. Hence, $(\overline{\mathbf{x}}, \overline{\mathbf{u}}, \overline{\mathbf{v}})$ is a saddle point of $L(\mathbf{x}, \mathbf{u}, \mathbf{v})$ according to the first part of this theorem. ■

Theorem 1.46 (Relationship between the saddlepoint criteria and the KKT conditions)
Consider problem (P) (1.20)–(1.23) *where the feasible set* (1.21)–(1.23) *is denoted by* Y. *Let* $f,\ g_i,\ i \in I_1,\ l_i,\ i \in I_2$ *be differentiable, and let* $\overline{\mathbf{x}} \in Y$ *satisfy the KKT conditions, that is, there exist* $\overline{\mathbf{u}} \geq \mathbf{0}$ *and* $\overline{\mathbf{v}}$ *such that*

$$\mathbf{f}'(\overline{\mathbf{x}}) + \sum_{i \in I_1} \overline{u}_i \mathbf{g}'_i(\overline{\mathbf{x}}) + \sum_{i \in I_2} \overline{v}_i \mathbf{l}'_i(\overline{\mathbf{x}}) = \mathbf{0}, \tag{1.27}$$

$$\sum_{i \in I_1} \overline{u}_i g_i(\overline{\mathbf{x}}) = 0. \tag{1.28}$$

Let $f,\ g_i,\ i \in I_1$ *be convex at* $\overline{\mathbf{x}}$ *and* $\overline{v}_i \neq 0$ *imply* l_i *is affine. Then* $(\overline{\mathbf{x}}, \overline{\mathbf{u}}, \overline{\mathbf{v}})$ *is a saddle point of the Lagrangian function* $L(\mathbf{x}, \mathbf{u}, \mathbf{v})$.

Conversely, if $(\overline{\mathbf{x}}, \overline{\mathbf{u}}, \overline{\mathbf{v}})$ *with* $\overline{\mathbf{x}} \in \operatorname{int} X$ *and* $\overline{\mathbf{u}} \geq \mathbf{0}$ *is a saddle point of* $L(\mathbf{x}, \mathbf{u}, \mathbf{v})$ *then* $\overline{\mathbf{x}}$ *is a feasible solution to the primal problem* (P) *and* $(\overline{\mathbf{x}}, \overline{\mathbf{u}}, \overline{\mathbf{v}})$ *satisfies the KKT conditions* (1.27)–(1.28).

Proof. Let $(\overline{\mathbf{x}}, \overline{\mathbf{u}}, \overline{\mathbf{v}})$, where $\overline{\mathbf{x}} \in Y$ and $\overline{\mathbf{u}} \geq \mathbf{0}$, satisfy the KKT conditions (1.27)–(1.28). Denote by I_2^0 the set of indices $i \in I_2$ with $\overline{v}_i = 0$. Since f and $g_i,\ i \in I_1$ are convex functions and l_i are affine functions for $i \in I_2$ with $\overline{v}_i \neq 0$ (that is, for $i \in I_2 \setminus I_2^0$) then

$$f(\mathbf{x}) - f(\overline{\mathbf{x}}) \geq \langle \mathbf{f}'(\overline{\mathbf{x}}), \mathbf{x} - \overline{\mathbf{x}} \rangle, \tag{1.29}$$

$$g_i(\mathbf{x}) - g_i(\overline{\mathbf{x}}) \geq \langle \mathbf{g}'_i(\overline{\mathbf{x}}), \mathbf{x} - \overline{\mathbf{x}} \rangle, \quad i \in I_1, \tag{1.30}$$

$$l_i(\mathbf{x}) - l_i(\overline{\mathbf{x}}) = \langle \mathbf{l}'_i(\overline{\mathbf{x}}), \mathbf{x} - \overline{\mathbf{x}} \rangle, \quad i \in I_2 \setminus I_2^0 \tag{1.31}$$

for all $\mathbf{x} \in X$ according to Theorem 1.30.

Multiplying (1.30) by $\overline{u}_i \geq 0,\ i \in I_1$, (1.31) by $\overline{v}_i$ for $i \in I_2 \setminus I_2^0$ and adding (1.29), we get

$$
\begin{aligned}
& f(\mathbf{x}) + \sum_{i \in I_1} \overline{u}_i g_i(\mathbf{x}) + \sum_{i \in I_2 \setminus I_2^0} \overline{v}_i l_i(\mathbf{x}) \\
& \geq f(\overline{\mathbf{x}}) + \sum_{i \in I_1} \overline{u}_i g_i(\overline{\mathbf{x}}) + \sum_{i \in I_2 \setminus I_2^0} \overline{v}_i l_i(\overline{\mathbf{x}}) \\
& + \langle \mathbf{f}'(\overline{\mathbf{x}}), \mathbf{x} - \overline{\mathbf{x}} \rangle + \sum_{i \in I_1} \overline{u}_i \langle \mathbf{g}_i'(\overline{\mathbf{x}}), \mathbf{x} - \overline{\mathbf{x}} \rangle + \sum_{i \in I_2 \setminus I_2^0} \overline{v}_i \langle \mathbf{l}_i'(\overline{\mathbf{x}}), \mathbf{x} - \overline{\mathbf{x}} \rangle
\end{aligned}
$$

for all $\mathbf{x} \in X$.

Adding $\sum_{i \in I_2^0} \overline{v}_i l_i(\mathbf{x})$ (= 0 because $\overline{v}_i = 0$ for $i \in I_2^0$ and $l_i(\mathbf{x}) = 0$ for $i \in I_2$) to the left-hand side and adding $\sum_{i \in I_2^0} \overline{v}_i l_i(\overline{\mathbf{x}})$ (= 0 because $\overline{v}_i = 0$ for $i \in I_2^0$ and $l_i(\overline{\mathbf{x}}) = 0$ for $i \in I_2$) and $\sum_{i \in I_2^0} \overline{v}_i \langle \mathbf{l}_i'(\overline{\mathbf{x}}), \mathbf{x} - \overline{\mathbf{x}} \rangle$ (= 0 because $\overline{v}_i = 0$ for $i \in I_2^0$) to the right-hand side of the above inequality, respectively, yield

$$
\begin{aligned}
& f(\mathbf{x}) + \sum_{i \in I_1} \overline{u}_i g_i(\mathbf{x}) + \sum_{i \in I_2} \overline{v}_i l_i(\mathbf{x}) \\
& \geq f(\overline{\mathbf{x}}) + \sum_{i \in I_1} \overline{u}_i g_i(\overline{\mathbf{x}}) + \sum_{i \in I_2} \overline{v}_i l_i(\overline{\mathbf{x}}) \\
& + \left\langle \mathbf{f}'(\overline{\mathbf{x}}) + \sum_{i \in I_1} \overline{u}_i \mathbf{g}_i'(\overline{\mathbf{x}}) + \sum_{i \in I_2} \overline{v}_i \mathbf{l}_i'(\overline{\mathbf{x}}),\ \mathbf{x} - \overline{\mathbf{x}} \right\rangle.
\end{aligned}
$$

Taking into account the definition of $L(\mathbf{x}, \mathbf{u}, \mathbf{v})$ and the KKT condition (1.27), this expression simplifies to

$$
L(\mathbf{x}, \overline{\mathbf{u}}, \overline{\mathbf{v}}) \geq L(\overline{\mathbf{x}}, \overline{\mathbf{u}}, \overline{\mathbf{v}}) \quad \text{for all } \mathbf{x} \in X.
$$

Since $g_i(\overline{\mathbf{x}}) \leq 0,\ i \in I_1,\ l_i(\overline{\mathbf{x}}) = 0,\ i \in I_2$ and $\sum_{i \in I_1} \overline{u}_i g_i(\overline{\mathbf{x}}) = 0$ it follows that

$$
\begin{aligned}
L(\overline{\mathbf{x}}, \mathbf{u}, \mathbf{v}) & \equiv f(\overline{\mathbf{x}}) + \sum_{i \in I_1} u_i g_i(\overline{\mathbf{x}}) + \sum_{i \in I_2} v_i l_i(\overline{\mathbf{x}}) \\
& \leq f(\overline{\mathbf{x}}) \\
& = f(\overline{\mathbf{x}}) + \sum_{i \in I_1} \overline{u}_i g_i(\overline{\mathbf{x}}) + \sum_{i \in I_2} \overline{v}_i l_i(\overline{\mathbf{x}}) \\
& \equiv L(\overline{\mathbf{x}}, \overline{\mathbf{u}}, \overline{\mathbf{v}})
\end{aligned} \tag{1.32}
$$

for all $(\mathbf{u}, \mathbf{v})$ such that $\mathbf{u} \geq \mathbf{0}$.

Therefore $(\overline{\mathbf{x}}, \overline{\mathbf{u}}, \overline{\mathbf{v}})$ is a saddle point of $L(\mathbf{x}, \mathbf{u}, \mathbf{v})$ by definition.

Conversely, let $(\overline{\mathbf{x}}, \overline{\mathbf{u}}, \overline{\mathbf{v}})$ with $\overline{\mathbf{x}} \in \text{int } X$ and $\overline{\mathbf{u}} \geq \mathbf{0}$ be a saddle point of $L(\mathbf{x}, \mathbf{u}, \mathbf{v})$. Then $L(\overline{\mathbf{x}}, \mathbf{u}, \mathbf{v}) \leq L(\overline{\mathbf{x}}, \overline{\mathbf{u}}, \overline{\mathbf{v}})$ for all $\mathbf{u} \geq \mathbf{0}$ by definition.

From this inequality, similar to the proof of Theorem 1.45 (see (1.26)), we have that $g_i(\overline{\mathbf{x}}) \leq 0,\ i \in I_1,\ l_i(\overline{\mathbf{x}}) = 0,\ i \in I_2$ and $\sum_{i \in I_1} \overline{u}_i g_i(\overline{\mathbf{x}}) = 0$ for $\overline{\mathbf{x}} \in X$. Therefore $\overline{\mathbf{x}}$ is a feasible solution to problem (P). Since $L(\overline{\mathbf{x}}, \overline{\mathbf{u}}, \overline{\mathbf{v}}) \leq L(\mathbf{x}, \overline{\mathbf{u}}, \overline{\mathbf{v}})$ for all $\mathbf{x} \in X$ (the second saddlepoint inequality) then $\overline{\mathbf{x}}$ is an optimal solution to the problem

$$\min\ L(\mathbf{x}, \overline{\mathbf{u}}, \overline{\mathbf{v}})$$

subject to

$$\mathbf{x} \in X.$$

Because $\overline{\mathbf{x}} \in \operatorname{int} X$ by assumption and $\overline{\mathbf{x}}$ is an optimal solution to this problem then $\frac{\partial L}{\partial \mathbf{x}}(\overline{\mathbf{x}}, \overline{\mathbf{u}}, \overline{\mathbf{v}}) = \mathbf{0}$, that is,

$$\mathbf{f}'(\overline{\mathbf{x}}) + \sum_{i \in I_1} \overline{u}_i \mathbf{g}_i'(\overline{\mathbf{x}}) + \sum_{i \in I_2} \overline{v}_i \mathbf{l}_i'(\overline{\mathbf{x}}) = \mathbf{0}.$$

Hence, the KKT conditions (1.27)–(1.28) are satisfied. ■

NOTES TO CHAPTER ONE

More information about concepts and results discussed in Chapter One *Preliminaries: Convex Analysis and Convex Programming* can be found in the books of Bazaraa, Sherali, and Shetty [5]; Kenderov, Hristov, and Dontchev [69]; Mangasarian [81]; Martos [83]; Pshenichnii [103]; Rockafellar [107, 108]; Rockafellar and Wets [110]; etc.

Theorem 1.41 is due to Karush 1939 (a weaker form, without nonnegativity of Lagrange multipliers), Kuhn and Tucker 1951, Uzawa 1958, and Karlin 1959. Sufficiency of Theorem 1.42 was proved by Kuhn and Tucker in 1951, and the necessity was proved by Fritz John in 1948. Slater's constraint qualification was formulated in 1950.

Part I
Separable Programming

Chapter 2
Introduction. Approximating the Separable Problem

In this chapter, some basic concepts and formulations concerning separable programs are introduced. Approximation of the separable problem by a linear program is considered and a restricted basis entry rule for solving the resulting linear program by the simplex method is discussed. Some numerical examples are also presented.

Definition 2.1 A function $f : X \to \mathbf{R},\ X \subseteq \mathbf{R}^n$ is said to be *separable* if $f(\mathbf{x}) = \sum_{j=1}^{n} f_j(x_j)$, that is, if $f(\mathbf{x}) \equiv f(x_1, \dots, x_n)$ can be expressed as the sum of n single-variable functions $f_1(x_1), \dots, f_n(x_n)$.

For example, the linear function

$$l(x_1, \dots, x_n) = \sum_{j=1}^{n} c_j x_j = c_1 x_1 + \cdots + c_n x_n,$$

where $c_1, \dots, c_n$ are constants, is separable.

Some nonlinear functions are not separable but can be made separable by appropriate substitutions.

For example, consider the problem

$$\min \{F(\mathbf{x}) = e^{x_1 + x_2}\}.$$

Set $f = e^{x_1 + x_2}$. Then $\ln f = x_1 + x_2$, and the problem becomes

$$\min \{F = f\}$$

subject to

$$\ln f = x_1 + x_2.$$

© The Author(s), under exclusive license to Springer Nature Switzerland AG 2021
S. M. Stefanov, *Separable Optimization*, Springer Optimization and Its Applications 177,
https://doi.org/10.1007/978-3-030-78401-0_2

This problem is separable.

Similarly, we can make separable the functions $e^{x_1^2+x_2}$, x_1x_2, $x_1^{x_2}$, etc.

When the objective function and the constraint functions are separable, the nonlinear program is called a *separable nonlinear program*.

An approximate solution can be obtained for any separable program by linear approximation and the simplex method of linear programming.

The new problem is obtained by replacing each nonlinear function with an approximating piecewise linear function.

At the beginning, consider a single-variable continuous function $f(x)$. The interval $[a, b]$ for x is partitioned into smaller intervals via *grid (mesh, breaking) points* $a = x_0 < x_1 < x_2 < \cdots < x_k = b$. The nonlinear function $f(x)$ is approximated in the interval $[x_l, x_{l+1}]$ as follows. Let x be a convex combination of x_l and x_{l+1}, that is,

$$x = \lambda x_l + (1 - \lambda)x_{l+1} \tag{2.1}$$

for some $\lambda \in [0, 1]$. Then

$$\widehat{f}(x) = \lambda f(x_l) + (1 - \lambda) f(x_{l+1}). \tag{2.2}$$

The grid points may or may not be equally spaced. As one would expect, the more the grid points, the better the approximation.

A point $x \in [x_l, x_{l+1}]$ can also be represented as a convex combination of two or more nonadjacent grid points.

More generally, the function f can be approximated over the interval $[a, b]$ via the grid points $x_0, x_1, \ldots, x_k$ by the piecewise linear function $\widehat{f}$ as follows

$$\widehat{f}(x) = \sum_{l=0}^{k} \lambda_l f(x_l), \quad \sum_{l=0}^{k} \lambda_l = 1, \ \lambda_l \geq 0, \ l = 0, 1, \ldots, k, \tag{2.3}$$

$$x = \sum_{l=0}^{k} \lambda_l x_l, \tag{2.4}$$

where at most two λ_l's are positive, and they must be adjacent. Sometimes λ_l is called a *weight associated with the grid point* x_l. Representation (2.3)–(2.4) is called the *λ-representation*, or the *λ-form approximation*. There is also an alternative representation, the *δ-form representation* (see, e.g., Bazaraa, Sherali, and Shetty [5]), which is not considered here.

Approximating the separable problem

Consider the separable nonlinear program

(SP)

$$\min \left\{ f(\mathbf{x}) = \sum_{j=1}^{n} f_j(x_j) \right\} \tag{2.5}$$

subject to

$$g_i(\mathbf{x}) \equiv \sum_{j=1}^{n} g_{ij}(x_j) \le \beta_i, \quad i = 1, \ldots, m \tag{2.6}$$

$$x_j \ge 0, \quad j = 1, \ldots, n. \tag{2.7}$$

Each variable x_j, for which either f_j or g_{ij} is nonlinear for some $i = 1, \ldots, m$, is replaced by a piecewise linear approximation (for example, λ-form or δ-form). Since some of f_j and g_{ij} may be linear in problem (SP), it is convenient to define the set $\mathcal{L}$ as follows:

$$\mathcal{L} = \{j : f_j \text{ and } g_{ij} \text{ are linear for } i = 1, \ldots, m\}. \tag{2.8}$$

For each $j \notin \mathcal{L}$, consider the interval $[a_j, b_j]$, $a_j, b_j \ge 0$. The grid points x_{lj}, $l = 0, \ldots, k_j$ of the jth variable x_j are defined as

$$a_j = x_{0j} < x_{1j} < \cdots < x_{k_j j} = b_j$$

and they may not be equidistant. Then the functions f_j and g_{ij}, $i = 1, \ldots, m$ can be approximated as follows

$$\widehat{f}_j(x_j) = \sum_{l=0}^{k_j} \lambda_{lj} f_j(x_{lj}), \quad j \notin \mathcal{L} \tag{2.9}$$

and

$$\widehat{g}_{ij}(x_j) = \sum_{l=0}^{k_j} \lambda_{lj} g_{ij}(x_{lj}), \quad i = 1, \ldots, m; \ j \notin \mathcal{L}, \tag{2.10}$$

respectively, where

$$\sum_{l=0}^{k_j} \lambda_{lj} = 1, \quad j \notin \mathcal{L} \tag{2.11}$$

$$\lambda_{lj} \ge 0, \quad l = 0, \ldots, k_j; \ j \notin \mathcal{L}. \tag{2.12}$$

Here λ_{lj}'s are weights associated with the lth grid point of the jth variable x_j.

Since f_j and g_{ij}, $i = 1, \dots, m$ are linear for $j \in \mathcal{L}$, their linear approximations are trivial:

$$\widehat{f}_j(x_j) = f_j(x_j), \quad \widehat{g}_{ij}(x_j) = g_{ij}(x_j), \quad i = 1, \dots, m; \; j \in \mathcal{L}. \tag{2.13}$$

Indeed, for $k_j = 1$ we have

$$x_j = \lambda x_{0j} + (1 - \lambda)x_{1j}, \quad \lambda \in [0, 1],$$

and since $f_j(x_j) = c_j x_j$, $g_{ij}(x_j) = a_{ij} x_j$, c_j's and a_{ij}'s are constants,

$$\begin{aligned} \widehat{f}_j(x_j) &= \lambda f_j(x_{0j}) + (1 - \lambda) f_j(x_{1j}) \\ &= \lambda c_j x_{0j} + (1 - \lambda) c_j x_{1j} = c_j x_j \stackrel{\text{def}}{=} f_j(x_j) \end{aligned}$$

and

$$\begin{aligned} \widehat{g}_{ij}(x_j) &= \lambda g_{ij}(x_{0j}) + (1 - \lambda) g_{ij}(x_{1j}) \\ &= \lambda a_{ij} x_{0j} + (1 - \lambda) a_{ij} x_{1j} = a_{ij} x_j \stackrel{\text{def}}{=} g_{ij}(x_j) \end{aligned}$$

for each $i = 1, \dots, m$ and $j \in \mathcal{L}$.

Then the *approximating separable problem* for (SP) is
(ASP)

$$\min \left\{ \widehat{f}(\mathbf{x}) = \sum_{j \in \mathcal{L}} f_j(x_j) + \sum_{j \notin \mathcal{L}} \widehat{f}_j(x_j) \right\} \tag{2.14}$$

subject to

$$\widehat{g}_i(\mathbf{x}) \equiv \sum_{j \in \mathcal{L}} g_{ij}(x_j) + \sum_{j \notin \mathcal{L}} \widehat{g}_{ij}(x_j) \le \beta_i, \quad i = 1, \dots, m \tag{2.15}$$

$$x_j \ge 0, \quad j = 1, \dots, n, \tag{2.16}$$

or, using definitions of $\widehat{f}_j$ and $\widehat{g}_{ij}$ for $j \notin \mathcal{L}$, we get the problem
$(LASP)$

$$\min \left\{ \widehat{f}(\mathbf{x}) = \sum_{j \in \mathcal{L}} f_j(x_j) + \sum_{j \notin \mathcal{L}} \sum_{l=0}^{k_j} \lambda_{lj} f_j(x_{lj}) \right\} \tag{2.17}$$

subject to

$$\widehat{g}_i(\mathbf{x}) \equiv \sum_{j\in\mathcal{L}} g_{ij}(x_j) + \sum_{j\notin\mathcal{L}} \sum_{l=0}^{k_j} \lambda_{lj} g_{ij}(x_{lj}) \le \beta_i, \quad i = 1, \ldots, m \tag{2.18}$$

$$\sum_{l=0}^{k_j} \lambda_{lj} = 1, \quad j \notin \mathcal{L} \tag{2.19}$$

$$\lambda_{lj} \ge 0, \quad l = 0, \ldots, k_j; \ j \notin \mathcal{L} \tag{2.20}$$

$$x_j \ge 0, \quad j \in \mathcal{L}. \tag{2.21}$$

Variables of the approximating problem $(LASP)$ are x_j, $j \in \mathcal{L}$ and λ_{lj}, $l = 0, \ldots, k_j$, $j \notin \mathcal{L}$. The objective function and constraint functions in problem $(LASP)$ are piecewise linear, and at most two adjacent λ_{lj}'s are positive for $j \notin \mathcal{L}$.

Problem $(LASP)$ differs from the linear program only because there is an additional constraint for λ_{lj}'s:

- for each $j \notin \mathcal{L}$, at most two λ_{lj}'s are positive and these positive λ_{lj}'s must be adjacent. (2.22)

Accuracy of the optimal solution depends on the length of intervals $[x_{l,j-1}, x_{lj}]$. The smaller the length, the greater the accuracy of the obtained solution (and therefore, the closer the approximate solution to the exact solution).

The *restricted basis entry rule* of the simplex method for solving problem $(LASP)$ is the rule that a nonbasic variable λ_{lj} is introduced into the basis only if it improves the objective function and if the new basis satisfies the additional constraint (2.22) for λ_{lj}'s. The strict optimality criterion of the simplex method is used to select the entering variable λ_{lj} only if it satisfies condition (2.22). Otherwise, the variable λ_{lj} which has the next best optimality indicator $\Delta_j = c_j - \mathbf{c}_B B^{-1}\mathbf{a}_j$ (where $\mathbf{a}_j$ is the jth column vector of the constraints, B is the basis matrix, and vector $\mathbf{c}_B$ consists of the objective function coefficients associated with the basic vector $\mathbf{x}_B$) is considered for entering the solution. The process is repeated until the optimality criterion is satisfied or until it is impossible to introduce a new λ_{lj} without violating the restricted basis entry rule. The last simplex tableau gives the approximate optimal solution to problem (SP).

SUMMARY of the simplex method with restricted basis entry rule for solving problem (SP) by solving problem $(LASP)$:

1. Each separable function is approximated by a piecewise linear function.
2. Construct the corresponding linear approximating separable problem $(LASP)$ (2.17)–(2.21).
3. Solve the linear problem (2.17)–(2.21) by the simplex method.

4. Find the optimal solution to the original problem (SP) and the corresponding optimum objective function value.

Separable problem (SP) (2.5)–(2.7) can also be approximated by using a mixed integer program whose variables satisfy the additional condition (2.22).

Let the number of grid points for the jth variable x_j be equal to $k_j + 1$ and let x_{lj} be its lth breaking value. Since $x_j \geq 0,\ j = 1, \dots, n$ then $x_{lj} \geq 0$ for all j's and l's. Let λ_{lj} be the weight associated with x_{lj}. Then the corresponding *mixed integer approximating problem* is

$(MIAP)$

$$\min \left\{ \widehat{f}(\mathbf{x}) = \sum_{j=1}^{n} \widehat{f}_j(x_j) = \sum_{j \in \mathcal{L}} f_j(x_j) + \sum_{j \notin \mathcal{L}} \sum_{l=0}^{k_j} \lambda_{lj} f_j(x_{lj}) \right\} \tag{2.23}$$

subject to

$$\sum_{j \in \mathcal{L}} g_{ij}(x_j) + \sum_{j \notin \mathcal{L}} \sum_{l=0}^{k_j} \lambda_{lj} g_{ij}(x_{lj}) \leq \beta_i, \quad i = 1, \dots, m \tag{2.24}$$

$$0 \leq \lambda_{0j} \leq y_{0j}, \quad j \notin \mathcal{L} \tag{2.25}$$

$$0 \leq \lambda_{lj} \leq y_{l-1,j} + y_{lj}, \quad l = 1, \dots, k_j - 1;\ j \notin \mathcal{L} \tag{2.26}$$

$$0 \leq \lambda_{k_j j} \leq y_{k_j - 1, j}, \quad j \notin \mathcal{L} \tag{2.27}$$

$$\sum_{l=0}^{k_j - 1} y_{lj} = 1, \quad j \notin \mathcal{L} \tag{2.28}$$

$$\sum_{l=0}^{k_j} \lambda_{lj} = 1, \quad j \notin \mathcal{L} \tag{2.29}$$

$$x_j = \sum_{l=0}^{k_j} \lambda_{lj} x_{lj}, \quad j \notin \mathcal{L} \tag{2.30}$$

$$y_{lj} = 0 \text{ or } 1, \quad l = 0, 1, \dots, k_j;\ j \notin \mathcal{L}. \tag{2.31}$$

Variables of the approximating problem $(MIAP)$ are $\lambda_{lj}, y_{lj}, l = 0, \dots, k_j, j \notin \mathcal{L}$, and $x_j,\ j \in \mathcal{L}$.

This formulation shows that any separable problem can be solved by a mixed integer programming problem. However, the number of constraints increases quite rapidly as the number of grid points increases. The computational implementation and effectiveness of this approach depend on the availability of reliable methods for solving large-scale mixed integer programming problems.

The mixed integer programming approach gives a global optimum to the approximate problem, whereas the simplex method with the restricted basis entry rule can only find a local optimum. Both approaches have the following disadvantage. The approximate problems may have additional extreme points that do not exist in the original separable problem. This depends on the linear approximation accuracy. In such cases, the approximate optimal solution may not be feasible for the original separable problem.

Consider again the separable problem (SP) (2.5)–(2.7).

When $f_j(x_j)$ are strictly convex (in case of minimization) or strictly concave (in case of maximization) and $g_{ij}(x_j)$ are convex, problem (SP) is a convex programming problem. In this case, the piecewise linear approximation is simpler than that in the general case.

Let x_j^l be the increment of the variable x_j in the interval $(x_{l-1,j}, x_{lj})$, $l = 1, \dots, k_j$ and let s_j^l be the slope of the corresponding line segment in the same interval. The function $f_j(x_j)$ is approximated as

$$f_j(x_j) \approx \sum_{l=1}^{k_j} s_j^l x_j^l + f_j(x_{0j}), \qquad x_j = \sum_{l=1}^{k_j} x_j^l$$

provided that

$$0 \le x_j^l \le x_{lj} - x_{l-1,j}, \quad l = 1, \dots, k_j.$$

Since $f_j(x_j)$ are strictly convex, according to Theorem 1.31 we have

$$s_j^1 < s_j^2 < \dots < s_j^{k_j}.$$

Therefore, $p < q$ implies x_j^p will enter the solution before x_j^q. The only limitation is that there is an upper bound $x_{lj} - x_{l-1,j}$ for x_j^l.

Analogously, we can approximate the convex constraint functions $g_{ij}(x_j)$. If s_{ij}^l denotes the slope of the lth line segment for $g_{ij}(x_j)$, then

$$g_{ij}(x_j) \approx \sum_{l=1}^{k_j} s_{ij}^l x_j^l + g_{ij}(x_{0j}).$$

Hence, the *approximating convex separable problem* is

$(ACSP)$

$$\min \left\{ f(\mathbf{x}) = \sum_{j=1}^{n} \left(\sum_{l=1}^{k_j} s_j^l x_j^l + f_j(x_{0j}) \right) \right\}$$

subject to

$$\sum_{j=1}^{n} \left(\sum_{l=1}^{k_j} s_{ij}^l x_j^l + g_{ij}(x_{0j}) \right) \le \beta_i, \quad i = 1, \dots, m$$

$$0 \le x_j^l \le x_{lj} - x_{l-1,j}, \quad l = 1, \dots, k_j, \ j = 1, \dots, n,$$

where

$$s_j^l = \frac{f_j(x_{lj}) - f_j(x_{l-1,j})}{x_{lj} - x_{l-1,j}}, \qquad s_{ij}^l = \frac{g_{ij}(x_{lj}) - g_{ij}(x_{l-1,j})}{x_{lj} - x_{l-1,j}}.$$

Problem $(ACSP)$ can be solved by the simplex method with upper bounded variables. It is not necessary to apply the restricted basis entry rule because, due to the convexity of the functions, the proper selection of variables is guaranteed.

Because of its importance, the case of convex functions $f_j(x_j)$ and $g_{ij}(x_j)$ is considered in Chapter Three.

Example 2.1 Solve the problem

$$\min \{ f(\mathbf{x}) = x_1^2 - 6x_1 - x_2 + 9 \}$$

subject to

$$\begin{aligned} 2x_1 + 3x_2 &\le 24 \\ x_1 + 2x_2 &\le 15 \\ 3x_1 + 2x_2 &\le 24 \\ x_2 &\le 4 \end{aligned}$$

$$x_1 \ge 0, \ x_2 \ge 0.$$

Function f is separable because f can be expressed as a sum of two single-variable functions: $f(\mathbf{x}) = f_1(x_1) + f_2(x_2)$, where $f_1(x_1) = x_1^2 - 6x_1 + 9$, $f_2(x_2) = -x_2$. Moreover, f is convex as a sum of two convex functions, and the feasible region X is convex. Therefore, a global minimum of f on X can be found by using the piecewise linear approximation.

Denote

$$g_{11}(x_1) = 2x_1, \quad g_{12}(x_2) = 3x_2, \quad b_1 = 24,$$

$$g_{21}(x_1) = x_1, \quad g_{22}(x_2) = 2x_2, \quad b_2 = 15,$$

$$g_{31}(x_1) = 3x_1, \quad g_{32}(x_2) = 2x_2, \quad b_3 = 24,$$

$$g_{41}(x_1) = 0, \quad g_{42}(x_2) = x_2, \quad b_4 = 4.$$

Only the objective function f is nonlinear, that is why we have to approximate only f. Since $f_2(x_2)$ is linear, it is sufficient to approximate only $f_1(x_1)$. In this case, $\mathcal{L} = \{2\}$ because $f_2(x_2)$ and $g_{i2},\ i = 1, 2, 3, 4$ are linear.

Since $x_1 \le 8$ when $x_2 = 0$, we can choose the breaking points in the closed segment $[0, 8]$, for example, as follows:

$$x_{l1} = l, \quad l = 0, 1, \dots, 8.$$

Then values of $f_1(x_1)$ at these points are

x_{l1}	0	1	2	3	4	5	6	7	8
$f_1(x_{l1})$	9	4	1	0	1	4	9	16	25

Using expressions of $\widehat{f_j}(x_j)$ and x_j we get

$$\widehat{f_1}(x_1) = 9\lambda_{01} + 4\lambda_{11} + \lambda_{21} + 0.\lambda_{31} + \lambda_{41} + 4\lambda_{51} + 9\lambda_{61} + 16\lambda_{71} + 25\lambda_{81},$$

$$x_1 = 0.\lambda_{01} + \lambda_{11} + 2\lambda_{21} + 3\lambda_{31} + 4\lambda_{41} + 5\lambda_{51} + 6\lambda_{61} + 7\lambda_{71} + 8\lambda_{81}.$$

Substitute these expressions in $\widehat{f}(\mathbf{x})$ and in constraints and convert the problem into standard linear programming form (Figure 2.1). Here $x_j,\ j = 3, 4, 5, 6$ are slack variables.

$$\max \{\widehat{f} = -9\lambda_{01} - 4\lambda_{11} - \lambda_{21} - \lambda_{41} - 4\lambda_{51} - 9\lambda_{61} - 16\lambda_{71} - 25\lambda_{81} + x_2\}$$

subject to

$$\begin{array}{llllllllllllllll}
 & 2\lambda_{11} & +4\lambda_{21} & +6\lambda_{31} & +8\lambda_{41} & +10\lambda_{51} & +12\lambda_{61} & +14\lambda_{71} & +16\lambda_{81} & +3x_2 & +x_3 & & & & = & 24 \\
 & \lambda_{11} & +2\lambda_{21} & +3\lambda_{31} & +4\lambda_{41} & +5\lambda_{51} & +6\lambda_{61} & +7\lambda_{71} & +8\lambda_{81} & +2x_2 & & +x_4 & & & = & 15 \\
 & 3\lambda_{11} & +6\lambda_{21} & +9\lambda_{31} & +12\lambda_{41} & +15\lambda_{51} & +18\lambda_{61} & +21\lambda_{71} & +24\lambda_{81} & +2x_2 & & & +x_5 & & = & 24 \\
 & & & & & & & & & x_2 & & & & +x_6 & = & 4 \\
\lambda_{01} & +\lambda_{11} & +\lambda_{21} & +\lambda_{31} & +\lambda_{41} & +\lambda_{51} & +\lambda_{61} & +\lambda_{71} & +\lambda_{81} & & & & & & = & 1
\end{array}$$

$$x_j \ge 0,\ j = 2, 3, 4, 5, 6; \quad \lambda_{l1} \ge 0,\ l = 0, 1, \dots, 8.$$

Fig. 2.1 Standard LP form for Example 2.1.

Since the standard LP form allows us to determine the starting basic feasible solution

$$\lambda_{01} = 1, \ x_3 = 24, \ x_4 = 15, \ x_5 = 24, \ x_6 = 4,$$

the simplex algorithm is applicable directly to this linear program.

Denote $\Delta_j = c_j - \mathbf{c}_B B^{-1}\mathbf{a}_j$, where $\mathbf{a}_j$ is the jth column vector of the constraints, B is the basis matrix, and $\mathbf{c}_B$ consists of the objective function coefficients associated with the basic vector.

Recall that the optimality criterion, using this definition of Δ_j, is $\Delta_j \leq 0$ for all indices j.

Using the simplex method, we obtain the table presented in Table 2.1.

Since $\Delta_j \leq 0$ for all variables in the third simplex tableau, then

$$(\lambda_{01}^*, \lambda_{11}^*, \lambda_{21}^*, \lambda_{31}^*, \lambda_{41}^*, \lambda_{51}^*, \lambda_{61}^*, \lambda_{71}^*, \lambda_{81}^*, x_2^*, x_3^*, x_4^*, x_5^*, x_6^*)$$

$$= (0, 0, 0, 1, 0, 0, 0, 0, 0, 4, 6, 4, 7, 0)$$

is an optimal solution to the linear problem. Moreover, this solution is unique because $\Delta_j < 0$ strictly for all nonbasic variables. Therefore

$$x_1^* \equiv \lambda_{11}^* + 2\lambda_{21}^* + 3\lambda_{31}^* + 4\lambda_{41}^* + 5\lambda_{51}^* + 6\lambda_{61}^* + 7\lambda_{71}^* + 8\lambda_{81}^* = 3.$$

Hence, the optimal solution to the original separable nonlinear problem is

$$\mathbf{x}^* = (x_1^*, x_2^*) = (3, 4).$$

This approximate solution turns out to be the exact optimal solution of the given problem as well. We have

$$f_{\min} = f(\mathbf{x}^*) = -4.$$

Problems.

2.1

$$\min \{f(\mathbf{x}) = x_1^2 - x_1 - x_2\}$$

subject to

$$x_1^2 + 2x_1 + 3x_2^2 \leq 3$$

$$x_1 \geq 0, \ x_2 \geq 0.$$

Answer. $\mathbf{x}^* \approx (0.2, 0.91852), \quad \widehat{f}_{\min} = -1.07852.$

2.2

$$\min \left\{ f(\mathbf{x}) = x_1^2 - 6x_1 + x_2^2 - 8x_2 - \frac{1}{2}x_3 \right\}$$

subject to

$$\begin{array}{ll} x_1 + x_2 + x_3 & \le 5 \\ x_1^2 - x_2 & \le 3 \end{array}$$

$$x_1 \ge 0,\ x_2 \ge 0,\ x_3 \ge 0.$$

Answer. $\mathbf{x}^* \approx (2, 3, 0)$, $\widehat{f}_{\min} = -22$; $f_{\min} = f(2, 3, 0) = -23$.

2.3

$$\max \{ f(\mathbf{x}) = x_1 + x_2^4 \}$$

subject to

$$3x_1 + 2x_2^2 \le 9$$

$$x_1 \ge 0,\ x_2 \ge 0.$$

Answer. $\mathbf{x}^* \approx (0, 2.1)$, $\widehat{f}_{\max} = 19.45$. Exact solution: $\mathbf{x}^*=(0, 2.12)$, $f_{\max} =$ 20.25.

2.4

$$\min \{ f(\mathbf{x}) = x_1^2 + x_2^2 + 5 \}$$

subject to

$$\begin{array}{ll} 3x_1^4 + x_2 & \le 243 \\ x_1 + 2x_2^2 & \le 32 \end{array}$$

$$x_1 \ge 0,\ x_2 \ge 0.$$

2.5

$$\min \{ f(\mathbf{x}) = (x_1 - 2)^2 + 4(x_2 - 6)^2 \}$$

subject to

$$6x_1 + 3(x_2 + 1)^2 \le 12$$

$$x_1 \ge 0,\ x_2 \ge 0.$$

Table 2.1 Simplex tableau for Example 2.1

Basis	$\mathbf{c}_B$	Solution	λ_{01}	λ_{11}	λ_{21}	λ_{31}	λ_{41}	λ_{51}	λ_{61}	λ_{71}	λ_{81}	x_2	x_3	x_4	x_5	x_6
			−9	−4	−1	0	−1	−4	−9	−16	−25	1	0	0	0	0
x_3	0	24	0	2	4	6	8	10	12	14	16	3	1	0	0	0
x_4	0	15	0	1	2	3	4	5	6	7	8	2	0	1	0	0
x_5	0	24	0	3	6	9	12	15	18	21	24	2	0	0	1	0
x_6	0	4	0	0	0	0	0	0	0	0	0	1	0	0	0	1
λ_{01}	−9	1	1	1	1	**1**	1	1	1	1	1	0	0	0	0	0
Δ		9	0	5	8	9	8	5	0	−7	−16	1	0	0	0	0
x_3	0	18	−6	−4	−2	0	2	4	6	8	10	3	1	0	0	0
x_4	0	12	−3	−2	−1	0	1	2	3	4	5	2	0	1	0	0
x_5	0	15	−9	−6	−3	0	3	6	9	12	15	2	0	0	1	0
x_6	0	4	0	0	0	0	0	0	0	0	0	**1**	0	0	0	1
λ_{31}	0	1	1	1	1	1	1	1	1	1	1	0	0	0	0	0
Δ		0	−9	−4	−1	0	−1	−4	−9	−16	−25	1	0	0	0	0
x_3	0	6	−6	−4	−2	0	2	4	6	8	10	0	1	0	0	−3
x_4	0	4	−3	−2	−1	0	1	2	3	4	5	0	0	1	0	−2
x_5	0	7	−9	−6	−3	0	3	6	9	12	15	0	0	0	1	−2
x_2	1	4	0	0	0	0	0	0	0	0	0	1	0	0	0	1
λ_{31}	0	1	1	1	1	1	1	1	1	1	1	0	0	0	0	0
Δ		−4	−9	−4	−1	0	−1	−4	−9	−16	−25	0	0	0	0	−1

Chapter 3
Convex Separable Programming

As it was pointed out in Chapter Two, if $f_j(x_j)$ are strictly convex and $g_{ij}(x_j)$ are convex for $i = 1, \dots, m$ and $j \notin \mathcal{L}$ in problem (SP) (2.5)–(2.7), the standard simplex method, discarding the restricted basis entry rule, is applicable to the approximating linear program $(LASP)$ (2.17)–(2.21).

Some results concerning the relationship between feasible and optimal solutions to problems (SP) (2.5)–(2.7), (ASP) (2.14)–(2.16), and $(LASP)$ and estimates on the accuracy of the piecewise linear approximation are presented below.

Theorem 3.1 (Properties of "approximating" variables in the approximating problem $(LASP)$)
Let (SP) (2.5)–(2.7) and $(LASP)$ (2.17)–(2.21) be the original separable problem and the approximating problem, defined in Chapter Two, respectively, and let the set $\mathcal{L}$ be defined by (2.8). Let f_j be strictly convex and g_{ij} be convex functions for $i = 1, \dots, m,\ j \notin \mathcal{L}$. If $\widehat{x}_j,\ j \in \mathcal{L}$ and $\widehat{\lambda}_{lj},\ l = 0, \dots, k_j,\ j \notin \mathcal{L}$ solve $(LASP)$, then for each $j \notin \mathcal{L}$, at most two $\widehat{\lambda}_{lj}$'s are positive and they are adjacent.

Proof. Suppose that there exist $\widehat{\lambda}_{pj} > 0$ and $\widehat{\lambda}_{qj} > 0$ such that x_{pj} and x_{qj} are not adjacent. Then there exists a grid point $x_{rj} \in (x_{pj}, x_{qj})$ with

$$x_{rj} = \lambda x_{pj} + (1 - \lambda)x_{qj}, \quad \lambda \in (0, 1).$$

Let $u_i \geq 0,\ i = 1, \dots, m$ be the optimal Lagrange multipliers associated with the constraints (2.18) and v_j be the optimal Lagrange multipliers associated with (2.19). The corresponding KKT conditions are

$$f_j(x_{pj}) + \sum_{i=1}^{m} u_i g_{ij}(x_{pj}) + v_j = 0 \tag{3.1}$$

© The Author(s), under exclusive license to Springer Nature Switzerland AG 2021

S. M. Stefanov, *Separable Optimization*, Springer Optimization and Its Applications 177,
https://doi.org/10.1007/978-3-030-78401-0_3

$$f_j(x_{qj}) + \sum_{i=1}^{m} u_i g_{ij}(x_{qj}) + v_j = 0 \tag{3.2}$$

$$f_j(x_{lj}) + \sum_{i=1}^{m} u_i g_{ij}(x_{lj}) + v_j \geq 0, \quad l = 0, \ldots, k_j. \tag{3.3}$$

However, since f_j are strictly convex and g_{ij} are convex, we get

$$\begin{aligned}
&f_j(x_{rj}) + \sum_{i=1}^{m} u_i g_{ij}(x_{rj}) + v_j \\
&< \lambda f_j(x_{pj}) + (1-\lambda) f_j(x_{qj}) \\
&+ \sum_{i=1}^{m} u_i [\lambda g_{ij}(x_{pj}) + (1-\lambda) g_{ij}(x_{qj})] + v_j \\
&\equiv \lambda \left[f_j(x_{pj}) + \sum_{i=1}^{m} u_i g_{ij}(x_{pj}) + v_j \right] \\
&+ (1-\lambda) \left[f_j(x_{qj}) + \sum_{i=1}^{m} u_i g_{ij}(x_{qj}) + v_j \right] = \lambda.0 + (1-\lambda).0 = 0,
\end{aligned}$$

which contradicts (3.3) for $l = r$. Therefore the assumption is incorrect, and x_{pj} and x_{qj} must be adjacent. ■

Theorem 3.2 (Feasible solution to separable problem (SP) from optimal solution to approximating problem $(LASP)$)
Under assumptions of Theorem 3.1, if $\widehat{x}_j = \sum_{l=0}^{k_j} \widehat{\lambda}_{lj} x_{lj},\ j \notin \mathcal{L}$, *then* $\widehat{\mathbf{x}} = (\widehat{x}_1, \ldots, \widehat{x}_n)$ *is a feasible solution to the original separable problem* (SP).

Proof. Using that g_{ij} are convex for $i = 1, \ldots, m,\ j \notin \mathcal{L}$ and $\widehat{x}_j,\ j \in \mathcal{L}, \widehat{\lambda}_{lj},\ l = 0, \ldots, k_j,\ j \notin \mathcal{L}$ solve $(LASP)$ (and, therefore, they are feasible for $(LASP)$), we get

$$\begin{aligned}
g_i(\widehat{\mathbf{x}}) &\equiv \sum_{j \in \mathcal{L}} g_{ij}(\widehat{x}_j) + \sum_{j \notin \mathcal{L}} g_{ij}(\widehat{x}_j) \\
&= \sum_{j \in \mathcal{L}} g_{ij}(\widehat{x}_j) + \sum_{j \notin \mathcal{L}} g_{ij} \left(\sum_{l=0}^{k_j} \widehat{\lambda}_{lj} x_{lj} \right) \\
&\leq \sum_{j \in \mathcal{L}} g_{ij}(\widehat{x}_j) + \sum_{j \notin \mathcal{L}} \sum_{l=0}^{k_j} \widehat{\lambda}_{lj} g_{ij}(x_{lj}) \leq \beta_i, \quad i = 1, \ldots, m.
\end{aligned}$$

Moreover, $\widehat{x}_j \geq 0$ for $j \in \mathcal{L}$ by the assumption that $\widehat{x}_j,\ j \in \mathcal{L},\ \widehat{\lambda}_{lj},\ l = 0, \ldots, k_j,$ $j \notin \mathcal{L}$ solve $(LASP)$, and since $\widehat{\lambda}_{lj} \geq 0,\ x_{lj} \geq 0,\ l = 0, \ldots, k_j,\ j \notin \mathcal{L}$, then $\widehat{x}_j = \sum_{l=0}^{k_j} \widehat{\lambda}_{lj} x_{lj} \geq 0$ for $j \notin \mathcal{L}$. Therefore, $\widehat{\mathbf{x}} = (\widehat{x}_1, \ldots, \widehat{x}_n)$ is feasible to the original separable problem (SP). ■

Statement of Theorem 3.2 means that, under some convexity assumptions, an optimal solution to $(LASP)$ is a feasible solution to the original problem (SP).

Theorem 3.3 (Inequalities for objective functions and constraint functions of (SP) and $(LASP)$)
Let f_j and g_{ij} be convex differentiable functions for $i = 1, \ldots, m$ and $j \notin \mathcal{L}$ in problems (SP) and $(LASP)$, and let $\widehat{f}_j$ and $\widehat{g}_{ij}$ be their piecewise linear approximations on the interval $[a_j, b_j]$, respectively. For $j \notin \mathcal{L}$ let η_j be such that $|f_j'(x_j)| \leq \eta_j$ for $x_j \in [a_j, b_j]$. For $j \notin \mathcal{L}$ and $i = 1, \ldots, m$ let κ_{ij} be such that $|g_{ij}'(x_j)| \leq \kappa_{ij}$ for $x_j \in [a_j, b_j]$. For $j \notin \mathcal{L}$, let ε_j be the maximum grid length used for the variable x_j.
Denote

$$\bar{\xi}_0 = 2 \sum_{j \notin \mathcal{L}} \eta_j \varepsilon_j, \quad \bar{\xi}_i = 2 \sum_{j \notin \mathcal{L}} \kappa_{ij} \varepsilon_j,\ i = 1, \ldots, m, \quad \xi = \max_{0 \leq i \leq m} \{\bar{\xi}_i\}.$$

Then

$$\widehat{f}(\mathbf{x}) \geq f(\mathbf{x}) \geq \widehat{f}(\mathbf{x}) - \xi,$$

$$\widehat{g}_i(\mathbf{x}) \geq g_i(\mathbf{x}) \geq \widehat{g}_i(\mathbf{x}) - \xi, \quad i = 1, \ldots, m.$$

Proof. We will prove that

$$\widehat{f}_j(x_j) \geq f_j(x_j) \geq \widehat{f}_j(x_j) - 2\eta_j \varepsilon_j, \quad j \notin \mathcal{L}.$$

Let $j \notin \mathcal{L}$ and $x_j \in [a_j, b_j]$. Then there exist grid points μ_k and μ_{k+1} such that $x_j \in [\mu_k, \mu_{k+1}]$. Therefore $x_j = \lambda \mu_k + (1 - \lambda)\mu_{k+1}$ for some $\lambda \in [0, 1]$. Hence

$$\begin{aligned} \widehat{f}_j(x_j) &= \lambda f_j(\mu_k) + (1 - \lambda) f_j(\mu_{k+1}) \\ &\geq f_j(\lambda \mu_k + (1 - \lambda)\mu_{k+1}) = f_j(x_j), \end{aligned}$$

the first inequality we want to prove. We have used the definition of $\widehat{f}_j$, the assumption that f_j is convex, $\lambda \in [0, 1]$, and the expression of x_j.

Since $\widehat{f}_j(x_j)$ is a linear piecewise approximation of $f_j(x_j)$ and μ_k, μ_{k+1} are grid points,

$$\frac{\widehat{f}_j(x_j) - f_j(\mu_k)}{x_j - \mu_k} = \frac{f_j(\mu_{k+1}) - f_j(\mu_k)}{\mu_{k+1} - \mu_k},$$

that is,

$$\widehat{f}_j(x_j) = f_j(\mu_k) + (x_j - \mu_k)\frac{f_j(\mu_{k+1}) - f_j(\mu_k)}{\mu_{k+1} - \mu_k}. \tag{3.4}$$

Using the property of convex differentiable functions (Theorem 1.30)

$$f(\mathbf{x}_1) - f(\mathbf{x}_2) \geq \langle \mathbf{f}'(\mathbf{x}_2), \mathbf{x}_1 - \mathbf{x}_2 \rangle,$$

we get

$$f_j(x_j) \geq f_j(\mu_k) + (x_j - \mu_k) f_j'(\mu_k). \tag{3.5}$$

From (3.4) and (3.5) it follows that

$$\widehat{f}_j(x_j) - f_j(x_j) \leq (x_j - \mu_k)\left[\frac{f_j(\mu_{k+1}) - f_j(\mu_k)}{\mu_{k+1} - \mu_k} - f_j'(\mu_k)\right]. \tag{3.6}$$

By using the Mean value theorem, there exists a $y \in [\mu_k, \mu_{k+1}]$ such that

$$\frac{f_j(\mu_{k+1}) - f_j(\mu_k)}{\mu_{k+1} - \mu_k} = f_j'(y).$$

By assumption,

$$\frac{f_j(\mu_{k+1}) - f_j(\mu_k)}{\mu_{k+1} - \mu_k} - f_j'(\mu_k) = f_j'(y) - f_j'(\mu_k) \leq 2\eta_j,$$

and $x_j - \mu_k \leq \varepsilon_j$ according to definition of ε_j. From (3.6) we get

$$\widehat{f}_j(x_j) - f_j(x_j) \leq 2\eta_j \varepsilon_j.$$

Therefore

$$\widehat{f}_j(x_j) \geq f_j(x_j) \geq \widehat{f}_j(x_j) - 2\eta_j \varepsilon_j, \quad j \notin \mathcal{L},\ x_j \in [a_j, b_j]. \tag{3.7}$$

From (3.7) we obtain

$$\begin{aligned}\sum_{j \notin \mathcal{L}} \widehat{f}_j(x_j) + \sum_{j \in \mathcal{L}} f_j(x_j) &\geq \sum_{j \notin \mathcal{L}} f_j(x_j) + \sum_{j \in \mathcal{L}} f_j(x_j) \\ &\geq \sum_{j \notin \mathcal{L}} \widehat{f}_j(x_j) + \sum_{j \in \mathcal{L}} f_j(x_j) - 2\sum_{j \notin \mathcal{L}} \eta_j \varepsilon_j,\end{aligned}$$

that is,

$$\widehat{f}(\mathbf{x}) \geq f(\mathbf{x}) \geq \widehat{f}(\mathbf{x}) - \overline{\xi}_0.$$

Similarly, we get

$$\widehat{g}_i(\mathbf{x}) \geq g_i(\mathbf{x}) \geq \widehat{g}_i(\mathbf{x}) - \overline{\xi}_i, \quad i = 1, \ldots, m.$$

By using the definition of ξ, we obtain the statement of Theorem 3.3. ■

Proposition 3.1 (Feasible solution to problem (SP) from optimal solutions to approximating problems (ASP) and $(LASP)$)
Let the separable problem (SP) and the set $\mathcal{L}$ be defined by (2.5)–(2.7) and (2.8), respectively, and let (ASP) (2.14)–(2.16) and $(LASP)$ (2.17)–(2.21) be the equivalent problems which approximate (SP). For $j \notin \mathcal{L}$, $i = 1, \ldots, m$ let $\widehat{g}_{ij}$ be the piecewise linear approximations of convex functions g_{ij}. Let $\widehat{x}_j$, $j \in \mathcal{L}$ and $\widehat{\lambda}_{lj}$, $l = 0, \ldots, k_j$, $j \notin \mathcal{L}$ be components of an optimal solution to problem $(LASP)$ such that the vector $\widehat{\mathbf{x}}$ with components $\widehat{x}_j$, $j \in \mathcal{L}$ and $\widehat{x}_j = \sum_{l=0}^{k_j} \widehat{\lambda}_{lj} x_{lj}$, $j \notin \mathcal{L}$ is an optimal solution to problem (ASP).

Then $\widehat{\mathbf{x}}$ is a feasible solution to problem (SP).

Proof. The vector $\widehat{\mathbf{x}}$ is feasible to problem (ASP), that is, $\widehat{g}_i(\widehat{\mathbf{x}}) \leq \beta_i$, $i = 1, \ldots, m$ and $\widehat{\mathbf{x}} \geq \mathbf{0}$, because $\widehat{\mathbf{x}}$ is an optimal solution to (ASP) by assumption. Using Theorem 3.3, we get

$$\widehat{g}_i(\widehat{\mathbf{x}}) \leq \beta_i \text{ implies } g_i(\widehat{\mathbf{x}}) \leq \widehat{g}_i(\widehat{\mathbf{x}}) \leq \beta_i, \; i = 1, \ldots, m,$$

and since $\widehat{\mathbf{x}} \geq \mathbf{0}$ then $\widehat{\mathbf{x}}$ is a feasible solution to problem (SP). ■

Theorem 3.4 (Error estimate for objective function values of (SP))
Consider problem (SP) (2.5)–(2.7). Let $\mathcal{L}$ be defined by (2.8). For $j \notin \mathcal{L}$, let $\widehat{f}_j$ and $\widehat{g}_{ij}$ be the piecewise linear approximations of f_j and g_{ij}, $i = 1, \ldots, m$, respectively. Let (ASP) (2.14)–(2.16) and $(LASP)$ (2.17)–(2.21) be the equivalent problems which approximate problem (SP). Let f_j and g_{ij}, $i = 1, \ldots, m$ be convex for $j \notin \mathcal{L}$ and $\mathbf{x}^*$ be an optimal solution to problem (SP). Let $\widehat{x}_j$, $j \in \mathcal{L}$ and $\widehat{\lambda}_{lj}$, $l = 0, \ldots, k_j$, $j \notin \mathcal{L}$ be components of an optimal solution to problem $(LASP)$ such that the vector $\widehat{\mathbf{x}}$ with components $\widehat{x}_j$, $j \in \mathcal{L}$ and $\widehat{x}_j = \sum_{l=0}^{k_j} \widehat{\lambda}_{lj} x_{lj}$, $j \notin \mathcal{L}$ is an optimal solution to problem (ASP). Let $\widehat{u}_i \geq 0$ be the optimal Lagrange multipliers associated with the constraints $\widehat{g}_i(\mathbf{x}) \leq \beta_i$, $i = 1, \ldots, m$.

Then $0 \leq f(\widehat{\mathbf{x}}) - f(\mathbf{x}^*) \leq \xi(1 + \sum_{i=1}^{m} \widehat{u}_i)$, where ξ is defined as in Theorem 3.3.

Proof. Since the linear piecewise approximation of a convex function is also convex, then $\widehat{f}_j$ and $\widehat{g}_{ij}$ are convex for $i = 1, \ldots, m$ and $j \notin \mathcal{L}$, and since the sum of convex functions is also convex (Theorem 1.17), then the objective function and constraint functions of problem (ASP) are convex. Therefore $(\widehat{\mathbf{x}}, \widehat{\mathbf{u}})$ satisfies the saddlepoint optimality criteria (Theorem 1.45) for problem (ASP), that is,

$$\widehat{f}(\widehat{\mathbf{x}}) \leq \widehat{f}(\mathbf{x}) + \langle \widehat{\mathbf{u}}, \widehat{\mathbf{g}}(\mathbf{x}) - \boldsymbol{\beta} \rangle \equiv \widehat{f}(\widehat{\mathbf{x}}) + \sum_{i=1}^{m} \widehat{u}_i (\widehat{g}_i(\mathbf{x}) - \beta_i), \quad \mathbf{x} \geq \mathbf{0}$$

where $\widehat{\mathbf{g}}(\mathbf{x}) = (\widehat{g}_1(\mathbf{x}), \ldots, \widehat{g}_m(\mathbf{x}))$, $\widehat{\mathbf{u}} = (\widehat{u}_1, \ldots, \widehat{u}_m)$, and $\boldsymbol{\beta} = (\beta_1, \ldots, \beta_m)$. Since $g_i(\mathbf{x}^*) \le \beta_i$, $i = 1, \ldots, m$ by assumption (constraints (2.15) of problem (ASP)), using Theorem 3.3, the second chain of inequalities, we get $\widehat{g}_i(\mathbf{x}^*) - \xi \le \beta_i$, $i = 1, \ldots, m$, that is, $\widehat{g}_i(\mathbf{x}^*) - \beta_i \le \xi$, $i = 1, \ldots, m$. Setting $\mathbf{x} = \mathbf{x}^*$ and using $\widehat{\mathbf{u}} \ge \mathbf{0}$, we obtain

$$\widehat{f}(\widehat{\mathbf{x}}) \le \widehat{f}(\mathbf{x}^*) + \xi \sum_{i=1}^{m} \widehat{u}_i .$$

As we have proved, $\widehat{\mathbf{x}}$ is feasible for problem (SP) (Proposition 3.1), and $\mathbf{x}^*$ is an optimal solution to (SP) by assumption. Therefore $f(\widehat{\mathbf{x}}) \ge f(\mathbf{x}^*)$, the left inequality we have to prove. From Theorem 3.3, $f(\mathbf{x}^*) \ge \widehat{f}(\mathbf{x}^*) - \xi$, hence $f(\widehat{\mathbf{x}}) \ge f(\mathbf{x}^*) \ge \widehat{f}(\mathbf{x}^*) - \xi$. Therefore,

$$f(\widehat{\mathbf{x}}) \ge f(\mathbf{x}^*) \ge \widehat{f}(\mathbf{x}^*) - \xi \ge \Big[\widehat{f}(\widehat{\mathbf{x}}) - \xi \sum_{i=1}^{m} \widehat{u}_i\Big] - \xi = \widehat{f}(\widehat{\mathbf{x}}) - \xi\Big[1 + \sum_{i=1}^{m} \widehat{u}_i\Big],$$

the right inequality we have to prove. ■

Obviously, if the grid length ε_j is reduced, ξ would be smaller (see definition of ξ, Theorem 3.3) and, therefore, a better approximation would be obtained.

Lagrange multipliers $\widehat{u}_i$, $i = 1, \ldots, m$, involved in Theorem 3.4, are available from the optimal simplex tableau for problem $(LASP)$. Then the estimate given by Theorem 3.4 can be evaluated.

Grid Point Generation

The accuracy of an approximating procedure for separable problem (SP) depends on the number of grid points for each variable. However, as the number of grid points increases, the number of variables in the approximating problem $(LASP)$ also increases. One possible approach to avoid this difficulty is to generate grid points when necessary.

Assumptions **A.**

Let x_{lj}, $l = 0, \ldots, k_j$, $j \notin \mathcal{L}$ be the grid points, where $\mathcal{L}$ is defined by (2.8).

Consider problem $(LASP)$ (2.17)–(2.21) *which approximates the original separable problem (SP)* (2.5)–(2.7) *using the grid points x_{lj}, $l = 0, \ldots, k_j$ for $j \notin \mathcal{L}$.*

Without loss of generality $f_j(x_j)$ are of the form $f_j(x_j) = c_j x_j$ and $g_{ij}(x_j)$ are of the form $g_{ij}(x_j) = a_{ij} x_j$ for $i = 1, \ldots, m$, $j \in \mathcal{L}$.

For $j \notin \mathcal{L}$, let f_j and g_{ij}, $i = 1, \ldots, m$ be convex functions.

Let $\widehat{x}_j$, $j \in \mathcal{L}$ and $\widehat{\lambda}_{lj}$, $l = 0, \ldots, k_j$, $j \notin \mathcal{L}$ be components of the optimal solution to problem $(LASP)$ and let $\widehat{f}$ be the corresponding objective function value.

Let $\widehat{u}_i \ge 0$, $i = 1, \ldots, m$ be the optimal Lagrange multipliers associated with the constraints (2.18) *and $\widehat{v}_j$, $j \notin \mathcal{L}$ be the Lagrange multipliers associated with the constraints* (2.19) *in problem $(LASP)$.*

We are interested in the question whether it is necessary to introduce an additional grid point for any of the variables x_j, $j \notin \mathcal{L}$ in order to obtain a better linear piecewise approximation so that, if this new point was considered in problem $(LASP)$, the minimum objective function value would decrease.

The Lagrangian for problem $(LASP)$ is

$$L(\mathbf{x}, \boldsymbol{\lambda}, \mathbf{u}, \mathbf{v}) = \sum_{j\in\mathcal{L}} f_j(x_j) + \sum_{j\notin\mathcal{L}} \sum_{l=0}^{k_j} \lambda_{lj} f_j(x_{lj})$$

$$+ \sum_{i=1}^{m} u_i \left[\sum_{j\in\mathcal{L}} g_{ij}(x_j) + \sum_{j\notin\mathcal{L}} \sum_{l=0}^{k_j} \lambda_{lj} g_{ij}(x_{lj}) - \beta_i \right]$$

$$+ \sum_{j\notin\mathcal{L}} v_j \left[\sum_{l=0}^{k_j} \lambda_{lj} - 1 \right], \tag{3.8}$$

where $x_j \geq 0,\ j \in \mathcal{L};\quad \lambda_{lj} \geq 0,\ l = 0, \ldots, k_j,\ j \notin \mathcal{L};\quad u_i \geq 0,\ i = 1, \ldots, m$.

The solution $\widehat{x}_j, \widehat{\lambda}_{lj}$ to problem $(LASP)$ and the optimal Lagrange multipliers $\widehat{u}_i, \widehat{v}_j$ satisfy the KKT conditions for this problem.

Taking into account that $f_j(x_j) = c_j x_j$ and $g_{ij}(x_j) = a_{ij} x_j$ for $j \in \mathcal{L}$, the KKT conditions for problem $(LASP)$ can be written as follows:

$$c_j + \sum_{i=1}^{m} u_i a_{ij} \geq 0, \quad j \in \mathcal{L} \tag{3.9}$$

$$x_j \left[c_j + \sum_{i=1}^{m} u_i a_{ij} \right] = 0, \quad j \in \mathcal{L} \tag{3.10}$$

$$f_j(x_{lj}) + \sum_{i=1}^{m} u_i g_{ij}(x_{lj}) + v_j \geq 0, \quad l = 0, \ldots, k_j,\ j \notin \mathcal{L} \tag{3.11}$$

$$\lambda_{lj} \left[f_j(x_{lj}) + \sum_{i=1}^{m} u_i g_{ij}(x_{lj}) + v_j \right] = 0, \quad l = 0, \ldots, k_j,\ j \notin \mathcal{L} \tag{3.12}$$

$$x_j \geq 0,\ j \in \mathcal{L}, \quad \lambda_{lj} \geq 0,\ l = 0, \ldots, k_j,\ j \notin \mathcal{L} \tag{3.13}$$

$$\sum_{j\in\mathcal{L}} a_{ij}x_j + \sum_{j\notin\mathcal{L}} \sum_{l=0}^{k_j} \lambda_{lj} g_{ij}(x_{lj}) - \beta_i \le 0, \quad i = 1, \dots, m \tag{3.14}$$

$$u_i \left[\sum_{j\in\mathcal{L}} a_{ij}x_j + \sum_{j\notin\mathcal{L}} \sum_{l=0}^{k_j} \lambda_{lj} g_{ij}(x_{lj}) - \beta_i \right] = 0, \quad i = 1, \dots, m \tag{3.15}$$

$$\sum_{l=0}^{k_j} \lambda_{lj} = 1, \quad j \notin \mathcal{L} \qquad (3.16) \equiv (2.19)$$

$$u_i \ge 0, \quad i = 1, \dots, m. \tag{3.17}$$

Multiplying both sides of condition (3.9) by $x_j \ge 0,\ j \in \mathcal{L}$, and using that $f_j(x_j) = c_j x_j$ and $g_{ij}(x_j) = a_{ij}x_j$ for $j \in \mathcal{L}$, we get

$$f_j(x_j) + \sum_{i=1}^{m} u_i g_{ij}(x_j) \ge 0 \quad \text{for } x_j \ge 0,\ j \in \mathcal{L}. \tag{3.9$'$}$$

Let $x_{l_0 j}$ be an additional grid point. If

$$f_j(x_{l_0 j}) + \sum_{i=1}^{m} \widehat{u}_i g_{ij}(x_{l_0 j}) + \widehat{v}_j \ge 0, \tag{3.18}$$

then, if we set $\widehat{\lambda}_{l_0 j} = 0$, the KKT conditions for problem $(LASP)$ will be satisfied.

Since we do not know where the new grid point should be located, we have to solve the following subproblem for each $j \notin \mathcal{L}$ in order to find out whether all x_j, satisfying $a_j \le x_j \le b_j,\ j \notin \mathcal{L}$, will satisfy (3.18).

Subproblem of (SP).

$$\min \left\{ h_j(x_j) = f_j(x_j) + \sum_{i=1}^{m} \widehat{u}_i g_{ij}(x_j) + \widehat{v}_j \right\}$$

subject to

$$a_j \le x_j \le b_j$$

where $[a_j, b_j]$ with $b_j > a_j \ge 0$ is the interval for $x_j,\ j \notin \mathcal{L}$.

If the minimum value of the objective function $h_j(x_j)$ is nonnegative for all $j \notin \mathcal{L}$, then we cannot find a new grid point violating (3.18). Theorem 3.5 below states that if this is the case, the current solution is optimal to the original problem (SP), and if the minimum value of the objective function is negative for some $j \notin \mathcal{L}$, we can obtain a better approximation to problem (SP).

Theorem 3.5 (Inequalities for objective function values of problems (SP) and $(LASP)$ and of the subproblem of (SP))
Let Assumptions **A** *be satisfied.*

For each $j \notin \mathcal{L}$, consider the Subproblem of (SP) formulated above. Let h_j^* be the optimal objective function value to this subproblem.

Then $\sum_{j \notin \mathcal{L}} (h_j^* - \widehat{v}_j) - \sum_{i=1}^{m} \widehat{u}_i \beta_i \leq \sum_{j=1}^{n} f_j(x_j^*) \leq \sum_{j=1}^{n} f_j(\widehat{x}_j) \leq \widehat{f}$

where $\widehat{x}_j = \sum_{l=0}^{k_j} \widehat{\lambda}_{lj} x_{lj}$, $j \notin \mathcal{L}$, and $\mathbf{x}^* = (x_1^*, \ldots, x_n^*)$ is an optimal solution to problem (SP).

Proof. Since $\widehat{u}_i$ and $\widehat{v}_j$ are the optimal Lagrange multipliers, associated with problem $(LASP)$, then the following KKT conditions (3.9) hold

$$c_j + \sum_{i=1}^{m} \widehat{u}_i a_{ij} \geq 0, \quad j \in \mathcal{L}.$$

Multiplying these inequalities by $x_j \geq 0$ and taking into account that $f_j(x_j) = c_j x_j$ and $g_{ij}(x_j) = a_{ij} x_j$ for $j \in \mathcal{L}$, we get

$$f_j(x_j) + \sum_{i=1}^{m} \widehat{u}_i g_{ij}(x_j) \geq 0 \quad \text{for } x_j \geq 0, \; j \in \mathcal{L}. \qquad (3.19) \equiv (3.9')$$

By definition of h_j^* we get

$$f_j(x_j) + \sum_{i=1}^{m} \widehat{u}_i g_{ij}(x_j) + \widehat{v}_j \geq h_j^*, \quad j \notin \mathcal{L}, \; a_j \leq x_j \leq b_j. \qquad (3.20)$$

Summing (3.19) for all $j \in \mathcal{L}$ and (3.20) for all $j \notin \mathcal{L}$ and subtracting $\sum_{i=1}^{m} \widehat{u}_i \beta_i$ from the resulting sum, we obtain

$$\sum_{j=1}^{n} f_j(x_j) + \sum_{i=1}^{m} \widehat{u}_i \left[\sum_{j=1}^{n} g_{ij}(x_j) - \beta_i \right] \geq \sum_{j \notin \mathcal{L}} (h_j^* - \widehat{v}_j) - \sum_{i=1}^{m} \widehat{u}_i \beta_i \qquad (3.21)$$

for $x_j \in [a_j, b_j]$. Using that

$$a_j \leq x_j^* \leq b_j, \quad \sum_{j=1}^{n} g_{ij}(x_j^*) \leq \beta_i, \quad \widehat{u}_i \geq 0,$$

from (3.21) it follows that

$$\sum_{j=1}^{n} f_j(x_j^*) \geq \sum_{j \notin \mathcal{L}} (h_j^* - \widehat{v}_j) - \sum_{i=1}^{m} \widehat{u}_i \beta_i,$$

the first inequality we have to prove.

According to Proposition 3.1, $\widehat{\mathbf{x}} = (\widehat{x}_1, \ldots, \widehat{x}_n)$ is a feasible solution to problem (SP), hence $\sum_{j=1}^{n} f_j(x_j^*) \leq \sum_{j=1}^{n} f_j(\widehat{x}_j)$, the second inequality we have to prove.

Using convexity of f_j for $j \notin \mathcal{L}$ and definition of $\widehat{f}$, we get

$$\begin{aligned}
\sum_{j=1}^{n} f_j(\widehat{x}_j) &= \sum_{j \in \mathcal{L}} f_j(\widehat{x}_j) + \sum_{j \notin \mathcal{L}} f_j(\widehat{x}_j) \\
&= \sum_{j \in \mathcal{L}} f_j(\widehat{x}_j) + \sum_{j \notin \mathcal{L}} f_j \left(\sum_{l=0}^{k_j} \widehat{\lambda}_{lj} x_{lj} \right) \\
&\leq \sum_{j \in \mathcal{L}} f_j(\widehat{x}_j) + \sum_{j \notin \mathcal{L}} \sum_{l=0}^{k_j} \widehat{\lambda}_{lj} f_j(x_{lj}) = \widehat{f},
\end{aligned}$$

the third inequality we have to prove. ■

Theorem 3.6 (Grid point generation in $(LASP)$)
Let Assumptions **A** *be satisfied and for each $j \notin \mathcal{L}$, let h_j^* be the optimal objective function value to the Subproblem of (SP). Then*

1. *If $h_j^* \geq 0$, $j \notin \mathcal{L}$, then $\widehat{\mathbf{x}} = (\widehat{x}_1, \ldots, \widehat{x}_n)$ is an optimal solution to problem (SP) and $\sum_{j=1}^{n} f_j(\widehat{x}_j) = \widehat{f}$.*

2. *Otherwise, if $h_j^* < 0$ for some $j \notin \mathcal{L}$ and $x_{l_0 j}$ is the corresponding optimal solution to the Subproblem of (SP), then, adding the grid point $x_{l_0 j}$ in problem $(LASP)$, we would obtain a new approximating problem of the form $(LASP)$ with a minimum objective function value not higher than $\widehat{f}$.*

Proof. Consider problem $(LASP)$. Using the complementary conditions among the KKT optimality conditions (that is, conditions (3.10), (3.12), (3.15)) with $x_j = \widehat{x}_j$, $u_i = \widehat{u}_i$, $v_j = \widehat{v}_j$, $\lambda_{lj} = \widehat{\lambda}_{lj}$ and the expressions $f_j(x_j) = c_j x_j$, $g_{ij}(x_j) = a_{ij} x_j$, $j \in \mathcal{L}$, we get

$$f_j(\widehat{x}_j) + \sum_{i=1}^{m} \widehat{u}_i g_{ij}(\widehat{x}_j) = 0, \quad j \in \mathcal{L} \tag{3.22}$$

$$\widehat{\lambda}_{lj}\left[f_j(x_{lj})+\sum_{i=1}^{m}\widehat{u}_i g_{ij}(x_{lj})+\widehat{v}_j\right]=0,\quad l=0,\ldots,k_j;\ j\notin\mathcal{L} \tag{3.23}$$

$$\widehat{u}_i\left[\sum_{j\in\mathcal{L}}g_{ij}(\widehat{x}_j)+\sum_{j\notin\mathcal{L}}\sum_{l=0}^{k_j}\widehat{\lambda}_{lj}g_{ij}(x_{lj})-\beta_i\right]=0,\quad i=1,\ldots,m. \tag{3.24}$$

Summing (3.22) for all $j\in\mathcal{L}$ and (3.23) for all $l=0,\ldots,k_j$ and $j\notin\mathcal{L}$, we obtain

$$\begin{aligned}&\left[\sum_{j\in\mathcal{L}}f_j(\widehat{x}_j)+\sum_{j\notin\mathcal{L}}\sum_{l=0}^{k_j}\widehat{\lambda}_{lj}f_j(x_{lj})\right]\\&+\sum_{i=1}^{m}\widehat{u}_i\left[\sum_{j\in\mathcal{L}}g_{ij}(\widehat{x}_j)+\sum_{j\notin\mathcal{L}}\sum_{l=0}^{k_j}\widehat{\lambda}_{lj}g_{ij}(x_{lj})\right]+\sum_{j\notin\mathcal{L}}\sum_{l=0}^{k_j}\widehat{\lambda}_{lj}\widehat{v}_j=0.\end{aligned} \tag{3.25}$$

The first term in (3.25) is equal to $\widehat{f}$ by definition, and the second one is equal to $\sum_{i=1}^{m}\widehat{u}_i\beta_i$ in accordance with (3.24). Since $\widehat{\lambda}_{lj}$'s are optimal to problem $(LASP)$, then $\sum_{l=0}^{k_j}\widehat{\lambda}_{lj}=1,\ j\notin\mathcal{L}$ (see (2.19)). Therefore

$$\widehat{f}+\sum_{i=1}^{m}\widehat{u}_i\beta_i+\sum_{j\notin\mathcal{L}}\widehat{v}_j=0. \tag{3.26}$$

From (3.26) and the first inequality of Theorem 3.5, written as follows $\sum_{i=1}^{m}\widehat{u}_i\beta_i+\sum_{j\notin\mathcal{L}}\widehat{v}_j\geq\sum_{j\notin\mathcal{L}}h_j^*-\sum_{j=1}^{n}f_j(x_j^*)$, we get

$$\widehat{f}+\sum_{j\notin\mathcal{L}}h_j^*\leq\sum_{j=1}^{n}f_j(x_j^*).$$

By the assumption of part 1. of Theorem 3.6, $h_j^*\geq 0$ for all $j\notin\mathcal{L}$. Then $\widehat{f}\leq\sum_{j=1}^{n}f_j(x_j^*)$ and from the second and third inequalities of Theorem 3.5,

$$\widehat{f}\leq\sum_{j=1}^{n}f_j(x_j^*)\leq\sum_{j=1}^{n}f_j(\widehat{x}_j)\leq\widehat{f}.$$

Therefore

$$\sum_{j=1}^{n}f_j(x_j^*)=\sum_{j=1}^{n}f_j(\widehat{x}_j)=\widehat{f}. \tag{3.27}$$

Since $\widehat{\mathbf{x}} = (\widehat{x}_1, \ldots, \widehat{x}_n)$ is a feasible solution to problem (SP) according to Proposition 3.1, then $\widehat{\mathbf{x}}$ is an optimal solution to (SP) as well according to (3.27), with optimal objective function value equal to $\widehat{f}$. The proof of part 1. of Theorem 3.6 is complete.

Let $x_{l_0 j}$ be the optimal solution to the Subproblem of (SP) with $h_j^* < 0$. Then

$$f_j(x_{l_0 j}) + \sum_{i=1}^{m} \widehat{u}_i g_{ij}(x_{l_0 j}) + \widehat{v}_j < 0.$$

However, if the grid point $x_{l_0 j}$ was included in the formulation of problem $(LASP)$, then one of the KKT conditions

$$f_j(x_{l_0 j}) + \sum_{i=1}^{m} \widehat{u}_i g_{ij}(x_{l_0 j}) + \widehat{v}_j \geq 0 \qquad (3.11')$$

would be violated, that is, $f_j(x_{l_0 j}) + \sum_{i=1}^{m} \widehat{u}_i g_{ij}(x_{l_0 j}) + \widehat{v}_j$ would be negative.

In order that (3.23) be satisfied for $l = l_0$, $\widehat{\lambda}_{l_0 j}$ must be equal to zero.

Hence, if we introduce $x_{l_0 j}$ in the basis, we would obtain an objective function value of problem $(LASP)$ which is not higher than $\widehat{f}$. ■

Chapter 4
Separable Programming: A Dynamic Programming Approach

This chapter is devoted to the dynamic programming approach for solving (separable) programs for both discrete and continuous cases. The Lagrange multiplier method for reducing the dimensionality of the problem is discussed, and some separable and other models are reviewed at the end of the chapter.

4.1 The Discrete Case

Consider the problem

$$\max \left\{ f(\mathbf{x}) = \sum_{j=1}^{n} f_j(x_j) \right\} \tag{4.1}$$

subject to

$$g_i(\mathbf{x}) \leq \beta_i, \quad i = 1, \dots, m \tag{4.2}$$

$$x_j \geq 0, \quad j = 1, \dots, n. \tag{4.3}$$

First of all, consider the simplest case when a single linear inequality constraint is involved

$$\max \left\{ f(\mathbf{x}) = \sum_{j=1}^{n} f_j(x_j) \right\} \tag{4.4}$$

subject to

$$\sum_{j=1}^{n} a_j x_j \leq b_n, \quad a_j > 0, j = 1, \dots, n \tag{4.5}$$

$$x_j \geq 0, j = 1, \dots, n. \tag{4.6}$$

© The Author(s), under exclusive license to Springer Nature Switzerland AG 2021

S. M. Stefanov, *Separable Optimization*, Springer Optimization and Its Applications 177,
https://doi.org/10.1007/978-3-030-78401-0_4

Suppose that x_j are integers and a_j, $j = 1, \ldots, n$; b_n are also integers.

If $f_j(x_j)$, $j = 1, \ldots, n$ are convex/concave, the Lagrange multiplier method can be applied for solving problem (4.4)–(4.6). However, if this is not the case, that is, if f has several local maxima, this method is not applicable because it finds only one local maximum.

Denote by f^* the maximum of $f(\mathbf{x})$ subject to (4.5)–(4.6).

Let x_n be selected and fixed among x_j, $j = 1, \ldots, n$. The maximum of f with respect to the remaining variables $x_1, \ldots, x_{n-1}$ depends on x_n because, due to separability, we have

$$\max_{x_1,\ldots,x_{n-1}} f(\mathbf{x}) = f_n(x_n) + \max_{x_1,\ldots,x_{n-1}} \sum_{j=1}^{n-1} f_j(x_j). \tag{4.7}$$

Since $f_n(x_n)$ depends only on x_n and does not depend on $x_1, \ldots, x_{n-1}$, we can write $f_n(x_n)$ before "max" in the right-hand side of (4.7).

For an arbitrary nonnegative integer value of x_n, the variables $x_1, \ldots, x_{n-1}$ must satisfy the constraint

$$\sum_{j=1}^{n-1} a_j x_j \le b_n - a_n x_n, \quad x_j \ge 0 \text{ and integer, } j = 1, \ldots, n-1.$$

Therefore

$$\Phi_{n-1}(b_{n-1}) \stackrel{\text{def}}{=} \max_{x_1,\ldots,x_{n-1}} \sum_{j=1}^{n-1} f_j(x_j), \tag{4.8}$$

where $b_{n-1} = b_n - a_n x_n$ depends on x_n. Then

$$f^* = \Phi_n(b_n) = \max_{x_1,\ldots,x_n} \sum_{j=1}^{n} f_j(x_j) = \max_{x_n \ge 0}[f_n(x_n) + \Phi_{n-1}(b_{n-1})]. \tag{4.9}$$

Here x_n assumes the values 0, 1, ..., $\left[\frac{b_n}{a_n}\right]$, where $\left[\frac{b_n}{a_n}\right]$ denotes the largest integer less than or equal to $\frac{b_n}{a_n}$. The equality $x_n = 1 + \left[\frac{b_n}{a_n}\right]$ is impossible because the constraint (4.5) with $x_j \ge 0$, $a_j > 0$ would be violated in this case.

The relation (4.9) can also be derived as follows:

$$\Phi_n(b_n) = \max_{\substack{\sum_{j=1}^{n} a_j x_j \le b_n \\ x_j \ge 0}} [f_n(x_n) + f_{n-1}(x_{n-1}) + \ldots + f_1(x_1)]$$

$$= \max_{0 \le x_n \le \left[\frac{b_n}{a_n}\right]} \left\{ \max_{\substack{\sum_{j=1}^{n-1} a_j x_j \le b_{n-1} \\ x_j \ge 0}} [f_n(x_n) + f_{n-1}(x_{n-1}) + \ldots + f_1(x_1)] \right\}$$

$$= \max_{0 \le x_n \le \left[\frac{b_n}{a_n}\right]} \left\{ f_n(x_n) + \max_{\substack{\sum_{j=1}^{n-1} a_j x_j \le b_{n-1} \\ x_j \ge 0}} [f_{n-1}(x_{n-1}) + \ldots + f_1(x_1)] \right\}$$

$$= \max_{0 \le x_n \le \left[\frac{b_n}{a_n}\right]} \Big[f_n(x_n) + \Phi_{n-1}(b_{n-1}) \Big],$$

where $b_{n-1} = b_n - a_n x_n$ and $\Phi_{n-1}(b_{n-1})$ is defined by (4.8).

If we know the function $\Phi_{n-1}(b_{n-1})$, then the problem of calculating f^* is reduced to maximizing the function $f_n(x_n) + \Phi_{n-1}(b_{n-1})$ which is a single-variable function of x_n.

That is why it is necessary to determine $\Phi_{n-1}(b_{n-1})$. For this reason, we have to find $x_1, \ldots, x_{n-1}$ which solve the problem

$$\max \left\{ \overline{f}(\mathbf{x}) = \sum_{j=1}^{n-1} f_j(x_j) \right\} \tag{4.10}$$

subject to

$$\sum_{j=1}^{n-1} a_j x_j \le b_{n-1}, \quad a_j > 0,\, j = 1, \ldots, n-1 \tag{4.11}$$

$$x_j \ge 0 \text{ and integer, } j = 1, \ldots, n-1. \tag{4.12}$$

This problem is similar to the original problem (4.4)–(4.6) but the number of variables is $n - 1$ instead of n.

Since

$$\Phi_{n-1}(b_{n-1}) \stackrel{\text{def}}{=} \max_{x_1,\ldots,x_{n-1}} \sum_{j=1}^{n-1} f_j(x_j)$$

subject to (4.11)–(4.12), using the same reasoning we get

$$\Phi_{n-1}(b_{n-1}) = \max_{x_{n-1} \ge 0} [f_{n-1}(x_{n-1}) + \Phi_{n-2}(b_{n-2})] \tag{4.13}$$

subject to

$$\sum_{j=1}^{n-2} a_j x_j \le b_{n-2}, \quad x_j \text{ nonnegative integers}, \; j = 1, \ldots, n-2, \tag{4.14}$$

where

$$\Phi_{n-2}(b_{n-2}) \overset{\text{def}}{=} \max_{x_1,\ldots,x_{n-2}} \sum_{j=1}^{n-2} f_j(x_j), \tag{4.15}$$

$$b_{n-2} = b_{n-1} - a_{n-1}x_{n-1}, \quad x_{n-1} \in \left\{0, 1, \ldots, \left[\frac{b_{n-1}}{a_{n-1}}\right]\right\}, \tag{4.16}$$

and so on, we determine Φ_{n-2} via Φ_{n-3}, Φ_{n-3} via Φ_{n-4}, . . ., until we find the function

$$\Phi_1(b_1) \overset{\text{def}}{=} \max f_1(x_1) \tag{4.17}$$

subject to

$$x_1 \in \left\{0, 1, \ldots, \left[\frac{b_1}{a_1}\right]\right\}, \quad b_1 = b_2 - a_2x_2, \quad 0 \le b_1 \le b_n. \tag{4.18}$$

Thus, the original n-variable problem is decomposed into a series of n similar single-variable problems (into n *stages* each of which comprises a single-variable subproblem). This is the advantage of the dynamic programming approach: to solve single-variable optimization subproblems instead of an n-variable problem.

This analysis shows the following. In order to obtain f^*, we have to construct the sequence of functions

$$\Phi_k(b_k) = \max_{x_1,\ldots,x_k} \sum_{j=1}^{k} f_j(x_j), \quad k = n, n-1, \ldots, 1, \tag{4.19}$$

where

$$b_k = b_n - \sum_{j=k+1}^{n} a_j x_j, \quad k = 1, \ldots, n-1, \tag{4.20}$$

such that $\Phi_1(b_1)$ is determined by (4.17), and all other Φ_k, via the recursion

$$\Phi_k(b_k) = \max_{x_k}[f_k(x_k) + \Phi_{k-1}(b_{k-1})]. \tag{4.21}$$

In particular, $\Phi_n(b_n) = \max f(\mathbf{x})$ subject to (4.5)–(4.6).

After that, we find f^* by using (4.9).

Variable x_k is called the *state* of the system at stage k and this state is connected with the information that links the stages together, such that optimal decisions for the remaining stages can be made regardless of how the decisions for the previous stages were made.

The relation (4.21) is called the *functional equation* (or *recurrence relation, forward recursion*) for the separable programming problem.

Principle of Optimality (*Richard Bellman*)

An optimal policy has the property that whatever the initial state and initial decisions are, the remaining decisions must constitute an optimal policy with regard to the state resulting from the first decision.

This principle means that future decisions for the remaining stages will constitute an optimal policy regardless of the policy adopted in previous stages.

We have calculated f^* but we do not know the point $\mathbf{x}^* = (x_1^*, \ldots, x_n^*)$ at which the function f attains its maximum value f^*. Since

$$b_k = b_n - \sum_{j=k+1}^{n} a_j x_j \quad \text{and} \quad 0 \leq \sum_{j=k+1}^{n} a_j x_j \leq b_n$$

($a_j > 0, x_j \geq 0$ for all $j = 1, \ldots, n$), then $0 \leq b_k \leq b_n$ for each $k = 1, \ldots, n-1$. Assume that b_n and a_j, $j = 1, \ldots, n$ are integers. This assumption and the assumption that x_j, $j = 1, \ldots, n$ are integers imply that all b_k are integers.

Solve the single-variable problem (4.17)–(4.18) of x_1 where x_1 assumes the integer values between 0 and $\left[\frac{b_1}{a_1}\right]$, $0 \leq b_1 \leq b_n$. Denote by $\widehat{x}_1(b_1)$ the optimal solution to (4.17)–(4.18). Results can be written as in Table 4.1.

Table 4.1 Results for problem (4.17)–(4.18), Step One

b_1	0	1	...	b_n
Optimum solution $\widehat{x}_1$	$\widehat{x}_1(0)$	$\widehat{x}_1(1)$	...	$\widehat{x}_1(b_n)$
$\max f_1(x_1)$	$\Phi_1(0)$	$\Phi_1(1)$	...	$\Phi_1(b_n)$

After that solve the problem

$$\Phi_2(b_2) = \max[f_2(x_2) + \Phi_1(b_1)]$$

subject to

$$0 \leq x_2 \leq \left[\frac{b_2}{a_2}\right], x_2 \text{ integer}$$

where $b_1 = b_2 - a_2 x_2$, $0 \leq b_2 < b_n$. Results are written in Table 4.2.

Table 4.2 Results for problem (4.17)–(4.18), Step Two

b_2	0	1	...	b_n
Optimum solution $\widehat{x}_2$	$\widehat{x}_2(0)$	$\widehat{x}_2(1)$	...	$\widehat{x}_2(b_n)$
$\max[f_2 + \Phi_1]$	$\Phi_2(0)$	$\Phi_2(1)$	...	$\Phi_2(b_n)$

Repeat analogous procedures for the remaining variables $x_3, \ldots, x_{n-1}$; corresponding results $\widehat{x}_k(b_k)$ and $\Phi_k(b_k)$ are written in subsequent tables.

From the last problem, we have to find

$$f^* = \max[f_n(x_n) + \Phi_{n-1}(b_{n-1})].$$

Since b_n is given, the table for $\widehat{x}_n$ consists of a single column. Among $\widehat{x}_k$, $k = 1, \ldots, n$ we have to choose one or more n–tuples $x_1^*, \ldots, x_n^*$ by using previously obtained tables.

We set $x_n^* = \widehat{x}_n$ because f^* is attained for $x_n = \widehat{x}_n$ regardless of values of $\widehat{x}_{n-1}, \ldots, \widehat{x}_1$. If there are several values of $\widehat{x}_n$, postoptimal analysis is made for each of these values.

Having known x_n^*, we calculate $b_{n-1}^* = b_n - a_n x_n^*$. Set $x_{n-1}^* = \widehat{x}_{n-1}(b_{n-1}^*)$.

Since we know x_n^* and x_{n-1}^*, we calculate $b_{n-2}^* = b_n - a_n x_n^* - a_{n-1}x_{n-1}^* = b_{n-1}^* - a_{n-1}x_{n-1}^*$ and set $x_{n-2}^* = \widehat{x}_{n-2}(b_{n-2}^*)$.

Using the same approach, we calculate x_k^*, $k = n-3, \ldots, 1$.

Each stage consists in determining the corresponding x_k^*, $k = n, \ldots, 1$; the state of the process is characterized by b_k (*state parameter*), and results are characterized by Φ_k (*state function*).

Compare the usual explicit enumeration and dynamic programming approach with respect to the number of arithmetic operations required. For the sake of simplicity, assume that $a_1 = a_2 = \cdots = a_n = 1$.

For the item-by-item examination, the number of possible solutions subject to integrality assumptions for the variables is equal to the number of ways we can allocate b_n states into n stages, that is,

$$C_{n+b_n-1}^{b_n} = \frac{(n+b_n-1)!}{b_n!(n-1)!}.$$

For example, if $n = 5$, $b_n = 25$, we have

$$C_{29}^{25} = \frac{29!}{25!4!} = 23751.$$

In the dynamic programming approach, in order to calculate $\Phi_k(b_k)$ for a fixed value of b_k we need $b_k + 1$ operations. Therefore, to fill in the table we have to perform

$$\sum_{b_k=0}^{b_n}(b_k + 1) = \frac{(b_k+1)(b_k+2)}{2!}$$

operations. Calculation of $\Phi_n(b_n)$ requires $b_n + 1$ operations. Hence, to find $\Phi_1, \ldots, \Phi_{n-1}$ and Φ_n, we need

$$N = (n-1)\frac{(b_n+1)(b_n+2)}{2!} + (b_n+1) = \frac{(b_n+1)[(n-1)(b_n+2)+2]}{2!}$$

operations. For $n = 5$, $b_n = 25$, this yields $N = 1430$ operations which is much less than $C_{29}^{25} = 23751$.

Problem (4.4)–(4.6) we have considered is known as the *knapsack problem*, the *cargo-loading problem*, or the *fly-away kit problem*. It can be formulated as follows.

There is a set of n items $V = \{1, \ldots, n\}$ and a knapsack with capacity b_n where b_n is a positive integer. Each item $j \in V$ has an associated value function $f_j(x_j)$ (usually $f_j(x_j)$ is linear, that is, $f_j(x_j) = c_j x_j$ where c_j is the value of a unit of item j) and a weight a_j, a_j are positive integers, $j = 1, \ldots, n$.

The problem is to find a maximum value subset of the set of items whose total weight does not exceed the knapsack capacity b_n.

Items j, $j = 1, \ldots, n$ can be interpreted as stages, x_j's as alternatives at stage j, b_j's as states at stage j (b_j's are the total weights for stages $j, j+1, \ldots, n$ combined), and function $\Phi_j(b_j)$ as the maximum return for the stages (items) $j, j+1, \ldots, n$ provided that the state is b_j.

Then $x_j = 0, 1, \ldots, \left[\frac{b_n}{a_j}\right]$.

The function $f_j(x_j)$ can be considered as the return from the jth activity due to respective allocations of x_j.

The *forward recursion* is

$$\Phi_k(b_k) = \max_{\substack{x_k = 0, \ldots, \left[\frac{b_n}{a_k}\right] \\ b_k = 0, \ldots, b_n}} [f_k(x_k) + \Phi_{k-1}(b_{k-1})]$$

$$= \max_{\substack{x_k = 0, \ldots, \left[\frac{b_n}{a_k}\right] \\ b_k = 0, \ldots, b_n}} [f_k(x_k) + \Phi_{k-1}(b_k - a_k x_k)], \quad k = 1, 2, \ldots, n,$$

$\Phi_0(b_0) \stackrel{\text{def}}{=} 0$ (knapsack problem with zero items).

The problem of the form (4.4)–(4.6) also describes the so-called *multi-item inventory control model with storage limitations*, considered in Section 4.8, and the *multi-facility location problem* in Chapter Thirteen. In Section 4.8, other models described by separable problems are also discussed.

4.2 Forward and Backward Recursion. Successive Decision-Making

Sometimes it is possible to calculate variables in dynamic programming in arbitrary order. However, in many problems variables must be determined in a given order during the time interval. Such problems are known as problems of *successive decision-making*.

In principle, there are two approaches to solve the latter class of problems.

1. *Forward approach*. The first step of the dynamic programming approach corresponds to the first (with respect to time period) decision made.

2. *Backward approach*. The last step of the dynamic programming approach corresponds to the first decision made.

Work Force Size Problem. The manager must determine the optimal number of laborers for each of n months provided that the work that must be done for each month is known.

Assume that the "ideal" number of laborers for the month j is m_j, $j = 1, \dots, n$. The manager can employ exactly m_j laborers in month j.

Assume that the work for month j can also be done by a smaller number of laborers, for example, assigning excessive amounts of work to each. Let x_j be the actual number of laborers employed in month j and let the cost of changing the number of laborers from month $j - 1$ to month j be described by the function $f_j(x_j - x_{j-1})$.

Depending on the sign of the difference $x_j - x_{j-1}$, function $f_j(x_j - x_{j-1})$ gives cost of hiring/firing. Obviously $f_j(0) = 0$ by definition. The difference between the actual number of laborers and m_j gives the cost $g_j(x_j - m_j)$ of maintaining an excess labor force $x_j - m_j$. Evidently $g_j(0) = 0$. Therefore the total costs for month j are $f_j(x_j - x_{j-1}) + g_j(x_j - m_j)$.

Let m_0 be the number of laborers at the beginning of the first month. Then the objective function of the problem under consideration is

$$f(\mathbf{x}) = \sum_{j=1}^{n} [f_j(x_j - x_{j-1}) + g_j(x_j - m_j)], \tag{4.22}$$

where $x_0 = m_0$ and this problem is with fixed initial value x_0.

The recursion is

$$f(\mathbf{x}) = \min_{x_1, \dots, x_n} \left\{ f_1(x_1 - m_0) + g_1(x_1 - m_1) + \sum_{j=2}^{n} [f_j(x_j - x_{j-1}) + g_j(x_j - m_j)] \right\},$$

$$\min_{x_2, \dots, x_n} \sum_{j=2}^{n} [f_j(x_j - x_{j-1}) + g_j(x_j - m_j)]$$

$$= \min_{x_2} \left\{ [f_2(x_2 - x_1) + g_2(x_2 - m_2)] + \min_{x_3,\dots,x_n} \sum_{j=3}^{n} [f_j(x_j - x_{j-1}) + g_j(x_j - m_j)] \right\}.$$

Solve the problem using the backward approach. Denote

$$\Phi_n(\gamma) = \min_{x_n} [f_n(x_n - \gamma) + g_n(x_n - m_n)], \tag{4.23}$$

where $\gamma = x_{n-1}$.

The *backward recursion* is

$$\Phi_k(\gamma) = \min_{x_k} [f_k(x_k - \gamma) + g_k(x_k - m_k) + \Phi_{k+1}(x_k)], \tag{4.24}$$

where $\Phi_k(\gamma)$ gives the minimum costs for months k through n in case that γ laborers are employed.

At the last step, we determine the optimum number x_1^* of workers for the first month provided that at the beginning of the first month there are $x_0 = m_0$ laborers. Having known x_1^*, we successively calculate $x_2^* = \widehat{x}_2(x_1^*)$, $x_3^* = \widehat{x}_3(x_2^*)$, and so on. In this case, the backward approach is more suitable because we know $x_0 = m_0$ only and do not know the number m_{n+1} of laborers for month $n + 1$.

Consider the case when both m_{n+1} and m_0 are given.

Since the final state m_{n+1} of the system is given, it is convenient to use the forward approach.

We have to find integers x_j that *minimize* the function

$$f(\mathbf{x}) = \sum_{j=1}^{n} [f_j(x_j - x_{j-1}) + g_j(x_j - m_j) + f_{n+1}(m_{n+1} - x_n)]. \tag{4.25}$$

This function can be written in the equivalent form

$$f(\mathbf{x}) = \min_{x_n} \left\{ f_{n+1}(m_{n+1} - x_n) + \min_{x_1,\dots,x_{n-1}} \sum_{j=1}^{n} [f_j(x_j - x_{j-1}) + g_j(m_j - x_j)] \right\}.$$

Define a sequence of state functions as follows:

$$\Phi_k(\gamma) = \min_{x_1,\dots,x_k} \left\{ f_{k+1}(\gamma - x_k) + g_{k+1}(\gamma - m_{k+1}) \right.$$
$$\left. + \sum_{j=1}^{k} [f_j(x_j - x_{j-1}) + g_j(m_j - x_j)] \right\}, \quad k = 1, 2, \dots, n,$$

where the minimum is taken over the nonnegative integers $x_1, \dots, x_k$.

Then

$$\Phi_1(\gamma) = \min[f_2(\gamma - x_1) + g_2(\gamma - m_2) + f_1(x_1 - m_0) + g_1(m_1 - x_1)]$$

and the forward recursion is

$$\Phi_k(\gamma) = \min_{x_k}[f_{k+1}(\gamma - x_k) + g_{k+1}(\gamma - m_{k+1}) + \Phi_{k-1}(x_k)]. \tag{4.26}$$

Function $\Phi_k(\gamma)$ gives the minimum costs during the first k months in case that γ laborers are employed in month $k+1$.

At the last step (with $k = n$), we set $\gamma = m_{n+1}$ and obtain the recursion

$$\Phi_n(m_{n+1}) = \min_{x_n}[f_{n+1}(m_{n+1} - x_n) + \Phi_{n-1}(x_n)]. \tag{4.27}$$

From (4.27) we determine the optimum value x_n^*. Using the table of the previous step $n-1$ with $\gamma = x_n^*$, we determine values $x_{n-1}^*, x_{n-2}^*, \ldots, x_1^*$.

SUMMARY.

Given a problem of successive decision-making during n periods of time. If the initial state is given, the problem is solved via the backward dynamic programming approach. If the final state of the system is given, the problem is solved by the forward dynamic programming approach. If both initial and final states are given, the problem can be solved by either the forward or backward approach, and both solutions coincide.

Example 4.1 Consider the work force size model with $n = 4$ periods and $m_0 = 2, m_1 = 2, m_2 = 5, m_3 = 3, m_4 = 1$.

Functions $f_j(x_j)$ and $g_j(x_j)$ are defined as follows:

$$f_j(x_j) = \begin{cases} 10(x_j - x_{j-1}), & \text{if } x_j > x_{j-1}, \\ 7(x_{j-1} - x_j), & \text{if } x_{j-1} > x_j, \end{cases}$$

$$f_j(0) = 0,$$

$$g_j(x_j) = \begin{cases} 8(x_j - m_j), & \text{if } x_j > m_j, \\ 11(m_j - x_j), & \text{if } m_j > x_j, \end{cases}$$

$$g_j(0) = 0.$$

Solution.

Since the initial state $x_0 = m_0 = 2$ is given, we use the backward dynamic programming approach.

The backward recursion is

$$\Phi_k(\gamma) = \min_{x_k}[f_k(x_k - \gamma) + g_k(x_k - m_k) + \Phi_{k+1}(x_k)],$$

where $\gamma = x_{k-1}$,

$$\Phi_4(\gamma) = \min_{x_4}[f_4(x_4 - \gamma) + g_4(x_4 - m_4)].$$

Step One. The recursion is

$$\Phi_4(\gamma) = \min_{x_4}[f_4(x_4 - \gamma) + g_4(x_4 - m_4)] = \min_{x_4} \Psi_4(x_4, \gamma).$$

In order to calculate $\Phi_4(\gamma)$, compute the following values of $\Psi_4(x_4, \gamma)$ and fill them in Table 4.3.

1.

$\gamma = 0, x_4 = 0,\quad \Psi_4(0, 0) = 0 + 11(1 - 0) = 11,$
$\gamma = 0, x_4 = 1,\quad \Psi_4(1, 0) = 10(1 - 0) + 0 = 10,$
$\gamma = 0, x_4 = 2,\quad \Psi_4(2, 0) = 10(2 - 0) + 8(2 - 1) = 28.$

2.

$\gamma = 1, x_4 = 0,\quad \Psi_4(0, 1) = 7(1 - 0) + 11(1 - 0) = 18,$
$\gamma = 1, x_4 = 1,\quad \Psi_4(1, 1) = 0 + 0 = 0.$

3.

$\gamma = 2, x_4 = 0,\quad \Psi_4(0, 2) = 7(2 - 0) + 11(1 - 0) = 25,$
$\gamma = 2, x_4 = 1,\quad \Psi_4(1, 2) = 7(2 - 1) + 0 = 7,$
$\gamma = 2, x_4 = 2,\quad \Psi_4(2, 2) = 7.0 + 8(2 - 1) = 8.$

4.

$\gamma = 3, x_4 = 0,\quad \Psi_4(0, 3) = 7(3 - 0) + 11(1 - 0) = 32,$
$\gamma = 3, x_4 = 1,\quad \Psi_4(1, 3) = 7(3 - 1) + 0 = 14,$
$\gamma = 3, x_4 = 2,\quad \Psi_4(2, 3) = 7(3 - 2) + 8(2 - 1) = 15.$

5.

$\gamma = 4, x_4 = 0,\quad \Psi_4(0, 4) = 7(4 - 0) + 11(1 - 0) = 39,$
$\gamma = 4, x_4 = 1,\quad \Psi_4(1, 4) = 7(4 - 1) + 0 = 21,$
$\gamma = 4, x_4 = 2,\quad \Psi_4(2, 4) = 7(4 - 2) + 8(2 - 1) = 22.$

6.

$\gamma = 5, x_4 = 0,\quad \Psi_4(0, 5) = 7(5 - 0) + 11(1 - 0) = 46,$
$\gamma = 5, x_4 = 1,\quad \Psi_4(1, 5) = 7(5 - 1) + 0 = 28,$
$\gamma = 5, x_4 = 2,\quad \Psi_4(2, 5) = 7(5 - 2) + 8(2 - 1) = 29.$

Table 4.3 Results for Example 4.1, Step One

$\gamma = x_3$	x_4	$\Psi_4(x_4, \gamma)$
0	0	11
	1	10
	2	28
1	0	18
	1	0
2	0	25
	1	7
	2	8

$\gamma = x_3$	x_4	$\Psi_4(x_4, \gamma)$
3	0	32
	1	14
	2	15
4	0	39
	1	21
	2	22
5	0	46
	1	28
	2	29

Write the values of min $\Psi_4(x_4, \gamma)$ in Table 4.4.

Table 4.4 Minimum values for Example 4.1, Step One

γ	x_4	$\Phi_4(\gamma)$
0	1	10
1	1	0
2	1	7
3	1	14
4	1	21
5	1	28

Since $f_j(x_j - x_{j-1})$ and $g_j(x_j - m_j)$ are convex (linear) functions of x_j for each j, then $\Psi_4(x_4, \gamma)$ is also a convex function of x_4 for all γ. Therefore, in order to determine $\Phi_4(\gamma)$, we have to find the first local minimum of $\Psi_4(x_4, \gamma)$ with respect to x_4; according to Theorem 1.36, it will be a global minimum as well. That is why we have calculated values of $\Psi_4(x_4, \gamma)$ only for the region of x_4 containing this minimum value.

Step Two. The backward recursion is

$$\Phi_3(\gamma) = \min_{x_3}[f_3(x_3 - \gamma) + g_3(x_3 - m_3) + \Phi_4(x_3)] = \min_{x_3} \Psi_3(x_3, \gamma)$$

where $\gamma = x_2$.

Similar to Step One, we obtain Tables 4.5 and 4.6.

Table 4.5 Results for Example 4.1, Step Two

$\gamma = x_2$	x_3	$\Psi_3(x_3, \gamma)$
0	0	40
	1	32
	2	38
1	0	50
	1	22
	2	28
2	0	57
	1	29
	2	18
	3	24

$\gamma = x_2$	x_3	$\Psi_3(x_3, \gamma)$
3	0	64
	1	36
	2	25
	3	14
	4	39
4	2	32
	3	21
	4	29
5	2	39
	3	28
	4	36

Table 4.6 Minimum values for Example 4.1, Step Two

$\gamma = x_2$	x_3	x_4	$\Phi_3(\gamma)$
0	1	1	32
1	1	1	22
2	2	1	18
3	3	1	14
4	3	1	21
5	3	1	28

Step Three. Calculate

$$\Phi_2(\gamma) = \min_{x_2}[f_2(x_2 - \gamma) + g_2(x_2 - m_2) + \Phi_3(x_2)] = \min_{x_2} \Psi_2(x_2, \gamma)$$

where $\gamma = x_1$, for all possible values of γ and fill the obtained results in Tables 4.7 and 4.8.

Table 4.7 Results for Example 4.1, Step Three

$\gamma = x_1$	x_2	$\Psi_2(x_2, \gamma)$
0	0	87
	1	76
	2	71
	3	66
	4	72
1	2	61
	3	56
	4	62
2	2	51
	3	46
	4	52

$\gamma = x_1$	x_2	$\Psi_2(x_2, \gamma)$
3	2	58
	3	36
	4	42
4	2	65
	3	43
	4	32
	5	38
5	4	39
	5	28
	6	46

Table 4.8 Minimum values for Example 4.1, Step Three

$\gamma = x_1$	x_2	x_3	x_4	$\Phi_2(\gamma)$
0	3	3	1	66
1	3	3	1	56
2	3	3	1	46
3	3	3	1	36
4	4	3	1	32
5	5	3	1	28

Step Four. Substituting the initial value $\gamma = x_0$ in the recursion

$$\Phi_1(\gamma = x_0) = \min_{x_1}[f_1(x_1 - x_0) + g_1(x_1 - m_1) + \Phi_2(x_1)] = \min_{x_1} \Psi_1(x_1, \gamma),$$

we find $\Phi_1(x_0 = 2)$ and the optimum value $x_1^* = 2$.

Table 4.9 Results for Example 4.1, Step Four

$\gamma = x_0$	x_1	$\Psi_1(x_1, \gamma)$
2	1	74
	2	46
	3	54

Using the values of Table 4.8 with $\gamma = x_1^* = 2$, we calculate

$$x_2^* = 3, \quad x_3^* = 3, \quad x_4^* = 1.$$

The minimum total cost is $\Phi_1(x_1^*) = \min \Psi_1(x_1, \gamma) = 46$ (Table 4.9).

4.3 The Continuous Case

The dynamic programming approach is very suitable for discrete programming problems. However, this approach can be adapted for separable programs with continuous variables. For this purpose, continuous variables are "quantized", and the step size of quantization (discretization) depends on the particular problem.

We can solve the "continuous" separable problem (4.4), (4.5), $x_j \geq 0,\ j = 1, \ldots, n$ as in the discrete case: find

$$\Phi_1(b_1) = \max f_1(x_1) \tag{4.28}$$

subject to

$$0 \leq b_1 \leq b_n \tag{4.29}$$

and after that

$$\begin{aligned}\Phi_k(b_k) &= \max_{x_k}[f_k(x_k) + \Phi_{k-1}(b_{k-1})] \\ &\equiv \max_{x_k}[f_k(x_k) + \Phi_{k-1}(b_k - a_k x_k)]\end{aligned} \tag{4.30}$$

subject to

$$0 \leq b_k \leq b_n, \quad x_k \geq 0, k = 2, \ldots, n. \tag{4.31}$$

From the last $\Phi_n(b_n)$, we obtain f^*. Equivalently,

$$\Phi_k(b_k) = \max \sum_{j=1}^{k} f_j(x_j) \tag{4.32}$$

subject to

$$\sum_{j=1}^{k} a_j x_j \leq b_k, \quad k = 1, \ldots, n. \tag{4.33}$$

The forward recursion is of the form

$$\begin{aligned}\Phi_k(b_k) = &\max_{0 \leq x_k \leq \frac{b_k}{a_k}} [f_k(x_k) + \Phi_{k-1}(b_{k-1})] \\ \equiv &\max_{0 \leq x_k \leq \frac{b_k}{a_k}} [f_k(x_k) + \Phi_{k-1}(b_k - a_k x_k)], \\ &k = 2, \ldots, n.\end{aligned} \tag{4.34}$$

This case differs from the discrete one in that variables $x_k, k = 1, \ldots, n$ assume values between 0 and $\frac{b_k}{a_k}$ instead of the *integer* values between 0 and $\left[\frac{b_k}{a_k}\right]$. Another difference consists in the type of methods by which we find x_k. In the discrete case, we can find any function Φ_k via item-by-item examination of the values of $f_k(x_k) + \Phi_{k-1}(b_{k-1})$, whereas analysis of the equation

$$\frac{d}{dx_k}[f_k + \Phi_{k-1}] = 0 \tag{4.35}$$

would be more suitable in the continuous case.

Denote

$$\Psi_k(x_k; b_k) = f_k(x_k) + \Phi_{k-1}(b_{k-1}) = f_k(x_k) + \Phi_{k-1}(b_k - a_k x_k). \tag{4.36}$$

Then

$$\Phi_k(b_k) = \max_{0 \le x_k \le \frac{b_k}{a_k}} \Psi_k(x_k; b_k). \tag{4.37}$$

If $\Phi_{k-1}(b_{k-1})$ is known, then values of $\Psi_k(x_k; b_k)$ can be calculated. To find $\Phi_k(b_k)$, we have to find all local maxima of Ψ_k and take the global one. For this purpose, values of $\Psi_k(x_k; b_k)$ are tabulated.

Functions $\Phi_k(b_k)$ must be tabulated for all feasible values because we do not know in advance which value we will need for calculating $\Phi_{k+1}(b_{k+1})$.

If all $f_j(x_j)$, $j = 1, \ldots, n$ are convex or concave functions, then calculations are simpler because if $f_j(x_j)$ are convex (concave), then $\Phi_k(b_k)$ is also convex (concave).

Case 1. Let $f_j(x_j)$, $j = 1, \ldots, n$ be convex. Since $\Phi_{k-1}(b_{k-1})$ is convex, then $\Phi_{k-1}(b_k - a_k x_k)$ is convex with respect to x_k. Therefore $\Psi_k(x_k; b_k) = f_k(x_k) + \Phi_{k-1}(b_k - a_k x_k)$ is also a convex function of x_k. We have to find $\max \Psi_k(x_k; b_k)$ with fixed b_k. Since maximum of a convex function on a compact set is attained at an extreme point (in our case, $x_k = 0$ and $x_k = \frac{b_k}{a_k}$ are such points), then we have to check only points $x_k = 0$ and $x_k = \frac{b_k}{a_k}$ in order to find

$$\Phi_k(b_k) = \max_{0 \le x_k \le \frac{b_k}{a_k}} \Psi_k(x_k; b_k).$$

Case 2. Let $f_j(x_j)$, $j = 1, \ldots, n$ be concave. Then $\Psi_k(x_k; b_k)$ is concave for fixed b_k. Hence, each local maximum is also a global one. To find

$$\Phi_k(b_k) = \max_{0 \le x_k \le \frac{b_k}{a_k}} \Psi_k(x_k; b_k)$$

we can use, for example, gradient-type method (for the gradient-type method of the steepest descent for a *minimization* problem, see Chapter Twelve).

4.4 Models Involving Two Types of Constraints. Problem of Dimensionality in Dynamic Programming

Up to now we have considered models with a state, represented by a single variable at any stage. For example, in the knapsack problem discussed in Section 4.1, the weight (capacity) was the only restriction on the cargo. However, there might be another restriction, for example, volume of the cargo. Such a model is two-dimensional and can be described as follows:

$$\max\left\{f(\mathbf{x}) = \sum_{j=1}^{n} f_j(x_j)\right\} \tag{4.38}$$

subject to

$$\sum_{j=1}^{n} a_{1j}x_j \le b_{1n} \tag{4.39}$$

$$\sum_{j=1}^{n} a_{2j}x_j \le b_{2n} \tag{4.40}$$

$$x_j \text{ nonnegative integers}, j = 1, \ldots, n. \tag{4.41}$$

Assume that $a_{1j}, a_{2j}, b_{1n}, b_{2n}$ are positive integers. Define

$$\Phi_k(b_{1k}, b_{2k}) = \max_{x_1,\ldots,x_k} \sum_{j=1}^{k} f_j(x_j) \tag{4.42}$$

subject to

$$\sum_{j=1}^{k} a_{1j}x_j \le b_{1k} \tag{4.43}$$

$$\sum_{j=1}^{k} a_{2j}x_j \le b_{2k} \tag{4.44}$$

$$x_j \ge 0 \text{ and integer}, \ j = 1, \ldots, k. \tag{4.45}$$

In this case, $m = 2$ in constraint (4.2) of problem (4.1)–(4.3). Here

$$b_{1k} = b_{1n} - \sum_{j=k+1}^{n} a_{1j}x_j, \quad b_{2k} = b_{2n} - \sum_{j=k+1}^{n} a_{2j}x_j,$$

and all x_k satisfy

$$0 \le x_k \le \left[\frac{b_{1k}}{a_{1k}}\right], \quad 0 \le x_k \le \left[\frac{b_{2k}}{a_{2k}}\right], \quad 0 \le b_{ik} \le b_{in}, i = 1, 2,$$

that is,

$$0 \le x_k \le \Delta_k \stackrel{\text{def}}{=} \min\left\{\left[\frac{b_{1k}}{a_{1k}}\right], \left[\frac{b_{2k}}{a_{2k}}\right]\right\}.$$

To determine $\Phi_1(b_1) = \max_{0 \le x_1 \le \Delta_1} f_1(x_1)$, we have to calculate

$$\Delta_1 = \min\left\{\left[\frac{b_{11}}{a_{11}}\right], \left[\frac{b_{21}}{a_{21}}\right]\right\}.$$

After that, using the same reasoning at each stage, we get the following forward recursion:

$$\begin{aligned} \Phi_k(b_{1k}, b_{2k}) &= \max_{0 \le x_k \le \Delta_k} [f_k(x_k) + \Phi_{k-1}(b_{1,k-1}, b_{2,k-1})] \\ &= \max_{0 \le x_k \le \Delta_k} [f_k(x_k) + \Phi_{k-1}(b_{1k} - a_{1k}x_k, b_{2k} - a_{2k}x_k)], \end{aligned} \tag{4.46}$$

where

$$\Delta_k = \min\left\{\left[\frac{b_{1k}}{a_{1k}}\right], \left[\frac{b_{2k}}{a_{2k}}\right]\right\}, \quad 0 \le b_{1k} \le b_{1n}, \quad 0 \le b_{2k} \le b_{2n}, \tag{4.47}$$

and determine the optimal solution $\widehat{x}_k(b_{1k}, b_{2k})$. At stage n, we get $\Phi_n(b_{1n}, b_{2n}) = f^*$, and $\widehat{x}_n(b_{1n}, b_{2n}) = x_n^*$.

Optimum values $x_{n-1}^*, \ldots, x_1^*$ of the remaining variables can be obtained by using tables of previous stages and the relations

$$\begin{aligned} x_{n-1}^* &= \widehat{x}_{n-1}(b_{1n} - a_{1n}x_n^*, b_{2n} - a_{2n}x_n^*), \\ x_{n-2}^* &= \widehat{x}_{n-2}(b_{1n} - a_{1n}x_n^* - a_{1,n-1}x_{n-1}^*, b_{2n} - a_{2n}x_n^* - a_{2,n-1}x_{n-1}^*), \\ &\vdots \\ x_{n-k}^* &= \widehat{x}_{n-k}(b_{1,n-k}^*, b_{2,n-k}^*), \end{aligned} \tag{4.48}$$

and so on.

In this case, $\Phi_k(b_{1k}, b_{2k})$, and $x_k(b_{1k}, b_{2k})$ are functions in two variables. All information about Φ_k and $\widehat{x}_k$ is written in Table 4.10.

The number N of values that must be stored in a table of this form for problem (4.1)–(4.3) is $N = 3\prod_{i=1}^{m} b_{in}$, where m is the number of constraints of the form (4.39)–(4.40) in (4.2). If $m = 2$, $b_{1n} = b_{2n} = 10$, we have to tabulate $\Phi_k(b_{1k}, b_{2k})$ in 100 points (Table 4.10), and $N = 300$. If $m = 3$, $b_{in} = 10^2, i = 1, 2, 3$, the corresponding function $\Phi_k(b_{1k}, b_{2k}, b_{3k})$ must be tabulated in 10^6 points and $N = 3.10^6$.

Table 4.10 Two-dimensional dynamic program

b_{1k} / b_{2k}	0	1	...	b_{1k}	...	b_{1n}
0						
1						
⋮						
b_{2k}				$\Phi_k, \widehat{x}_k, \Delta_k$		
⋮						
b_{2n}						

When $m = 4$, $b_{in} = 10^2$, $i = 1, 2, 3, 4$, then Φ_k must be tabulated in 10^8 points and $N = 3.10^8$. Moreover, methods for finding $\widehat{x}_k$ and Δ_k become more complex. Thus, the more the state variables are, the more computations at each stage must be done. Without transforming the original problem, in practice it is very difficult to solve a problem with more than three state parameters b_{in} by the dynamic programming approach because of the enormous number of computations. Because of this reason, Richard Bellman has used the phrase *curse of dimensionality*.

4.5 Two-dimensional Allocation Problem

Consider a version of problem (4.38)–(4.41) in which two types of resources are involved, and let the return $f_j(x_j, y_j)$ be a function of both resources. The corresponding program is

$$\max \left\{ f(\mathbf{x}, \mathbf{y}) = \sum_{j=1}^{n} f_j(x_j, y_j) \right\} \tag{4.49}$$

subject to

$$\sum_{j=1}^{n} a_{1j} x_j \le b_{1n} \tag{4.50}$$

$$\sum_{j=1}^{n} a_{2j} y_j \le b_{2n} \tag{4.51}$$

$$x_j, y_j \text{ nonnegative integers}, j = 1, \ldots, n. \tag{4.52}$$

The state function is defined as follows:

$$\Phi_k(b_{1k}, b_{2k}) = \max_{\substack{x_1, \ldots, x_k \\ y_1, \ldots, y_k}} \sum_{j=1}^{k} f_j(x_j, y_j) \tag{4.53}$$

subject to

$$\sum_{j=1}^{k} a_{1j} x_j \le b_{1k} \tag{4.54}$$

$$\sum_{j=1}^{k} a_{2j} y_j \le b_{2k} \tag{4.55}$$

$$x_j, y_j \text{ nonnegative integers}, j = 1, \ldots, k. \tag{4.56}$$

The forward recursion is

$$\begin{aligned} \Phi_k(b_{1k}, b_{2k}) &= \max_{x_k, y_k} [f_k(x_k, y_k) + \Phi_{k-1}(b_{1,k-1}, b_{2,k-1})] \\ &= \max_{0 \le x_k \le \left[\frac{b_{1k}}{a_{1k}}\right]} \max_{0 \le y_k \le \left[\frac{b_{2k}}{a_{2k}}\right]} \Phi_k(b_{1k}, b_{2k}) \\ &= \max_{x_k, y_k} [f_k(x_k, y_k) + \Phi_{k-1}(b_{1k} - a_{1k} x_k, b_{2k} - a_{2k} y_k)], \\ & k = 2, \ldots, n, \end{aligned} \tag{4.57}$$

where

$$b_{1,k-1} = b_{1k} - a_{1k} x_k, \quad b_{2,k-1} = b_{2k} - a_{2k} y_k, \quad 0 \le b_{ik} \le b_{in}, i = 1, 2 \tag{4.58}$$

and the maximum with respect to x_k and y_k is found via item-by-item examination of all possible values.

In particular,

$$f^* = \max_{x_n, y_n} [f_n(x_n, y_n) + \Phi_{n-1}(b_{1,n-1}, b_{2,n-1})] \tag{4.59}$$

where

$$\Phi_{n-1}(b_{1,n-1}, b_{2,n-1}) = \max_{\substack{x_1, \ldots, x_{n-1} \\ y_1, \ldots, y_{n-1}}} \sum_{j=1}^{n-1} f_j(x_j, y_j). \tag{4.60}$$

The problem of dimensionality is also important in this case. For example, if both x_k and y_k can assume 100 different values, then, in order to find $\Phi_k(b_{1k}, b_{2k})$ with fixed b_{1k} and b_{2k}, we have to examine item-by-item 10^4 values of the function

$$\begin{aligned}\Psi_k(x_k, y_k; b_{1k}, b_{2k}) &= f_k(x_k, y_k) + \Phi_{k-1}(b_{1,k-1}, b_{2,k-1}) \\ &= f_k(x_k, y_k) + \Phi_{k-1}(b_{1k} - a_{1k}x_k, b_{2k} - a_{2k}y_k).\end{aligned} \tag{4.61}$$

We have considered separable programs subject to *linear* inequality constraints because of the simplicity of formulas, estimations, and intervals for $x_k, b_k, k = 1, \ldots, n-1$, etc. When the inequality constraints are defined by separable *non-linear* functions, some difficulties concerning description and estimation of b_k, Φ_k and $\widehat{x}_k$ arise.

4.6 Application of Lagrange Multiplier Method for Reducing the Dimensionality of the Problem

Lagrange multiplier method can be employed for reducing the dimensionality of problems solved by dynamic programming methods.

Consider a separable problem subject to equality constraints:

$$\max \left\{ f(\mathbf{x}) = \sum_{j=1}^{n} f_j(x_j) \right\} \tag{4.62}$$

subject to

$$\sum_{j=1}^{n} a_{1j}x_j = b_{1n} \tag{4.63}$$

$$\sum_{j=1}^{n} a_{2j}x_j = b_{2n} \tag{4.64}$$

$$x_j \geq 0, \quad j = 1, \ldots, n. \tag{4.65}$$

Transform this problem into a problem with a single constraint

$$\max_{x_1,\ldots,x_n} \left\{ h(\mathbf{x}) = \sum_{j=1}^{n} f_j(x_j) - \lambda \sum_{j=1}^{n} a_{2j}x_j \right\} \tag{4.66}$$

subject to

$$\sum_{j=1}^{n} a_{1j}x_j = b_{1n}, \quad x_j \geq 0, j = 1, \ldots, n. \tag{4.67}$$

Problem (4.66)–(4.67) has a single state parameter λ, the Lagrange multiplier associated with the constraint (4.64). The value of λ is not known in advance. It is varied until the constraint (4.64) is satisfied.

The optimal solution to problem (4.66)–(4.67) depends on λ:

$$x_j^* = x_j^*(\lambda),\ j = 1, \ldots, n.$$

If $x_j^*(\lambda)$ satisfies (4.64), then $x_j^*(\lambda)$ is an optimal solution to problem (4.62)–(4.65) as well. Otherwise, if $\sum_{j=1}^n a_{2j} x_j^*(\lambda) > b_{2n}$, then λ must be increased (price of the second resource must be increased because we would use more of this resource than available), else if $\sum_{j=1}^n a_{2j} x_j^*(\lambda) < b_{2n}$, then λ must be decreased.

To find λ, the method of successive approximations is used. If two optimal solutions $\mathbf{x}_1^*(\lambda_1)$, $\mathbf{x}_2^*(\lambda_2)$ associated with the values λ_1 and λ_2 of λ, respectively, have been tried, then using linear interpolation, we determine

$$\lambda_3 = \frac{\lambda_2 - \lambda_1}{b_{2n}^2 - b_{2n}^1}(b_{2n} - b_{2n}^1) + \lambda_1,$$

where $b_{2n}^1 = \sum_{j=1}^n a_{2j} x_j^*(\lambda_1)$, $b_{2n}^2 = \sum_{j=1}^n a_{2j} x_j^*(\lambda_2)$.

Having known λ_1, λ_2, and λ_3, a quadratic interpolation can be used to determine λ_4, and so on.

If the optimal solution $\mathbf{x}^*$ to problem (4.66)–(4.67) satisfies (4.64), then $\mathbf{x}^*$ is an optimal solution to problem (4.62)–(4.65) as well.

Denote by $\mathbf{x}^0$ an optimal solution to problem (4.62)–(4.65). Because $\mathbf{x}^*$ is an optimal solution to problem (4.66)–(4.67), then

$$h(\mathbf{x}^*) = f(\mathbf{x}^*) - \lambda \sum_{j=1}^n a_{2j} x_j^* \geq f(\mathbf{x}^0) - \lambda \sum_{j=1}^n a_{2j} x_j^0 = h(\mathbf{x}^0).$$

Since $\sum_{j=1}^n a_{2j} x_j^* = \sum_{j=1}^n a_{2j} x_j^0$ by the assumption ($\mathbf{x}^*$ also satisfies (4.64)) then $f(\mathbf{x}^*) \geq f(\mathbf{x}^0)$.

However, $\mathbf{x}^0$ is an optimal solution to problem (4.62)–(4.65) and $\mathbf{x}^*$ satisfies (4.64) and (4.67) = (4.63) $\cup$ (4.65). Then $f(\mathbf{x}^0) \geq f(\mathbf{x}^*)$. From the two inequalities it follows that $f(\mathbf{x}^*) = f(\mathbf{x}^0)$, that is, $\mathbf{x}^*$ is also an optimal solution to problem (4.62)–(4.65).

The forward recursion for problem (4.66)–(4.67) is

$$\begin{aligned}\Phi_k(b_k) &= \max_{0 \leq x_k \leq \frac{b_{1k}}{a_{1k}}} [f_k(x_k) - \lambda a_{2k} x_k + \Phi_{k-1}(b_{1,k-1})] \\ &= \max_{0 \leq x_k \leq \frac{b_{1k}}{a_{1k}}} [f_k(x_k) - \lambda a_{2k} x_k + \Phi_{k-1}(b_{1k} - a_{1k} x_k)].\end{aligned}$$

It can be shown that $\sum_{j=1}^{n} a_{2j}x_j(\lambda)$ decreases monotonically as λ increases from 0 to $+\infty$.

Denote $w(\lambda) = \sum_{j=1}^{n} a_{2j}x_j^*(\lambda)$, and let $\mu > \lambda > 0$. Then, using the optimality of $x_j^*(\lambda)$ for $h(\mathbf{x})$ (4.66), we get

$$\sum_{j=1}^{n} f_j(x_j^*(\lambda)) - \lambda w(\lambda) \geq \sum_{j=1}^{n} f_j(x_j^*(\mu)) - \lambda w(\mu),$$

$$\sum_{j=1}^{n} f_j(x_j^*(\mu)) - \mu w(\mu) \geq \sum_{j=1}^{n} f_j(x_j^*(\lambda)) - \mu w(\lambda).$$

Hence

$$\sum_{j=1}^{n} f_j(x_j^*(\lambda)) - \lambda w(\lambda) + \sum_{j=1}^{n} f_j(x_j^*(\mu)) - \mu w(\mu)$$

$$\geq \sum_{j=1}^{n} f_j(x_j^*(\mu)) - \lambda w(\mu) + \sum_{j=1}^{n} f_j(x_j^*(\lambda)) - \mu w(\lambda).$$

Then

$$(\mu - \lambda)w(\lambda) \geq (\mu - \lambda)w(\mu)$$

and taking into account that $\mu > \lambda$, we get $w(\lambda) \geq w(\mu)$, that is, $\sum_{j=1}^{n} a_{2j}x_j(\lambda)$ is a monotonically decreasing function of λ.

Using this property, it is easier to find λ in particular problems.

Consider the more general case

$$\max \left\{ f(\mathbf{x}) = \sum_{j=1}^{n} f_j(x_j) \right\} \tag{4.68}$$

subject to

$$\sum_{j=1}^{n} a_{ij}x_j \leq b_i, \quad i = 1, \ldots, m \tag{4.69}$$

$$x_j \geq 0, \quad j = 1, \ldots, n. \tag{4.70}$$

Introduce p Lagrange multipliers $\lambda_1, \ldots, \lambda_p$ $(p < m)$ and pose the new problem as follows:

$$\max_{x_1,\ldots,x_n} \left\{ h(\mathbf{x}) = \sum_{j=1}^{n} f_j(x_j) - \sum_{i=1}^{p} \lambda_i \sum_{j=1}^{n} a_{ij}x_j \right\}$$

$$\equiv \max_{x_1,\dots,x_n} \left\{ h(\mathbf{x}) = \sum_{j=1}^{n} \left[f_j(x_j) - \sum_{i=1}^{p} \lambda_i a_{ij} x_j \right] \right\} \tag{4.71}$$

subject to

$$\sum_{j=1}^{n} a_{ij} x_j \le b_i, \quad i = p+1, \dots, m, \tag{4.72}$$

where (4.72) are the remaining $m - p$ constraints, not involved in (4.71).

We have reduced the problem with m constraints (m state variables) to a problem of determining a sequence of functions of $m - p$ state variables together with a search over the p-dimensional space of λ_i, $i = 1, \dots, p$, where these λ_i's are the Lagrange multipliers associated with the first p constraints in (4.69). Thus, the original problem is reduced to a sequence of problems of smaller dimensions. Value of p is chosen depending on the particular problem to be solved.

4.7 Application of Dynamic Programming Approach to the Transportation Problem with Nonlinear Separable Objective Function

Consider the transportation model with two sources (depots) A_1, A_2 and n destinations (sinks, demand points) $B_1, B_2, \dots, B_n$, and amounts of supply a_1, a_2 at sources and amounts of demand $b_1, b_2, \dots, b_n$ at destinations, respectively.

Denote by $x_{ij}, i = 1, 2;\ j = 1, \dots, n$ the quantity sent from $A_i, i = 1, 2$ to $B_j, j = 1, \dots, n$; $X = (x_{ij})_{2\times n}$, and by $f_{ij}(x_{ij})$ the unit transportation cost from A_i to $B_j, i = 1, 2;\ j = 1, \dots, n$.

Then the transportation model can be formulated as follows: determine the unknowns x_{ij} so as to *minimize* the total transportation cost

$$c(X) = \sum_{i=1}^{2} \sum_{j=1}^{n} f_{ij}(x_{ij}) \tag{4.73}$$

subject to the supply and demand restrictions

$$\sum_{j=1}^{n} x_{ij} = a_i, \quad i = 1, 2 \tag{4.74}$$

$$\sum_{i=1}^{2} x_{ij} = b_j, \quad j = 1, \dots, n \tag{4.75}$$

$$x_{ij} \ge 0, \quad i = 1, 2;\ j = 1, \dots, n. \tag{4.76}$$

The relation (4.74) requires that the total quantity shipped from any source must equal the supply there, and (4.75) requires that the total quantity shipped to any destination must equal the demand at this site.

In this model, any transshipment is not allowed, that is, we cannot transport from one source to another and then to a destination, or from one source to a destination and then to another destination.

If $f_{ij}(x_{ij}) = c_{ij}x_{ij}$, where c_{ij} is the transportation cost per unit of the item from A_i to B_j, then (4.73)–(4.76) is the *linear transportation problem*.

However, we assume that $f_{ij}(x_{ij})$ are *nonlinear* functions.

Although the number of constraints is $n + 2$, a state function can be described by one parameter only.

Since $x_{2j} = b_j - x_{1j}$ according to (4.75), it is sufficient to determine x_{1j} for each $j = 1, \ldots, n$.

Because a_1 and a_2 are the amounts of supply at A_1 and A_2, respectively, and $a_1 + a_2 = \sum_{j=1}^{n} b_j$ (we assume that the model is *balanced*, that is, the total supply $\sum_{i=1}^{2} a_i$ is equal to the total demand $\sum_{j=1}^{n} b_j$) then $a_2 = \sum_{j=1}^{n} b_j - a_1$. Hence a_1 is the only state parameter.

The more general problem where the supply exceeds the demand or the demand exceeds the supply is not considered here.

Define the sequence of functions

$$\Phi_k(a_1) = \min_{\{x_{ij}\}} \sum_{i=1}^{2} \sum_{j=1}^{n} f_{ij}(x_{ij}), \tag{4.77}$$

where we take the minimum over all nonnegative x_{ij} which satisfy conditions

$$\sum_{j=1}^{k} x_{1j} = a_1,$$

$$\sum_{j=1}^{k} x_{2j} = \sum_{j=1}^{k} b_j - a_1,$$

$$x_{1j} + x_{2j} = b_j, \quad j = 1, \ldots, k.$$

The forward recursion is then of the form

$$\Phi_k(a_1) = \min_{0 \le x_{1k} \le \min\{a_1, b_k\}} [f_{1k}(x_{1k}) + f_{2k}(b_k - x_{1k}) + \Phi_{k-1}(a_1 - x_{1k})]. \tag{4.78}$$

Function $\Phi_k(a_1)$ should be tabulated for all $a_1 = 0, 1, \ldots, \sum_{j=1}^{n} b_j$. At the last step, when $k = n$, we find x_{1n}^* and $x_{2n}^* = b_n - x_{1n}^*$. All other values x_{1k}^* can be calculated from the relations

$$x^*_{1,n-i} = \widehat{x}_{1,n-i}\left(a_1 - \sum_{k=0}^{i-1} x^*_{1,n-k}\right), \quad i = 1, \ldots, n-1.$$

Consider the similar transportation problem with *three* sources A_1, A_2, and A_3 and n destinations. In this case, the state function is defined as follows:

$$\Phi_k(a_1, a_2) = \min_{\{x_{ij}\}} \sum_{i=1}^{3} \sum_{j=1}^{n} f_{ij}(x_{ij}), \quad k = 1, \ldots, n, \tag{4.79}$$

where we take the minimum over all nonnegative x_{ij} which satisfy conditions

$$\sum_{j=1}^{k} x_{1j} = a_1,$$
$$\sum_{j=1}^{k} x_{2j} = a_2,$$
$$\sum_{j=1}^{k} x_{3j} = \sum_{j=1}^{k} b_j - a_1 - a_2,$$
$$x_{1j} + x_{2j} + x_{3j} = b_j, \quad j = 1, \ldots, k.$$

The forward recursion is then of the form

$$\Phi_k(a_1, a_2) = \min_{x_{1k}, x_{2k}} [f_{1k}(x_{1k}) + f_{2k}(x_{2k}) + f_{3k}(b_k - x_{1k} - x_{2k}) + \Phi_{k-1}(a_1 - x_{1k}, a_2 - x_{2k})], \tag{4.80}$$

where the minimum is taken over all x_{1k} and x_{2k} which satisfy conditions

$$0 \le x_{1k} \le \min\{a_1, b_k\},$$
$$0 \le x_{2k} \le \min\{a_2, b_k\},$$
$$x_{1k} + x_{2k} \le b_k,$$
$$x_{3k} = b_k - x_{1k} - x_{2k} \le \sum_{j=1}^{k} b_j - a_1 - a_2,$$

and the last condition can be written as

$$x_{1k} + x_{2k} \ge a_1 + a_2 - \sum_{j=1}^{k-1} b_j.$$

The Lagrange multiplier method can be used for reducing the dimensionality of the problem.

Assume that there are unlimited amounts at the second and third sources A_2 and A_3 and that λ and 1 are the unit transportation costs associated with the amounts carried from A_2 and A_3, respectively.

Consider the problem

$$\min_{\{x_{ij}\}} \left\{ \sum_{i=1}^{3} \sum_{j=1}^{n} f_{ij}(x_{ij}) + \lambda \sum_{j=1}^{n} x_{2j} + \sum_{j=1}^{n} x_{3j} \right\} \tag{4.81}$$

subject to

$$\begin{aligned} &\sum_{j=1}^{n} x_{1j} = a_1 \\ &x_{1j} + x_{2j} + x_{3j} = b_j, \quad j = 1, \ldots, n \\ &x_{ij} \geq 0, \quad i = 1, 2, 3; \; j = 1, \ldots, n \end{aligned} \tag{4.82}$$

for different values of λ. We vary λ until we get $\sum_{j=1}^{n} x_{2j} = a_2$ for the minimizing values of x_{ij}'s. In this case, the constraint $\sum_{j=1}^{n} x_{3j} = a_3$ will also be satisfied.

Define the state function as

$$\Phi_k(\gamma) = \min_{\{x_{ij}\}} \sum_{i=1}^{3} \sum_{j=1}^{k} [f_{ij}(x_{ij}) + \lambda x_{2j} + x_{3j}], \tag{4.83}$$

where

$$\begin{aligned} &\sum_{j=1}^{k} x_{1j} = \gamma, \\ &x_{1j} + x_{2j} + x_{3j} = b_j, \quad j = 1, \ldots, k, \\ &x_{ij} \geq 0, \quad i = 1, 2, 3; \; j = 1, \ldots, k. \end{aligned}$$

The forward recursion is then

$$\begin{aligned} \Phi_k(\gamma) = \min_{x_{1k}, x_{2k}} &[f_{1k}(x_{1k}) + f_{2k}(x_{2k}) + f_{3k}(b_k - x_{1k} - x_{2k}) \\ &+ \lambda x_{2k} + x_{3k} + \Phi_{k-1}(\gamma - x_{1k})], \quad k = 2, \ldots, n, \end{aligned} \tag{4.84}$$

where x_{1k}, x_{2k}, and x_{3k} satisfy the constraints $x_{ik} \geq 0$, $i = 1, 2, 3$; $x_{1k} + x_{2k} + x_{3k} = b_k$.

Thus, a transportation problem with three sources and n destinations is transformed into a sequence of n-stage problems with a single state parameter.

4.8 Review of Some Separable Inventory and Other Models

In this section, some inventory and other models are formulated, which are connected with separable programming. For completeness, some simpler models are also included. The probabilistic inventory model is also considered in Chapter Thirteen in connection with the implementation of stochastic quasigradient methods. Methods studied in this book are applicable to models under consideration.

Inventory control problems arise when the inventory of goods must be set up and maintained in order to satisfy their demand during a time interval. A reasonable inventory should be maintained because, on the one hand, the production process is continuous and, on the other hand, orders are usually periodic; the production process is regular whereas the consumption is irregular; there are random factors in the demand between replenishment moments, order quantity, ordering cycle length, etc.

Demand can be satisfied by a single setup of inventory of goods for the period or by replenishment for each unit time interval. The *oversupply* (with respect to the unit time interval) usually corresponds to the first possibility, and *shortfall* (with respect to the whole time interval) usually corresponds to the second possibility described above.

The oversupply leads to a surplus at some locations and a shortage at other locations. The oversupply requires higher (with respect to unit time interval) costs for storage but shortage arises rarely and we have to reorder new quantities rarely.

In case of a shortfall, holding costs are lower but the risk of deficiency rises and this may lead to the irregular production process, interruption of regular relations between companies, etc.

In both extreme cases, there is considerable economic loss. The inventory control problem determines the inventory level that balances these two cases.

Each inventory policy consists of two important elements: reorder time and order quantity, that is, it answers the questions, when and how much to order?

The decision about the *economic order quantity* (EOQ) and the moment of order is based on minimizing the total costs subject to some constraints.

If the inventory system requires *periodic review* (for example, every day, week, month, or year), the time of receiving the new order is at the beginning of each period.

If the inventory system requires *continuous review*, a new order is placed when the inventory level is decreased down to a level called *reorder point*.

The *total inventory cost* is the sum of the following costs:

- *purchasing cost*—this is the price per unit of the item.
- *setup cost*—this is the charge for placing the order. Usually this cost is fixed and does not depend on the order quantity. Sometimes the setup cost is proportional to the number or quantity of the items stored.
- *holding cost*—this is the cost for maintaining the inventory in stock. It includes, for example, cost of storage, maintenance, and handling.
- *shortage cost*—this is the penalty in case of insufficient quantity of goods. This cost consists mainly of the loss of income.

The *constraints* are on maximum available storage area (capacity of the warehouse), maximum weight, maximum cost, average cost, replenishment number per time interval, order quantity, etc.

A classification of inventory models is given below.

- according to demand:
  - deterministic (known with certainty):
    - static (with constant demand over time),
    - dynamic (demand over time varies),
  - probabilistic (described by a probability distribution):
    - with continuous review,
    - with periodic review (single-period, multiperiod models),
  - undetermined,
- according to the number of warehouses (stores): models with one and with several warehouses,
- according to number of products: single-item and multi-item models,
- according to order and use of resources: with continuous review and with periodic review (single-period and multiperiod models),
- according to replenishment type: instantaneous, with fixed delay, with random delay,

  etc.

4.8.1 Deterministic Static Models

The Classic Model

Consider some of the simplest inventory models with constant demand rate a and limit inventory level r.

Let

T be the total ordering cycle length (time)
y be the order quantity (replenishment rate)
w be the maximum shortage
s be the setup cost
p be the unit holding cost per unit time
q be the unit shortage cost (the penalty for unit of item j in case of deficiency of this item).

Assume that p and q are proportional to the average inventory level and shortage quantity, respectively, and to the length of the time interval during which this inventory is maintained (this shortage lasts, respectively).

Then the total cost is

$$c(y) = s + p \int_0^{t_1+t_2} y(t)dt - q \int_{t_1+t_2}^{T} y(t)dt \tag{4.85}$$

where

$$y(t) = \begin{cases} (r-a)t,\ 0 \le t \le t_1 \\ y - a(t-t_1),\ t_1 < t \le t_1 + t_2 + t_3 \\ -w + (r-a)(t - t_1 - t_2 - t_3),\ t_1 + t_2 + t_3 < t \le T, \end{cases} \tag{4.86}$$

$$w = \frac{(T - (t_1 + t_2))y}{t_1 + t_2} = \frac{Ty}{t_1 + t_2} - y. \tag{4.87}$$

Function $y(t)$ (4.86) is a linear increasing function with slope $r - a$ for $0 \le t \le t_1$, a linear decreasing function with slope $-a$ for $t_1 < t \le t_1 + t_2 + t_3$ and a linear increasing function with slope $r - a$ for $t_1 + t_2 + t_3 < t \le T$. Function $y(t)$ has positive values for $0 \le t \le t_1 + t_2$ and negative values for $t_1 + t_2 \le t \le T$.

Here

$$t_1 = \frac{y}{r-a}, \quad t_2 = \frac{y}{a}, \quad t_3 = \frac{(r-a)T}{r} - \frac{y}{a},$$
$$t_4 \stackrel{\text{def}}{=} T - (t_1 + t_2 + t_3) = \frac{a}{r}T - \frac{y}{r-a}.$$

Then

$$w = \frac{a}{r}(r-a)T - y. \tag{4.88}$$

Taking into account (4.86), (4.85) becomes

$$c(y) = s + \frac{pry^2}{2a(r-a)} + \frac{qr}{2a(r-a)}\left[\frac{a}{r}(r-a)T - y\right]^2. \tag{4.89}$$

Then the average cost per unit time is

$$c_{\text{av}}(y) = \frac{c(y)}{T} = \frac{1}{T}\left[s + \frac{(p+q)ry^2}{2a(r-a)}\right] + \frac{qa(r-a)}{2r}T - qy. \tag{4.90}$$

The optimum values of the order quantity y^* and the total cycle length T^* are determined by minimizing the function $c_{\text{av}}(y)$ with respect to y and T. Necessary conditions for minimum of $c_{\text{av}}(y)$ (assuming that y and T are continuous) are

$$\frac{\partial c_{\text{av}}(y)}{\partial y} \equiv \frac{(p+q)ry}{Ta(r-a)} - q = 0, \tag{4.91}$$

$$\frac{\partial c_{\text{av}}(y)}{\partial T} \equiv -\frac{1}{T^2}\left[s + \frac{(p+q)ry^2}{2a(r-a)}\right] + \frac{qa(r-a)}{2r} = 0. \tag{4.92}$$

Since $c_{\text{av}}(y)$ is a convex function of y and T (see, for example, Theorem 1.32), then these conditions are also sufficient. The solution of system (4.91)–(4.92) is

$$y^* = \sqrt{\frac{2asq(r-a)}{rp(p+q)}} \equiv \sqrt{\frac{2as(1-\frac{a}{r})}{p(1+\frac{p}{q})}}, \tag{4.93}$$

$$T^* = \sqrt{\frac{2sr(p+q)}{apq(r-a)}} \equiv \sqrt{\frac{2s(1+\frac{p}{q})}{ap(1-\frac{a}{r})}}. \tag{4.94}$$

The minimum value of $c_{\mathrm{av}}(y)$ is then

$$c^*_{\mathrm{av}} = c_{\mathrm{av}}(y^*) = \sqrt{\frac{2aspq(r-a)}{r(p+q)}} \equiv \sqrt{\frac{2asp(1-\frac{a}{r})}{1+\frac{p}{q}}}. \tag{4.95}$$

From (4.95) we can obtain the following important cases.

Case 1. No shortage model.

Then $q \to \infty$, therefore $\frac{p}{q} \to 0$ and (4.93)–(4.95) give

$$y^* = \sqrt{\frac{2as(1-\frac{a}{r})}{p}}, \quad T^* = \sqrt{\frac{2s}{ap(1-\frac{a}{r})}}, \quad c^*_{\mathrm{av}} = \sqrt{2asp\left(1-\frac{a}{r}\right)}. \tag{4.96}$$

Case 2. Instantaneous order replenishment model.

Then $r \to \infty$, therefore $\frac{a}{r} \to 0$ and (4.93)–(4.95) give

$$y^* = \sqrt{\frac{2as}{p(1+\frac{p}{q})}}, \quad T^* = \sqrt{\frac{2s(1+\frac{p}{q})}{ap}}, \quad c^*_{\mathrm{av}} = \sqrt{\frac{2asp}{1+\frac{p}{q}}}. \tag{4.97}$$

Case 3. Instantaneous order replenishment model with no shortage.

Then $q \to \infty$ and $r \to \infty$, therefore $\frac{p}{q} \to 0$, $\frac{a}{r} \to 0$ and

$$y^* = \sqrt{\frac{2as}{p}}, \quad T^* = \sqrt{\frac{2s}{ap}}, \quad c^*_{\mathrm{av}} = \sqrt{2asp}. \tag{4.98}$$

Formulas (4.98) are called *Wilson's economic lot size formulas*.

A Model with Price Breaks

In this model, price of the inventory item depends on the size of the order, that is, an inventory item can be purchased at a discount if the size y of the order exceeds some order rate y_0.

Then the unit purchasing price $f(y)$ is given by the function

$$f(y) = \begin{cases} 0, \text{ if } y = 0 \\ s + b_1 y, \text{ if } 0 < y \le y_0 \\ s + b_2 y, \text{ if } y > y_0 \end{cases} \tag{4.99}$$

where s is the constant cost of the order and $b_1 > b_2$.

Consider the instantaneous order replenishment model with no shortage.
As above, let p be the unit holding cost per unit time and a be the demand rate. Then total cost for the whole time interval T is

$$c(y) = \frac{1}{2}pyT + f(y),$$

and total cost per unit time is

$$c_{\text{av}}(y) = \frac{c(y)}{T} = \frac{1}{2}py + \frac{1}{T}f(y) = \begin{cases} c_{\text{av}}^{(1)}(y) = \frac{1}{2}py + \frac{sa}{y} + b_1 a, \ y \le y_0 \\ c_{\text{av}}^{(2)}(y) = \frac{1}{2}py + \frac{sa}{y} + b_2 a, \ y > y_0 \end{cases} \tag{4.100}$$

We have used that $T = \frac{y}{a}$; see (4.98).

Obviously function $c_{\text{av}}(y)$ has a break point at $y = y_0$. Since the two functions in (4.100) differ only by the constant $|(b_1 - b_2)a|$, their minimum points coincide. Minimum of $c_{\text{av}}(y)$ is attained either at a point for which $\frac{\partial c_{\text{av}}(y)}{\partial y} = 0$ or at the break point $y = y_0$.

Calculate

$$\frac{\partial c_{\text{av}}(y)}{\partial y} \equiv \frac{1}{2}p - \frac{sa}{y^2} = 0,$$

whence it follows that $y_m = \sqrt{\frac{2sa}{p}}$.

Consider both possible cases: $y_m > y_0$ and $y_m < y_0$.

1) Let $y_m > y_0$. Then the optimum order quantity is $y^* = y_m$.

2) Let $y_m < y_0$. Calculate y_μ $(y_\mu > y_m)$ such that $c_{\text{av}}^{(1)}(y_m) = c_{\text{av}}^{(2)}(y_\mu)$. Therefore

$$\frac{1}{2}py_m + \frac{sa}{y_m} + b_1 a = \frac{1}{2}py_\mu + \frac{sa}{y_\mu} + b_2 a,$$

whence, using the expression of y_m, we get

$$\sqrt{2psa} + b_1 a = \frac{1}{2}py_\mu + \frac{sa}{y_\mu} + b_2 a.$$

The order size is $y = y_0$ if and only if $c_{\text{av}}^{(2)}(y_0) < c_{\text{av}}^{(1)}(y_m) = c_{\text{av}}^{(2)}(y_\mu)$, that is, if and only if

$$\frac{1}{2}py_0 + \frac{sa}{y_0} + b_2 a < \sqrt{2psa} + b_1 a,$$

whence it follows that

$$b_1 - b_2 > \frac{1}{a}\left(\frac{1}{2}py_0 + \frac{sa}{y_0} - \sqrt{2psa}\right). \tag{4.101}$$

Since $c''_{\text{av}}(y) = \frac{2sa}{y^3} > 0$ for $y > 0, s > 0, a > 0$, then $c_{\text{av}}(y)$ is a strictly convex function according to Theorem 1.32, and therefore $c'_{\text{av}}(y)$ is a monotone increasing function. Hence, $c_{\text{av}}^{(2)}(y_\mu) > c_{\text{av}}^{(2)}(y_0)$ implies $y_\mu > y_0$, and since we consider the case $y_m < y_0$, then $y_m < y_0 < y_\mu$. Therefore $y^* = y_0$ in this case.

If $y_0 > y_\mu (> y_m)$, then according to convexity of $c_{\text{av}}^{(1)}$ and $c_{\text{av}}^{(2)}$ we get

$$c_{\text{av}}^{(1)}(y_0) \geq c_{\text{av}}^{(1)}(y_\mu) \geq c_{\text{av}}^{(1)}(y_m)$$

and

$$c_{\text{av}}^{(2)}(y_0) \geq c_{\text{av}}^{(2)}(y_\mu) = c_{\text{av}}^{(1)}(y_m).$$

Therefore the optimum order quantity y^*, which minimizes $c_{\text{av}}(y)$ in this case, is $y^* = y_m$.

Summarize the obtained results:

$$y^* = \begin{cases} y_m = \sqrt{\frac{2sa}{p}}, & y_m > y_0 \text{ or } y_0 > y_\mu \\ y_0, & y_m < y_0 < y_\mu. \end{cases}$$

Multi-item Model with Limited Capacity of the Warehouse

Consider the generalization of the classic model with instantaneous order replenishment and no shortage, with $n \geq 2$ items and given maximum storage area for all n items.

For each item j, $j = 1, \ldots, n$ define

y_j order quantity
a_j demand rate
s_j setup cost
p_j unit holding cost
d_j storage area per inventory unit
α maximum storage area (the storage capacity of the warehouse)

We have to minimize the total cost subject to storage limitation and positivity of order quantities, that is, the mathematical problem is

$$\min \left\{ c(\mathbf{y}) \equiv c(y_1, \ldots, y_n) = \sum_{j=1}^{n} \left(\frac{s_j a_j}{y_j} + \frac{p_j y_j}{2} \right) \right\} \tag{4.102}$$

subject to

$$\sum_{j=1}^{n} d_j y_j \leq \alpha \tag{4.103}$$

$$y_j \geq 0, \quad j = 1, \ldots, n. \tag{4.104}$$

We are able to compute the unconstrained optimum order quantities for $c(\mathbf{y})$ (4.102) (also cf. (4.98))

$$y_j^* = \sqrt{\frac{2s_j a_j}{p_j}}, \quad j = 1, \ldots, n. \tag{4.105}$$

If all y_j^* satisfy (4.103), these y_j^*'s are optimal. Otherwise, if $\sum_{j=1}^n d_j y_j^* > \alpha$, then (4.103) must be satisfied as an equality $\sum_{j=1}^n d_j y_j^* = \alpha$ and the Lagrange multiplier method can be applied in this case. The Lagrangian for problem (4.102)–(4.104) is

$$L(\mathbf{y}, \lambda) = \sum_{j=1}^{n} \left(\frac{s_j a_j}{y_j} + \frac{p_j y_j}{2} \right) + \lambda \left(\sum_{j=1}^{n} d_j y_j - \alpha \right)$$

(see Definition 1.24) where $\lambda \geq 0$ is the Lagrange multiplier associated with the inequality constraint (4.103).

Since $L(\mathbf{y}, \lambda)$ is a convex function of $\mathbf{y}$ and λ, optimum values $\mathbf{y}^*$, λ^* satisfy the following necessary and sufficient conditions

$$\frac{\partial L}{\partial y_j} \equiv -\frac{s_j a_j}{y_j^2} + \frac{p_j}{2} + \lambda d_j = 0, \quad j = 1, \ldots, n, \tag{4.106}$$

$$\frac{\partial L}{\partial \lambda} \equiv \sum_{j=1}^{n} d_j y_j - \alpha = 0. \tag{4.107}$$

Conditions (4.106) give

$$y_j^* = \sqrt{\frac{2s_j a_j}{p_j + 2\lambda^* d_j}}, \quad j = 1, \ldots, n, \tag{4.108}$$

and (4.107) states that the inequality constraint (4.103) is satisfied with equality.

When $\lambda^* = 0$, that is, when there are no constraints (no storage limitation), expressions (4.105) and (4.108) coincide.

4.8.2 Dynamic Inventory Models

Consider a deterministic inventory model with periodic review which is *dynamic*, that is, the demand for each period is deterministic but it may vary from a period to the next one, and let the *finite* number of periods be of equal length.

Setup Model with No Shortage

For each period j, $j = 1, \ldots, n$, let

y_j be the order quantity
p_j be the unit holding cost from period j to period $j + 1$
a_j be the demand rate for period j
x_j be the inventory at the beginning of period j, remained from period $j - 1$
s_j be the setup cost for period j
$g_j(y_j)$ be the marginal production cost function
$h_j(x_{j+1})$ be the minimum inventory cost for maintaining the surplus quantity x_{j+1} at the end of period j

Inventory level at the end of period j is $x_{j+1} = y_j + x_j - a_j$.
Then the production and holding cost function for period j is

$$C_j(y_j, x_j) = \begin{cases} p_j x_{j+1}, \ y_j = 0 \\ s_j + g_j(y_j) + p_j x_{j+1}, \ y_j > 0. \end{cases}$$

The total cost function for all n periods is

$$c(\mathbf{y}, \mathbf{x}) = \sum_{j=1}^{n} c_j(y_j, x_j) = \sum_{j=1}^{n} [C_j(y_j, x_j) + h_j(x_{j+1})], \tag{4.109}$$

where $x_{j+1} = y_j + x_j - a_j$. We must find x_j's and y_j's which *minimize* function (4.109) subject to $x_{j+1} = y_j + x_j - a_j$. For this purpose, we use the forward recursion dynamic programming approach.

The state at stage j is x_{j+1}. Note that the concept of *stage* introduced in Section 4.1 coincides with the concept of *period* here.

Since the remaining inventory level x_{j+1} can satisfy the demand for all the remaining periods, then $0 \leq x_{j+1} \leq a_{j+1} + \cdots + a_n$.

Define the sequence of state functions as follows:

$$\begin{aligned} \Phi_k(x_{k+1}) &= \min_{\{y_j\}} \sum_{j=1}^{k} c_j(y_j, x_j) \\ &= \min_{\{y_j\}} \sum_{j=1}^{k} [C_j(y_j, x_j) + h_j(y_j + x_j - a_j)], \end{aligned} \tag{4.110}$$

where $x_{k+1} = y_k + x_k - a_k$.

The forward recursion is then

$$\Phi_k(x_{k+1}) = \min_{0 \leq y_k \leq a_k + x_{k+1}} [c_k(y_k, x_k) + \Phi_{k-1}(x_k)]$$

$$= \min_{0 \leq y_k \leq a_k + x_{k+1}} [c_k(y_k, x_{k+1} + a_k - y_k) + \Phi_{k-1}(x_{k+1} + a_k - y_k)]. \tag{4.111}$$

When $k = 1$, we have

$$\Phi_1(x_2) = \min_{0 \le y_1 \le a_1 + x_2} [C_1(y_1, x_1) + h_1(y_1 + x_1 - a_1)]$$

subject to $x_2 = y_1 + x_1 - a_1$.

Determining successively $\Phi_1(x_2), \Phi_2(x_3), \dots,$ at the last step we determine $\Phi_n(x_{n+1})$ and $y_n^* = \widehat{y}_n(x_{n+1})$. Then

$$y_k^* = \widehat{y}_k\left(x_{n+1} + \sum_{j=k+1}^{n} (a_j - y_j^*)\right).$$

Consider the special case when all $C_j(y_j, x_j)$ are concave functions of y_j and x_j, and h_j are linear, that is,

$$h_j(y_j + x_j - a_j) = h_j(x_{j+1}) = h_j x_{j+1}, \quad h_j = const., \quad j = 1, \dots, n.$$

Then the total cost function is

$$c(\mathbf{y}, \mathbf{x}) = \sum_{j=1}^{n} [C_j(y_j, x_j) + h_j x_{j+1}] \overset{\text{def}}{=} \sum_{j=1}^{n} c_j(y_j, x_j).$$

Let x_1 and x_{n+1} be given, purchase be at the beginning of each period, and demands for the current period must be satisfied.

Find $\mathbf{x} = (x_1, \dots, x_n)$, $\mathbf{y} = (y_1, \dots, y_n)$ that *minimize* the total cost function subject to $x_{j+1} = y_j + x_j - a_j$, that is, solve the problem

$$\min_{\{y_j\}} \left\{ c(\mathbf{y}, \mathbf{x}) = \sum_{j=1}^{n} c_j(y_j, x_j) = \sum_{j=1}^{n} [C_j(y_j, x_j) + h_j x_{j+1}] \right\} \tag{4.112}$$

subject to

$$x_{j+1} = y_j + x_j - a_j, \quad j = 1, \dots, n. \tag{4.113}$$

Since we have to minimize the sum of concave functions $c_j(y_j, x_j)$ (4.112), the optimal solution is an extreme point of the polyhedron (4.113).

Because the number of constraints is n and problem (4.112)–(4.113) has $2n$ variables, then the optimal solution contains at most n nonzero components. From (4.113) it follows that x_j's and y_j's cannot be equal to zero simultaneously. Then

$$x_j^* y_j^* = 0, \quad j = 1, \dots, n \tag{4.114}$$

where $(\mathbf{y}^*, \mathbf{x}^*) = (y_1^*, \dots, y_n^*, x_1^*, \dots, x_n^*)$ is the optimal solution to (4.112)–(4.113).

Condition (4.114), considering (4.113), is equivalent to the following two conditions:

$$x_j^* = 0 \text{ implies } y_j^* > 0; \qquad x_j^* > 0 \text{ implies } y_j^* = 0. \tag{4.115}$$

This means that we do not order a new quantity y_j in case there is inventory $x_j > 0$. Therefore, $y_j^* = 0$, or $y_j^* = a_j$, or $y_j^* = a_j + a_{j+1}$, and so on; that is, the order quantity is equal to the demand for all periods.

Since $x_{n+1} = 0$, we use the forward approach to solve the problem. Let $\Phi_k(x_{k+1})$ be the minimum cost during k periods provided that inventory at the beginning of period $k+1$ is x_{k+1}.

The forward recursion is

$$\Phi_k(x_{k+1}) = \min_{\{y_k\}}[c_k(y_k, x_k) + \Phi_{k-1}(x_{k+1} + a_k - y_k)], \tag{4.116}$$

where $x_{k+1} = y_k + x_k - a_k$, $k = 1, \ldots, n$.

Since $x_k^* y_k^* = 0$ and $\sum_{j=1}^n c_j(y_j, x_j)$ is concave, then minimum in (4.116) is attained at $y_k = 0$ or at $y_k = x_{k+1} + a_k$. Therefore

$$\Phi_k(x_{k+1}) = \min \begin{cases} c_k(0, x_{k+1}) + \Phi_{k-1}(x_{k+1} + a_k), & y_k = 0. \\ c_k(x_{k+1} + a_k, x_{k+1}) + \Phi_{k-1}(0), & y_k = x_{k+1} + a_k. \end{cases} \tag{4.117}$$

Similarly,

$$\Phi_{k-1}(x_{k+1} + a_k)$$
$$= \min \begin{cases} c_{k-1}(0, x_{k+1} + a_k) + \Phi_{k-2}(x_{k+1} + a_k + a_{k-1}), & y_k = 0. \\ c_{k-1}(x_{k+1} + a_k + a_{k-1}, x_{k+1} + a_k) + \Phi_{k-2}(0), & y_k = x_{k+1} + a_k. \end{cases} \tag{4.118}$$

Analogously we get

$$\Phi_k(x_{k+1}) = \min_{1 \le i \le k} F_k(i) \tag{4.119}$$

where

$$F_k(i) \stackrel{\text{def}}{=} \Phi_{i-1}(0) + c_i\left(x_{i+1} + \sum_{l=i}^{k} a_l, x_{i+1} + \sum_{l=i+1}^{k} a_l\right)$$
$$+ \sum_{j=i+1}^{k} c_j\left(0, x_{j+1} + \sum_{l=j+1}^{k} a_l\right) \tag{4.120}$$

is the cost function for k periods provided that the last order is at the beginning of period i.

When production cost function for period j is

$$C_j(y_j) = \begin{cases} 0, \ y_j = 0 \\ s_j + g y_j, \ y_j > 0, \end{cases}$$

we have another special case of the inventory model under consideration.

Define the function

$$\chi_j(y_j) = \begin{cases} 0, \ y_j = 0 \\ 1, \ y_j > 0. \end{cases} \tag{4.121}$$

Then

$$\sum_{j=1}^{n} C_j(y_j) = \sum_{j=1}^{n} s_j \chi_j(y_j) + g \sum_{j=1}^{n} y_j$$

$$= \sum_{j=1}^{n} s_j \chi_j(y_j) + g \sum_{j=1}^{n} (x_{j+1} - x_j + a_j)$$

$$= \sum_{j=1}^{n} s_j \chi_j(y_j) + g \sum_{j=1}^{n} (a_j + (x_{n+1} - x_1)),$$

and since $g \sum_{j=1}^{n} (a_j + (x_{n+1} - x_1)) = const.$, then the corresponding inventory model is described by the following linear program:

$$\min \left\{ \sum_{j=1}^{n} [s_j \chi_j(y_j) + h_j x_{j+1}] \right\} \tag{4.122}$$

subject to

$$y_j + x_j - x_{j+1} = a_j, \quad j = 1, \ldots, n \tag{4.123}$$

$$x_j y_j = 0, \quad j = 1, \ldots, n, \tag{4.124}$$

where $\chi_j(y_j)$ is defined by (4.121).

From (4.120), using (4.121)–(4.124) and $x_{k+1} = 0$, it follows that

$$F_k(i) = \Phi_{i-1}(0) + s_i + \sum_{j=i}^{k-1} h_j + \sum_{l=j+1}^{k} a_l, \tag{4.125}$$

$$\Phi_k(0) = \min_{1 \le i \le k} F_k(i). \tag{4.126}$$

Example 4.2 A company will receive goods during a seven-month period. Demands are

$$a_1 = 90, a_2 = 125, a_3 = 140, a_4 = 100, a_5 = 45, a_6 = 60, a_7 = 130,$$

respectively. Unit holding costs per month are $h_j = 2$, $j = 1, \ldots, 7$. Setup costs are $s_j = 300$, $j = 1, \ldots, 7$ and these costs occur only at the beginning of months when goods are produced/placed. Assume that the model is with instantaneous order replenishment.

Determine in which periods the company must order goods, and how much it must order for these periods.

Solution.

Let inventory at the beginning of the first period (month) be $x_1 = 0$.
The model describing the situation is

$$\min_{\{x_j\},\{y_j\}} \left\{c(\mathbf{y}, \mathbf{x}) = 300 \sum_{j=1}^{7} \chi_j(y_j) + 2 \sum_{j=1}^{7} x_{j+1}\right\}$$

subject to

$$\begin{aligned} y_1 - x_2 &= a_1 = 90 \\ y_j + x_j - x_{j+1} &= a_j,\ j = 2, \ldots, 6 \\ y_7 + x_7 &= a_7 = 130. \end{aligned}$$

Using (4.125) and (4.126), we get $\Phi_1(0) = \min\{s_1\} = 300$,

$$\Phi_2(0) = \min \begin{cases} s_2 + \Phi_1(0) = 300 + 300 = 600 \\ s_1 + 2a_2 = 300 + 2 \times 125 = 550. \end{cases}$$

Therefore $\Phi_2(0) = \min\{600, 550\} = 550$.

Hence, only when there are two periods ($k = 2$) and inventory at the end of a period is equal to zero, it is reasonable to have a single setup which satisfies the demands of both periods.

When $k = 3$, from (4.125) and (4.126) we obtain

$$\Phi_3(0) = \min \begin{cases} s_3 + \Phi_2(0) = 300 + 550 = 850 \\ s_2 + 2a_3 + \Phi_1(0) = 300 + 2 \times 140 + 300 = 880 \\ s_1 + 2a_3 + 2(a_2 + a_3) = 300 + 2 \times 140 + 2(125 + 140) = 1110. \end{cases}$$

Hence, $\Phi_3(0) = \min\{850, 880, 1110\} = 850$.

These and all other results are given in Table 4.11. Numbers in square brackets give numbers of periods for which the demand is satisfied at the beginning of the first period written in these square brackets. Boldface numbers give $\Phi_k(0) = \min_{1 \le i \le k} F_k(i)$.

From this table, it follows that the optimal solution to the problem under consideration is

$y_1^* = a_1 + a_2 = 90 + 125 = 215$
$y_2^* = 0$
$y_3^* = a_3 + a_4 = 140 + 100 = 240$
$y_4^* = 0$
$y_5^* = a_5 + a_6 = 45 + 60 = 105$
$y_6^* = 0$
$y_7^* = a_7 = 130$
$c(\mathbf{y}^*, \mathbf{x}^*) = \Phi_7(0) = 1770.$

Table 4.11 Results for Example 4.2

Period k	1	2	3	4	5	6	7
Demand a_k	90	125	140	100	45	60	130
$F_k(1)$	300	**550**	1110				
$F_k(2)$		600	880				
$F_k(3)$			**850**	**1050**	**1230**	1590	
$F_k(4)$				1150	1240	1480	
$F_k(5)$					1350	**1470**	1990
$F_k(6)$						1530	1790
$F_k(7)$							**1770**
$\Phi_k(0)$	300	550	850	1050	1230	1470	1770
	[1]	[1, 2]	[1, 2]	[1, 2]	[1, 2]	[1, 2]	[1, 2]
			[3]	[3, 4]	[3, 4, 5]	[3, 4]	[3, 4]
						[5, 6]	[5, 6]
							[7]

4.8.3 Probabilistic Inventory Models. No-Setup Models

Demand in probabilistic inventory models is described by a probability distribution. Among probabilistic models, we consider only the so-called *periodic review models.*

Single-Period Model

Single-period models are characterized by a single order of item(s) in order to satisfy the demand during the period.

Let

y be the order quantity
b be the purchasing or production cost per unit
p be the unit holding cost during the period
q be the unit shortage cost
a be the probabilistic demand rate during the period
$f(a)$ be the probability density function
x be the amount available before the order is placed
r be the revenue per unit of the item

We have to find the optimum value of the order quantity y that *minimizes* the total cost (purchasing/production cost, holding cost, and shortage cost).

In this model, we assume that demand occurs at the beginning of the period, immediately after the order is placed.

Let the probabilistic demand a be *continuous.*

Mathematical expectation of the total cost for the period is

$$\mathbf{E}c(y) = b(y - x) + p \int_0^y (y - a)f(a)da + q \int_y^\infty (a - y)f(a)da. \tag{4.127}$$

Since $\mathbf{E}c(y)$ is a convex function of y, its minimum is unique, and the necessary and sufficient condition for y^* to be a minimum point for $\mathbf{E}c(y)$ is

$$[\mathbf{E}c(y)]' \equiv b + p \int_0^y f(a)da - q \int_y^\infty f(a)da = 0,$$

that is,

$$b + pP\{a \le y\} - q(1 - P\{a \le y\}) = 0.$$

Hence

$$P\{a \le y^*\} = \frac{q - b}{p + q}$$

where $P\{a \le y\}$ denotes the probability with which $a \le y$. Ratio $\frac{q-b}{p+q}$ sometimes is called the *critical ratio*. If $\frac{q-b}{p+q} \ge 0$, that is, if $q \ge b$ (p and q are positive), then y^* is well-defined. Otherwise, if $q < b$, this case is not considered because the purchasing/production cost is higher than the unit shortage cost in case the item is not available.

Let the probabilistic demand a be *discrete*.

Then the expectation of the total cost for the period is

$$\mathbf{E}c(y) = b(y - x) + p \sum_{a=0}^{y} (y - a)f(a) + q \sum_{a=y+1}^{\infty} (a - y)f(a).$$

In this case, the necessary and sufficient optimality conditions are

$$\mathbf{E}c(y - 1) \ge \mathbf{E}c(y) \quad \text{and} \quad \mathbf{E}c(y + 1) \ge \mathbf{E}c(y).$$

Therefore,

$$P\{a \le y^* - 1\} \le \frac{q - b}{p + q} \le P\{a \le y^*\}.$$

Multiperiod Model

Consider a no-setup model that includes n periods. Also, let the model under consideration be without replenishment delay, the demand a for each period be described by a stationary probability density function $f(a)$, and the demand that is not satisfied can be accumulated for exactly one period. In addition to these assumptions, let the model be with a "compound discount", that is, if $\delta < 1$ is the discount rate per period, then the quantity Y after n periods becomes $\delta^n Y$.

Let x_j be the inventory available before the order be placed in period j. Denote by $\Phi_j(x_j)$ the expectation of the maximum profit for periods $j, j + 1, \ldots, n$. Then

$$
\begin{aligned}
\Phi_j(x_j) = \max_{y_j \geq x_j} \Bigg\{ &- b(y_j - x_j) + \int_0^{y_j} [ra - p(y_j - a)] f(a) da \\
&+ \int_{y_j}^{\infty} [ry_j + \delta r(a - y_j) - q(a - y_j)] f(a) da \\
&+ \delta \int_0^{\infty} \Phi_{j+1}(y_j - a) f(a) da \Bigg\}, \quad j = 1, \ldots, n,
\end{aligned} \tag{4.128}
$$

where $\Phi_{n+1}(y_n - a) = 0$. Problem (4.128) is a dynamic programming problem.

Since the unsatisfied demand is accumulated, the value of x_j may be negative. In the third summand of (4.128), the term $\delta r(a - y_j)$ is included because the unsatisfied demand in period j, which must be satisfied during the next period $j + 1$, is $a - y_j$.

In the case of an infinite number of periods, denote by x and y the inventory levels for each period before and after orders are received, respectively. Then (4.128) becomes

$$
\begin{aligned}
\Phi(x) = \max_{y \geq x} \Bigg\{ &- b(y - x) + \int_0^{y} [ra - p(y - a)] f(a) da \\
&+ \int_{y}^{\infty} [ry + \delta r(a - y) - q(a - y)] f(a) da \\
&+ \delta \int_0^{\infty} \Phi(y - a) f(a) da \Bigg\}.
\end{aligned} \tag{4.129}
$$

Since (4.129) is a *maximization* concave problem, the necessary condition for optimality turns out to be a sufficient condition as well.

The optimum value y^* of y can be calculated from the condition

$$
\begin{aligned}
\frac{\partial \Phi}{\partial y} \equiv &-b - p \int_0^{y} f(a) da + \int_{y}^{\infty} [(1 - \delta) r + q] f(a) da \\
&+ \delta \int_0^{\infty} \frac{\partial \Phi(y - a)}{\partial y} f(a) da = 0.
\end{aligned}
$$

4.8.4 Investment Models

Deterministic Investment Problem

Consider the following situation. We can invest the amounts $m_1, \ldots, m_n$ at the beginning of each of n periods. There are two opportunities: a bank pays an interest rate r_1 and another bank pays an interest rate r_2. These interest rates are compounded annually. Once an investment is deposited in a bank, it must remain there until the end of the n-year period. In order to encourage investments, both banks pay bonuses on new deposits. Each bonus is a percentage of the amount invested in this bank, it is paid at the end of the year in which the investment is made, and it may be reinvested in either bank in the next year. The bonus percentages vary from one year to another. For year j, let bonus percentages be q_{j1} for the first bank and q_{j2} for the second bank.

Find the best (optimal) investment scheme for the next n years.

Reformulate this problem in terms of dynamic programming.

Let

j be the stage (period), $j = 1, \ldots, n$ (this is the year)

$A_j^{(1)}$ and $A_j^{(2)}$ be the alternatives at stage j (the amounts invested in the first and in the second bank, respectively)

x_j be the state at stage j (this is the amount available for investment at the beginning of year j)

$\Phi_j(x_j)$ be the optimal value of investments for years $j, j+1, \ldots, n$ provided that the amount x_j is available at the beginning of year j

By definition, $A_j^{(2)} = x_j - A_j^{(1)}$. Then

$$\begin{aligned} x_1 &= m_1, \\ x_j &= m_j + q_{j-1,1}A_{j-1}^{(1)} + q_{j-1,2}(x_{j-1} - A_{j-1}^{(1)}) \\ &= m_j + (q_{j-1,1} - q_{j-1,2})A_{j-1}^{(1)} + q_{j-1,2}x_{j-1}, \quad j = 2, \ldots, n. \end{aligned}$$

The investment amount x_j includes only new deposited money and any bonus for investments made in year $j-1$.

Let s_j be the accumulated amount at the end of year j provided that $A_j^{(1)}$ and $A_j^{(2)} = x_j - A_j^{(1)}$ are investments in year j in the first and in the second bank, respectively.

Then the problem can be formulated as follows:

$$\max \left\{ c = \sum_{j=1}^{n} s_j \right\} \tag{4.130}$$

subject to

$$\begin{aligned} s_j &= A_j^{(1)}\delta_1^{n+1-j} + (x_j - A_j^{(1)})\delta_2^{n+1-j} \\ &= (\delta_1^{n+1-j} - \delta_2^{n+1-j})A_j^{(1)} + \delta_2^{n+1-j}x_j, \quad j = 1, \ldots, n-1 \end{aligned} \tag{4.131}$$

$$s_n = (\delta_1 + q_{n1} - \delta_2 - q_{n2})A_n^{(1)} + (\delta_2 + q_{n2})x_n, \tag{4.132}$$

where $\delta_i = 1 + r_i, i = 1, 2$.

The backward recursion is

$$\Phi_k(x_k) = \max_{0 \le A_k^{(1)} \le x_k} [s_k + \Phi_{k+1}(x_{k+1})], \quad k = 1, \ldots, n-1, \tag{4.133}$$

$$\Phi_{n+1}(x_{n+1}) = 0. \tag{4.134}$$

Probabilistic Investment Problem

Consider a probabilistic version of the model discussed above.

An amount of up to \$$m$ may be invested during the next n years. The investment plan is to buy goods at the beginning of the year and sell them at the end of the same year. The money earned (or part of it) may be reinvested at the beginning of the next year. Return on the investment depends on l market conditions which may be favorable or unfavorable. For condition i, $i = 1, \dots, l$, return r_i with probability P_i is expected.

How to invest the amount of \$$m$ in order to obtain the highest accumulation at the end of the n-year period?

Since investing money in such a way involves a risk that is described by random parameters, this investment model is probabilistic.

Let

x_j be the amount (in dollars) available for deposit at the beginning of year j
y_j be the actual amount invested at the beginning of year j
$\Phi_j(x_j)$ be the maximum expected amount for years $j, j+1, \dots, n$ provided that amount x_j is available at the beginning of year j

By definition, $x_1 = m$ and $y_j \le x_j$.

Introduce the concepts of dynamic programming as follows:

j is the stage (this is the year j), $j = 1, \dots, n$
y_j are the alternatives at stage j, $j = 1, \dots, n$
x_j is the state at stage j, $j = 1, \dots, n$

Then

$$x_{j+1} = (1 + r_i)y_j + (x_j - y_j) \equiv r_i y_j + x_j, \quad i = 1, \dots, l$$

for market condition i.

Since the probability for the market condition i is P_i, $i = 1, \dots, l$, the backward recursion is

$$\Phi_k(x_k) = \max_{0 \le y_k \le x_k} \left\{ \sum_{i=1}^{l} P_i \Phi_{k+1}(x_k + r_i y_k) \right\}, \quad k = 1, \dots, n, \tag{4.135}$$

and $\Phi_n(x_{n+1}) = x_{n+1}$ because we do not invest after the n-year period. Hence, for $k = n$ we get

$$\begin{aligned} \Phi_n(x_n) &= \max_{0 \le y_n \le x_n} \left\{ \sum_{i=1}^{l} P_i.(x_n + r_i y_n) \right\} \\ &= \sum_{i=1}^{l} P_i.(x_n + r_i x_n) = x_n \sum_{i=1}^{l} P_i.(1 + r_i). \end{aligned} \tag{4.136}$$

We have used that, due to linearity of function $\sum_{i=1}^{l} P_i.(x_n + r_i y_n)$ with respect to y_n and nonnegativity of its slope $\sum_{i=1}^{l} P_i r_i$ (because $P_i \geq 0, r_i \geq 0$), the maximum function value is attained at $y_n^* = x_n$.

NOTES TO PART ONE

The simplex method with restricted basis entry rule for solving the approximated separable programming problems is discussed, for example, in the works of Charnes and Cooper [23], Dantzig, Johnson and White [34], Markowitz and Manne [82], Meyer [86, 87] (for two-segment approximations), Miller [89], and Wolfe [173].

The grid generation scheme of Wolfe can be found in [173].

In works of Meyer [87] and Thakur [157], bounds on error (upon early termination) are considered.

Separable programming is also discussed in the books of Bazaraa, Sherali, and Shetty [5], Taha [155], etc.

The principle of optimality is formulated, for example, in the book of Bellman and Dreyfus [7].

Dynamic programming is considered in the books of Bellman [6], Bellman and Dreyfus [7], Bertsekas [9], Denardo [37], Dreyfus and Law [39], Luus [80], Sniedovich [117], Taha [155], Zaichenko [179], etc. Some aspects of dynamic programming are also discussed by Clarke [26]. In [80], the author suggested an iterative version of dynamic programming in order to overcome the "curse of dimensionality".

More details about deterministic and probabilistic inventory models can be found, for example, in the books of Bellman and Dreyfus [7], Taha [155], Zaichenko [179], etc.

Part II
Convex Separable Programming with Bounds on the Variables

Chapter 5
Statement of the Main Problem. Basic Result

Consider the following convex separable programming problem
(C)

$$\min \left\{ c(\mathbf{x}) = \sum_{j \in J} c_j(x_j) \right\} \tag{5.1}$$

subject to

$$\sum_{j \in J} d_j(x_j) \leq \alpha \tag{5.2}$$

$$a_j \leq x_j \leq b_j, \quad j \in J, \tag{5.3}$$

where $c_j(x_j)$ and $d_j(x_j)$ are twice differentiable convex functions, defined on the open convex sets X_j in $\mathbf{R}$, $j \in J$, respectively; $d_j'(x_j) > 0$ for every $j \in J$, $\mathbf{x} = (x_j)_{j \in J}$, and $J \stackrel{\text{def}}{=} \{1, ..., n\}$.

The feasible region X (5.2)–(5.3) is an intersection of the convex set (5.2) (Theorem 1.21) and the box (5.3) of dimension $|J| = n$. Therefore X is a convex set (Theorem 1.1). In particular, when $d_j(x_j)$ are linear functions, that is, when $d_j(x_j) = d_j x_j$ (a linear function is both convex and concave, Example 1.2), then (5.2) is a half-space and the assumption $d_j'(x_j) > 0$ becomes $d_j > 0, \ j \in J$.

Since the derivative of a convex differentiable function is a monotone nondecreasing function (Theorem 1.31), the condition $d_j'(x_j) > 0$ for $a_j \leq x_j \leq b_j, j \in J$ is satisfied if and only if $d_j'(a_j) > 0$, when $a_j > -\infty$, or if and only if $\lim_{t \to -\infty} d_j'(t) > 0$, when $a_j = -\infty$. Set $d_j'(b_j) = \lim_{t \to \infty} d_j'(t)$ when $b_j = +\infty$.

The assumption that $c_j(x_j)$ and $d_j(x_j), \ j \in J$, are convex functions is reasonable because the class of convex programming problems is the largest class of nonlinear programming problems for which the saddle point optimality criteria are both necessary and sufficient conditions and for which any local minimum turns out to be

© The Author(s), under exclusive license to Springer Nature Switzerland AG 2021

S. M. Stefanov, *Separable Optimization*, Springer Optimization and Its Applications 177,
https://doi.org/10.1007/978-3-030-78401-0_5

a global minimum. Moreover, the class of convex functions is the largest class of functions such that their subdifferentials (subgradient sets) are *monotone* mappings. These properties are employed in the approach suggested in this part of the book.

Furthermore, problem (5.1)–(5.3) arises in many cases, especially when $d_j(x_j)$ are linear functions, for example, in production planning and scheduling, in the allocation of financial resources, in the allocation of promotional resources among competing activities, in the theory of search, in subgradient optimization, in facility location, in the implementation of projection methods when the feasible region is of the form (5.2)–(5.3), etc. That is why we need efficient algorithms for solving such a problem.

Let the following assumptions be satisfied.

5.a) $a_j \le b_j$ for all $j \in J$. If $a_k = b_k$ for some $k \in J$, then the value $x_k := a_k = b_k$ is determined in advance.

5.b) $\sum_{j\in J} d_j(a_j) \le \alpha$. Otherwise the constraints (5.2)–(5.3) are inconsistent and $X = \emptyset$. In addition to this assumption, we suppose that $\alpha \le \sum_{j\in J} d_j(b_j)$ in some cases which are specified below.

Assume that *Slater's constraint qualification* is satisfied, that is, there exists a point $\overline{\mathbf{x}} = (\overline{x}_1, \dots, \overline{x}_n) \in X$ such that $\sum_{j\in J} d_j(\overline{x}_j) < \alpha$.

Let $h_j^{\le}$ be the value of x_j for which $c'_j(h_j^{\le}) = 0,\ j \in J$. If such a point does not exist, since $c'_j(x_j)$ is a monotone nondecreasing function ($c_j(x_j)$ is convex), we adopt $h_j^{\le} = -\infty$.

The Lagrangian for problem (5.1)–(5.3) is

$$L(\mathbf{x}, \mathbf{u}, \mathbf{v}, \lambda)$$

$$= \sum_{j\in J} c_j(x_j) + \lambda\Big(\sum_{j\in J} d_j(x_j) - \alpha\Big) + \sum_{j\in J} u_j(a_j - x_j) + \sum_{j\in J} v_j(x_j - b_j),$$

where $\lambda \in \mathbf{R}^1_+$; $\mathbf{u},\ \mathbf{v} \in \mathbf{R}^n_+$, and $\mathbf{R}^n_+$ consists of all vectors with n real nonnegative components.

Theorem 5.1 (Characterization of the optimal solution to problem (C))
A feasible solution $\mathbf{x}^* = (x_j^*)_{j\in J} \in X$ *is an optimal solution to problem* (C) (5.1)–(5.3) *if and only if there exists a* $\lambda \in \mathbf{R}^1_+$ *such that*

$$x_j^* = a_j, \quad j \in J_a^\lambda \stackrel{\text{def}}{=} \left\{ j \in J : \lambda \ge -\frac{c'_j(a_j)}{d'_j(a_j)} \right\}, \tag{5.4}$$

$$x_j^* = b_j, \quad j \in J_b^\lambda \stackrel{\text{def}}{=} \left\{ j \in J : \lambda \le -\frac{c'_j(b_j)}{d'_j(b_j)} \right\}, \tag{5.5}$$

$$x_j^* : \lambda d_j'(x_j^*) = -c_j'(x_j^*),$$

$$j \in J^\lambda \stackrel{\text{def}}{=} \left\{ j \in J : -\frac{c_j'(b_j)}{d_j'(b_j)} \le \lambda \le -\frac{c_j'(a_j)}{d_j'(a_j)} \right\}. \tag{5.6}$$

When $c(\mathbf{x})$ is strictly convex, inequalities in (5.6) *are strict.*

Proof. The Karush-Kuhn-Tucker (KKT) necessary and sufficient optimality conditions for the minimum solution $\mathbf{x}^*$ are

$$c_j'(x_j^*) + \lambda d_j'(x_j^*) - u_j + v_j = 0\,, \quad j \in J \tag{5.7}$$

$$u_j(a_j - x_j^*) = 0, \quad j \in J \tag{5.8}$$

$$v_j(x_j^* - b_j) = 0, \quad j \in J \tag{5.9}$$

$$\lambda\left(\sum_{j\in J} d_j(x_j^*) - \alpha\right) = 0, \quad \lambda \in \mathbf{R}_+^1 \tag{5.10}$$

$$\sum_{j\in J} d_j(x_j^*) \le \alpha \tag{5.11}$$

$$a_j \le x_j^* \le b_j, \quad j \in J \tag{5.12}$$

$$u_j \in \mathbf{R}_+^1,\ v_j \in \mathbf{R}_+^1\,, \quad j \in J. \tag{5.13}$$

Here λ, u_j, v_j $(j \in J)$ are the Lagrange multipliers associated with the constraints (5.2), $a_j \le x_j,\ x_j \le b_j$ $(j \in J)$, respectively. If $a_j = -\infty$ or $b_j = +\infty$ for some $j \in J$, we do not consider the corresponding condition (5.8) [condition (5.9), respectively] and multiplier u_j [multiplier v_j, respectively].

Necessity. Let $\mathbf{x}^* = (x_j^*)_{j\in J}$ be an optimal solution to problem (C) (5.1)–(5.3). Then there exist constants $\lambda, u_j, v_j,\ j \in J$ such that KKT conditions (5.7)–(5.13) are satisfied.

Consider both possible cases for $\lambda \ge 0$.

1) Let $\lambda > 0$. Then system (5.7)–(5.13) becomes (5.7), (5.8), (5.9), (5.12), (5.13), and

$$\sum_{j\in J} d_j(x_j^*) = \alpha, \tag{5.14}$$

that is, the inequality constraint (5.2) is satisfied as an equality for $x_j^*,\ j \in J$ in this case.

a) If $x_j^* = a_j$, then $u_j \ge 0$, $v_j = 0$ according to (5.8), (5.9), and (5.13). Therefore (5.7) and (5.13) imply $c_j'(x_j^*) = u_j - \lambda d_j'(x_j^*) \ge -\lambda d_j'(x_j^*)$. Since $d_j'(x_j^*) > 0$ by assumption,

$$\lambda \geq -\frac{c'_j(x_j^*)}{d'_j(x_j^*)} \equiv -\frac{c'_j(a_j)}{d'_j(a_j)}.$$

b) If $x_j^* = b_j$, then $u_j = 0,\ v_j \geq 0$ according to (5.8), (5.9), and (5.13). Therefore (5.7) and (5.13) imply $c'_j(x_j^*) = -v_j - \lambda d'_j(x_j^*) \leq -\lambda d'_j(x_j^*)$. Hence

$$\lambda \leq -\frac{c'_j(x_j^*)}{d'_j(x_j^*)} \equiv -\frac{c'_j(b_j)}{d'_j(b_j)}.$$

c) If $a_j < x_j^* < b_j$, then $u_j = v_j = 0$ according to (5.8) and (5.9). Therefore (5.7) implies $-c'_j(x_j^*) = \lambda d'_j(x_j^*)$. Since $c_j(x_j),\ d_j(x_j),\ j \in J$ are single-variable convex differentiable functions, $c'_j(x_j),\ d'_j(x_j)$ are monotone nondecreasing functions of $x_j,\ j \in J$ (Theorem 1.31), that is,

$$[c'_j(x_j^1) - c'_j(x_j^2)](x_j^1 - x_j^2) \geq 0 \quad \forall\, x_j^1, x_j^2,\ j \in J, \tag{5.15}$$

$$[d'_j(x_j^1) - d'_j(x_j^2)](x_j^1 - x_j^2) \geq 0 \quad \forall\, x_j^1, x_j^2,\ j \in J. \tag{5.16}$$

Since $d'_j(x_j^*) > 0,\ j \in J,\ \lambda > 0$ by the assumption, then $-c'_j(x_j^*) > 0$. From

$$a_j < x_j^* < b_j$$

in the considered case and from (5.15), (5.16) it follows that

$$c'_j(a_j) \leq c'_j(x_j^*) \leq c'_j(b_j), \quad j \in J,$$
$$d'_j(a_j) \leq d'_j(x_j^*) \leq d'_j(b_j), \quad j \in J.$$

Using that $-c'_j(x_j^*) > 0,\ d'_j(x_j^*) > 0,\ \lambda = -\frac{c'_j(x_j^*)}{d'_j(x_j^*)},\ \frac{1}{d'_j(x_j)}$ is a monotone nonincreasing function as a reciprocal of the derivative of the convex function $d_j(x_j)$ with $d'_j(x_j) > 0$, we obtain

$$\lambda = -\frac{c'_j(x_j^*)}{d'_j(x_j^*)} \leq -\frac{c'_j(x_j^*)}{d'_j(a_j)} \leq -\frac{c'_j(a_j)}{d'_j(a_j)},$$

$$\lambda = -\frac{c'_j(x_j^*)}{d'_j(x_j^*)} \geq -\frac{c'_j(x_j^*)}{d'_j(b_j)} \geq -\frac{c'_j(b_j)}{d'_j(b_j)},$$

that is,

$$-\frac{c'_j(b_j)}{d'_j(b_j)} \leq \lambda \leq -\frac{c'_j(a_j)}{d'_j(a_j)}.$$

When $c(\mathbf{x})$ is strictly convex, these inequalities are strict

$$-\frac{c_j'(b_j)}{d_j'(b_j)} < \lambda < -\frac{c_j'(a_j)}{d_j'(a_j)}.$$

2) Let $\lambda = 0$. Then system (5.7)–(5.13) becomes

$$c_j'(x_j^*) - u_j + v_j = 0, \quad j \in J \tag{5.7'}$$

and (5.8), (5.9), (5.11), (5.12), and (5.13).

a) If $x_j^* = a_j$, then $u_j \geq 0$, $v_j = 0$ according to (5.8), (5.9), and (5.13). Therefore (5.7′) implies $c_j'(a_j) \equiv c_j'(x_j^*) = u_j \geq 0$. Multiplying both sides of this inequality by $-\frac{1}{d_j'(a_j)}$ (< 0 by the assumption), we obtain

$$-\frac{c_j'(a_j)}{d_j'(a_j)} \leq 0 \equiv \lambda.$$

b) If $x_j^* = b_j$, then $u_j = 0$, $v_j \geq 0$ in accordance with (5.8), (5.9), and (5.13). Therefore (5.7′) implies $c_j'(b_j) \equiv c_j'(x_j^*) = -v_j \leq 0$. Multiplying this inequality by $-\frac{1}{d_j'(b_j)} < 0$, we get

$$-\frac{c_j'(b_j)}{d_j'(b_j)} \geq 0 \equiv \lambda.$$

c) If $a_j < x_j^* < b_j$, then $u_j = v_j = 0$ according to (5.8), (5.9). Therefore (5.7′) implies $c_j'(x_j^*) = 0$, that is, $x_j^* = h_j^{\leq}$. Since $b_j > x_j^*, x_j^* > a_j$, $j \in J$ by the assumption, then from (5.15) it follows that

$$c_j'(b_j) \geq c_j'(x_j^*) = 0, \quad 0 = c_j'(x_j^*) \geq c_j'(a_j).$$

Multiplying the first inequality by $-\frac{1}{d_j'(b_j)} < 0$ and the second inequality by $-\frac{1}{d_j'(a_j)} < 0$, we obtain

$$-\frac{c_j'(b_j)}{d_j'(b_j)} \leq 0 \equiv \lambda, \quad \lambda \equiv 0 \leq -\frac{c_j'(a_j)}{d_j'(a_j)}.$$

Therefore

$$-\frac{c_j'(b_j)}{d_j'(b_j)} \leq \lambda \leq -\frac{c_j'(a_j)}{d_j'(a_j)}.$$

When $c(\mathbf{x})$ is strictly convex, these inequalities are strict

$$-\frac{c_j'(b_j)}{d_j'(b_j)} < \lambda < -\frac{c_j'(a_j)}{d_j'(a_j)}.$$

In order to describe cases a), b), and c) for both 1) and 2), it is convenient to introduce the following index sets: J_a^λ, J_b^λ, J^λ defined by (5.4), (5.5), and (5.6), respectively. Obviously $J_a^\lambda \cup J_b^\lambda \cup J^\lambda = J$. The "necessity" part of Theorem 5.1 is proved.

Sufficiency. Conversely, let $\mathbf{x}^* \in X$ and components of $\mathbf{x}^*$ satisfy (5.4), (5.5), and (5.6), where $\lambda \in \mathbf{R}_+^1$.

1) If $\lambda > 0$ then $c_j'(x_j^*) < 0,\ j \in J^\lambda$ according to (5.6) with $d_j'(x_j^*) > 0$. Set:

$$\lambda = -\frac{c_j'(x_j^*)}{d_j'(x_j^*)} = \lambda(\mathbf{x}^*) \quad (> 0) \quad \text{from (5.6) such that}$$

$$\sum_{j \in J_a^\lambda} d_j(a_j) + \sum_{j \in J_b^\lambda} d_j(b_j) + \sum_{j \in J^\lambda} d_j(x_j^*) = \alpha;$$

$$u_j = v_j = 0 \qquad \text{for } j \in J^\lambda;$$

$$u_j = c_j'(a_j) + \lambda d_j'(a_j) \ (\geq 0 \text{ according to (5.4)}), \quad v_j = 0 \quad \text{for } j \in J_a^\lambda;$$

$$u_j = 0,\ v_j = -c_j'(b_j) - \lambda d_j'(b_j) \ (\geq 0 \text{ according to (5.5)}) \quad \text{for } j \in J_b^\lambda.$$

By using these expressions, it is easy to check that conditions (5.7), (5.8), (5.9), (5.10), and (5.13) are satisfied, and conditions (5.11) and (5.12) are also satisfied according to the assumption $\mathbf{x}^* \in X$.

2) If $\lambda = 0$, then $c_j'(x_j^*) = 0,\ j \in J^\lambda$ according to (5.6), and

$$J^{\lambda=0} = \left\{ j \in J : -\frac{c_j'(b_j)}{d_j'(b_j)} \leq 0 \leq -\frac{c_j'(a_j)}{d_j'(a_j)} \right\}.$$

Since $d_j'(x_j) > 0$ for each x_j, then $c_j'(b_j) \geq 0,\ c_j'(a_j) \leq 0,\ j \in J^0$. Therefore there exists some $x_j^* = h_j^{\leq} \in [a_j, b_j]$ such that $c_j'(x_j^*) = 0$ according to the Darboux theorem (see Appendix A, Theorem A.4).
Set:

$$\lambda = -\frac{c_j'(x_j^*)}{d_j'(x_j^*)} \quad (= 0); \quad u_j = v_j = 0 \qquad \text{for } j \in J^{\lambda=0};$$

$$u_j = c_j'(a_j) + \lambda d_j'(a_j) = c_j'(a_j) \quad (\geq 0), \quad v_j = 0 \qquad \text{for } j \in J_a^{\lambda=0};$$

$$u_j = 0, \quad v_j = -c_j'(b_j) - \lambda d_j'(b_j) = -c_j'(b_j) \quad (\geq 0) \quad \text{for } j \in J_b^{\lambda=0}.$$

Obviously conditions (5.7), (5.8), (5.9), and (5.13) are satisfied; conditions (5.11) and (5.12) are also satisfied according to the assumption $\mathbf{x}^* \in X$, and condition (5.10) obviously is satisfied for $\lambda = 0$.

In both cases 1) and 2) of the "sufficiency" part of the proof, x_j^*, λ, u_j, v_j, $j \in J$ satisfy KKT conditions (5.7)–(5.13) which are necessary and sufficient conditions for a feasible solution to be an optimal solution to the convex minimization problem. Therefore $\mathbf{x}^*$ is an optimal solution to problem (C) (5.1)–(5.3). When $c(\mathbf{x}) = \sum_{j \in J} c_j(x_j)$ is a strictly convex function, this optimal solution is *unique* (Theorem 1.38). ■

Remark 5.1 *Since* $\lambda \geq 0,\ u_j \geq 0,\ v_j \geq 0,\ j \in J\ (|J| = n)$ *and since the complementary conditions* (5.8), (5.9), *and* (5.10) *must be satisfied, in order to find* $x_j^*,\ j \in J$ *from the system* (5.7)–(5.13), *we have to consider all possible cases for* λ, u_j, v_j*: all* λ, u_j, v_j *equal to* 0*; all* λ, u_j, v_j *different from* 0*; some of them equal to* 0*; some of them different from* 0*. The number of these cases is* $2^{2|J|+1} = 2^{2n+1}$*, where* $2n+1$ *is the number of* $\lambda, u_j, v_j,\ j \in J$*. Obviously this is an enormous number of cases even with small values of n, for example,* $n = 50$ *or* $n = 100$*. Moreover, in each case we have to solve a large-scale system of nonlinear equations.*

In view of this discussion, the importance of Theorem 5.1 consists in the fact that it describes components of the optimal solution to problem (C) *only through the Lagrange multiplier* λ *associated with the inequality constraint* (5.2).

If we knew the optimal value of λ*, then by the use of Theorem 5.1 we could directly obtain the optimal solution to problem* (C) *(or establish that it does not have an optimal solution). However, we do not know this optimal value of* λ *in advance. That is why we define an iterative process with respect to the Lagrange multiplier* λ *and we prove the convergence of this process in Chapter Seven* The Algorithms.

From $d_j'(x_j) > 0$*, from* (5.15), (5.16), *and* $a_j \leq b_j,\ j \in J$*, it follows that*

$$ub_j \stackrel{\text{def}}{=} -\frac{c_j'(b_j)}{d_j'(b_j)} \leq -\frac{c_j'(b_j)}{d_j'(a_j)} \leq -\frac{c_j'(a_j)}{d_j'(a_j)} \stackrel{\text{def}}{=} la_j,\ j \in J$$

for the expressions by means of which we define the index sets $J_a^\lambda, J_b^\lambda, J^\lambda$.

The problem of how to ensure a feasible solution to problem (C) (5.1)–(5.3), which is an assumption of Theorem 5.1, is discussed in Chapter Seven *The Algorithms.*

Chapter 6
Version One: Linear Equality Constraints

6.1 Single Linear Equality Constraint

Denote by $(C^=)$ a version of problem (C) where instead of (5.2), we have a *linear equality constraint*:
$(C^=)$

$$\min \left\{ c(\mathbf{x}) = \sum_{j \in J} c_j(x_j) \right\} \tag{6.1}$$

subject to

$$\sum_{j \in J} d_j x_j = \alpha \tag{6.2}$$

$$a_j \le x_j \le b_j, \quad j \in J, \tag{6.3}$$

where $c_j(x_j)$ are twice differentiable convex functions, defined on the open convex sets X_j in $\mathbf{R}$, $j \in J$, respectively, $d_j > 0$, $j \in J$, and $J \stackrel{\text{def}}{=} \{1, ..., n\}$.

Assumptions:

6.a) $a_j \le b_j$ for each $j \in J$.

6.b) $\sum_{j \in J} d_j a_j \le \alpha \le \sum_{j \in J} d_j b_j$. Otherwise the constraints (6.2)–(6.3) are inconsistent and $X^L = \emptyset$, where X^L is defined by (6.2)–(6.3).

Let $h_j^=$, $j \in J$ be the value of x_j for which $c_j'(h_j^=) = 0$. If a finite point with this property does not exist, this means that the function $c_j(x_j)$ does not change the type of its monotonicity, that is, $c_j(x_j)$ is a nondecreasing or nonincreasing function in the interval $(-\infty, +\infty)$. That is why in the case there does not exist a finite point $h_j^=$ such that $c_j'(h_j^=) = 0$, we adopt

i) $h_j^= := -\infty$, if $c_j(x_j)$ is a nondecreasing function;

ii) $h_j^= := +\infty$, if $c_j(x_j)$ is a nonincreasing function.

Since optimality conditions in convex programming are both necessary and sufficient only if the equality constraints are linear, we do not consider a problem of

© The Author(s), under exclusive license to Springer Nature Switzerland AG 2021
S. M. Stefanov, *Separable Optimization*, Springer Optimization and Its Applications 177,
https://doi.org/10.1007/978-3-030-78401-0_6

the form (6.1), (6.2′) $\sum_{j\in J} d_j(x_j) = \alpha$, (6.3), where $d_j(x_j)$ are *convex nonlinear* differentiable functions with $d'_j(x_j) > 0,\ j \in J$.

For problem $(C^=)$, we can prove the following Theorem 6.1 which is similar to Theorem 5.1. However, since $d'_j(x_j) = d_j = const. > 0$ and since the Lagrange multiplier λ associated with (6.2) is not involved in a complementary condition of the form (5.10), the proof of Theorem 6.1 is simpler than the proof of Theorem 5.1.

For the sake of simplicity, throughout Theorem 6.1 we use the same notation $\lambda, u_j, v_j, j \in J$ for the Lagrange multipliers associated with (6.2), $a_j \le x_j$, $x_j \le b_j$, respectively, as we have used it in Theorem 5.1.

Theorem 6.1 (Characterization of the optimal solution to problem $(C^=)$) *A feasible solution* $\mathbf{x}^* = (x_j^*)_{j\in J} \in X^L$ *is an optimal solution to problem* $(C^=)$ *if and only if there exists a* $\lambda \in \mathbf{R}^1$ *such that*

$$x_j^* = a_j, \quad j \in J_a^\lambda \stackrel{\text{def}}{=} \left\{ j \in J : \lambda \ge -\frac{c'_j(a_j)}{d_j} \right\}, \tag{6.4}$$

$$x_j^* = b_j, \quad j \in J_b^\lambda \stackrel{\text{def}}{=} \left\{ j \in J : \lambda \le -\frac{c'_j(b_j)}{d_j} \right\}, \tag{6.5}$$

$$x_j^* : \lambda d_j = -c'_j(x_j^*), \quad j \in J^\lambda \stackrel{\text{def}}{=} \left\{ j \in J : -\frac{c'_j(b_j)}{d_j} \le \lambda \le -\frac{c'_j(a_j)}{d_j} \right\}. \tag{6.6}$$

When $c(\mathbf{x})$ *is strictly convex, inequalities in* (6.6) *are strict.*

Proof. The Lagrangian for problem $(C^=)$ is

$$L(\mathbf{x}, \mathbf{u}, \mathbf{v}, \lambda)$$

$$= \sum_{j\in J} c_j(x_j) + \lambda\Big(\sum_{j\in J} d_j x_j - \alpha\Big) + \sum_{j\in J} u_j(a_j - x_j) + \sum_{j\in J} v_j(x_j - b_j),$$

where $\lambda \in \mathbf{R}^1$; $\mathbf{u}, \mathbf{v} \in \mathbf{R}^n_+$. The KKT necessary and sufficient optimality conditions for problem $(C^=)$ are

$$c'_j(x_j^*) + \lambda d_j - u_j + v_j = 0\,, \quad j \in J \tag{6.7}$$

$$u_j(a_j - x_j^*) = 0\,, \quad j \in J \tag{6.8}$$

$$v_j(x_j^* - b_j) = 0\,, \quad j \in J \tag{6.9}$$

$$\lambda \in \mathbf{R}^1;\ u_j \in \mathbf{R}^1_+,\ v_j \in \mathbf{R}^1_+\,, \quad j \in J \tag{6.10}$$

$$\sum_{j\in J} d_j x_j^* = \alpha \tag{6.11}$$

$$a_j \le x_j^* \le b_j, \quad j \in J. \tag{6.12}$$

Necessity. Let $\mathbf{x}^* = (x_j^*)_{j \in J}$ be an optimal solution to problem $(C^=)$.

a) If $x_j^* = a_j$, then $u_j \geq 0$, $v_j = 0$ according to (6.8), (6.9), and (6.10). Therefore (6.7) implies $c_j'(x_j^*) = u_j - \lambda d_j \geq -\lambda d_j$. Since $d_j > 0$,

$$\lambda \geq -\frac{c_j'(x_j^*)}{d_j} \equiv -\frac{c_j'(a_j)}{d_j}.$$

b) If $x_j^* = b_j$, then $u_j = 0$, $v_j \geq 0$ according to (6.8), (6.9), and (6.10). Therefore (6.7) implies $c_j'(x_j^*) = -v_j - \lambda d_j \leq -\lambda d_j$. Hence

$$\lambda \leq -\frac{c_j'(x_j^*)}{d_j} \equiv -\frac{c_j'(b_j)}{d_j}.$$

c) If $a_j < x_j^* < b_j$, then $u_j = v_j = 0$ according to (6.8), (6.9). Therefore (6.7) implies $c_j'(x_j^*) = -\lambda d_j$. Since $c_j(x_j)$ are convex differentiable functions, $c_j'(x_j)$ are nondecreasing functions, $j \in J$. Because $b_j > x_j^* > a_j$, $j \in J$ by assumption, from the monotonicity of $c_j'(x_j)$ it follows that

$$c_j'(b_j) \geq c_j'(x_j^*) \geq c_j'(a_j), \quad j \in J.$$

Multiplying this chain of inequalities by $-\frac{1}{d_j}$ and using that $d_j > 0$, $\lambda = -\frac{c_j'(x_j^*)}{d_j}$, we obtain

$$-\frac{c_j'(b_j)}{d_j} \leq \lambda \leq -\frac{c_j'(a_j)}{d_j}.$$

When $c(\mathbf{x})$ is strictly convex, these inequalities are strict.

To describe cases a), b), and c), we introduce the following index sets: J_a^λ, J_b^λ, J^λ defined by (6.4), (6.5), and (6.6), respectively. Obviously $J_a^\lambda \cup J_b^\lambda \cup J^\lambda = J$. The "necessity" part of Theorem 6.1 is proved.

Sufficiency. Conversely, let $\mathbf{x}^* \in X^L$ and components of $\mathbf{x}^*$ satisfy (6.4), (6.5), and (6.6). Set:

$$\lambda = -\frac{c_j'(x_j^*)}{d_j}; \quad u_j = v_j = 0 \qquad \text{for } j \in J^\lambda;$$

$$u_j = c_j'(a_j) + \lambda d_j \quad (\geq 0), \quad v_j = 0 \qquad \text{for } j \in J_a^\lambda;$$

$$u_j = 0, \quad v_j = -c_j'(b_j) - \lambda d_j \quad (\geq 0) \qquad \text{for } j \in J_b^\lambda.$$

Clearly $\mathbf{x}^*, \lambda, u_j, v_j, j \in J$ satisfy KKT conditions (6.7)–(6.12). Consequently, $\mathbf{x}^*$ is an optimal solution to problem $(C^=)$ (6.1)–(6.3). When $c(\mathbf{x}) = \sum_{j \in J} c_j(x_j)$ is a strictly convex function, this optimal solution is *unique* (Theorem 1.38). ■

The importance of Theorem 6.1 consists in the fact that it describes the optimal solution to problem $(C^{=})$ only through the Lagrange multiplier λ associated with the single equality constraint (6.2).

As in the fourth paragraph of Remark 5.1, we can obtain that

$$-\frac{c_j'(b_j)}{d_j} \le -\frac{c_j'(a_j)}{d_j}.$$

The problem of how to ensure a feasible solution to problem $(C^{=})$ is discussed after the proof of Theorem 7.2 in Section 7.5 *Algorithm 2 (for Problem* $(C^{=})$*) and Its Convergence.*

6.2 Several Linear Equality Constraints

Denote by $(C_m^{=})$ the problem

$(C_m^{=})$

$$\min \left\{ c(\mathbf{x}) = \sum_{j \in J} c_j(x_j) \right\} \tag{6.13}$$

subject to

$$D\mathbf{x} = \boldsymbol{\alpha} \tag{6.14}$$

$$\mathbf{a} \le \mathbf{x} \le \mathbf{b}, \tag{6.15}$$

where $c_j(x_j)$ are differentiable strictly convex functions, defined on the open convex sets X_j in $\mathbf{R}$, $j \in J \stackrel{\text{def}}{=} \{1, ..., n\}$, respectively, $D = (d_{ij}) \in \mathbf{R}^{m \times n}$, $\boldsymbol{\alpha} \in \mathbf{R}^m$, $\mathbf{a} = (a_1, \ldots, a_n)$, $\mathbf{b} = (b_1, \ldots, b_n) \in \mathbf{R}^n$.

The feasible region (6.14)–(6.15) is an intersection of m hyperplanes (6.14) and the box (6.15).

Problem $(C^{=})$, considered in Section 6.1, is a special case of problem $(C_m^{=})$ with $m = 1$. We consider problem $(C^{=})$ separately because Theorem 6.1 and the algorithm suggested for $(C^{=})$ in Chapter Seven are used as a part of the algorithm for another problem in Chapter Eight.

Denote by $P_c(D, \boldsymbol{\alpha}, \mathbf{a}, \mathbf{b})$ the solution to problem $(C_m^{=})$. Since $c(\mathbf{x})$ is strictly convex as a sum of strictly convex functions (Theorem 1.17) then $P_c(D, \boldsymbol{\alpha}, \mathbf{a}, \mathbf{b})$ is uniquely defined, that is, there is at most one minimum which is both local and global (Theorem 1.36, Theorem 1.38).

Denote $\mathbf{y} = [\mathbf{x}]_{\mathbf{a}}^{\mathbf{b}}$ where $y_j = \min\{\max\{x_j, a_j\}, b_j\}$ for each $j \in J$.

The KKT conditions for $\mathbf{x}^* \in \mathbf{R}^n$ to be a minimum solution of $(C_m^{=})$ are

$$D\mathbf{x}^* = \boldsymbol{\alpha} \tag{6.16}$$

$$\mathbf{a} \leq \mathbf{x}^* \tag{6.17}$$

$$\mathbf{x}^* \leq \mathbf{b} \tag{6.18}$$

$$\mathbf{c}'(\mathbf{x}^*) + D^T\boldsymbol{\lambda} - \mathbf{u} + \mathbf{v} = \mathbf{0} \tag{6.19}$$

$$u_j(a_j - x_j^*) = 0, \quad j \in J \tag{6.20}$$

$$v_j(x_j^* - b_j) = 0, \quad j \in J \tag{6.21}$$

$$\mathbf{u} \geq \mathbf{0} \tag{6.22}$$

$$\mathbf{v} \geq \mathbf{0}, \tag{6.23}$$

where $\boldsymbol{\lambda} \in \mathbf{R}^m$, $\mathbf{u}, \mathbf{v} \in \mathbf{R}_+^n$ are the Lagrange multipliers associated with (6.14) and the two inequalities of (6.15), respectively.

The map $\mathbf{c}' \equiv \nabla c : \mathbf{R}^n \to \mathbf{R}^n$ is strictly monotone increasing since c is a strictly convex function. Therefore $(\nabla c)^{-1} : \mathbf{R}^n \to \mathbf{R}^n$ is well-defined.

Theorem 6.2 (Characterization of the optimal solution to problem $(C_m^=)$; Berman, Kovoor and Pardalos [8])
Let $c : \mathbf{R}^n \to \mathbf{R}$ be separable, differentiable, and strictly convex. Then

$$\{P_c(D, \boldsymbol{\alpha}, \mathbf{a}, \mathbf{b})\} = \left\{ (\mathbf{c}')^{-1}\Big[- D^T\mathbf{t} \Big]_{\mathbf{c}'(\mathbf{a})}^{\mathbf{c}'(\mathbf{b})} : \mathbf{t} \in \mathbf{R}^m \right\} \tag{6.24}$$

where $D, \boldsymbol{\alpha}, \mathbf{a}, \mathbf{b}$ are defined above.

Proof. Relation (6.24) is proved by two-way inclusion.

i) Let $\mathbf{x}^* = P_c(D, \boldsymbol{\alpha}, \mathbf{a}, \mathbf{b})$ for some $\boldsymbol{\alpha} \in \mathbf{R}^m$. Then there exist $\boldsymbol{\lambda} \in \mathbf{R}^m, \mathbf{u}, \mathbf{v} \in \mathbf{R}_+^n$ satisfying the KKT conditions (6.16)–(6.23) together with this $\mathbf{x}^*$.

From (6.19) it follows that

$$D^T\boldsymbol{\lambda} = -\mathbf{c}'(\mathbf{x}^*) + \mathbf{u} - \mathbf{v},$$

that is,

$$\langle \mathbf{D}_j, \boldsymbol{\lambda} \rangle = -c_j'(x_j^*) + u_j - v_j \tag{6.25}$$

for each $j \in J$.

If $\langle \mathbf{D}_j, \boldsymbol{\lambda} \rangle > -c_j'(x_j^*)$, then $u_j > v_j \geq 0$, hence $x_j^* = a_j$ according to (6.20), that is,

$$\langle \mathbf{D}_j, \boldsymbol{\lambda} \rangle > -c_j'(x_j^*) \quad \text{implies} \quad x_j^* = a_j. \tag{6.26}$$

Similarly, if $\langle \mathbf{D}_j, \boldsymbol{\lambda} \rangle < -c_j'(x_j^*)$, then $v_j > u_j \geq 0$, hence $x_j^* = b_j$ according to (6.21), that is,

$$\langle \mathbf{D}_j, \boldsymbol{\lambda} \rangle < -c_j'(x_j^*) \quad \text{implies} \quad x_j^* = b_j. \tag{6.27}$$

Since $a_j \le b_j,\ j \in J$, by assumption, we have three cases to consider.

Case 1. $\langle \mathbf{D}_j, \boldsymbol{\lambda} \rangle > -c'_j(a_j)$.

Then $\langle \mathbf{D}_j, \boldsymbol{\lambda} \rangle > -c'_j(x^*_j)$ according to (6.17) and the monotonicity of c'_j, hence $x^*_j = a_j$ in accordance with (6.26).

Case 2. $\langle \mathbf{D}_j, \boldsymbol{\lambda} \rangle < -c'_j(b_j)$.

Then $\langle \mathbf{D}_j, \boldsymbol{\lambda} \rangle < -c'_j(x^*_j)$ according to (6.18) and the monotonicity of c'_j, hence $x^*_j = b_j$ in accordance with (6.27).

Case 3. $-c'_j(b_j) \le \langle \mathbf{D}_j, \boldsymbol{\lambda} \rangle \le -c'_j(a_j)$.

If $\langle \mathbf{D}_j, \boldsymbol{\lambda} \rangle < -c'_j(x^*_j)$, then $x^*_j = b_j$ according to (6.27). Therefore $\langle \mathbf{D}_j, \boldsymbol{\lambda} \rangle \ge -c'_j(x^*_j)$ because $\langle \mathbf{D}_j, \boldsymbol{\lambda} \rangle \ge -c'_j(b_j)$ by the assumption of Case 3, a contradiction. Similarly, if we assume that $\langle \mathbf{D}_j, \boldsymbol{\lambda} \rangle > -c'_j(x^*_j)$ strictly, this would imply $x^*_j = a_j$ according to (6.26) and $\langle \mathbf{D}_j, \boldsymbol{\lambda} \rangle \le -c'_j(x^*_j)$, a contradiction.

Then $\langle \mathbf{D}_j, \boldsymbol{\lambda} \rangle = -c'_j(x^*_j)$, therefore $x^*_j = (c'_j)^{-1}(-\langle \mathbf{D}_j, \boldsymbol{\lambda} \rangle)$.

In the three cases considered, we have

$$x^*_j = (c'_j)^{-1}\Big[-\langle \mathbf{D}_j, \boldsymbol{\lambda} \rangle\Big]_{c'_j(a_j)}^{c'_j(b_j)}.$$

Hence $\mathbf{x}^* = (\mathbf{c}')^{-1}\Big[-D^T\boldsymbol{\lambda}\Big]_{\mathbf{c}'(\mathbf{a})}^{\mathbf{c}'(\mathbf{b})}$, that is,

$$\{P_c(D, \boldsymbol{\alpha}, \mathbf{a}, \mathbf{b})\} \subseteq \left\{(\mathbf{c}')^{-1}\Big[-D^T\mathbf{t}\Big]_{\mathbf{c}'(\mathbf{a})}^{\mathbf{c}'(\mathbf{b})} : \mathbf{t} \in \mathbb{R}^m\right\}.$$

ii) Conversely, suppose that $\mathbf{x}^* \in \mathbb{R}^n$ and $\mathbf{x}^* = (\mathbf{c}')^{-1}\Big[-D^T\mathbf{t}\Big]_{\mathbf{c}'(\mathbf{a})}^{\mathbf{c}'(\mathbf{b})}$ for some $\mathbf{t} \in \mathbb{R}^m$.

Set:

$\boldsymbol{\alpha} = D(\mathbf{c}')^{-1}\Big[-D^T\mathbf{t}\Big]_{\mathbf{c}'(\mathbf{a})}^{\mathbf{c}'(\mathbf{b})}$

$\boldsymbol{\lambda} = \mathbf{t}$

$\mathbf{u} = \mathbf{c}'(\mathbf{a}) + D^T\mathbf{t}$

$\mathbf{v} = -\mathbf{c}'(\mathbf{b}) - D^T\mathbf{t}$.

We have to prove that $\mathbf{x}^*, \boldsymbol{\alpha}, \boldsymbol{\lambda}, \mathbf{u}, \mathbf{v}$ satisfy the KKT conditions (6.16)–(6.23).

Obviously $\mathbf{x}^*$ and $\boldsymbol{\alpha}$ satisfy (6.16) $\equiv$ (6.14), $\mathbf{x}^*$ satisfies (6.17) and (6.18) (these are (6.15)) according to the definition of $[\mathbf{x}]_{\mathbf{a}}^{\mathbf{b}}$ and monotonicity of $\mathbf{c}'$.

In order to verify (6.19)–(6.23), we consider each $j \in J$. There are three possible cases.

Case 1. $\langle \mathbf{D}_j, \mathbf{t} \rangle > -c'_j(a_j)$.

Then $c'_j(a_j) + \langle \mathbf{D}_j, \mathbf{t} \rangle > 0$, and since $a_j \le b_j$ then $-c'_j(b_j) - \langle \mathbf{D}_j, \mathbf{t} \rangle < 0$.

Therefore $x^*_j = a_j,\ \boldsymbol{\lambda} = \mathbf{t},\ u_j = c'_j(a_j) + \langle \mathbf{D}_j, \mathbf{t} \rangle,\ v_j = 0$.

Case 2. $\langle \mathbf{D}_j, \mathbf{t} \rangle < -c'_j(b_j)$.
Then $-c'_j(b_j) - \langle \mathbf{D}_j, \mathbf{t} \rangle > 0$, and since $a_j \le b_j$ then $c'_j(a_j) + \langle \mathbf{D}_j, \mathbf{t} \rangle < 0$.
Therefore $x_j^* = b_j, \ \boldsymbol{\lambda} = \mathbf{t}, \ u_j = 0, \ v_j = -c'_j(b_j) - \langle \mathbf{D}_j, \mathbf{t} \rangle$.

Case 3. $-c'_j(b_j) \le \langle \mathbf{D}_j, \mathbf{t} \rangle \le -c'_j(a_j)$.
Then $-c'_j(b_j) - \langle \mathbf{D}_j, \mathbf{t} \rangle \le 0, \quad \langle \mathbf{D}_j, \mathbf{t} \rangle + c'_j(a_j) \le 0$.
Therefore $x_j^* = (\mathbf{c}')_j^{-1}(-\langle \mathbf{D}_j, \mathbf{t} \rangle), \ \boldsymbol{\lambda} = \mathbf{t}, \ u_j = v_j = 0$.

Obviously in each of the three cases, $x_j^*, u_j, v_j (j \in J), \boldsymbol{\lambda}$ satisfy (6.19)–(6.23) as well.

Therefore $\mathbf{x}^*, \boldsymbol{\alpha}, \boldsymbol{\lambda}, \mathbf{u}, \mathbf{v}$ satisfy the KKT conditions (6.16)–(6.23), hence $\mathbf{x}^* \in P_c(D, \boldsymbol{\alpha}, \mathbf{a}, \mathbf{b})$.

The two-way inclusion, proved in cases i) and ii), implies (6.24). ■

Define the functions $\mathbf{x} : \mathbb{R}^m \to \mathbb{R}^n, \ \boldsymbol{\alpha} : \mathbb{R}^m \to \mathbb{R}^m$ by

$$\mathbf{x}(\mathbf{t}) = (\mathbf{c}')^{-1}\Big[- D^T\mathbf{t}\Big]_{\mathbf{c}'(\mathbf{a})}^{\mathbf{c}'(\mathbf{b})}, \tag{6.28}$$

$$\boldsymbol{\alpha}(\mathbf{t}) = D(\mathbf{c}')^{-1}\Big[- D^T\mathbf{t}\Big]_{\mathbf{c}'(\mathbf{a})}^{\mathbf{c}'(\mathbf{b})}. \tag{6.29}$$

Then the following Corollary holds.

Corollary 6.1 (Further characterization of the optimal solution and optimal Lagrange multipliers to problem $(C_m^=)$; Berman, Kovoor and Pardalos [8])
Vectors $\mathbf{x}^* \in \mathbb{R}^n, \ \boldsymbol{\alpha}^* \in \mathbb{R}^m$ *satisfy* $\mathbf{x}^* = P_c(D, \boldsymbol{\alpha}^*, \mathbf{a}, \mathbf{b})$ *if and only if there exists* $\mathbf{t}^* \in \mathbb{R}^m$ *such that*

$$\mathbf{x}(\mathbf{t}^*) = \mathbf{x}^* \tag{6.30}$$

$$\boldsymbol{\alpha}(\mathbf{t}^*) = \boldsymbol{\alpha}^*. \tag{6.31}$$

Proof of Corollary 6.1 follows from the statement of problem $(C_m^=)$ and (6.24).

From Corollary 6.1 it follows that $\mathbf{x}^* = P_c(D, \boldsymbol{\alpha}^*, \mathbf{a}, \mathbf{b})$ can be solved with respect to $\mathbf{x}^*$ for given $\boldsymbol{\alpha}^*$ by first solving (6.31) for $\mathbf{t}^*$ and then calculating $\mathbf{x}^*$ by using (6.30).

Let S be the set of solutions to (6.31) for a particular value of $\boldsymbol{\alpha}^*$:

$$S = \{\mathbf{t} \in \mathbb{R}^m : \ \boldsymbol{\alpha}(\mathbf{t}) = \boldsymbol{\alpha}^*\}. \tag{6.32}$$

According to (6.29), each component of $\boldsymbol{\alpha}(\mathbf{t})$ is a linear combination of the same set of terms. Each term $(\mathbf{c}')_j^{-1}\Big[- D^T\mathbf{t}\Big]_{\mathbf{c}'_j(a_j)}^{\mathbf{c}'_j(b_j)}$ is a smooth function of $\mathbf{t}$ except on the pair of break hyperplanes

$$A_j = \{\mathbf{t} \in \mathbb{R}^m : \ \langle \mathbf{D}_j, \mathbf{t} \rangle = -c'_j(a_j)\},$$

$$B_j = \{\mathbf{t} \in \mathbf{R}^m : \langle \mathbf{D}_j, \mathbf{t} \rangle = -c'_j(b_j)\}.$$

These break hyperplanes are generalizations of break points considered in Chapter Seven, Sections 7.1 *Analysis of the Optimal Solution to Problem* (C) and 7.4*Analysis of the Optimal Solution to Problem* $(C^=)$.

In [8], a linear time algorithm based on the construction of hyperplane arrangements is proposed. This algorithm is not considered here.

Chapter 7
The Algorithms

7.1 Analysis of the Optimal Solution to Problem (C)

Before the formal statement of the algorithm suggested, some properties of the optimal solution to problem (C) are discussed which turn out to be useful.

Using (5.4), (5.5), and (5.6), condition (5.10) can be written in the form

$$\lambda\Big(\sum_{j\in J_a^\lambda} d_j(a_j) + \sum_{j\in J_b^\lambda} d_j(b_j) + \sum_{j\in J^\lambda} d_j(x_j^*) - \alpha\Big) = 0, \quad \lambda \geq 0. \tag{5.10'}$$

Since the optimal solution $\mathbf{x}^*$ to problem (C) (5.1)–(5.3) depends on λ, consider the components of $\mathbf{x}^*$ as functions of λ for different $\lambda \in \mathbf{R}^1_+$:

$$x_j(\lambda) = \begin{cases} a_j, & \lambda \geq -\frac{c_j'(a_j)}{d_j'(a_j)} \qquad (j \in J_a^\lambda), \\ b_j, & \lambda \leq -\frac{c_j'(b_j)}{d_j'(b_j)} \qquad (j \in J_b^\lambda), \\ x_j^* : \; c_j'(x_j^*) + \lambda d_j'(x_j^*) = 0, & -\frac{c_j'(b_j)}{d_j'(b_j)} \leq \lambda \leq -\frac{c_j'(a_j)}{d_j'(a_j)} \qquad (j \in J^\lambda). \end{cases} \tag{7.1}$$

Functions $x_j(\lambda),\ j \in J$ are piecewise, monotone, piecewise differentiable functions of λ (with two break points at $\lambda = -\frac{c_j'(a_j)}{d_j'(a_j)}$ and $\lambda = -\frac{c_j'(b_j)}{d_j'(b_j)}$, $j \in J$).

Let

$$\delta(\lambda) \stackrel{\text{def}}{=} \sum_{j\in J_a^\lambda} d_j(a_j) + \sum_{j\in J_b^\lambda} d_j(b_j) + \sum_{j\in J^\lambda} d_j(x_j(\lambda)) - \alpha. \tag{7.2}$$

According to (7.1) and $u_j = v_j = 0,\ j \in J^\lambda$, condition (5.7) becomes

$$c_j'(x_j(\lambda)) + \lambda d_j'(x_j(\lambda)) = 0, \quad j \in J^\lambda. \tag{7.3}$$

© The Author(s), under exclusive license to Springer Nature Switzerland AG 2021

S. M. Stefanov, *Separable Optimization*, Springer Optimization and Its Applications 177,
https://doi.org/10.1007/978-3-030-78401-0_7

If we differentiate (7.3) with respect to λ (using that $c_j''(x_j)$, $d_j''(x_j)$ exist by assumption for all $j \in J^\lambda$; $x_j'(\lambda)$ exist for all $j \in J^\lambda$ because $x_j(\lambda)$ are defined by $x_j(\lambda) = x_j^*$ such that $c_j'(x_j^*) + \lambda d_j'(x_j^*) = 0$ for such $j \in J^\lambda$), we obtain

$$c_j''(x_j(\lambda))x_j'(\lambda) + d_j'(x_j(\lambda)) + \lambda d_j''(x_j(\lambda))x_j'(\lambda) = 0, \quad j \in J^\lambda. \tag{7.4}$$

Therefore

$$x_j'(\lambda) = -\frac{d_j'(x_j(\lambda))}{\lambda d_j''(x_j(\lambda)) + c_j''(x_j(\lambda))}, \quad j \in J^\lambda \tag{7.5}$$

and since $c_j''(x_j) \geq 0, \;\; d_j''(x_j) \geq 0$ as second derivatives of convex differentiable functions, $d_j'(x_j) > 0$ by the assumption, $\lambda \geq 0$, then $x_j'(\lambda) < 0, \; j \in J^\lambda$. (If we assume that $\lambda d_j''(x_j(\lambda)) + c_j''(x_j(\lambda)) = 0$, then $d_j'(x_j(\lambda)) = 0, \; j \in J^\lambda$ according to (7.4). However, $d_j'(x_j) > 0, \; j \in J$ by the assumption, a contradiction.) Consequently,

$$\delta'(\lambda) \equiv \sum_{j \in J^\lambda} d_j'(x_j(\lambda))x_j'(\lambda) < 0 \tag{7.6}$$

when $J^\lambda \neq \emptyset$, and $\delta'(\lambda) = 0$ when $J^\lambda = \emptyset$. Hence $\delta(\lambda)$ is *a monotone nonincreasing function* of $\lambda, \; \lambda \in \mathbf{R}^1_+$. Therefore $\max_{\lambda \geq 0} \delta(\lambda)$ is attained at the minimal admissible value of λ, that is, at $\lambda = 0$.

Case 1. If $\delta(0) > 0$, in order that (5.10′) and (5.11) be satisfied, there exists a $\lambda^* > 0$ such that $\delta(\lambda^*) = 0$, that is,

$$\sum_{j \in J} d_j(x_j^*) = \alpha. \tag{7.7}$$

This means that the inequality constraint (5.2) is satisfied as equality for λ^* in this case.

Case 2. If $\delta(0) < 0$, then $\delta(\lambda) < 0$ for all $\lambda \geq 0$, and the maximum of $\delta(\lambda)$ with $\lambda \geq 0$ is $\delta(0) = \max_{\lambda \geq 0} \delta(\lambda)$ and it is attained at $\lambda = 0$ in this case. In order that (5.10′) be satisfied, λ must be equal to 0. Therefore $x_j^* = h_j^{\leq}, \; j \in J^{\lambda=0}$ according to (5.6) and definition of $h_j^{\leq}$.

Case 3. In the special case when $\delta(0) = 0$, the maximum $\delta(0) = \max_{\lambda \geq 0} \delta(\lambda)$ of $\delta(\lambda)$ is also attained at the minimal admissible value of λ, that is, at $\lambda = 0$, because $\delta(\lambda)$ is a monotone nonincreasing function.

As we have seen, for the optimal value of λ we have $\lambda \geq 0$ in all possible cases, as the KKT condition (5.10) requires. We have shown that in Case 1 we need an algorithm for finding λ^* which satisfies the KKT conditions (5.7) –(5.13) and such that λ^* satisfies (5.11) with an equality. In order that this be satisfied, the set

$$X^= \stackrel{\text{def}}{=} \left\{ \mathbf{x} \in \mathbf{R}^n : \sum_{j \in J} d_j(x_j) = \alpha, \; a_j \leq x_j \leq b_j, \; j \in J \right\}$$

must be nonempty. That is why we have required $\alpha \le \sum_{j\in J} d_j(b_j)$ in some cases in addition to the assumption $\sum_{j\in J} d_j(a_j) \le \alpha$ (see assumption 5.b). We have also used this in the proof of Theorem 5.1, the "sufficiency" part, when $\lambda > 0$.

From the equation $\delta(\lambda) = 0$, where $\delta(\lambda)$ is defined by (7.2), we are able to determine λ as an implicit function of $\mathbf{x}$,

$$\lambda = \lambda(\mathbf{x}), \tag{7.8}$$

because $\delta'(\lambda) < 0$ when $J^\lambda \neq \emptyset$ according to (7.6) (it is important that $\delta'(\lambda) \neq 0$). If it is possible to obtain a closed form expression of λ (for example, this is always possible when $d_j(x_j) = d_j x_j,\ j \in J$), we use it in the algorithm suggested for solving problem (C). It turns out that for our purposes, without loss of generality, we can assume that $\delta'(\lambda) \neq 0$, that is, $\delta(\lambda)$ depends on λ, which means that $J^\lambda \neq \emptyset$ (see the third paragraph of Remark 7.1 below).

At iteration k of the implementation of algorithms, denote by $\lambda^{(k)}$ the value of the Lagrange multiplier associated with the constraint (5.2) [constraint (6.2)], by $\alpha^{(k)}$ the right-hand side of constraint (5.2) [constraint (6.2)], by $J^{(k)}$, $J_a^{\lambda(k)}$, $J_b^{\lambda(k)}$, and $J^{\lambda(k)}$ the current sets J, J_a^λ, J_b^λ, J^λ, respectively.

The following algorithm for solving problem (C) with strictly convex differentiable functions $c_j(x_j),\ j \in J$ is based on Theorem 5.1.

7.2 Statement of Algorithm 1 (for Problem (C))

Step 0. (Initialization) $J := \{1, \dots, n\}$,
$k := 0,\ \alpha^{(0)} := \alpha,\ n^{(0)} := n$,
$J^{(0)} := J,\ J_a^\lambda := \emptyset,\ J_b^\lambda := \emptyset$,
initialize $h_j^{\le},\ j \in J$.
If $\sum_{j\in J} d_j(a_j) \le \alpha$ go to Step 1 else go to Step 9.

Step 1. Construct the sets J_a^0, J_b^0, J^0 (for $\lambda = 0$). Calculate

$$\delta(0) := \sum_{j\in J_a^0} d_j(a_j) + \sum_{j\in J_b^0} d_j(b_j) + \sum_{j\in J^0} d_j(h_j^{\le}) - \alpha.$$

If $\delta(0) \le 0$ then $\lambda := 0$, go to Step 8
else if $\delta(0) > 0$ then
if $\alpha \le \sum_{j\in J} d_j(b_j)$ then go to Step 2
else if $\alpha > \sum_{j\in J} d_j(b_j)$ go to Step 9 (there does not exist $\lambda^* > 0$ such that $\delta(\lambda^*) = 0$).

Step 2. $J^{\lambda(k)} := J^{(k)}$.
Calculate $\lambda^{(k)}$ by using the explicit expression of λ (when it is available), determined from the equality $\sum_{j\in J^{\lambda(k)}} d_j(x_j) = \alpha^{(k)}$, where $x_j,\ j \in J^{\lambda(k)}$ are given by (5.6) with $\lambda = \lambda^{(k)}$. Go to Step 3.

Step 3. Construct the sets $J_a^{\lambda(k)}$, $J_b^{\lambda(k)}$, $J^{\lambda(k)}$ through (5.4), (5.5), and (5.6) (with $J^{(k)}$ instead of J) and find their cardinalities $|J_a^{\lambda(k)}|$, $|J_b^{\lambda(k)}|$, $|J^{\lambda(k)}|$, respectively. Go to Step 4.

Step 4. Calculate

$$\delta(\lambda^{(k)}) := \sum_{j \in J_a^{\lambda(k)}} d_j(a_j) + \sum_{j \in J_b^{\lambda(k)}} d_j(b_j) + \sum_{j \in J^{\lambda(k)}} d_j(x_j^{(k)}) - \alpha^{(k)},$$

where $x_j^{(k)}$, $j \in J^{\lambda(k)}$ are determined from (5.6) with $\lambda = \lambda^{(k)}$.
Go to Step 5.

Step 5. If $\delta(\lambda^{(k)}) = 0$ or $J^{\lambda(k)} = \emptyset$ then
$\lambda := \lambda^{(k)}$, $J_a^{\lambda} := J_a^{\lambda} \cup J_a^{\lambda(k)}$, $J_b^{\lambda} := J_b^{\lambda} \cup J_b^{\lambda(k)}$, $J^{\lambda} := J^{\lambda(k)}$,
go to Step 8
else if $\delta(\lambda^{(k)}) > 0$ go to Step 6
else if $\delta(\lambda^{(k)}) < 0$ go to Step 7.

Step 6. $x_j^* := a_j$ for $j \in J_a^{\lambda(k)}$,
$\alpha^{(k+1)} := \alpha^{(k)} - \sum_{j \in J_a^{\lambda(k)}} d_j(a_j)$,
$J^{(k+1)} := J^{(k)} \setminus J_a^{\lambda(k)}$, $n^{(k+1)} := n^{(k)} - |J_a^{\lambda(k)}|$,
$J_a^{\lambda} := J_a^{\lambda} \cup J_a^{\lambda(k)}$, $k := k + 1$.
Go to Step 2.

Step 7. $x_j^* := b_j$ for $j \in J_b^{\lambda(k)}$,
$\alpha^{(k+1)} := \alpha^{(k)} - \sum_{j \in J_b^{\lambda(k)}} d_j(b_j)$,
$J^{(k+1)} := J^{(k)} \setminus J_b^{\lambda(k)}$, $n^{(k+1)} := n^{(k)} - |J_b^{\lambda(k)}|$,
$J_b^{\lambda} := J_b^{\lambda} \cup J_b^{\lambda(k)}$, $k := k + 1$.
Go to Step 2.

Step 8. $x_j^* := a_j$ for $j \in J_a^{\lambda}$, $x_j^* := b_j$ for $j \in J_b^{\lambda}$,
assign x_j^* the values, determined from (5.6) for $j \in J^{\lambda}$.
Go to Step 10.

Step 9. Problem (C) has no optimal solution because $X = \emptyset$ or there does not exist a $\lambda > 0$ satisfying KKT necessary and sufficient conditions for problem (C).

Step 10. End.

7.3 Convergence and Computational Complexity of Algorithm 1

Theorem 7.1 (Convergence of Algorithm 1)
Let $\{\lambda^{(k)}\}$ be the sequence generated by Algorithm 1. Then
i) *if $\delta(\lambda^{(k)}) > 0$ then $\lambda^{(k)} \le \lambda^{(k+1)}$;*
ii) *if $\delta(\lambda^{(k)}) < 0$ then $\lambda^{(k)} \ge \lambda^{(k+1)}$.*

Proof. Denote by $x_j^{(k)}$ the components of $\mathbf{x}^{(k)} = (x_j)_{j \in J^{(k)}}$ at iteration k of the implementation of Algorithm 1.

Taking into consideration (7.6), Case 1, Case 2, and Case 3 of Section 7.1, and Step 1 (sign of $\delta(0)$) and Step 2 of Algorithm 1, it follows that $\lambda^{(k)} \geq 0$ for every k. Since $x_j^{(k)}$ are determined from (5.6) with $\lambda = \lambda^{(k)},\ x_j = x_j^{(k)}$:

$$\lambda^{(k)} d_j'(x_j^{(k)}) + c_j'(x_j^{(k)}) = 0,$$

substituted in

$$\sum_{j \in J^{\lambda(k)}} d_j(x_j^{(k)}) = \alpha^{(k)}$$

at Step 2 of Algorithm 1, and since $\lambda^{(k)} \geq 0,\ d_j'(x_j^{(k)}) > 0$, then $c_j'(x_j^{(k)}) \leq 0$, that is, $-c_j'(x_j^{(k)}) \geq 0$.

i) Let $\delta(\lambda^{(k)}) > 0$. Using Step 6 of Algorithm 1 (which is performed namely when $\delta(\lambda^{(k)}) > 0$), we obtain

$$\sum_{j \in J^{\lambda(k+1)}} d_j(x_j^{(k)}) = \sum_{j \in J^{(k+1)}} d_j(x_j^{(k)}) = \sum_{j \in J^{(k)} \setminus J_a^{\lambda(k)}} d_j(x_j^{(k)})$$

$$= \alpha^{(k)} - \sum_{j \in J_a^{\lambda(k)}} d_j(x_j^{(k)}). \tag{7.9}$$

We want to prove that $x_j^{(k)} \leq a_j$ for $j \in J_a^{\lambda(k)}$. If we assume the contrary, that $x_j^{(k)} > a_j$ for $j \in J_a^{\lambda(k)}$, then $d_j'(x_j^{(k)}) \geq d_j'(a_j)$ according to (5.16), and since $d_j'(x_j^{(k)}) > 0$, then $\frac{d_j'(a_j)}{d_j'(x_j^{(k)})} \in (0, 1]$. From the definition of $J_a^{\lambda(k)}$ we have

$$-\frac{c_j'(a_j)}{d_j'(a_j)} \leq \lambda^{(k)} = -\frac{c_j'(x_j^{(k)})}{d_j'(x_j^{(k)})}, \tag{7.10}$$

and therefore $-c_j'(a_j) \leq -c_j'(x_j^{(k)}) \frac{d_j'(a_j)}{d_j'(x_j^{(k)})}$. Using

$$-c_j'(x_j^{(k)}) \geq 0, \quad d_j'(x_j^{(k)}) > 0 \quad \text{and} \quad \frac{d_j'(a_j)}{d_j'(x_j^{(k)})} \in (0, 1]$$

by assumption, it follows that $-c_j'(a_j) \leq -c_j'(x_j^{(k)})$. Hence $x_j^{(k)} \leq a_j$ according to (5.15), a contradiction with the assumption that $a_j < x_j^{(k)}$. Therefore the assumption that $a_j < x_j^{(k)},\ j \in J_a^{\lambda(k)}$ is incorrect.

From (7.9), using that $d_j(x_j)$ is an increasing function ($d'_j(x_j) > 0$) and Step 6, we obtain

$$\sum_{j \in J^{\lambda(k+1)}} d_j(x_j^{(k)}) = \alpha^{(k)} - \sum_{j \in J_a^{\lambda(k)}} d_j(x_j^{(k)}) \geq \alpha^{(k)} - \sum_{j \in J_a^{\lambda(k)}} d_j(a_j)$$

$$= \alpha^{(k+1)} = \sum_{j \in J^{\lambda(k+1)}} d_j(x_j^{(k+1)}).$$

Therefore there exists at least one $j_0 \in J^{\lambda(k+1)}$ such that $d_{j_0}(x_{j_0}^{(k)}) \geq d_{j_0}(x_{j_0}^{(k+1)})$, and since $d_j(x_j)$ is an increasing function, $x_{j_0}^{(k)} \geq x_{j_0}^{(k+1)}$. Then

$$\lambda^{(k)} = -\frac{c'_{j_0}(x_{j_0}^{(k)})}{d'_{j_0}(x_{j_0}^{(k)})} \leq -\frac{c'_{j_0}(x_{j_0}^{(k+1)})}{d'_{j_0}(x_{j_0}^{(k)})} \leq -\frac{c'_{j_0}(x_{j_0}^{(k+1)})}{d'_{j_0}(x_{j_0}^{(k+1)})} = \lambda^{(k+1)}.$$

We have used that the relationship between $\lambda^{(k)}$ and $x_j^{(k)}$ is given by (5.6) for $j \in J^{\lambda(k)}$ according to Step 2 of Algorithm 1, $-c'_{j_0}(x_{j_0})$ is a monotone nonincreasing function according to Theorem 1.31, $-c'_j(x_j^{(k)}) \geq 0,\ j \in J^{\lambda(k)}$ according to (5.6) with $\lambda^{(k)} \geq 0$ and $d'_j(x_j^{(k)}) > 0$, and that $\frac{1}{d'_j(x_j)}$ is a monotone nonincreasing function as a reciprocal of the derivative of the convex function $d_j(x_j)$ with $d'_j(x_j) > 0$.

The proof of part ii) is similar to the proof of part i).

ii) Let $\delta(\lambda^{(k)}) < 0$. From Step 7 of Algorithm 1, we get

$$\sum_{j \in J^{\lambda(k+1)}} d_j(x_j^{(k)}) = \sum_{j \in J^{(k+1)}} d_j(x_j^{(k)}) = \sum_{j \in J^{(k)} \setminus J_b^{\lambda(k)}} d_j(x_j^{(k)})$$

$$= \alpha^{(k)} - \sum_{j \in J_b^{\lambda(k)}} d_j(x_j^{(k)}). \tag{7.11}$$

Our purpose is to prove that $j \in J_b^{\lambda(k)}$ implies $x_j^{(k)} \geq b_j$. Assume the contrary that $x_j^{(k)} < b_j$ for $j \in J_b^{\lambda(k)}$. From (5.16) it follows that $d'_j(x_j^{(k)}) \leq d'_j(b_j)$, and since $d'_j(x_j^{(k)}) > 0$, then $\frac{d'_j(b_j)}{d'_j(x_j^{(k)})} \geq 1 > 0$. Taking into account the definition of $J_b^{\lambda(k)}$, we obtain

$$-\frac{c'_j(b_j)}{d'_j(b_j)} \geq \lambda^{(k)} = -\frac{c'_j(x_j^{(k)})}{d'_j(x_j^{(k)})},$$

and since $d'_j(x_j^{(k)}) > 0$, then $-c'_j(b_j) \geq -c'_j(x_j^{(k)})\frac{d'_j(b_j)}{d'_j(x_j^{(k)})}$. From

$$-c'_j(x_j^{(k)}) \geq 0, \quad d'_j(x_j^{(k)}) > 0 \quad \text{and} \quad \frac{d'_j(b_j)}{d'_j(x_j^{(k)})} \geq 1 \ (> 0)$$

it follows that $-c'_j(b_j) \geq -c'_j(x_j^{(k)})$. Consequently, $x_j^{(k)} \geq b_j,\ j \in J_b^{\lambda(k)}$ according to (5.15), which contradicts the assumption that $x_j^{(k)} < b_j$ for $j \in J_b^{\lambda(k)}$. Therefore this assumption is incorrect.

From (7.11), using $d'_j(x_j) > 0$, Step 7, and $x_j^{(k)} \geq b_j,\ j \in J_b^{\lambda(k)}$, we obtain

$$\sum_{j \in J^{\lambda(k+1)}} d_j(x_j^{(k)}) = \alpha^{(k)} - \sum_{j \in J_b^{\lambda(k)}} d_j(x_j^{(k)}) \leq \alpha^{(k)} - \sum_{j \in J_b^{\lambda(k)}} d_j(b_j)$$

$$= \alpha^{(k+1)} = \sum_{j \in J^{\lambda(k+1)}} d_j(x_j^{(k+1)}).$$

Therefore there exists at least one $j_0 \in J^{\lambda(k+1)}$ such that $d_{j_0}(x_{j_0}^{(k)}) \leq d_{j_0}(x_{j_0}^{(k+1)})$, and since $d_j(x_j)$ is an increasing function, $x_{j_0}^{(k)} \leq x_{j_0}^{(k+1)}$. Therefore

$$\lambda^{(k)} = -\frac{c'_{j_0}(x_{j_0}^{(k)})}{d'_{j_0}(x_{j_0}^{(k)})} \geq -\frac{c'_{j_0}(x_{j_0}^{(k+1)})}{d'_{j_0}(x_{j_0}^{(k)})} \geq -\frac{c'_{j_0}(x_{j_0}^{(k+1)})}{d'_{j_0}(x_{j_0}^{(k+1)})} = \lambda^{(k+1)},$$

where we have used again the facts pointed out at the end of the proof of part i). ■

Remark 7.1 *Since we do not know the optimal value of λ which is involved in Theorem 5.1, we approximate the value of λ until we obtain the optimal value of λ at the last iteration of algorithm performance. In order to determine the current value $\lambda^{(k)}$ of λ at each iteration (including the initial value), we assume that $J^{\lambda(k)} = J^{(k)}$ at the beginning of the corresponding iteration (Step 2).*

Theorem 7.1, definitions of J_a^{λ} (5.4), J_b^{λ} (5.5), J^{λ} (5.6), and Steps 6 and 7 of Algorithm 1 allow us to state that the values of $\lambda^{(k)}$, $k = 0, 1, \ldots$, calculated at Step 2, are such that if we do not reduce index sets $J^{(k)}$ at Steps 6 and 7 of Algorithm 1 at each iteration, then $j \in J_a^{\lambda(k)}$ would imply $j \in J_a^{\lambda(k+1)}$; $j \in J_b^{\lambda(k)}$ would imply $j \in J_b^{\lambda(k+1)}$; and that $j \in J^{\lambda(k+1)}$ implies $j \in J^{\lambda(k)}$, that is, we would have $J_a^{\lambda(k)} \subseteq J_a^{\lambda(k+1)}$, $J_b^{\lambda(k)} \subseteq J_b^{\lambda(k+1)}$, and we have $J^{\lambda(k)} \supseteq J^{\lambda(k+1)}$. This means that if j belongs to current set $J_a^{\lambda(k)}$, then j belongs to the "optimal" set J_a^{λ} according to Theorem 7.1 and definition (5.4); the same holds true about the sets $J_b^{\lambda(k)}$ and J_b^{λ} (5.5). Therefore $\lambda^{(k)}$ converges to the optimal value λ from Theorem 5.1 and index sets $J_a^{\lambda(k)}$, $J_b^{\lambda(k)}$, $J^{\lambda(k)}$ "converge" to the optimal index sets J_a^{λ}, J_b^{λ}, J^{λ}, respectively. This means that the calculation of λ, the operations $x_j^* := a_j,\ j \in J_a^{\lambda(k)}$ (Step 6), $x_j^* := b_j,\ j \in J_b^{\lambda(k)}$ (Step 7), and the construction of index sets J_a^{λ}, J_b^{λ}, J^{λ} are in accordance with Theorem 5.1. The final sets J_a^{λ}, J_b^{λ}, J^{λ} are constructed at Step 1 or at Step 5 (when $\delta(\lambda^{(k)}) = 0$ or $J^{\lambda(k)} = \emptyset$) and Steps 6, 7 of iteration k where k is the number of the last iteration of algorithm performance.

Since at the beginning of Algorithm 1 we have $J^{\lambda(0)} := J$ *(Steps 0, 2) and since* $J^{\lambda(k)} \supseteq J^{\lambda(k+1)}$*, then* $J^{\lambda(k)} \neq \emptyset$ *for all* $k \leq k_0$ *where* k_0 *is some nonnegative integer. If we obtain* $J^{\lambda(k_0)} = \emptyset$*, this would mean that* $J_a^{\lambda(k_0)} \cup J_b^{\lambda(k_0)} = J$*, that is, the problem has been already solved at iteration* k_0*, and* $\delta(\lambda^{(k_0)}) = const$.

As we have seen in the proof of Theorem 7.1, Algorithm 1 guarantees that $\lambda \geq 0$ *for problem (C) as Theorem 5.1 requires. In the proof of Theorem 7.1, we have essentially used that* $\lambda^{(k)} \geq 0$ *in order to conclude that* $-c_j'(x_j^{(k)}) \geq 0$.

At each iteration, Algorithm 1 determines the value of at least one variable (Steps 6, 7, and 8) and at each iteration, we solve a problem of the type (C) but *of lower dimension* (Steps 2–7). Therefore Algorithm 1 is finite and it converges with at most $n = |J|$ iterations, that is, the iteration complexity of Algorithm 1 is $\mathcal{O}(n)$.

Step 0 takes time $\mathcal{O}(n)$. Step 1 (construction of sets J_a^0, J_b^0, J^0, calculation of $\delta(0)$, and checking whether X is empty) also takes time $\mathcal{O}(n)$. The calculation of $x_j^{(k)}$, $j \in J$ and $\lambda^{(k)}$ requires $\mathcal{O}(n)$ time (Step 2). Step 3 takes $\mathcal{O}(n)$ time because of the construction of $J_a^{\lambda(k)}$, $J_b^{\lambda(k)}$, $J^{\lambda(k)}$. Step 4 also requires $\mathcal{O}(n)$ time and Step 5 requires constant time. Each of Steps 6, 7, and 8 takes time which is bounded by $\mathcal{O}(n)$ because at these steps, we assign some of x_j's the optimal values, and since the number of all x_j's is n, Steps 6, 7, and 8 take time $\mathcal{O}(n)$. Hence Algorithm 1 has $\mathcal{O}(n^2)$ running time, and it belongs to the class of strongly polynomially bounded algorithms.

As computational experiments show, the number of iterations of the algorithm performance is not only at most n but also is much less than n for large n. In fact, this number does not depend on n but only on the three index sets defined by (5.4), (5.5), and (5.6). In practice, Algorithm 1 has $\mathcal{O}(n)$ running time.

Consider the **feasibility** of $\mathbf{x}^* = (x_j^*)_{j \in J}$, generated by Algorithm 1.

Components $x_j^* = a_j,\ j \in J_a^\lambda$ and $x_j^* = b_j,\ j \in J_b^\lambda$ obviously satisfy constraints (5.3).

Let $j \in J^\lambda$. Suppose that $x_j^* \leq a_j$ or $x_j^* \geq b_j$ for $j \in J^\lambda$. By monotonicity of $c_j'(x_j)$ and $d_j'(x_j)$, it follows that $c_j'(x_j^*) \leq c_j'(a_j),\ d_j'(x_j^*) \leq d_j'(a_j)$ or $c_j'(x_j^*) \geq c_j'(b_j),\ d_j'(x_j^*) \geq d_j'(b_j)$, respectively. Hence, using that $d_j'(x_j) > 0$ by assumption, $-c_j'(x_j) \geq 0,\ j \in J^\lambda$ (see proof of Theorem 7.1), we get

$$\lambda = -\frac{c_j'(x_j^*)}{d_j'(x_j^*)} \geq -\frac{c_j'(a_j)}{d_j'(x_j^*)} \geq -\frac{c_j'(a_j)}{d_j'(a_j)}$$

or

$$\lambda = -\frac{c_j'(x_j^*)}{d_j'(x_j^*)} \leq -\frac{c_j'(b_j)}{d_j'(x_j^*)} \leq -\frac{c_j'(b_j)}{d_j'(b_j)},$$

that is, $j \in J_a^\lambda$ or $j \in J_b^\lambda$, respectively, a contradiction. Therefore the assumption is incorrect, and $x_j^* \in (a_j, b_j),\ j \in J^\lambda$.

Consequently all $x_j^*,\ j \in J$ satisfy (5.3).

We have proved in Section 7.1 that if $\delta(\lambda)_{|\lambda=0} > 0$, then there exists a $\lambda^* > 0$ such that $\delta(\lambda^*) = 0$ (Case 1). Since at Step 2 we determine $\lambda^{(k)}$ from $\sum_{j\in J^{\lambda(k)}} d_j(x_j^{(k)}) = \alpha^{(k)}$ for each k, then (5.2) is satisfied with an equality in this case. Otherwise, if $\delta(0) < 0$, we set $\lambda = 0$ (Step 1) and we have $\sum_{j\in J} d_j(x_j(0)) - \alpha \equiv \delta(0) < 0$, that is, (5.2) is also satisfied in this case but as a strict inequality. When $\delta(0) = 0$, since $\delta(\lambda)$ is monotone nonincreasing, then (5.2) is also satisfied as a strict inequality.

Therefore Algorithm 1 generates $\mathbf{x}^*$ which is feasible for problem (C), which is an assumption of Theorem 5.1.

7.4 Analysis of the Optimal Solution to Problem $(C^=)$

The following discussion, similar to that of Section 7.1, is useful for the algorithm suggested for problem $(C^=)$.

Since the optimal solution $\mathbf{x}^*$ to problem $(C^=)$ depends on λ, consider the components of $\mathbf{x}^*$ as functions of λ for different $\lambda \in \mathbf{R}^1$:

$$x_j(\lambda) = \begin{cases} a_j, & \lambda \geq -\frac{c_j'(a_j)}{d_j} \quad (j \in J_a^\lambda), \\ b_j, & \lambda \leq -\frac{c_j'(b_j)}{d_j} \quad (j \in J_b^\lambda), \\ x_j^* : \ c_j'(x_j^*) + \lambda d_j = 0, & -\frac{c_j'(b_j)}{d_j} \leq \lambda \leq -\frac{c_j'(a_j)}{d_j} \quad (j \in J^\lambda). \end{cases}$$

Let

$$\delta(\lambda) \stackrel{\text{def}}{=} \sum_{j\in J_a^\lambda} d_j a_j + \sum_{j\in J_b^\lambda} d_j b_j + \sum_{j\in J^\lambda} d_j x_j(\lambda) - \alpha.$$

Since $u_j = v_j = 0,\ j \in J^\lambda$, condition (6.7) becomes

$$c_j'(x_j(\lambda)) + \lambda d_j = 0, \quad j \in J^\lambda. \tag{7.12}$$

If we differentiate both sides of each of these expressions with respect to λ (note that $c_j''(x_j)$ exist for all $j \in J^\lambda$ by assumption; $x_j'(\lambda)$ exist, $j \in J^\lambda$ because $x_j(\lambda)$ are defined by $x_j(\lambda) = x_j^* : \ c_j'(x_j^*) + \lambda d_j = 0$ for each $j \in J^\lambda$), we get

$$c_j''(x_j(\lambda))x_j'(\lambda) + d_j = 0, \quad j \in J^\lambda. \tag{7.13}$$

Therefore

$$x_j'(\lambda) = -\frac{d_j}{c_j''(x_j(\lambda))}, \quad j \in J^\lambda \tag{7.14}$$

and since $c_j''(x_j) \geq 0,\ j \in J$ as the second derivatives of convex differentiable functions, $d_j > 0$ by the assumption, then $x_j'(\lambda) < 0,\ j \in J^\lambda$. (If we assume that $c_j''(x_j(\lambda)) = 0$, then $d_j = 0$ according to (7.13). However, $d_j > 0,\ j \in J$ by the assumption, a contradiction.) Therefore

$$\delta'(\lambda) \equiv \sum_{j \in J^\lambda} d_j x'_j(\lambda) < 0 \tag{7.15}$$

when $J^\lambda \neq \emptyset$, and $\delta'(\lambda) = 0$ when $J^\lambda = \emptyset$. Hence $\delta(\lambda)$ is *a monotone nonincreasing function* of λ, $\lambda \in \mathbf{R}^1$.

From the equation $\delta(\lambda) = 0$, where $\delta(\lambda)$ is defined above, we are always able to determine λ as an implicit function of $\mathbf{x} : \lambda = \lambda(\mathbf{x})$ because $\delta'(\lambda) < 0$ when $J^\lambda \neq \emptyset$ according to (7.15) (it is important that $\delta'(\lambda) \neq 0$). Moreover, since $\delta(\lambda)$ is a linear function of $\mathbf{x}(\lambda)$, it is always possible to obtain a closed form expression of λ. It turns out that, without loss of generality, we can assume that $\delta'(\lambda) \neq 0$, that is, that $\delta(\lambda)$ depends on λ, hence, $J^\lambda \neq \emptyset$ (see the third paragraph of Remark 7.1).

According to Theorem 6.1 and the preliminary analysis, we can suggest the following algorithm for solving problem $(C^=)$ with strictly convex differentiable functions $c_j(x_j)$.

7.5 Algorithm 2 (for Problem $(C^=)$) and Its Convergence

Step 1. (Initialization) $J := \{1, \ldots, n\}$,
$k := 0,\ \alpha^{(0)} := \alpha,\ n^{(0)} := n$,
$J^{(0)} := J,\ J_a^\lambda := \emptyset,\ J_b^\lambda := \emptyset$,
initialize $h_j^=,\ j \in J$.
If $\sum_{j \in J} d_j a_j \leq \alpha \leq \sum_{j \in J} d_j b_j$ go to Step 2, else go to Step 9.

Step 2. $J^{\lambda(k)} := J^{(k)}$.
Calculate $\lambda^{(k)}$ by using the explicit expression of λ, determined from the equality constraint $\sum_{j \in J^{\lambda(k)}} d_j x_j = \alpha^{(k)}$, where $x_j,\ j \in J^{\lambda(k)}$ are given by (6.6) with $\lambda = \lambda^{(k)}$. Go to Step 3.

Step 3. Construct the sets $J_a^{\lambda(k)},\ J_b^{\lambda(k)},\ J^{\lambda(k)}$ through (6.4), (6.5), and (6.6) (with $J^{(k)}$ instead of J) and find their cardinalities $|J_a^{\lambda(k)}|$, $|J_b^{\lambda(k)}|$, $|J^{\lambda(k)}|$, respectively. Go to Step 4.

Step 4. Calculate

$$\delta(\lambda^{(k)}) := \sum_{j \in J_a^{\lambda(k)}} d_j a_j + \sum_{j \in J_b^{\lambda(k)}} d_j b_j + \sum_{j \in J^{\lambda(k)}} d_j x_j^{(k)} - \alpha^{(k)},$$

where $x_j^{(k)},\ j \in J^{\lambda(k)}$ are determined from (6.6) with $\lambda = \lambda^{(k)}$.
Go to Step 5.

Step 5. If $\delta(\lambda^{(k)}) = 0$ or $J^{\lambda(k)} = \emptyset$
then $\lambda := \lambda^{(k)},\ J_a^\lambda := J_a^\lambda \cup J_a^{\lambda(k)},\ J_b^\lambda := J_b^\lambda \cup J_b^{\lambda(k)},\ J^\lambda := J^{\lambda(k)}$,
go to Step 8

else if $\delta(\lambda^{(k)}) > 0$ go to Step 6
else if $\delta(\lambda^{(k)}) < 0$ go to Step 7.
Step 6. $x_j^* := a_j$ for $j \in J_a^{\lambda(k)}$,
$\alpha^{(k+1)} := \alpha^{(k)} - \sum_{j \in J_a^{\lambda(k)}} d_j a_j$,
$J^{(k+1)} := J^{(k)} \setminus J_a^{\lambda(k)}$, $n^{(k+1)} := n^{(k)} - |J_a^{\lambda(k)}|$,
$J_a^\lambda := J_a^\lambda \cup J_a^{\lambda(k)}$, $k := k + 1$.
Go to Step 2.
Step 7. $x_j^* := b_j$ for $j \in J_b^{\lambda(k)}$,
$\alpha^{(k+1)} := \alpha^{(k)} - \sum_{j \in J_b^{\lambda(k)}} d_j b_j$,
$J^{(k+1)} := J^{(k)} \setminus J_b^{\lambda(k)}$, $n^{(k+1)} := n^{(k)} - |J_b^{\lambda(k)}|$,
$J_b^\lambda := J_b^\lambda \cup J_b^{\lambda(k)}$, $k := k + 1$.
Go to Step 2.
Step 8. $x_j^* := a_j$ for $j \in J_a^\lambda$, $x_j^* := b_j$ for $j \in J_b^\lambda$,
assign x_j^* the values, determined from (6.6) for $j \in J^\lambda$.
Go to Step 10.
Step 9. The problem has no optimal solution because $X^L = \emptyset$.
Step 10. End.

Remark 7.2 *To avoid a possible "endless loop" in programming Algorithms 1 and 2, the criterion of Step 5 to go to Step 8 at iteration k usually is not $\delta(\lambda^{(k)}) = 0$ but $\delta(\lambda^{(k)}) \in [-\varepsilon, \varepsilon]$ where $\varepsilon > 0$ is some given tolerance value up to which the equality $\delta(\lambda^*) = 0$ may (for Algorithm 1) or must (for Algorithm 2) be satisfied.*

Since Steps 5, 6, 7, and 8 of Algorithm 1 and Algorithm 2 are the same, we are able to prove a theorem, analogous to Theorem 7.1, which states the "convergence" of $\lambda^{(k)}$, $J^{\lambda(k)}$, $J_a^{\lambda(k)}$, $J_b^{\lambda(k)}$ to the "optimal" λ, J^λ, J_a^λ, J_b^λ, respectively, for Algorithm 2.

Theorem 7.2 (Convergence of Algorithm 2)
Let $\{\lambda^{(k)}\}$ be the sequence generated by Algorithm 2. Then
i) *if $\delta(\lambda^{(k)}) > 0$ then $\lambda^{(k)} \le \lambda^{(k+1)}$;*
ii) *if $\delta(\lambda^{(k)}) < 0$ then $\lambda^{(k)} \ge \lambda^{(k+1)}$.*

Proof of Theorem 7.2 repeats in part the proof of Theorem 7.1.

Proof. Denote by $x_j^{(k)}$ the components of $\mathbf{x}^{(k)} = (x_j)_{j \in J^{(k)}}$ at iteration k of the implementation of Algorithm 2.

i) Let $\delta(\lambda^{(k)}) > 0$. According to Step 6 of Algorithm 2 (which is performed when $\delta(\lambda^{(k)}) > 0$), we have

$$\sum_{j \in J^{\lambda(k+1)}} d_j x_j^{(k)} = \sum_{j \in J^{(k+1)}} d_j x_j^{(k)} = \sum_{j \in J^{(k)} \setminus J_a^{\lambda(k)}} d_j x_j^{(k)}$$

$$= \alpha^{(k)} - \sum_{j \in J_a^{\lambda(k)}} d_j x_j^{(k)}. \tag{7.16}$$

Let $j \in J_a^{\lambda(k)}$. According to the definition of $J_a^{\lambda(k)}$ (6.4), we have

$$-\frac{c'_j(a_j)}{d_j} \le \lambda^{(k)} = -\frac{c'_j(x_j^{(k)})}{d_j}.$$

Multiplying this inequality by $-d_j < 0$, we conclude that $c'_j(a_j) \ge c'_j(x_j^{(k)})$. Hence $a_j \ge x_j^{(k)},\ j \in J_a^{\lambda(k)}$ in accordance with (5.15).

Taking into consideration (7.16), $d_j > 0$, $a_j \ge x_j^{(k)},\ j \in J_a^{\lambda(k)}$, and Step 6, we obtain

$$\sum_{j\in J^{\lambda(k+1)}} d_j x_j^{(k)} = \alpha^{(k)} - \sum_{j\in J_a^{\lambda(k)}} d_j x_j^{(k)} \ge \alpha^{(k)} - \sum_{j\in J_a^{\lambda(k)}} d_j a_j$$

$$= \alpha^{(k+1)} = \sum_{j\in J^{\lambda(k+1)}} d_j x_j^{(k+1)}.$$

Since $d_j > 0,\ j \in J$, there exists at least one $j_0 \in J^{\lambda(k+1)}$ such that $x_{j_0}^{(k)} \ge x_{j_0}^{(k+1)}$. Then

$$\lambda^{(k)} = -\frac{c'_{j_0}(x_{j_0}^{(k)})}{d_{j_0}} \le -\frac{c'_{j_0}(x_{j_0}^{(k+1)})}{d_{j_0}} = \lambda^{(k+1)},$$

where we have used the relationship (6.6) between $\lambda^{(k)}$ and $x_j^{(k)}$ for $j \in J^{\lambda(k)}$ according to Step 2 of Algorithm 2, the fact that $-c'_{j_0}(x_{j_0})$ is a monotone nonincreasing function according to Theorem 1.31, and $d_j > 0,\ j \in J$.

The proof of part ii) is similar to the proof of part i).

ii) Let $\delta(\lambda^{(k)}) < 0$. According to the definition of $J_b^{\lambda(k)}$ (6.5), we obtain

$$-\frac{c'_j(b_j)}{d_j} \ge \lambda^{(k)} = -\frac{c'_j(x_j^{(k)})}{d_j}.$$

Therefore $c'_j(b_j) \le c'_j(x_j^{(k)})$. Hence $b_j \le x_j^{(k)},\ j \in J_b^{\lambda(k)}$ according to (5.15).

From Step 7 of Algorithm 2, we get

$$\sum_{j\in J^{\lambda(k+1)}} d_j x_j^{(k)} = \sum_{j\in J^{(k+1)}} d_j x_j^{(k)} = \sum_{j\in J^{(k)}\setminus J_b^{\lambda(k)}} d_j x_j^{(k)}$$

$$= \alpha^{(k)} - \sum_{j\in J_b^{\lambda(k)}} d_j x_j^{(k)} \qquad (7.17)$$

$$\le \alpha^{(k)} - \sum_{j\in J_b^{\lambda(k)}} d_j b_j = \alpha^{(k+1)} = \sum_{j\in J^{\lambda(k+1)}} d_j x_j^{(k+1)}.$$

Since $d_j > 0$ for every $j \in J$, there exists at least one $j_0 \in J^{\lambda(k+1)}$ such that $x_{j_0}^{(k)} \le x_{j_0}^{(k+1)}$. Then, taking into account (6.6), $d_{j_0} > 0$ and (5.15) we obtain

$$\lambda^{(k)} = -\frac{c'_{j_0}(x_{j_0}^{(k)})}{d_{j_0}} \ge -\frac{c'_{j_0}(x_{j_0}^{(k+1)})}{d_{j_0}} = \lambda^{(k+1)}. \qquad \blacksquare$$

A remark analogous to Remark 7.1 (but for an arbitrary $\lambda \in \mathbf{R}^1$) holds for Algorithm 2.

Algorithm 2 belongs to the class of so-called "active set" algorithms.

Using the same reasoning as in Section 7.3, we can show that Algorithm 2 belongs to the class of $\mathcal{O}(n)$ algorithms.

Feasibility of $\mathbf{x}^* = (x_j^*)_{j \in J}$.

Values $x_j^* = a_j,\ j \in J_a^\lambda$ and $x_j^* = b_j,\ j \in J_b^\lambda$ obviously satisfy (6.3). From

$$-\frac{c'_j(b_j)}{d_j} < \lambda = -\frac{c'_j(x_j^*)}{d_j} < -\frac{c'_j(a_j)}{d_j}, \quad j \in J^\lambda$$

and $d_j > 0$, it follows that

$$c'_j(a_j) < c'_j(x_j^*) < c'_j(b_j), \quad j \in J^\lambda.$$

Therefore $a_j \le x_j^* \le b_j$ for all $j \in J^\lambda$ according to monotonicity of $c'_j(x_j)$; see (5.15).

In the sequel, since at each iteration $\lambda^{(k)}$ is determined from the "current" equality constraint (6.2) (Step 2 of Algorithm 2) and since $x_j,\ j \in J$ are determined in accordance with $\lambda^{(k)}$ at each iteration (Steps 5, 6, 7, and 8 of Algorithm 2) then $\mathbf{x}^*$ satisfies (6.2) as well. Therefore $\mathbf{x}^*$, obtained by Algorithm 2, is feasible for problem ($C^=$).

7.6 Commentary

Some of the main characteristics of the approach, suggested for problems (C) and ($C^=$) in this book, are the following.

Since the method, proposed for problem (C), uses values of the first derivatives of functions $c_j(x_j),\ j \in J$, we can consider it as a *first-order method.* Also, this method is a *saddle point method* or, more precisely, a *dual variable saddle point method* because it is based on convergence with respect to the Lagrange multiplier (dual variable) λ associated with the single constraint (5.2).

At Step 2 of Algorithm 1, we use the expression of $\lambda^{(k)}$, determined from the equality $\delta(\lambda^{(k)}) = 0$, where x_j^* are from (5.6), $j \in J^{\lambda(k)} = J^{(k)}$. As it was proved, under the assumptions of problem (C) we can always determine $\lambda = \lambda(\mathbf{x}^*)$ from $\delta(\lambda) = 0$

as an implicit function of $\mathbf{x}^*$; see (7.8). For example, when $d_j(x_j),\ j \in J$ are linear, the explicit expression of λ is always available for Algorithm 1. Other examples of functions for which it is possible to obtain closed form expressions of λ are given in Chapter Eleven. Of course, there are also other functions $c_j(x_j),\ d_j(x_j),\ j \in J$, for which the approach, suggested here, is applicable and gives good results.

Analogous commentary is valid for the method suggested for solving problem $(C^=)$.

Some additional concepts concerning Lagrangians and their saddle points are considered in Chapter Nine.

When the (optimal) Lagrange multiplier λ^* associated with (5.2) is known, then problem (C) (5.1)– (5.3) can be replaced by the following separable convex optimization problem

$$\min \left\{ \sum_{j \in J} [c_j(x_j) + \lambda^* d_j(x_j)] - \lambda^* \alpha \right\}$$

subject to

$$\mathbf{x} \in A \stackrel{\text{def}}{=} \{\mathbf{x} \in \mathbf{R}^n : a_j \le x_j \le b_j,\ j \in J\}.$$

The problem, dual to problem (C), is

$$\max\ \Psi(\lambda)$$

subject to

$$\lambda \in \mathbf{R}^1_+$$

where

$$\Psi(\lambda) = \min_{\mathbf{x} \in A} \left\{ \sum_{j \in J} [c_j(x_j) + \lambda d_j(x_j)] - \lambda \alpha \right\}.$$

Problem $(C^=)$ can be considered similarly, with $d_j(x_j) = d_j x_j$ and $\lambda \in \mathbf{R}^1$.

Thus, using the Lagrangian duality and Theorem 5.1 for problem (C) (Theorem 6.1 for problem $(C^=)$, respectively), we have replaced the *multivariate* problem (C) (problem $(C^=)$, respectively) of $\mathbf{x} \in \mathbf{R}^n$ by the *single-variable* optimization problem for finding $\lambda \in \mathbf{R}^1_+$ ($\lambda \in \mathbf{R}^1$, respectively).

Chapter 8
Version Two: Linear Constraint of the Form "≥"

8.1 Statement of Problem ($C^{\geq}$) and Results

Consider the following version ($C^{\geq}$) of problem (C):
($C^{\geq}$)

$$\min \left\{ c(\mathbf{x}) = \sum_{j \in J} c_j(x_j) \right\} \tag{8.1}$$

subject to

$$\sum_{j \in J} d_j x_j \geq \alpha \tag{8.2}$$

$$a_j \leq x_j \leq b_j, \quad j \in J, \tag{8.3}$$

where $c_j(x_j)$ are twice differentiable convex functions, defined on the open convex sets X_j in $\mathbf{R}$, $j \in J$, respectively, $d_j > 0$, $j \in J$, and $J \stackrel{\text{def}}{=} \{1, ..., n\}$.

Assumptions:

8.a) $a_j \leq b_j$ for all $j \in J$.

8.b) $\alpha \leq \sum_{j \in J} d_j b_j$. Otherwise the constraints (8.2)–(8.3) are inconsistent and $X^{\geq} = \emptyset$, where $X^{\geq}$ is defined by (8.2) –(8.3).

The convex programming approach, similar to that applied to problem (C), is not applicable to the more general problem (8.1), (8.2a), and (8.3) where

$$\sum_{j \in J} d_j(x_j) \geq \alpha \tag{8.2a}$$

and $d_j(x_j)$ are convex differentiable functions with $d'_j(x_j) > 0, \ j \in J$, because in convex programming, only inequality constraints of the form $g_i(\mathbf{x}) \leq 0, \ i \in I$ with $g_i(\mathbf{x}), \ i \in I$ convex functions are considered.

© The Author(s), under exclusive license to Springer Nature Switzerland AG 2021

S. M. Stefanov, *Separable Optimization*, Springer Optimization and Its Applications 177,
https://doi.org/10.1007/978-3-030-78401-0_8

Let $h_j^{\geq}$, $j \in J$ be the value of x_j for which $c_j'(h_j^{\geq}) = 0$. If such a point does not exist, we adopt $h_j^{\geq} = +\infty$ for problem $(C^{\geq})$.

Problems (C), $(C^{=})$, and $(C^{\geq})$ can be considered, to a certain extent, as continuous versions of the famous packing problem, partitioning problem, and covering problem in discrete optimization, respectively.

Rewrite $(C^{\geq})$ in the form

$$\min \left\{ c(\mathbf{x}) = \sum_{j \in J} c_j(x_j) \right\} \tag{8.1}$$

subject to

$$-\sum_{j \in J} d_j x_j \leq -\alpha \tag{8.2'}$$

$$a_j \leq x_j \leq b_j, \quad j \in J. \tag{8.3}$$

Since the linear function $d(\mathbf{x}) \stackrel{\text{def}}{=} -\sum_{j \in J} d_j x_j + \alpha$ is both convex and concave (Example 1.2), then $(C^{\geq})$ is also a convex optimization problem.

Let λ and $\lambda^{\geq}$ be the Lagrange multipliers associated with constraint (6.2) [for problem $(C^{=})$] and with constraint (8.2′) [for problem $(C^{\geq})$], respectively, and x_j^*, $x_j^{\geq}$, $j \in J$ be the components of the optimal solutions to problems $(C^{=})$, $(C^{\geq})$, respectively. For the sake of simplicity, we will use u_j, v_j $(j \in J)$ instead of $u_j^{\geq}$, $v_j^{\geq}$ $(j \in J)$ for the Lagrange multipliers associated with $a_j \leq x_j$, $x_j \leq b_j$ $(j \in J)$ in (8.3), respectively.

The Lagrangian for problem $(C^{\geq})$ is

$$L(\mathbf{x}, \mathbf{u}, \mathbf{v}, \lambda^{\geq})$$

$$= \sum_{j \in J} c_j(x_j) + \lambda^{\geq}\left(-\sum_{j \in J} d_j x_j + \alpha \right) + \sum_{j \in J} u_j(a_j - x_j) + \sum_{j \in J} v_j(x_j - b_j),$$

and the KKT conditions for problem $(C^{\geq})$ are

$$c_j'(x_j^{\geq}) - \lambda^{\geq} d_j - u_j + v_j = 0, \quad j \in J \tag{8.4}$$

$$u_j(a_j - x_j^{\geq}) = 0, \quad j \in J \tag{8.5}$$

$$v_j(x_j^{\geq} - b_j) = 0, \quad j \in J \tag{8.6}$$

$$\lambda^{\geq}\left(\alpha - \sum_{j \in J} d_j x_j^{\geq} \right) = 0, \quad \lambda^{\geq} \in \mathbf{R}_+^1 \tag{8.7}$$

$$-\sum_{j\in J} d_j x_j^{\geq} \leq -\alpha \tag{8.8}$$

$$a_j \leq x_j^{\geq} \leq b_j, \quad j \in J \tag{8.9}$$

$$u_j \in \mathbf{R}^1_+, \ v_j \in \mathbf{R}^1_+, \quad j \in J. \tag{8.10}$$

We can replace (8.4) and (8.7) by

$$c_j'(x_j^{\geq}) + \lambda^{\geq} d_j - u_j + v_j = 0\,, \quad j \in J \tag{8.4'}$$

and

$$\lambda^{\geq}\left(\sum_{j\in J} d_j x_j^{\geq} - \alpha\right) = 0, \quad \lambda^{\geq} \in \mathbf{R}^1_-, \ d_j > 0, \tag{8.7'}$$

respectively, where we have redenoted $\lambda^{\geq} := -\lambda^{\geq}$. That is why $\lambda^{\geq} \in \mathbf{R}^1_+$ in (8.4) and (8.7), and $\lambda^{\geq} \in \mathbf{R}^1_-$ in (8.4′) and (8.7′). Here (8.4′), (8.5), (8.6), (8.9), and (8.10) (withλ instead of $\lambda^{\geq}$) are among the KKT conditions for problem ($C^{=}$).

Theorem 8.1 (Sufficient condition for optimal solution to problem ($C^{\geq}$))
i) *If* $\lambda = -\frac{c_j'(x_j^*)}{d_j} \leq 0$ *then* $x_j^*, \ j \in J$, *solve problem* ($C^{\geq}$) *as well.*
ii) *If* $\lambda = -\frac{c_j'(x_j^*)}{d_j} > 0$ *then* $x_j^{\geq}, \ j \in J$, *defined as follows:*
$x_j^{\geq} = b_j, \quad j \in J_b^{\lambda}$;
$x_j^{\geq} = \min\{b_j, h_j^{\geq}\}, \quad j \in J^{\lambda}$;
$x_j^{\geq} = \min\{b_j, h_j^{\geq}\}$ *for all* $j \in J_a^{\lambda}$ *such that* $c_j'(a_j) < 0$;
$x_j^{\geq} = a_j$ *for all* $j \in J_a^{\lambda}$ *such that* $c_j'(a_j) \geq 0$
solve problem ($C^{\geq}$).

Proof. i) Let $\lambda = -\frac{c_j'(x_j^*)}{d_j} \leq 0$. Since $x_j^*, \ j \in J$ satisfy KKT conditions (6.7)–(6.12) for problem ($C^{=}$) as components of the optimal solution to ($C^{=}$), then (8.4′), (8.5), (8.6), (8.7′), (8.8) with equality (and therefore (8.7)), (8.9), and (8.10) are satisfied as well (with λ instead of $\lambda^{\geq}$). Since they are the KKT necessary and sufficient conditions for problem ($C^{\geq}$), then $x_j^*, \ j \in J$ solve problem ($C^{\geq}$) as well.

ii) Let $\lambda = -\frac{c_j'(x_j^*)}{d_j} > 0$. Since $\mathbf{x}^* = (x_j^*)_{j\in J}$ is an optimal solution to problem ($C^{=}$) by the assumption, KKT conditions (6.7)–(6.12) for problem ($C^{=}$) are satisfied. If $\mathbf{x}^{\geq} := (x_j^{\geq})_{j\in J}$ is an optimal solution to problem ($C^{\geq}$), then $\mathbf{x}^{\geq}$ satisfies (8.4′), (8.5), (8.6), (8.7′), (8.8), (8.9), and (8.10). Since $\lambda > 0$, then λ cannot play the role of $\lambda^{\geq}$ in (8.4′) and (8.7′) because $\lambda^{\geq}$ must be a nonpositive real number in (8.4′) and (8.7′). Therefore $x_j^*, \ j \in J$, which satisfy KKT conditions (6.7)–(6.12) for problem ($C^{=}$), including (6.11), cannot play the roles of $x_j^{\geq}, \ j \in J$ in (8.4′), (8.5), (8.6), (8.7′), (8.8), (8.9), and (8.10). Hence, in the general case the equality

$$\sum_{j \in J} d_j x_j = \alpha$$

is not satisfied for $x_j = x_j^{\geq}$. Therefore, in order that (8.7′) be satisfied, $\lambda^{\geq}$ must be equal to 0. This conclusion helps us to prove the theorem.

Let $\mathbf{x}^{\geq} := (x_j^{\geq})_{j \in J}$ be defined as in part ii) of the statement of Theorem 8.1.

Set: $\lambda^{\geq} = 0$;

1) $u_j = 0,\ v_j = -c_j'(b_j)$ (≥ 0 since $\lambda > 0,\ d_j > 0$ and according to definition of J_b^λ (6.5)) for $j \in J_b^\lambda$;

2) $u_j = v_j = 0$ for $j \in J_a^\lambda : c_j'(a_j) < 0$ and $j \in J^\lambda$ such that $h_j^{\geq} < b_j$;

3) $u_j = 0,\ v_j = -c_j'(b_j)\ (\geq 0)$ for $j \in J^\lambda$ such that $h_j^{\geq} \geq b_j$;

4) $u_j = c_j'(a_j) \geq 0,\ v_j = 0$ for $j \in J_a^\lambda$ such that $c_j'(a_j) \geq 0$.

In case 2), we have $c_j'(a_j) < 0 \equiv c_j'(h_j^{\geq})$, therefore $a_j \leq h_j^{\geq} = x_j^{\geq}$ according to (5.15) and definition of $x_j^{\geq}$ in this case. In case 3), since $b_j \leq h_j^{\geq}$, then $c_j'(b_j) \leq c_j'(h_j^{\geq}) = 0$ according to (5.15) and therefore $v_j := -c_j'(b_j) \geq 0$. Consequently, conditions (8.9) and (8.10) are satisfied for all j according to cases 1)–4).

As we have proved, (8.7′) is satisfied with $\lambda^{\geq} = 0$. Since the equality constraint (6.2) $\sum_{j \in J} d_j x_j = \alpha$ is satisfied for the optimal solution $\mathbf{x}^*$ to problem $(C^=)$, since the components of $\mathbf{x}^{\geq}$, defined in the formulation of Theorem 8.1, ii) are such that some of them are the same as the corresponding components of $\mathbf{x}^*$, since some of components of $\mathbf{x}^{\geq}$, namely those for $j \in J_a^\lambda$ with $c_j'(a_j) < 0$, are greater than the corresponding components $x_j^* = a_j,\ j \in J_a^\lambda$ of $\mathbf{x}^*$, and since $d_j > 0,\ j \in J$, then obviously the inequality condition (8.8) holds for $\mathbf{x}^{\geq}$. It is easy to check that other conditions (8.4′), (8.5), and (8.6) are also satisfied. Thus, $x_j^{\geq},\ j \in J$ defined above satisfy the KKT conditions for problem $(C^{\geq})$. Therefore $\mathbf{x}^{\geq}$ is an optimal solution to $(C^{\geq})$. When $c(\mathbf{x}) = \sum_{j \in J} c_j(x_j)$ is a strictly convex function, this solution is *unique* (Theorem 1.38). ■

Theorem 7 ([11]) follows from Theorem 8.1 with $d_j = 1,\ j \in J$.

According to Theorem 8.1, optimal solution to problem $(C^{\geq})$ is obtained by using the optimal solution and optimal value of the Lagrange multiplier λ for problem $(C^=)$. That is why we suppose that $\sum_{j \in J} d_j a_j \leq \alpha$ in addition to assumption 8.b) (see Step 1 of Algorithm 3 below) as we assumed this in assumption 6.b).

Algorithm 3 for solving problem $(C^{\geq})$ with strictly convex differentiable functions $c_j(x_j)$ is based on Theorem 8.1 and Algorithm 2.

8.2 Algorithm 3 (for Problem $(C^{\geq})$)

Step 1. (Initialization) $J := \{1, \ldots, n\}$,
$k := 0,\ \alpha^{(0)} := \alpha,\ n^{(0)} := n$,
$J^{(0)} := J,\ J_a^{\lambda} := \emptyset,\ J_b^{\lambda} := \emptyset$,
initialize $h_j^{\geq},\ j \in J$.
If $\sum_{j \in J} d_j a_j \leq \alpha \leq \sum_{j \in J} d_j b_j$ then go to Step 2, else go to Step 9.

Steps 2–7. are the same as Steps 2–7 of Algorithm 2, respectively.

Step 8. If $\lambda \leq 0$ then $x_j^{\geq} := a_j$ for $j \in J_a^{\lambda}$, $x_j^{\geq} := b_j$ for $j \in J_b^{\lambda}$,
assign $x_j^{\geq}$ the values, determined through (6.6) for $j \in J^{\lambda}$,
go to Step 10
else if $\lambda > 0$ then
$x_j^{\geq} := b_j$ for $j \in J_b^{\lambda}$,
$x_j^{\geq} := \min\{b_j, h_j^{\geq}\}$ for $j \in J^{\lambda}$,
if $j \in J_a^{\lambda}$ and $c'_j(a_j) < 0$ then $x_j^{\geq} := \min\{b_j, h_j^{\geq}\}$
else if $j \in J_a^{\lambda}$ and $c'_j(a_j) \geq 0$ then $x_j^{\geq} := a_j$,
go to Step 10.

Step 9. Problem $(C^{\geq})$ has no optimal solution because $X^{\geq} = \emptyset$ or
there do not exist $x_j^* \in [a_j, b_j],\ j \in J$ such that $\sum_{j \in J} d_j x_j^* = \alpha$.

Step 10. End.

Since Algorithm 3 is based on Theorem 8.1 and Algorithm 2 and since the "iterative" Steps 2–7 of Algorithm 2 and Algorithm 3 are the same, then the "convergence" of Algorithm 3 follows from Theorem 7.2 as well. Because of the same reason, the computational complexity of Algorithm 3 is the same as that of Algorithm 2.

Chapter 9
Well-Posedness of Optimization Problems

9.1 Well-Posedness and Calmness of Optimization Problems

Questions of the existence of solutions and how they depend on problem's parameters are usually important for many problems of mathematics, not only in optimization. The term *well-posedness* refers to the existence and uniqueness of a solution, and to its continuous behavior with respect to data perturbations, which is referred to as *stability*. In general, a problem is said to be *stable* if

$$\varepsilon(\delta) \to 0 \quad \text{when} \quad \delta \to 0,$$

where δ is a given tolerance of the problem's data, $\varepsilon(\delta)$ is the accuracy with which the solution can be determined, and $\varepsilon(\delta)$ is a continuous function of δ. Besides these conditions, accompanying robustness properties in the convergence of a sequence of approximate solutions are also required.

Problems which are not well-posed are called *ill-posed*, or, sometimes, *improperly posed*.

9.1.1 Tychonov and Hadamard Well-Posedness. Well-Posedness in the Generalized Sense

Recall (Definition 1.18) that f is a *proper function* if $f(\mathbf{x}) < \infty$ for at least one $\mathbf{x} \in \mathbf{R}^n$ and $f(\mathbf{x}) > -\infty$ for all $\mathbf{x} \in \mathbf{R}^n$, or, in other words, if

$$\operatorname{dom} f \stackrel{\text{def}}{=} \{\mathbf{x} \in \mathbf{R}^n : \ f(\mathbf{x}) < \infty\}$$

© The Author(s), under exclusive license to Springer Nature Switzerland AG 2021

S. M. Stefanov, *Separable Optimization*, Springer Optimization and Its Applications 177,
https://doi.org/10.1007/978-3-030-78401-0_9

is a nonempty set on which $f(\mathbf{x}) > -\infty$, where dom f is the *effective domain* of f (Definition 1.16). Otherwise f is called *improper*.

Definition 9.1 Let X be a space with either a topology or a convergence structure associated, and let $f : X \to \overline{\mathbf{R}} \equiv \mathbf{R} \cup \{+\infty\}$ be a proper extended real-valued function. Consider the problem

$$\min\ f(\mathbf{x}) \tag{9.1}$$

subject to

$$\mathbf{x} \in X. \tag{9.2}$$

The problem (9.1)–(9.2) is *Tychonov well-posed* if and only if f has a unique global minimum point on X toward which every minimizing sequence converges.

An equivalent definition is as follows: problem (9.1)–(9.2) is *Tychonov well-posed* if and only if there exists a unique $\mathbf{x}_0 \in X$ such that $f(\mathbf{x}_0) \le f(\mathbf{x})$ for all $\mathbf{x} \in X$ and

$$f(\mathbf{x}_n) \to f(\mathbf{x}_0) \quad \text{implies} \quad \mathbf{x}_n \to \mathbf{x}_0.$$

There are two ways to cope with ill-posedness.

The first one is to change the statement of the problem.

The second one is so-called *Tychonov regularization*. A parametric functional is constructed such that if it approaches 0, the solution of the "regularized" problem converges to the exact solution of the original problem.

Let $X \subseteq \mathbf{R}^n$. Consider the problem

$$\min_{\mathbf{x}\in X} f(\mathbf{x}) = f(\mathbf{x}^*). \tag{9.3}$$

Associate the following problem with problem (9.3)

$$\min_{\mathbf{x}\in X} [f(\mathbf{x}) + n_k(\mathbf{x})] = z_k(\mathbf{x}^*(n_k(\mathbf{x}))), \tag{9.4}$$

where $n_k(\mathbf{x})$ is the perturbation in the input data and $\mathbf{x}^*(n_k(\mathbf{x}))$ is an optimal solution to the perturbed problem.

Let

$$\int_{\mathbf{R}^n} n_k^2(\mathbf{x})d\mathbf{x} \le c_k^2 < \infty.$$

If

$$|z_k(\mathbf{x}^*(n_k(\mathbf{x}))) - f(\mathbf{x}^*)| \to 0$$

when $c_k \to 0$, then problem (9.3) is stable with respect to perturbation $n_k(\mathbf{x})$.

A parametric function $F(\mathbf{x}, \Delta, f(\mathbf{x}))$ with a parameter Δ is called a *regularizing function* for problem (9.3) with respect to perturbation $n_k(\mathbf{x})$ if the following conditions are satisfied:

1. $F(\mathbf{x}, \Delta, f(\mathbf{x}))$ is defined for all $\mathbf{x} \in X$ and $\Delta > 0$.
2. If $\mathbf{x}^*(n_k(\mathbf{x}), \Delta)$ is an optimal solution to problem

$$\min_{\mathbf{x} \in X} F(\mathbf{x}, \Delta, z_k(\mathbf{x})) = F(\mathbf{x}^*(n_k(\mathbf{x}), \Delta), \Delta, z_k(\mathbf{x})),$$

then there exists a function $\Delta_k = \Delta_k(c_k)$ such that

$$F(\mathbf{x}^*(n_k(\mathbf{x}), \Delta_k), \Delta_k, z_k(\mathbf{x})) \to f(\mathbf{x}^*)$$

when $c_k \to 0$.

Following A. N. Tychonov, an ill-posed problem is said to be *regularizable* if there exists at least one regularizing function for it.

The concept of Tychonov well-posedness can be extended to problems without the uniqueness of the optimal solution.

Definition 9.2 Let X be a space with either a topology or a convergence structure associated, and $f : X \to \overline{\mathbf{R}} \equiv \mathbf{R} \cup \{+\infty\}$ be a proper real-valued function. The problem (9.1)–(9.2) is said to be *well-posed in the generalized sense* if and only if $\arg\min_{\mathbf{x} \in X} f(\mathbf{x}) \neq \emptyset$ and every sequence $\{\mathbf{u}_n\} \subset X$ such that $f(\mathbf{u}_n) \to \inf \{f(\mathbf{x}) : \mathbf{x} \in X\}$ has some subsequence $\{\mathbf{v}_n\} \to \mathbf{u}$ with $\mathbf{u} \in \arg\min_{\mathbf{x} \in X} f(\mathbf{x})$.

Problem (9.1)–(9.2) is Tychonov well-posed if and only if it is well-posed in the generalized sense and $\arg\min_{\mathbf{x} \in X} f(\mathbf{x})$ is a singleton.

Hadamard well-posedness is primarily connected with problems of mathematical physics (boundary value problems for partial differential equations) and can be extended to mathematical programming problems. This topic is not discussed here.

As recent studies in the calculus of variations, optimal control, and numerical methods of optimization show, uniqueness and continuity are often too restrictive to be adopted as the standards of well-posedness. It turns out that practical concepts concerning well-posedness are some forms of semicontinuity in the problem's data and solution mapping, along with potential multivaluedness in this mapping.

9.1.2 *Calmness in the Sense of Clarke*

Let X be a Banach space.

Definition 9.3 Let Y be a subset of X. A function $f : Y \to \mathbf{R}$ is said to satisfy a *Lipschitz condition* on Y provided that, for some nonnegative scalar K, the following inequality holds true

$$|f(\mathbf{y}) - f(\mathbf{y}')| \leq K \|\mathbf{y} - \mathbf{y}'\|$$

for all points $\mathbf{y}, \mathbf{y}' \in Y$; this is also referred to as a *Lipschitz condition of rank K*. We say that f is *Lipschitz* (of rank K) *near* $\mathbf{x}$ if for some $\varepsilon > 0$, f satisfies a Lipschitz condition (of rank K) on the set $\mathbf{x} + \varepsilon B$ (that is, within an $\varepsilon-$neighborhood of $\mathbf{x}$) where B is the open unit ball around $\mathbf{0}$.

A function f which satisfies a Lipschitz condition sometimes is said to be *Lipschitz continuous*.

Consider the general mathematical programming problem
(P)

$$\min\ f(\mathbf{x})$$

subject to

$$\begin{aligned} &g_i(\mathbf{x}) \leq 0, \quad i = 1, \ldots, t \\ &h_j(\mathbf{x}) = 0, \quad j = 1, \ldots, m \\ &\mathbf{x} \in C, \qquad C \subseteq X \end{aligned}$$

where g_i, h_j are real-valued functions on X.

Let $\mathbf{g}$ and $\mathbf{h}$ be the functions $\mathbf{g} = [g_1, \ldots, g_t] : X \to \mathbf{R}^t$, $\mathbf{h} = [h_1, \ldots, h_m] : X \to \mathbf{R}^m$.

Let problem (P) be imbedded in a parameterized family $P(\mathbf{p}, \mathbf{q})$ of mathematical programs where $\mathbf{p} \in \mathbf{R}^t$, $\mathbf{q} \in \mathbf{R}^m$:
$P(\mathbf{p}, \mathbf{q})$

$$\min\ f(\mathbf{x})$$

subject to

$$\begin{aligned} &\mathbf{g}(\mathbf{x}) + \mathbf{p} \leq \mathbf{0} \\ &\mathbf{h}(\mathbf{x}) + \mathbf{q} = \mathbf{0} \\ &\mathbf{x} \in C. \end{aligned}$$

Denote by A the feasible region of problem $P(\mathbf{p}, \mathbf{q})$.

Definition 9.4 (Clarke [27]) The *value function* $V : \mathbf{R}^t \times \mathbf{R}^m \to \mathbf{R} \cup \{\pm\infty\}$ is defined via $V(\mathbf{p}, \mathbf{q}) = \inf \{P(\mathbf{p}, \mathbf{q})\}$ (that is, the value of the problem $P(\mathbf{p}, \mathbf{q})$). If there are no feasible points for $P(\mathbf{p}, \mathbf{q})$, then the infimum is over the empty set, and $V(\mathbf{p}, \mathbf{q})$ is assigned the value $+\infty$.

Definition 9.5 (Clarke [27]) Let $\mathbf{x}$ solve problem (P). Then (P) is said to be *calm* at $\mathbf{x}$ provided that there exist positive ε and M such that for all $(\mathbf{p}, \mathbf{q}) \in \varepsilon B$, for all $\mathbf{x}' \in \mathbf{x} + \varepsilon B$ which are feasible for $P(\mathbf{p}, \mathbf{q})$, we have

$$f(\mathbf{x}') - f(\mathbf{x}) + M\|(\mathbf{p}, \mathbf{q})\| \geq 0,$$

where B is the open unit ball in X and $\|(\mathbf{p}, \mathbf{q})\|$ is the Euclidean norm of $(\mathbf{p}, \mathbf{q})$.

Let U be an open convex subset of X.

Theorem 9.1 (Lipschitz condition from boundedness of a convex function; Roberts and Varberg [104], Clarke [27])
Let f be a convex function, bounded above on a neighborhood of some point of U. Then, for any $\mathbf{x}$ in U, f is Lipschitz near $\mathbf{x}$.

Recall that the *limit superior* of a bounded sequence $\{x_n\}$ in $\mathbf{R}^1$, denoted by $\limsup \{x_n\}$ or $\overline{\lim} \{x_n\}$, equals the infimum of all numbers $q \in \mathbf{R}^1$ for which at most a finite number of elements of $\{x_n\}$ (strictly) exceed q. Similarly, the *limit inferior* of $\{x_n\}$ is given by $\liminf \{x_n\} \equiv \underline{\lim} \{x_n\} \equiv \sup \{q :$ at most a finite number of elements of $\{x_n\}$ are (strictly) less than $q\}$.

A bounded sequence always has a unique limit superior and limit inferior.

Theorem 9.2 (Calmness; Clarke [27])
Let $V(\mathbf{0}, \mathbf{0})$ be finite and suppose that

$$\liminf_{(\mathbf{p},\mathbf{q})\to(\mathbf{0},\mathbf{0})} \frac{V(\mathbf{p}, \mathbf{q}) - V(\mathbf{0}, \mathbf{0})}{\|(\mathbf{p}, \mathbf{q})\|} > -\infty$$

(this is true in particular if V is Lipschitz near $(\mathbf{0}, \mathbf{0})$). Then, for any solution $\mathbf{x}$ to problem (P), (P) is calm at $\mathbf{x}$.

Sometimes problem (P) is said to be *calm* provided V satisfies the hypothesis of Theorem 9.2.

Slater-Type Conditions. Suppose that problem (P) has no equality constraints (that is, $m = 0$), that the functions g_i, $i = 1, \ldots, t$ are convex, and that C is a convex set. Recall (Section 1.5) that *Slater's condition* (*Slater's constraint qualification*) then is as follows: there exists a point $\overline{\mathbf{x}}$ in C such that $g_i(\overline{\mathbf{x}}) < 0$, $i = 1, \ldots, t$ ($\overline{\mathbf{x}}$ is called a *strictly feasible point*).

For $\mathbf{p} \in \mathbf{R}^t$, let $V(\mathbf{p})$ be the infimum in the problem $P(\mathbf{p})$ in which the constraints $g_i(\mathbf{x}) \leq 0$ of problem (P) are replaced by $g_i(\mathbf{x}) + p_i \leq 0$.

Theorem 9.3 (Lipschitz property of the value function from Slater's condition; Clarke [27])
If C is bounded and f is Lipschitz on C, then Slater's condition (that is, the existence of a strictly feasible point) implies that V is Lipschitz near $\mathbf{0}$.

Theorems 9.2 and 9.3 mean that Slater's constraint qualification implies calmness of problem $P(\mathbf{p})$ in this case.

Theorem 9.4 (Calmness of a problem subject to inequality constraints; Clarke [27])
Let problem (P) incorporate only inequality constraints $g_i(\mathbf{x}) \leq 0$ and the abstract constraint $\mathbf{x} \in C$, and suppose that the value function $V(\mathbf{p})$ is finite for $\mathbf{p}$ near $\mathbf{0}$. Then for almost all $\mathbf{p}$ in a neighborhood of $\mathbf{0}$, the problem $P(\mathbf{p})$ is calm.

Remark 9.1 *In the case of problem (P), in which equality constraints exist, it is a consequence ([27]) of Ekeland's theorem (Theorem C.2, Appendix C) that $P(\mathbf{p}, \mathbf{q})$ is calm for all $(\mathbf{p}, \mathbf{q})$ in a dense subset of any open set upon which V is bounded and lower semicontinuous.*

Consider the following way of perturbing problem (P) ([27]):

$P(\boldsymbol{\alpha})$

$$\min \{f(\mathbf{x}, \boldsymbol{\alpha}) : \mathbf{g}(\mathbf{x}, \boldsymbol{\alpha}) \leq \mathbf{0},\ \mathbf{h}(\mathbf{x}, \boldsymbol{\alpha}) = \mathbf{0},\ (\mathbf{x}, \boldsymbol{\alpha}) \in D\}$$

where $\boldsymbol{\alpha}$ is a vector of k real components. The value function V then would be a function of $\boldsymbol{\alpha}$: $V(\boldsymbol{\alpha}) = \inf P(\boldsymbol{\alpha})$.

Problem $P(\boldsymbol{\alpha})$ is a special case of problem $P(\mathbf{p}, \mathbf{q})$ with $k = t + m$, $\boldsymbol{\alpha} = (\mathbf{p}, \mathbf{q})$, $f(\mathbf{x}, \boldsymbol{\alpha}) = f(\mathbf{x})$, $\mathbf{g}(\mathbf{x}, \boldsymbol{\alpha}) = \mathbf{g}(\mathbf{x}) + \mathbf{p}$, $\mathbf{h}(\mathbf{x}, \boldsymbol{\alpha}) = \mathbf{h}(\mathbf{x}) + \mathbf{q}$, and $D = C \times \mathbf{R}^{t+m}$. R. T. Rockafellar has observed that, at least when the dependence of $f, \mathbf{g}$ and $\mathbf{h}$ on $\boldsymbol{\alpha}$ is locally Lipschitz, we can consider problem $P(\mathbf{p}, \mathbf{q})$ with $k = t + m$, $\boldsymbol{\alpha} = (\mathbf{p}, \mathbf{q})$, $f(\mathbf{x}, \boldsymbol{\alpha}) = f(\mathbf{x})$, $\mathbf{g}(\mathbf{x}, \boldsymbol{\alpha}) = \mathbf{g}(\mathbf{x}) + \mathbf{p}$, $\mathbf{h}(\mathbf{x}, \boldsymbol{\alpha}) = \mathbf{h}(\mathbf{x}) + \mathbf{q}$, and $D = C \times \mathbf{R}^{t+m}$ rather than problem $P(\boldsymbol{\alpha})$. Hence, the methods and results considered above can be applied to perturbed family $P(\boldsymbol{\alpha})$ as well.

Constraint qualifications (regularity conditions) can be classified into two categories: on the one hand, Mangasarian-Fromowitz and Slater-type conditions and their extensions, and, on the other hand, constraint qualifications called *calmness*. It turns out that calmness is the weakest of these conditions, since it is implied by all the others (see, for example, Theorem 9.2 and Theorem 9.3).

9.1.3 Well-Posedness of Problems (C), $(C^{=})$, *and* $(C^{\geq})$

Existence of solutions

The question of the existence of solutions to problems (C), $(C^{=})$, and $(C^{\geq})$ has been discussed in Theorems 5.1, 6.1, and 8.1, respectively. Steps 0, 1, and 9 of Algorithm 1, Steps 1 and 9 of Algorithms 2 and 3, respectively, refer to these results.

Uniqueness of solution

The question of the uniqueness of the optimal solution to problems under consideration is also important.

If $c(\mathbf{x}) \equiv \sum_{j \in J} c_j(x_j)$ (5.1) [(6.1), (8.1), respectively] is a strictly convex function, then problem (C) [problem $(C^{=})$, $(C^{\geq})$, respectively] has a unique optimal solution in the feasible region X [feasible region X^L, $X^{\geq}$, respectively] in case problem (C) [problem $(C^{=})$, $(C^{\geq})$, respectively] has feasible solutions (Theorem 1.38), that is, x_j^*, $j \in J^{\lambda}$ are uniquely determined from (5.6) [from (6.6), from Theorem 8.1, respectively] in $[a_j, b_j]$ in this case. If the parameters a_j, b_j, etc. of particular problems of the form (C) [$(C^{=})$, $(C^{\geq})$] are generated in intervals where the functions $c_j(x_j)$ are strictly convex, then problem (C) [problem $(C^{=})$, $(C^{\geq})$, respectively], if it has feasible solutions, has a *unique* optimal solution.

In the general case, if functions $c_j(x_j)$ are convex but not necessarily strictly convex, then, as it is known, a convex programming problem has more than one optimal solution and the set of optimal solutions to such a problem is convex (Theorem 1.37). Further, the optimal value of the objective function is the same for all optimal solutions to problem (C) [problem $(C^{=})$, $(C^{\geq})$, respectively] if it has more than one optimal solution. If, for example, (5.6) [(6.6), respectively] is a linear equation of x_j^*, then $x_j^*,\ j \in J^\lambda$ are also uniquely determined from (5.6) [from (6.6), respectively].

Calmness of the problems (of the optimal solutions)

Let $(C(\mathbf{p})),\ (C^=(\mathbf{p}, q)),\ (C^\geq(\mathbf{p}))$ be the parameterized families of mathematical programs associated with problems (C), $(C^=)$, $(C^\geq)$, respectively.

Feasible regions of problems (C) and $(C^\geq)$ are nonempty by the assumption, which is satisfied when $\sum_{j \in J} d_j(a_j) \leq \alpha$ and $\sum_{j \in J} d_j b_j \geq \alpha$, respectively. Without loss of generality, feasible regions

$X(\mathbf{p})$

$$\sum_{j \in J} d_j(x_j) + p_0 \leq \alpha$$

$$a_j \leq x_j + p_j \leq b_j, \quad j \in J$$

and $X^\geq(\mathbf{p})$

$$\sum_{j \in J} d_j x_j + p_0 \geq \alpha$$

$$a_j \leq x_j + p_j \leq b_j, \quad j \in J$$

of problems $(C(\mathbf{p}))$ and $(C^\geq(\mathbf{p}))$, respectively, are also nonempty in a neighborhood of $\mathbf{p} = \mathbf{0}$.

Since the value function $V(\mathbf{p})$ associated with problems $(C(\mathbf{p}))$ and $(C^\geq(\mathbf{p}))$ is finite near $\mathbf{0}$ (according to Definition 9.4 and the assumption that the corresponding feasible set is nonempty), then both problems are calm according to Theorem 9.4.

An alternative proof of calmness of problem (C) is the following.

The objective function $c(\mathbf{x})$ of problem (C) is convex (and, therefore, Lipschitz in accordance with Theorem 9.1) and Slater's constraint qualification is satisfied by assumption. From Theorem 9.3 it follows that the value function $V(\mathbf{p})$ is Lipschitz, and problem (C) is calm at any solution $\mathbf{x}^*$ of problem (C) according to Theorem 9.2.

Consider the parameterized family $(C^=(\mathbf{p}, q))$ in which problem $(C^=)$ is imbedded:

$(C^=(\mathbf{p}, q))$

$$\min \left\{ c(\mathbf{x}) = \sum_{j \in J} c_j(x_j) \right\}$$

subject to

$$\mathbf{x} \in X^L(\mathbf{p}, q),$$

where $X^L(\mathbf{p}, q)$ is defined as

$$\sum_{j\in J} d_j x_j + q = \alpha$$

$$a_j \le x_j + p_j \le b_j, \quad j \in J.$$

As has been pointed out, $X^L \ne \emptyset$ if

$$\sum_{j\in J} d_j a_j \le \alpha \le \sum_{j\in J} d_j b_j,$$

whereas $X^L(\mathbf{p}, q) \ne \emptyset$ if

$$\sum_{j\in J} d_j(a_j - p_j) \le \alpha - q \le \sum_{j\in J} d_j(b_j - p_j).$$

Without loss of generality, assume that there exists a $(\mathbf{p}, q)$ such that $X^L(\mathbf{p}, q) \ne \emptyset$. This is satisfied, for example, when $\sum_{j\in J} d_j p_j = q$ in addition to the requirement $X^L \ne \emptyset$. Then the value function $V(\mathbf{p}, q)$, associated with problem $(C^{=}(\mathbf{p}, q))$, is finite by Definition 9.4.

Theorem 9.5 (Convexity of the infimum of a convex function subject to linear equality constraints)
Let f be a convex function and S be a convex set in $\mathbf{R}^n$. Then function

$$h(\mathbf{y}) \overset{\text{def}}{=} \inf_{\mathbf{x}} \{f(\mathbf{x}) : A\mathbf{x} = \mathbf{y},\ A_{m\times n},\ \mathbf{x} \in S,\ \mathbf{y} \in \mathbf{R}^m\}$$

is convex.

Proof. Let $\mathbf{x} = \lambda\mathbf{x}_1 + (1-\lambda)\mathbf{x}_2,\ \lambda \in [0, 1],\ \mathbf{x}_1, \mathbf{x}_2 \in S,\ \mathbf{y}_1, \mathbf{y}_2 \in \mathbf{R}^m$. Therefore $\mathbf{x} \in S$ because S is convex by assumption. Then

$$\begin{aligned}
& h(\lambda\mathbf{y}_1 + (1-\lambda)\mathbf{y}_2) \overset{\text{def}}{=} \inf_{\mathbf{x}} \{f(\mathbf{x}) : A\mathbf{x} = \lambda\mathbf{y}_1 + (1-\lambda)\mathbf{y}_2\} \\
&= \inf_{\mathbf{x}_1,\mathbf{x}_2} \{f(\lambda\mathbf{x}_1 + (1-\lambda)\mathbf{x}_2) : A(\lambda\mathbf{x}_1 + (1-\lambda)\mathbf{x}_2) = \lambda\mathbf{y}_1 + (1-\lambda)\mathbf{y}_2\} \\
&\le \inf_{\mathbf{x}_1,\mathbf{x}_2} \{f(\lambda\mathbf{x}_1 + (1-\lambda)\mathbf{x}_2) : A\mathbf{x}_1 = \mathbf{y}_1,\ A\mathbf{x}_2 = \mathbf{y}_2\} \\
&\le \inf_{\mathbf{x}_1,\mathbf{x}_2} \{\lambda f(\mathbf{x}_1) + (1-\lambda) f(\mathbf{x}_2) : A\mathbf{x}_1 = \mathbf{y}_1,\ A\mathbf{x}_2 = \mathbf{y}_2\} \\
&= \inf_{\mathbf{x}_1} \{\lambda f(\mathbf{x}_1) : A\mathbf{x}_1 = \mathbf{y}_1\} + \inf_{\mathbf{x}_2} \{(1-\lambda) f(\mathbf{x}_2) : A\mathbf{x}_2 = \mathbf{y}_2\} \\
&\overset{\text{def}}{=} \lambda h(\mathbf{y}_1) + (1-\lambda) h(\mathbf{y}_2).
\end{aligned}$$

We have used that f is a convex function, the property that

$$\{\mathbf{x}_1, \mathbf{x}_2 \in S : A\mathbf{x}_1 = \mathbf{y}_1, A\mathbf{x}_2 = \mathbf{y}_2\}$$
$$\subseteq \{\mathbf{x}_1, \mathbf{x}_2 \in S : \lambda A\mathbf{x}_1 + (1 - \lambda)A\mathbf{x}_2 = \lambda\mathbf{y}_1 + (1 - \lambda)\mathbf{y}_2\}$$

and the fact that $X \subseteq Y$ implies $\inf_{\mathbf{x}\in Y} f(\mathbf{x}) \leq \inf_{\mathbf{x}\in X} f(\mathbf{x})$.

Therefore $h(\mathbf{y})$ is convex by definition. ■

In problem $(C^=(\mathbf{p}, q))$, matrix A of Theorem 9.5 consists of a single row, that is, $m = 1$, and convex set S is the n-dimensional box

$$S = \{\mathbf{x} \in \mathbf{R}^n : a_j - p_j \leq x_j \leq b_j - p_j, \ j \in J\}.$$

The value function associated with problem $(C^=(\mathbf{p}, q))$ is

$$V(\mathbf{p}, q) = \inf_{\mathbf{x}\in X^L(\mathbf{p},q)} f(\mathbf{x}).$$

From Theorem 9.5 and the assumption that $X^L(\mathbf{p}, q) \neq \emptyset$ it follows that $V(\mathbf{p}, q)$ is convex and finite, respectively, and from Theorem 9.1 it follows that it is Lipschitz. Then, problem $(C^=(\mathbf{p}, q))$ is calm according to Theorem 9.2.

In the general case, if the mathematical program is not convex and equality constraints exist, we can use the approach of Remark 9.1.

Besides the well-posedness of the optimization problems, the stability of methods for solving these problems is also important.

9.2 On the Stability of the Set of Saddle Points of the Lagrangian

9.2.1 The Concept of Stability of Saddle Points of the Lagrangian

Let $\Phi(\mathbf{x}, \mathbf{y})$ be a convex function of $\mathbf{x} \in X$ and a concave function of $\mathbf{y} \in Y$, where X and Y are convex and closed sets.

Recall the definition of a saddle point (Definition 1.25).

A point $(\widehat{\mathbf{x}}, \widehat{\mathbf{y}})$ is said to be a *saddle point* of function $\Phi(\mathbf{x}, \mathbf{y})$, $\mathbf{x} \in X$, $\mathbf{y} \in Y$ if the following inequalities hold

$$\Phi(\widehat{\mathbf{x}}, \mathbf{y}) \leq \Phi(\widehat{\mathbf{x}}, \widehat{\mathbf{y}}) \leq \Phi(\mathbf{x}, \widehat{\mathbf{y}})$$

for all $\mathbf{x} \in X, \mathbf{y} \in Y$, that is, if

$$\Phi(\widehat{\mathbf{x}}, \widehat{\mathbf{y}}) = \min_{\mathbf{x}\in X} \max_{\mathbf{y}\in Y} \Phi(\mathbf{x}, \mathbf{y}) = \max_{\mathbf{y}\in Y} \min_{\mathbf{x}\in X} \Phi(\mathbf{x}, \mathbf{y}).$$

This means that $\Phi(\mathbf{x}, \mathbf{y})$ attains at the saddle point $(\widehat{\mathbf{x}}, \widehat{\mathbf{y}})$ its maximum with respect to $\mathbf{y}$ for fixed $\widehat{\mathbf{x}}$, and $\Phi(\mathbf{x}, \mathbf{y})$ attains at $(\widehat{\mathbf{x}}, \widehat{\mathbf{y}})$ its minimum with respect to $\mathbf{x}$ for fixed $\widehat{\mathbf{y}}$.

Set

$$\chi(\mathbf{x}) = \max_{\mathbf{y}\in Y} \Phi(\mathbf{x}, \mathbf{y}), \quad \psi(\mathbf{y}) = \min_{\mathbf{x}\in X} \Phi(\mathbf{x}, \mathbf{y}).$$

Denote by X^* and Y^* the sets of optimal solutions to the optimization problems

$$\min_{\mathbf{x}\in X} \chi(\mathbf{x})$$

and

$$\max_{\mathbf{y}\in Y} \psi(\mathbf{y}),$$

respectively, that is,

$$X^* \stackrel{\text{def}}{=} \{\mathbf{x} : \chi(\mathbf{x}) = \min_{\mathbf{x}\in X} \max_{\mathbf{y}\in Y} \Phi(\mathbf{x}, \mathbf{y})\} \equiv \{\mathbf{x} : \max_{\mathbf{y}\in Y} \Phi(\mathbf{x}, \mathbf{y}) = \Phi(\widehat{\mathbf{x}}, \widehat{\mathbf{y}})\},$$

$$Y^* \stackrel{\text{def}}{=} \{\mathbf{y} : \psi(\mathbf{y}) = \max_{\mathbf{y}\in Y} \min_{\mathbf{x}\in X} \Phi(\mathbf{x}, \mathbf{y})\} \equiv \{\mathbf{y} : \min_{\mathbf{x}\in X} \Phi(\mathbf{x}, \mathbf{y}) = \Phi(\widehat{\mathbf{x}}, \widehat{\mathbf{y}})\}.$$

Let X^*, Y^* be bounded sets. Then

$$\chi(\mathbf{x}^*) = \psi(\mathbf{y}^*), \ \mathbf{x}^* \in X^*, \ \mathbf{y}^* \in Y^*,$$

that is,

$$\chi(\mathbf{x}^*) \stackrel{\text{def}}{=} \max_{\mathbf{y}\in Y} \Phi(\mathbf{x}^*, \mathbf{y}) = \psi(\mathbf{y}^*) \stackrel{\text{def}}{=} \min_{\mathbf{x}\in X} \Phi(\mathbf{x}, \mathbf{y}^*) = \Phi(\widehat{\mathbf{x}}, \widehat{\mathbf{y}}).$$

This means that $X^* \times Y^*$ is the set of saddle points of $\Phi(\mathbf{x}, \mathbf{y})$ and

$$\Phi(\widehat{\mathbf{x}}, \widehat{\mathbf{y}}) = \Phi(\mathbf{x}^*, \mathbf{y}^*).$$

Consider the sets

$$X_{\mathbf{y}^*} \stackrel{\text{def}}{=} \{\mathbf{x} : \Phi(\mathbf{x}, \mathbf{y}^*) = \Phi(\mathbf{x}^*, \mathbf{y}^*)\},$$

$$Y_{\mathbf{x}^*} \stackrel{\text{def}}{=} \{\mathbf{y} : \Phi(\mathbf{x}^*, \mathbf{y}) = \Phi(\mathbf{x}^*, \mathbf{y}^*)\};$$

that is, $X_{\mathbf{y}^*}$ and $Y_{\mathbf{x}^*}$ denote the sets of arguments of $\Phi(\mathbf{x}, \mathbf{y})$ with $\mathbf{y} = \mathbf{y}^*$ and $\mathbf{x} = \mathbf{x}^*$, respectively, for which the value of $\Phi(\mathbf{x}, \mathbf{y})$ is equal to its value at the saddle point.

In the general case, $X^* \subseteq X_{\mathbf{y}^*}$, $Y^* \subseteq Y_{\mathbf{x}^*}$, that is, the sets $X_{\mathbf{y}^*}$, $Y_{\mathbf{x}^*}$ contain sets X^*, Y^*, respectively.

Definition 9.6 If $X^* = X_{\mathbf{y}^*}$ and $Y^* = Y_{\mathbf{x}^*}$, then the set of saddle points of $\Phi(\mathbf{x}, \mathbf{y})$ is said to be *stable*.

If the set of saddle points of $\Phi(\mathbf{x}, \mathbf{y})$ is stable, then from

$$\lim_{k\to\infty} \Phi(\mathbf{x}^{(k)}, \mathbf{y}^{*(k)}) = \Phi(\mathbf{x}^*, \mathbf{y}^*), \quad \mathbf{x}^{(k)} \in X, \ \mathbf{y}^{*(k)} \in Y$$

it follows that

$$\delta(\mathbf{x}^{(k)}, X^*) \to 0, \ k \to \infty,$$

and from

$$\lim_{k\to\infty} \Phi(\mathbf{x}^{*(k)}, \mathbf{y}^{(k)}) = \Phi(\mathbf{x}^*, \mathbf{y}^*), \quad \mathbf{x}^{*(k)} \in X, \ \mathbf{y}^{(k)} \in Y$$

it follows that

$$\delta(\mathbf{y}^{(k)}, Y^*) \to 0, \ k \to \infty,$$

where $\delta(\mathbf{x}, X) \stackrel{\text{def}}{=} \min_{\mathbf{z}\in X} \|\mathbf{x} - \mathbf{z}\|$ is the distance from $\mathbf{x}$ to the set X. The implications written above mean that the convergence of $\Phi(\mathbf{x}, \mathbf{y}^*)$ to $\Phi(\mathbf{x}^*, \mathbf{y}^*)$ with respect to $\mathbf{x}^{(k)}$ and convergence of $\Phi(\mathbf{x}^*, \mathbf{y})$ to $\Phi(\mathbf{x}^*, \mathbf{y}^*)$ with respect to $\mathbf{y}^{(k)}$ imply the convergence of sequence $(\{\mathbf{x}^{(k)}\}, \{\mathbf{y}^{(k)}\})$ to an element $(\mathbf{x}^*, \mathbf{y}^*)$ of the set $X^* \times Y^*$ of saddle points of $\Phi(\mathbf{x}, \mathbf{y})$.

The concept of stability, introduced by Definition 9.6, is important for constructing iterative gradient algorithms for finding saddle points of the Lagrangian, associated with an optimization problem.

The set of saddle points of the Lagrangian, associated with the problem

$$\min_{\mathbf{x}\in S} f(\mathbf{x}), \tag{9.5}$$

$$S = \{\mathbf{x} \in \mathbf{R}^n : \varphi_i(\mathbf{x}) \le 0, \ i = 1, \ldots, m, \ \mathbf{x} \in X\}, \tag{9.6}$$

is not stable according to Definition 9.6. Concerning the dual variables, this can be proved as follows.

Let rth constraint of (9.6) be satisfied as an equality at $\mathbf{x}^*$, that is,

$$\varphi_r(\mathbf{x}^*) = 0$$

for some r, $1 \le r \le m$. Then the Lagrangian $L(\mathbf{x}^*, \boldsymbol{\lambda})$ of problem (9.5)–(9.6) does not depend on λ_r and therefore $L(\mathbf{x}^*, \boldsymbol{\lambda}^*) = L(\mathbf{x}^*, \boldsymbol{\lambda})$ is satisfied for every λ_r. Hence, it is impossible to determine λ_r by using the relation $L(\mathbf{x}^*, \boldsymbol{\lambda}^*) = L(\mathbf{x}^*, \boldsymbol{\lambda})$.

In order to avoid this difficulty, so-called *modified Lagrangians* are used instead of the "classical" Lagrangian. Modified Lagrangians are usually *nonlinear* functions of $\boldsymbol{\lambda}$ and the set of their saddle points is stable and it coincides, under some assumptions, with the set of saddle points of the "classical" Lagrangian for the same problem. This is important to ensure convergence of iterative gradient algorithms (see, for example, Gol'shtein [51]).

9.2.2 Concerning Stability of the Set of Saddle Points for the Methods Suggested for Problems (C), $(C^{=})$, $(C^{\geq})$

Consider problem (C) [problem $(C^{=})$, $(C^{\geq})$, respectively]. Obviously the Lagrange multiplier λ associated with the constraint (5.2) [constraint (6.2), (8.2), respectively] is not involved in the equality

$$L(\mathbf{x}^*, \mathbf{u}^*, \mathbf{v}^*, \lambda^*) = L(\mathbf{x}^*, \mathbf{u}, \mathbf{v}, \lambda)$$

when $\delta(\lambda^*) = 0$, that is, when $\sum_{j \in J} d_j(x_j) = \alpha$. For problem (C), λ^* (> 0) is either determined *uniquely* from $\delta(\lambda^*) = 0$ when $\delta(0) > 0$ or we set $\lambda^* := 0$ when $\delta(0) \leq 0$. Although the set of saddle points of the Lagrangian $L(\mathbf{x}, \mathbf{u}, \mathbf{v}, \lambda)$, associated with problem (C) [problem $(C^{=})$, $(C^{\geq})$, respectively], is not stable in the sense of Gol'shtein, the specificity of the approach suggested in Part Two of this book (Algorithms 1, 2, and 3 are not of gradient type and λ^* is determined uniquely in all cases for the three problems under consideration) overcomes this "weakness" of the classical Lagrangian. That is why it is not necessary to use modified Lagrangians for problems (C), $(C^{=})$, and $(C^{\geq})$.

On the one hand, we need a closed form expression of λ at Step 2 of the algorithms suggested. However, it is this feature of the algorithms that allows us to use classical Lagrangians instead of modified Lagrangians in the proposed approach. Moreover, the method for finding λ, and, therefore, for finding x_j^*, $j \in J$ in the corresponding problem, is *exact* although it is an iterative method.

As it usually happens, the disadvantage in one aspect turns out to be an advantage in another aspect, and vice versa.

All conclusions in this chapter have been drawn under the assumption that the objective function $c(\mathbf{x})$ and the constraint function(s) $d_i(\mathbf{x})$ are *nondegenerate*, that is, $\mathbf{c}'(\mathbf{x}^*) \not\equiv \mathbf{0}$, $\mathbf{d}_i'(\mathbf{x}^*) \not\equiv \mathbf{0}$; otherwise the application of the Karush-Kuhn-Tucker theorem with differentiability is void of meaning.

Some optimality criteria for degenerate mathematical programs are given, for example, in the book of Karmanov [67], pp. 269 - 278.

Chapter 10
Extensions

10.1 Theoretical Aspects

If it is allowed for problem (C): i) $d_j'(x_j) \equiv 0$ or ii) $d_j'(x_j) \not\equiv 0$ but $d_j'(a_j) = 0$ and/or $d_j'(b_j) = 0$ for some $j \in J$ in (5.2), then for such j's, we cannot construct the expressions $-\frac{c_j'(a_j)}{d_j'(a_j)}$ and/or $-\frac{c_j'(b_j)}{d_j'(b_j)}$, by means of which the sets J_a^λ (5.4), J_b^λ (5.5), J^λ (5.6) are defined. In case i) we have $d_j(x_j) =: d_j = const.$ and x_j's are not involved in (5.2) for such j's.

Denote

$$\begin{aligned} Z0 &= \{j \in J : d_j'(x_j) \equiv 0\}, \\ ZA &= \{j \in J \setminus Z0 : d_j'(a_j) = 0\}, \\ ZB &= \{j \in J \setminus Z0 : d_j'(b_j) = 0\}. \end{aligned}$$

Here "0" denotes the "computer zero". In particular, when $J = Z0$ and $d_j(x_j) =: d_j = 0,\ j \in J$, $\alpha = 0$, then X is defined only by (5.3).

Theorem 10.1 (Characterization of the optimal solution to problem (C): an extended version)
Problem (C) can be decomposed into two subproblems:
$(C1)$ *for $j \in Z0$, and*
$(C2)$ *for $j \in J \setminus Z0$ with $\alpha := \alpha - \sum_{j \in Z0} d_j(x_j^*) \equiv \alpha - \sum_{j \in Z0} d_j$.*
The optimal solution to subproblem $(C1)$ is

$$x_j^* = \begin{cases} a_j, & j \in Z0 \text{ and } h_j^{\leq} \leq a_j, \\ b_j, & j \in Z0 \text{ and } h_j^{\leq} \geq b_j, \\ h_j^{\leq}, & j \in Z0 \text{ and } a_j < h_j^{\leq} < b_j, \end{cases} \tag{10.1}$$

© The Author(s), under exclusive license to Springer Nature Switzerland AG 2021

S. M. Stefanov, *Separable Optimization*, Springer Optimization and Its Applications 177,
https://doi.org/10.1007/978-3-030-78401-0_10

that is, subproblem $(C1)$ *is decomposed into* $n_0 \equiv |Z0|$ *independent subproblems. The optimal solution to* $(C2)$ *is given by* (5.4), (5.5), (5.6) *with* $J := J \setminus Z0$, $\alpha := \alpha - \sum_{j\in Z0} d_j(x_j^*)$, *where we adopt* $\dfrac{c_j'(a_j)}{d_j'(a_j)} = \lim_{t\to a_j} \dfrac{c_j'(t)}{d_j'(t)}$ *if* $j \in ZA$, *and we adopt* $\dfrac{c_j'(b_j)}{d_j'(b_j)} = \lim_{t\to b_j} \dfrac{c_j'(t)}{d_j'(t)}$ *when* $j \in ZB$.

It is possible some of the limits above be equal to $-\infty$ or $+\infty$.

Proof of Theorem 10.1 repeats in part the proof of Theorem 5.1.

Proof. *Necessity.* Let $\mathbf{x}^* = (x_j^*)_{j\in J}$ be an optimal solution to (C).

1) Let $j \in Z0$, that is, $d_j'(x_j) \equiv 0$. Then the KKT conditions are

$$c_j'(x_j^*) - u_j + v_j = 0, \ j \in Z0 \quad \text{from (5.7)}$$
$$\text{and } (5.8) - (5.13).$$

a) If $x_j^* = a_j$, then $u_j \geq 0$, $v_j = 0$ according to (5.8), (5.9), (5.13). From (5.7) and definition of $h_j^{\leq}$ it follows that $c_j'(x_j^*) = u_j \geq 0 \equiv c_j'(h_j^{\leq})$. Since $c_j'(x_j)$ is a monotone nondecreasing function of x_j for each $j \in J$, then $x_j^* \equiv a_j \geq h_j^{\leq}$.

b) If $x_j^* = b_j$, then $u_j = 0$, $v_j \geq 0$ according to (5.8), (5.9), (5.13). Therefore (5.7) implies that $c_j'(x_j^*) = -v_j \leq 0 \equiv c_j'(h_j^{\leq})$. Using that $c_j'(x_j)$ is a monotone nondecreasing function of x_j for each $j \in J$, we obtain $x_j^* \equiv b_j \leq h_j^{\leq}$.

c) If $a_j < x_j^* < b_j$, then $u_j = v_j = 0$ according to (5.8), (5.9). Therefore (5.7) implies that $c_j'(x_j^*) = 0$, that is, $x_j^* = h_j^{\leq}$ according to definition of $h_j^{\leq}$.

2) Components of the optimal solution to subproblem $(C2)$ are obtained by using the same approach as that of the proof of Theorem 5.1 but with the reduced index set $J := J \setminus Z0$ and reduced right-hand side of (5.2) $\alpha := \alpha - \sum_{j\in Z0} d_j(x_j^*) \equiv \alpha - \sum_{j\in Z0} d_j$.

Sufficiency. Conversely, let $\mathbf{x}^* \in X$ and components of $\mathbf{x}^*$ satisfy: (10.1) for $j \in Z0$, and (5.4), (5.5), (5.6) with $J := J \setminus Z0$ and $\alpha := \alpha - \sum_{j\in Z0} d_j(x_j^*) \equiv \alpha - \sum_{j\in Z0} d_j$ for $j \in J \setminus Z0$. Set:

$$\begin{array}{ll} u_j = v_j = 0 & \text{for } a_j < x_j^* < b_j, \ j \in Z0; \\ u_j = c_j'(a_j) \ \ (\geq 0), \ v_j = 0 & \text{for } x_j^* = a_j, \ j \in Z0; \\ u_j = 0, \ v_j = -c_j'(b_j) \ \ (\geq 0) & \text{for } x_j^* = b_j, \ j \in Z0. \end{array}$$

If $\lambda > 0$, set:

$$\begin{array}{ll} \lambda = -\dfrac{c_j'(x_j^*)}{d_j'(x_j^*)} = \lambda(\mathbf{x}^*) \ \ (> 0) & \text{from (5.6), (7.8);} \\ u_j = v_j = 0 & \text{for } a_j < x_j^* < b_j, \ j \in J \setminus Z0; \\ u_j = c_j'(a_j) + \lambda d_j'(a_j) \ \ (\geq 0), \quad v_j = 0 & \text{for } x_j^* = a_j, \ j \in J \setminus Z0; \\ u_j = 0, \quad v_j = -c_j'(b_j) - \lambda d_j'(b_j) \ \ (\geq 0) & \text{for } x_j^* = b_j, \ j \in J \setminus Z0. \end{array}$$

If $\lambda = 0$, set:

$$\begin{aligned} &u_j = v_j = 0 && \text{for } a_j < x_j^* < b_j,\ j \in J \setminus Z0; \\ &u_j = c_j'(a_j) \quad (\geq 0), \quad v_j = 0 && \text{for } x_j^* = a_j,\ j \in J \setminus Z0; \\ &u_j = 0, \quad v_j = -c_j'(b_j) \quad (\geq 0) && \text{for } x_j^* = b_j,\ j \in J \setminus Z0. \end{aligned}$$

It can be verified that $\mathbf{x}^*$, λ, u_j, v_j, $j \in J$ satisfy the KKT conditions for problem (C). Then $\mathbf{x}^*$ with components: (10.1) for $j \in Z0$, and (5.4), (5.5), (5.6) with $\alpha := \alpha - \sum_{j \in Z0} d_j(x_j^*) \equiv \alpha - \sum_{j \in Z0} d_j$ for $j \in J \setminus Z0$ is an optimal solution to problem $(C) = (C1) \cup (C2)$. ■

Thus, with the use of Theorem 10.1, we can express components x_j^*, $j \in Z0$ of the optimal solution to problem (C) *without* the necessity of constructing expressions $-\frac{c_j'(a_j)}{d_j'(a_j)}$ with $d_j'(a_j) = 0$ and $-\frac{c_j'(b_j)}{d_j'(b_j)}$ with $d_j'(b_j) = 0$.

A result, similar to Theorem 10.1, holds for problem $(C^=)$ as well (with $Z0 = \{j \in J : d_j = 0\}$, $ZA = ZB = \emptyset$).

Theorem 10.2 (Characterization of the optimal solution to problem $(C^=)$: an extended version)
Problem $(C^=)$ can be decomposed into two subproblems:
$(C1^=)$ *for* $j \in Z0 := \{j \in J : d_j = 0\}$ *and*
$(C2^=)$ *for* $j \in J \setminus Z0$.
The optimal solution to subproblem $(C1^=)$ is

$$x_j^* = \begin{cases} a_j, & j \in Z0 \text{ and } h_j^= \leq a_j \\ b_j, & j \in Z0 \text{ and } h_j^= \geq b_j \\ h_j^=, & j \in Z0 \text{ and } a_j < h_j^= < b_j. \end{cases} \tag{10.2}$$

The optimal solution to subproblem $(C2^=)$ is given by (6.4), (6.5), (6.6) *with* $J := J \setminus Z0$.

Proof. *Necessity.* Let $\mathbf{x}^* = (x_j^*)_{j \in J}$ be an optimal solution to $(C^=)$.

1) Let $j \in Z0$, that is, $d_j = 0$ for this j. The KKT conditions are:
$c_j'(x_j^*) - u_j + v_j = 0,\ j \in Z0$ from (6.7)
and (6.8) – (6.12).

a) If $x_j^* = a_j$, then $u_j \geq 0$, $v_j = 0$ according to (6.8), (6.9), (6.10). From (6.7) and definition of $h_j^=$ we obtain that $c_j'(x_j^*) = u_j \geq 0 \equiv c_j'(h_j^=)$. Therefore, $x_j^* = a_j \geq h_j^=$ according to (5.15).

b) If $x_j^* = b_j$, then $u_j = 0$, $v_j \geq 0$ according to (6.8), (6.9), (6.10). Hence, (6.7) implies $c_j'(x_j^*) = -v_j \leq 0 \equiv c_j'(h_j^=)$. Using (5.15), we obtain $x_j^* = b_j \leq h_j^=$.

c) If $a_j < x_j^* < b_j$, then $u_j = v_j = 0$ according to (6.8), (6.9). Hence, (6.7) implies $c_j'(x_j^*) = 0$, that is, $x_j^* = h_j^=$ according to definition of $h_j^=$.

2) Components of the optimal solution to $(C2^=)$ are obtained by using the approach from the proof of Theorem 6.1, necessity part, with $J := J \setminus Z0$.

Sufficiency. Conversely, let $\mathbf{x}^* \in X^L$ and components of $\mathbf{x}^*$ satisfy: (10.2) for $j \in Z0$, and (6.4), (6.5), (6.6) for $j \in J \setminus Z0$. Set:

$$
\begin{array}{ll}
u_j = v_j = 0 & \text{for } a_j < x_j^* < b_j,\ j \in Z0; \\
u_j = c_j'(a_j)\ \ (\geq 0),\ v_j = 0 & \text{for } x_j^* = a_j,\ j \in Z0; \\
u_j = 0,\ v_j = -c_j'(b_j)\ \ (\geq 0) & \text{for } x_j^* = b_j,\ j \in Z0; \\
\lambda = -\dfrac{c_j'(x_j^*)}{d_j};\ \ u_j = v_j = 0 & \text{for } a_j < x_j^* < b_j,\ j \in J \setminus Z0; \\
u_j = c_j'(a_j) + \lambda d_j\ \ (\geq 0),\ \ v_j = 0 & \text{for } x_j^* = a_j,\ j \in J \setminus Z0; \\
u_j = 0,\ \ v_j = -c_j'(b_j) - \lambda d_j\ \ (\geq 0) & \text{for } x_j^* = b_j,\ j \in J \setminus Z0.
\end{array}
$$

Since $\mathbf{x}^*$, λ, u_j, v_j, $j \in J$ satisfy the KKT conditions for problem $(C^=)$, then $\mathbf{x}^*$ with components: (10.2) for $j \in Z0$, and (6.4), (6.5), (6.6) for $j \in J \setminus Z0$, is an optimal solution to problem $(C^=) = (C1^=) \cup (C2^=)$. ■

Since Theorem 8.1 and Algorithm 3 are based on the sets of indices J_a^λ, J_b^λ, J^λ of problem $(C^=)$, Theorem 10.2 solves the problem of decomposition of problem $(C^\geq)$ as well.

With the use of set $Z0$, we can deduce the following about the check whether the feasible region is empty or nonempty when $J = Z0$.

About problem (C).

When $J = Z0$, $a_j < b_j$, $j \in J$, $\sum_{j \in J} d_j \leq \alpha$, the feasible region of problem (C) is always nonempty and it is not necessary to check anything else in this case.

About problems $(C^=)$ **and** $(C^\geq)$.

When $J = Z0$, $a_j < b_j$, $j \in J$, $\alpha = 0$, the feasible regions of problems $(C^=)$ and $(C^\geq)$ are always nonempty and it is not necessary to check anything else in this case.

10.2 Computational Aspects

Algorithms 1, 2, and 3 are also applicable in cases when $a_j = -\infty$ for some $j \in J$ and/or $b_j = \infty$ for some $j \in J$. However, if we use the computer values of $-\infty$ and $+\infty$ at the first step of the algorithms to check whether the corresponding feasible region is empty or nonempty and at Step 3 in the expressions $-\frac{c_j'(x_j)}{d_j'(x_j)}$ or $-\frac{c_j'(x_j)}{d_j}$ with $x_j = -\infty$ and/or $x_j = +\infty$, by means of which we construct index sets J_a^λ, J_b^λ, J^λ, this can sometimes lead to arithmetic overflow. If we use other values of $-\infty$ and $+\infty$ with smaller absolute values than those of the computer values of $-\infty$ and $+\infty$, this would lead to inconvenience and dependence on the data of the particular problems. To avoid these difficulties and to take into account the above discussion, it is useful to do the following.

Construct the following sets of indices:

$$
\begin{aligned}
SVN &= \{j \in J \setminus Z0 : a_j > -\infty,\ \ b_j < +\infty\}, \\
SV1 &= \{j \in J \setminus Z0 : a_j > -\infty,\ \ b_j = +\infty\}, \\
SV2 &= \{j \in J \setminus Z0 : a_j = -\infty,\ \ b_j < +\infty\}, \\
SV &= \{j \in J \setminus Z0 : a_j = -\infty,\ \ b_j = +\infty\}.
\end{aligned}
\tag{10.3}
$$

It is obvious that $Z0 \cup SV \cup SV1 \cup SV2 \cup SVN = J$, that is, the set $J \setminus Z0$ is partitioned into the four subsets SVN, $SV1$, $SV2$, SV, defined above.

When programming the algorithms, we use computer values of $-\infty$ and $+\infty$ for constructing the sets SVN, $SV1$, $SV2$, SV.

In order to construct the sets J_a^λ, J_b^λ, J^λ without the necessity of calculating the values $-\frac{c_j'(x_j)}{d_j'(x_j)}$ or $-\frac{c_j'(x_j)}{d_j}$ with $x_j = -\infty$ or $+\infty$, except for the sets J, $Z0$, SV, $SV1$, $SV2$, SVN, we need some subsidiary sets defined as follows:

About Problem (C)

For SVN:

$$J^{\lambda\, SVN} = \left\{ j \in SVN : -\frac{c_j'(b_j)}{d_j'(b_j)} < \lambda < -\frac{c_j'(a_j)}{d_j'(a_j)} \right\},$$

$$J_a^{\lambda\, SVN} = \left\{ j \in SVN : \lambda \geq -\frac{c_j'(a_j)}{d_j'(a_j)} \right\},$$

$$J_b^{\lambda\, SVN} = \left\{ j \in SVN : \lambda \leq -\frac{c_j'(b_j)}{d_j'(b_j)} \right\};$$

for $SV1$:

$$J^{\lambda\, SV1} = \left\{ j \in SV1 : \lambda < -\frac{c_j'(a_j)}{d_j'(a_j)} \right\}, \tag{10.4}$$

$$J_a^{\lambda\, SV1} = \left\{ j \in SV1 : \lambda \geq -\frac{c_j'(a_j)}{d_j'(a_j)} \right\};$$

for $SV2$:

$$J^{\lambda\, SV2} = \left\{ j \in SV2 : \lambda > -\frac{c_j'(b_j)}{d_j'(b_j)} \right\},$$

$$J_b^{\lambda\, SV2} = \left\{ j \in SV2 : \lambda \leq -\frac{c_j'(b_j)}{d_j'(b_j)} \right\};$$

for SV:

$$J^{\lambda\, SV} = SV.$$

About Problems $(C^{=})$ *and* $(C^{\geq})$

Subsidiary index sets are defined as in (10.4) but since $d_j(x_j) = d_j x_j,\ j \in J$ in problems $(C^{=})$ and $(C^{\geq})$ then $d'_j(a_j) = d'_j(b_j) = d_j,\ j \in J$.

Then

$$\begin{aligned} J^{\lambda} &:= J^{\lambda\, SVN} \cup J^{\lambda\, SV1} \cup J^{\lambda\, SV2} \cup J^{\lambda\, SV}, \\ J_a^{\lambda} &:= J_a^{\lambda\, SVN} \cup J_a^{\lambda\, SV1}, \\ J_b^{\lambda} &:= J_b^{\lambda\, SVN} \cup J_b^{\lambda\, SV2}. \end{aligned} \tag{10.5}$$

The index sets J^{λ}, J_a^{λ}, J_b^{λ}, constructed in such a way, are used as the corresponding index sets with the same names in Algorithms 1, 2, 3.

The relationships among index sets are presented in Figure 10.1.

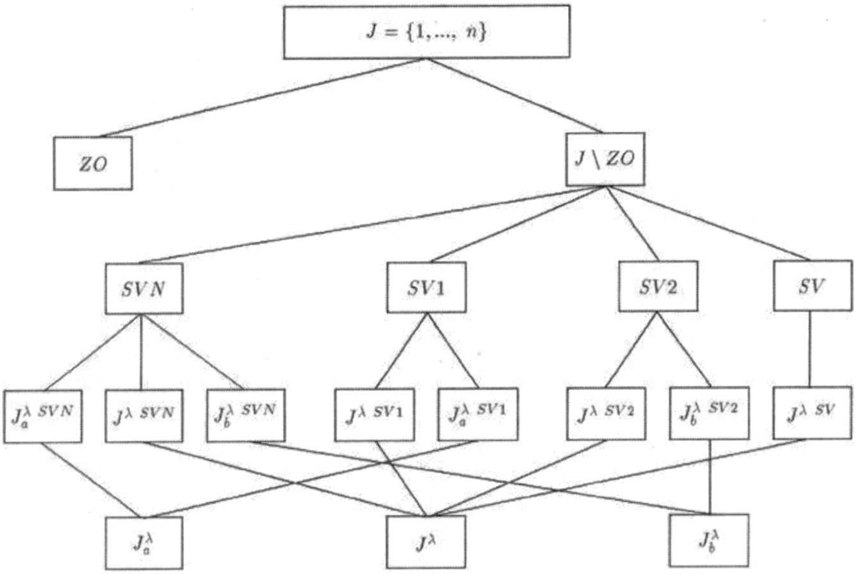

Fig. 10.1 Relationships among index sets for problems (C), $(C^{=})$, $(C^{\geq})$.

The reason to construct the sets (10.3) and (10.4) is the following.

If $j \in SVN$, then none of the a_j's is equal to $-\infty$ and none of the b_j's is equal to $+\infty$. That is why there are no computational difficulties, described above, with the expressions involving such indices j.

If $j \in SV1$, that is, if $a_j > -\infty$ and $b_j = +\infty$, then $v_j = 0,\ j \in SV1$ according to (5.9) for problem (C) (according to (6.9) for $(C^{=})$ and according to (8.6) for $(C^{\geq})$). Then $j \in J_a^{\lambda}$ or $j \in J^{\lambda}$ using the same reasoning as in a) and c) from the proofs (necessity part) of Theorem 5.1 and Theorem 6.1. Therefore, it is sufficient to consider only sets of the form J_a^{λ} and J^{λ} for $SV1$, and we have denoted these sets by $J_a^{\lambda\, SV1}$ and $J^{\lambda\, SV1}$ in (10.4), respectively.

Similarly, if $j \in SV2$ then $a_j = -\infty$ and $b_j < +\infty$ for these j's. Then $u_j = 0$ according to (5.8) for (C) (according to (6.8) for $(C^{=})$ and according to (8.5) for

$(C^{\geq})$). Hence, $j \in J_b^{\lambda}$ or $j \in J^{\lambda}$ according to cases b) and c) of the proofs (necessity part) of Theorem 5.1 and Theorem 6.1. That is why it is sufficient to consider only sets of the form J_b^{λ} and J^{λ} for $SV2$. We have denoted these sets by $J_b^{\lambda\, SV2}$ and $J^{\lambda\, SV2}$ in (10.4), respectively.

If $j \in SV$, then $a_j = -\infty$ and $b_j = +\infty$. Therefore, $u_j = v_j = 0$ according to (5.8) and (5.9) for (C), according to (6.8) and (6.9) for $(C^{=})$, and according to (8.5) and (8.6) for $(C^{\geq})$. Hence, $j \in J^{\lambda}$ according to case c) of the proofs (necessity part) of Theorem 5.1 and Theorem 6.1. Therefore, $j \in SV$ implies $j \in J^{\lambda}$, and we have denoted by $J^{\lambda\, SV}$ the set $\left\{ j \in SV : -\frac{c_j'(b_j)}{d_j'(b_j)} < \lambda < -\frac{c_j'(a_j)}{d_j'(a_j)} \right\}$ in (10.4) in this case.

Since Theorem 8.1 (Sufficient condition for solution to $(C^{\geq})$) is based on the index sets for problem $(C^{=})$ from Theorem 6.1, these conclusions are valid for problem $(C^{\geq})$ as well.

Because $d_j'(x_j) > 0, \ j \in J$ for problem (C) $(d_j > 0, \ j \in J$ for problems $(C^{=})$ and $(C^{\geq})$), then $d_j(x_j)$ is a monotone nondecreasing function for each $j \in J$. This helps us to draw the following conclusions.

About Problem (C)

i) If $SVN \cup SV1 = J \setminus Z0$, that is, if all a_j's are finite but some of (or all) b_j's are equal to $+\infty$ for the variables which are involved in (5.2), then it is sufficient to check whether

$$\sum_{j \in J \setminus Z0} d_j(a_j) \leq \alpha$$

and it is not necessary to check whether $\alpha \leq \sum_{j \in J \setminus Z0} d_j(b_j)$ in this case;

ii) Else if $SV2 \cup SV \neq \emptyset$, that is, if there exists at least one variable x_j which is involved in (5.2) with $a_j = -\infty$, then:

if $c_j(x_j) \geq 0, \ j \in J$ and $SVN \cup SV1 \neq \emptyset$ then it is sufficient to check whether

$$\sum_{j \in SVN \cup SV1} d_j(a_j) \leq \alpha$$

else $X \neq \emptyset$ and it is not necessary to check anything else in this case.

Similarly, we can consider problems $(C^{=})$ and $(C^{\geq})$.

About Problem $(C^{=})$

i) If $SVN = J \setminus Z0$, that is, if a_j's and b_j's are finite for all $j \in J$ which are involved in (6.2) with $d_j \neq 0$, then the check whether the corresponding feasible region is nonempty is

$$\sum_{j \in SVN} d_j a_j \leq \alpha \leq \sum_{j \in SVN} d_j b_j?$$

ii) Else if $SV1 \cup SVN = J \setminus Z0$, that is, if all a_j's are finite but some of (or all) b_j's are equal to $+\infty$ for the variables which are involved in (6.2), then the check is whether

$$\sum_{j \in J \setminus Z0} d_j a_j \leq \alpha$$

and it is not necessary to check whether $\alpha \leq \sum_{j \in J \setminus Z0} d_j b_j$ in this case;

iii) Else if $SV2 \cup SVN = J \setminus Z0$, that is, if all b_j's are finite but some of (or all) a_j's are equal to $-\infty$ for variables which are involved in (6.2), then the check is whether

$$\alpha \leq \sum_{j \in J \setminus Z0} d_j b_j$$

and it is not necessary to check whether $\sum_{j \in J \setminus Z0} d_j a_j \leq \alpha$ in this case;

iv) Else if $SV \neq \emptyset$, that is, when there exists at least one variable x_j which is involved in (6.2) with $a_j = -\infty$ and $b_j = +\infty$, then $X^L \neq \emptyset$ and it is not necessary to check anything else in this case.

About Problem $(C^{\geq})$

i) If $SV2 \cup SVN = J \setminus Z0$, that is, if all b_j's are finite but some of (or all) a_j's are equal to $-\infty$ for variables which are involved in (8.2), then it is sufficient to check whether

$$\alpha \leq \sum_{j \in J \setminus Z0} d_j b_j$$

and it is not necessary to check whether $\sum_{j \in J \setminus Z0} d_j a_j \leq \alpha$ in this case;

ii) Else if $SV1 \cup SVN = J \setminus Z0$, that is, if all a_j's are finite but some of (or all) b_j's are equal to $+\infty$ for the variables which are involved in (8.2), then we have to check whether $\sum_{j \in J \setminus Z0} d_j a_j \leq \alpha$, otherwise there does not exist $x_j^* \in [a_j, b_j],\ j \in J$, such that $\sum_{j \in J \setminus Z0} d_j x_j^* = \alpha$.

iii) Else if $SV1 \cup SV \neq \emptyset$, that is, if there exists at least one variable x_j which is involved in (8.2) with $b_j = +\infty$, then $X^{\geq} \neq \emptyset$ and it is not necessary to check anything else in this case.

Using the previous results of this chapter, Steps 0, 1, and 3 of Algorithm 1, and Steps 1 and 3 of Algorithms 2 and 3 can be modified as follows, respectively.

About Algorithm 1.

Step 0[1]. (Initialization) $J := \{1, \ldots, n\}$,
$k := 0,\ \alpha^{(0)} := \alpha,\ n^{(0)} := n$,
$J^{(0)} := J,\ J_a^{\lambda} := \emptyset,\ J_b^{\lambda} := \emptyset$,
initialize $h_j^{\leq},\ j \in J$.
Construct the set $Z0$. If $j \in Z0$ then:
if $h_j^{\leq} \leq a_j$ then $x_j^* := a_j$

else if $h_j^{\leq} \geq b_j$ then $x_j^* := b_j$
else if $a_j < h_j^{\leq} < b_j$ then $x_j^* := h_j^{\leq}$.
If $J = Z0$ and $\alpha = 0$ then go to Step 10
else if $J = Z0$ and $\alpha \neq 0$ then go to Step 9.
Set $J := J \setminus Z0,\ J^{(0)} := J,\ n^{(0)} := n - |Z0|$,
$\alpha^{(0)} := \alpha - \sum_{j \in Z0} d_j(x_j^*) \equiv \alpha - \sum_{j \in Z0} d_j$.
Construct the sets SVN, $SV1$, $SV2$, SV.
If $SVN \cup SV1 = J$ then
if $\sum_{j \in J} d_j(a_j) \leq \alpha$ go to Step 1
else go to Step 9 (feasible region X is empty)
else if $SV2 \cup SV \neq \emptyset$ then
if $c_j(x_j) \geq 0,\ j \in J$ and $SVN \cup SV1 \neq \emptyset$ then
if $\sum_{j \in SVN \cup SV1} d_j(a_j) \leq \alpha$ go to Step 1
else go to Step 9 (feasible region is empty)
else go to Step 1 (feasible region is always nonempty).

Step 1[1]. Construct the sets $J^{0\,SVN}$, $J_a^{0\,SVN}$, $J_b^{0\,SVN}$, $J^{0\,SV1}$, $J_a^{0\,SV1}$, $J^{0\,SV2}$, $J_b^{0\,SV2}$, $J^{0\,SV}$ (for $\lambda = 0$).
Construct the sets J_a^0, J_b^0, J^0 through (10.5). Calculate

$$\delta(0) := \sum_{j \in J_a^0} d_j(a_j) + \sum_{j \in J_b^0} d_j(b_j) + \sum_{j \in J^0} d_j(h_j^{\leq}) - \alpha.$$

If $\delta(0) \leq 0$ then $\lambda := 0$, go to Step 8
else if $\delta(0) > 0$ then:
if $SV2 \cup SVN = J$ then
if $\alpha \leq \sum_{j \in J} d_j(b_j)$ go to Step 2
else go to Step 9 (there does not exist a $\lambda^* > 0$ such that $\delta(\lambda^*) = 0$)
else if $SV1 \cup SV \neq \emptyset$ go to Step 2 (there exists a $\lambda^* > 0$ such that $\delta(\lambda^*) = 0$).

Step 3[1]. Construct the sets $J^{\lambda\,SVN}$, $J_a^{\lambda\,SVN}$, $J_b^{\lambda\,SVN}$, $J^{\lambda\,SV1}$, $J_a^{\lambda\,SV1}$, $J^{\lambda\,SV2}$, $J_b^{\lambda\,SV2}$, $J^{\lambda\,SV}$ (with $J^{(k)}$ instead of J).
Construct the sets $J_a^{\lambda(k)}$, $J_b^{\lambda(k)}$, $J^{\lambda(k)}$ by using (10.5) and find their cardinalities $|J_a^{\lambda(k)}|$, $|J_b^{\lambda(k)}|$, $|J^{\lambda(k)}|$, respectively. Go to Step 4.

About Algorithm 2.

Step 1[2]. (Initialization) $J := \{1, \ldots, n\}$,
$k := 0,\ \alpha^{(0)} := \alpha,\ n^{(0)} := n$,
$J^{(0)} := J,\ J_a^{\lambda} := \emptyset,\ J_b^{\lambda} := \emptyset$,
initialize $h_j^{=},\ j \in J$.
Construct the set $Z0$. If $j \in Z0$ then:
if $h_j^{=} \leq a_j$ then $x_j^* := a_j$
else if $h_j^{=} \geq b_j$ then $x_j^* := b_j$
else if $a_j < h_j^{=} < b_j$ then $x_j^* := h_j^{=}$.

If $J = Z0$ and $\alpha = 0$ then go to Step 10
 else if $J = Z0$ and $\alpha \neq 0$ go to Step 9.
Set $J := J \setminus Z0,\ J^{(0)} := J,\ n^{(0)} := n - |Z0|$,
$\alpha^{(0)} := \alpha - \sum_{j \in Z0} d_j x_j^* \equiv \alpha$.
Construct the sets SVN, $SV1$, $SV2$, SV.
If $SVN = J$ then
 if $\sum_{j \in J} d_j a_j \leq \alpha \leq \sum_{j \in J} d_j b_j$ then go to Step 2
 else go to Step 9 (feasible region X^L is empty)
else if $SV1 \cup SVN = J$ then
 if $\sum_{j \in J} d_j a_j \leq \alpha$ then go to Step 2
 else go to Step 9 (feasible region X^L is empty)
else if $SV2 \cup SVN = J$ then
 if $\alpha \leq \sum_{j \in J} d_j b_j$ go to Step 2
 else go to Step 9 (feasible region X^L is empty)
else if $SV \neq \emptyset$ go to Step 2 (feasible region X^L is always nonempty).

Step 3^2. is the same as **Step 3^1**.

About Algorithm 3.

Step 1^3. (Initialization) $J := \{1, \ldots, n\}$,
$k := 0,\ \alpha^{(0)} := \alpha,\ n^{(0)} := n$,
$J^{(0)} := J,\ J_a^\lambda := \emptyset,\ J_b^\lambda := \emptyset$,
initialize $h_j^{\geq},\ j \in J$.
Construct the set $Z0$. If $j \in Z0$ then:
 if $h_j^{\geq} \leq a_j$ then $x_j^* := a_j$
 else if $h_j^{\geq} \geq b_j$ then $x_j^* := b_j$
 else if $a_j < h_j^{\geq} < b_j$ then $x_j^* := h_j^{\geq}$.
If $J = Z0$ and $\alpha = 0$ then go to Step 10
 else if $J = Z0$ and $\alpha \neq 0$ go to Step 9.
Set $J := J \setminus Z0,\ J^{(0)} := J,\ n^{(0)} := n - |Z0|$,
$\alpha^{(0)} := \alpha - \sum_{j \in Z0} d_j x_j^* \equiv \alpha$.
Construct the sets SVN, $SV1$, $SV2$, SV.
If $SV2 \cup SVN = J$ then
 if $\alpha > \sum_{j \in J} d_j b_j$ then go to Step 9 (feasible region $X^{\geq}$ is empty)
 else if $\alpha \leq \sum_{j \in J} d_j b_j$ or $SV1 \cup SV \neq \emptyset$ then feasible region
 $X^{\geq}$ is nonempty.
If $SV1 \cup SVN = J$ then
 if $\sum_{j \in J} d_j a_j \leq \alpha$ go to Step 2
 else go to Step 9 (there do not exist $x_j^* \in [a_j, b_j],\ j \in J$
 such that $\sum_{j \in J} d_j x_j^* = \alpha$)
else if $SV1 \cup SV \neq \emptyset$ go to Step 2.

Step 3^3. is the same as **Step 3^1**.

Modifications of the algorithms connected with theoretical and computational aspects do not influence their computational complexity, discussed in Sections 7.3, 7.5, and 8.2, because these modifications do not affect the "iterative" steps of the algorithms suggested.

Chapter 11
Applications and Computational Experiments

In this chapter, we present some important types of functions for problems (C), $(C^{=})$, and $(C^{\geq})$, considered in the previous chapters of Part Two of this book, and the explicit expressions which are necessary for implementing the suggested algorithms. We also present computational results obtained by applying these algorithms to problems under consideration.

11.1 Some Important Types of Functions for Problems (C), $(C^{=})$, and $(C^{\geq})$

Problem 11.1

$$c_j(x_j) = \frac{s_j}{x_j}, \quad s_j \neq 0, \; x_j \neq 0, \tag{11.1.0}$$

$$d_j(x_j) = d_j x_j^p, \quad d_j > 0, \; p \geq 1, \; x_j > 0.$$

Since $c''_j(x_j) = \frac{2s_j}{x_j^3}$, then $c_j(x_j)$ are convex for $s_j x_j \geq 0$, and since $s_j \neq 0, \; x_j \neq 0$, then $c_j(x_j)$ are strictly convex for $s_j x_j > 0, \; j \in J$ (Theorem 1.32).

As it is known, $d_j(x_j) = d_j x_j^p$ are convex functions for $d_j \geq 0, \; p \geq 1$ and $x_j \geq 0$, and strictly convex for $d_j > 0, \; p > 1$ and $x_j > 0$ (Example 1.5, Chapter One). Since $d'_j(x_j) = p d_j x_j^{p-1}$ and since $d_j > 0, \; p \geq 1$, then the requirement $d'_j(x_j) > 0$ in this case is: $d_j > 0, \; x_j > 0, \; j \in J$; and since $s_j x_j > 0, \; x_j > 0$, then $s_j > 0, \; j \in J$.

According to Theorem 5.1, components of the optimal solution to problem (C) with $c_j(x_j) = \frac{s_j}{x_j}$ and $d_j(x_j) = d_j x_j^p$ are

© The Author(s), under exclusive license to Springer Nature Switzerland AG 2021

S. M. Stefanov, *Separable Optimization*, Springer Optimization and Its Applications 177,
https://doi.org/10.1007/978-3-030-78401-0_11

$$x_j^* = a_j, \quad j \in J_a^\lambda = \left\{ j \in J : \lambda \geq \frac{s_j}{pd_j a_j^{p+1}} \right\}, \tag{11.1.1}$$

$$x_j^* = b_j, \quad j \in J_b^\lambda = \left\{ j \in J : \lambda \leq \frac{s_j}{pd_j b_j^{p+1}} \right\}, \tag{11.1.2}$$

$$x_j^* = \sqrt[p+1]{\frac{s_j}{p\lambda d_j}}, \quad j \in J^\lambda = \left\{ j \in J : \frac{s_j}{pd_j b_j^{p+1}} \leq \lambda \leq \frac{s_j}{pd_j a_j^{p+1}} \right\}. \tag{11.1.3}$$

We assume that $0 \notin [a_j, b_j], \ j \in J$, in particular, $a_j, \ b_j \neq 0$ because 0 is not a feasible point for $c_j(x_j) = \frac{s_j}{x_j}$. According to Theorem 5.1, $\lambda \geq 0$. Indeed, since $p\lambda d_j (x_j^*)^{p+1} = s_j$, where $d_j > 0, \ x_j^* > 0$; s_j and x_j^* have the same signs ($s_j x_j > 0$), $s_j \neq 0$ (otherwise $c_j(x_j) \equiv 0$ for $j \in J$), $p \geq 1$, then $\lambda > 0$. Since $\dfrac{s_j}{pd_j a_j^{p+1}} > 0$, $\dfrac{s_j}{pd_j b_j^{p+1}} > 0$ under the assumptions, then if $\lambda := 0$ at Step 1 of Algorithm 1, obviously $J^{\lambda=0} = \emptyset$ and it would not be necessary to compute $x_j^*, \ j \in J^0$ with (11.1.3). An analogous remark is valid for other examples of functions when $\lambda := 0$ and λ is involved in the denominator of the expression of $x_j^*, \ j \in J^\lambda$.

$$\delta(\lambda) := \sum_{j \in J_a^\lambda} d_j a_j^p + \sum_{j \in J_b^\lambda} d_j b_j^p + \sum_{j \in J^\lambda} \sqrt[p+1]{\frac{d_j s_j^p}{\lambda^p p^p}} - \alpha. \tag{11.1.4}$$

Current value of λ

$$\lambda^{(k)} := \left(\alpha^{(k)}\right)^{-(1+\frac{1}{p})} \left(\sum_{j \in J^{\lambda(k)}} \sqrt[p+1]{\frac{d_j s_j^p}{p^p}} \right)^{1+\frac{1}{p}}, \quad k = 0, 1, \dots . \tag{11.1.5}$$

Explicit expression of the optimal value λ^*

$$\lambda^* = \left(\alpha - \sum_{j \in J_a^\lambda} d_j a_j^p - \sum_{j \in J_b^\lambda} d_j b_j^p \right)^{-(1+\frac{1}{p})} \left(\sum_{j \in J^\lambda} \sqrt[p+1]{\frac{d_j s_j^p}{p^p}} \right)^{1+\frac{1}{p}}, \tag{11.1.6}$$

$$h_j^{\leq} := -\infty, \ h_j^{=} := +\infty, \ h_j^{\geq} := +\infty. \tag{11.1.7}$$

Problem 11.2

$$c_j(x_j) = -s_j \ln m_j x_j, \quad s_j > 0,\ m_j > 0,\ x_j > 0, \tag{11.2.0}$$

$$d_j(x_j) = d_j x_j^p, \quad d_j > 0,\ p \geq 1,\ x_j > 0.$$

Since $c_j''(x_j) \equiv \frac{s_j}{x_j^2} > 0$, then $c_j(x_j)$ are strictly convex functions (Theorem 1.32) defined for $x_j > 0,\ m_j > 0,\ j \in J$.

According to Theorem 5.1, components of the optimal solution to problem (C) with $c_j(x_j) = -s_j \ln m_j x_j$ and $d_j(x_j) = d_j x_j^p$ are

$$x_j^* = a_j, \quad j \in J_a^\lambda = \left\{ j \in J : \lambda \geq \frac{s_j}{p d_j a_j^p} \right\}, \tag{11.2.1}$$

$$x_j^* = b_j, \quad j \in J_b^\lambda = \left\{ j \in J : \lambda \leq \frac{s_j}{p d_j b_j^p} \right\}, \tag{11.2.2}$$

$$x_j^* = \sqrt[p]{\frac{s_j}{p\lambda d_j}}, \quad j \in J^\lambda = \left\{ j \in J : \frac{s_j}{p d_j b_j^p} \leq \lambda \leq \frac{s_j}{p d_j a_j^p} \right\}. \tag{11.2.3}$$

Here $\lambda > 0$ because $p\lambda d_j (x_j^*)^p = s_j,\ j \in J^\lambda$ and $d_j > 0,\ x_j > 0,\ p \geq 1 > 0,\ s_j > 0$ by assumption.

$$\delta(\lambda) := \sum_{j \in J_a^\lambda} d_j a_j^p + \sum_{j \in J_b^\lambda} d_j b_j^p + \frac{1}{p\lambda} \sum_{j \in J^\lambda} s_j - \alpha, \tag{11.2.4}$$

$$\lambda^{(k)} := \frac{1}{p\alpha^{(k)}} \sum_{j \in J^{\lambda(k)}} s_j, \quad k = 0, 1, \ldots, \tag{11.2.5}$$

$$\lambda^* = \frac{1}{p} \left(\alpha - \sum_{j \in J_a^\lambda} d_j a_j^p - \sum_{j \in J_b^\lambda} d_j b_j^p \right)^{-1} \sum_{j \in J^\lambda} s_j, \tag{11.2.6}$$

$$h_j^{\leq} := -\infty,\ h_j^{=} := +\infty,\ h_j^{\geq} := +\infty. \tag{11.2.7}$$

Problem 11.3

$$c_j(x_j) = c_j x_j^q, \quad c_j > 0,\ q > 1,\ x_j > 0, \tag{11.3.0}$$

$$d_j(x_j) = d_j x_j^p, \quad d_j > 0,\ p > 1,\ x_j > 0.$$

Since $c_j''(x_j) \equiv c_j q(q-1)x_j^{q-2} > 0$ for $c_j > 0,\ q > 1,\ x_j > 0$, then $c_j(x_j),\ j \in J$ are strictly convex functions (Theorem 1.32). Similarly, under the assumptions, $d_j(x_j),\ j \in J$ are also strictly convex functions.

According to Theorem 5.1, $\mathbf{x}^*$ is an optimal solution to problem (C) with $c_j(x_j) = c_j x_j^q$ and $d_j(x_j) = d_j x_j^p$ if and only if there exists a $\lambda \in \mathbf{R}^1_+$ such that

$$x_j^* = a_j, \quad j \in J_a^\lambda = \left\{ j \in J : \lambda \geq -\frac{qc_j a_j^{q-p}}{pd_j} \right\}, \tag{11.3.1}$$

$$x_j^* = b_j, \quad j \in J_b^\lambda = \left\{ j \in J : \lambda \leq -\frac{qc_j b_j^{q-p}}{pd_j} \right\}. \tag{11.3.2}$$

Since $c_j > 0,\ d_j > 0,\ x_j > 0$ (therefore $a_j > 0,\ b_j > 0$), $p > 1 > 0$, $q > 1 > 0$, then

$$-\frac{qc_j a_j^{q-p}}{pd_j} < 0, \quad -\frac{qc_j b_j^{q-p}}{pd_j} < 0,$$

and since λ must be nonnegative, obviously $J_a^0 = J,\ J_b^0 = J^0 = \emptyset$. Then

$$\delta(0) := \sum_{j \in J_a^0} d_j(a_j) + \sum_{j \in J_b^0} d_j(b_j) + \sum_{j \in J^0} d_j(h_j^{\leq}) - \alpha = \sum_{j \in J_a^0 = J} d_j a_j^p - \alpha \leq 0$$

according to assumption 5.a). Therefore, $\lambda := 0$ at Step 1 and $J_a^0 = J$, $J_b^0 = J^0 = \emptyset$, where $J^0 = \left\{ j \in J : -\frac{qc_j b_j^{q-p}}{pd_j} \leq 0 \leq -\frac{qc_j a_j^{q-p}}{pd_j} \right\}$. Thus, it is *not necessary* to use the expressions of $x_j^*,\ j \in J^\lambda$, of $\delta(\lambda)$ and of $\lambda^{(k)}$.

$$\lambda^* = 0,$$

$$h_j^{\leq} = h_j^{=} = h_j^{\geq} = 0.$$

Problem 11.4

$$c_j(x_j) = -\frac{s_j(x_j + c_j)}{x_j + m_j}, \quad m_j > c_j,\ s_j > 0, \tag{11.4.0}$$

$$d_j(x_j) = d_j x_j, \quad d_j > 0.$$

Since $c_j''(x_j) = \frac{2s_j(m_j - c_j)}{(x_j+m_j)^3}$, then $c_j(x_j)$ are strictly convex functions for $x_j > -m_j,\ j \in J$ (Theorem 1.32).

The optimal solution to problem $(C^=)$ with $c_j(x_j)$ (11.4.0) and $d_j(x_j) = d_j x_j$ according to Theorem 6.1 has components

$$x_j^* = a_j, \quad j \in J_a^\lambda = \left\{ j \in J : \lambda \geq \frac{s_j(m_j - c_j)}{d_j(a_j + m_j)^2} \right\}, \tag{11.4.1}$$

$$x_j^* = b_j, \quad j \in J_b^\lambda = \left\{ j \in J : \lambda \leq \frac{s_j(m_j - c_j)}{d_j(b_j + m_j)^2} \right\}, \tag{11.4.2}$$

$$x_j^* = \sqrt{\frac{s_j(m_j - c_j)}{\lambda d_j}} - m_j,$$

$$j \in J^\lambda = \left\{ j \in J : \frac{s_j(m_j - c_j)}{d_j(b_j + m_j)^2} \leq \lambda \leq \frac{s_j(m_j - c_j)}{d_j(a_j + m_j)^2} \right\}. \tag{11.4.3}$$

$$\delta(\lambda) := \sum_{j \in J_a^\lambda} d_j a_j + \sum_{j \in J_b^\lambda} d_j b_j - \sum_{j \in J^\lambda} d_j m_j + \sum_{j \in J^\lambda} \sqrt{\frac{d_j s_j(m_j - c_j)}{\lambda}} - \alpha, \tag{11.4.4}$$

$$\lambda^{(k)} := \left(\alpha^{(k)} + \sum_{j \in J^{\lambda(k)}} d_j m_j \right)^{-2} \left(\sum_{j \in J^{\lambda(k)}} \sqrt{d_j s_j(m_j - c_j)} \right)^2, \tag{11.4.5}$$

$$k = 0, 1, \ldots,$$

$$\lambda^* = \left(\alpha - \sum_{j \in J_a^\lambda} d_j a_j - \sum_{j \in J_b^\lambda} d_j b_j + \sum_{j \in J^\lambda} d_j m_j \right)^{-2}$$

$$\times \left(\sum_{j \in J^\lambda} \sqrt{d_j s_j(m_j - c_j)} \right)^2, \tag{11.4.6}$$

$$h_j^{\leq} := -\infty,\ h_j^{=} := +\infty,\ h_j^{\geq} := +\infty. \tag{11.4.7}$$

Problem 11.5

$$c_j(x_j) = \frac{1}{2}c_j.(x_j - \widehat{x}_j)^2, \quad c_j > 0, \tag{11.5.0}$$

$$d_j(x_j) = d_j x_j, \quad d_j > 0.$$

Here $\widehat{x}_j,\ j \in J$ are known. Since $c''_j(x_j) = c_j > 0$, then $c_j(x_j),\ j \in J$ are strictly convex functions (Theorem 1.32).

According to Theorem 6.1, components of the optimal solution to problem $(C^=)$ with $c_j(x_j) = \frac{1}{2}c_j.(x_j - \widehat{x}_j)^2$ and $d_j(x_j) = d_j x_j$ are

$$x_j^* = a_j, \quad j \in J_a^\lambda = \left\{ j \in J : \lambda \geq \frac{c_j.(\widehat{x}_j - a_j)}{d_j} \right\}, \tag{11.5.1}$$

$$x_j^* = b_j, \quad j \in J_b^\lambda = \left\{ j \in J : \lambda \leq \frac{c_j.(\widehat{x}_j - b_j)}{d_j} \right\}, \tag{11.5.2}$$

$$x_j^* = \widehat{x}_j - \frac{\lambda d_j}{c_j},$$

$$j \in J^\lambda = \left\{ j \in J : \frac{c_j.(\widehat{x}_j - b_j)}{d_j} \leq \lambda \leq \frac{c_j.(\widehat{x}_j - a_j)}{d_j} \right\}. \tag{11.5.3}$$

$$\delta(\lambda) := \sum_{j \in J_a^\lambda} d_j a_j + \sum_{j \in J_b^\lambda} d_j b_j + \sum_{j \in J^\lambda} d_j \widehat{x}_j - \lambda \sum_{j \in J^\lambda} \frac{d_j^2}{c_j} - \alpha, \tag{11.5.4}$$

$$\lambda^{(k)} := \left(\sum_{j \in J^{\lambda(k)}} \frac{d_j^2}{c_j} \right)^{-1} \left(\sum_{j \in J^{\lambda(k)}} d_j \widehat{x}_j - \alpha^{(k)} \right), \tag{11.5.5}$$

$$k = 0, 1, \ldots,$$

$$\lambda^* = \left(\sum_{j \in J^\lambda} \frac{d_j^2}{c_j} \right)^{-1} \left(\sum_{j \in J_a^\lambda} d_j a_j + \sum_{j \in J_b^\lambda} d_j b_j + \sum_{j \in J^\lambda} d_j \widehat{x}_j - \alpha \right), \tag{11.5.6}$$

$$h_j^{\leq} = h_j^{=} = h_j^{\geq} = \widehat{x}_j. \tag{11.5.7}$$

Problem (6.1)–(6.3) with $c_j(x_j)$ defined by (11.5.0) and $c_j = 1$ is equivalent to projecting the point $\widehat{\mathbf{x}} = (\widehat{x}_1, \ldots, \widehat{x}_n)$ onto the feasible region (6.2)–(6.3).

The following two examples are of this type.

Example 11.5a Find the projection of $\widehat{\mathbf{x}} = (55, 12, 15, 85, 30)$ onto a set defined as follows:

$$\begin{aligned}
&x_1 + x_2 + 2x_3 + 3x_4 + x_5 = 200\\
&0 \leq x_1 \leq 50\\
&0 \leq x_2 \leq 7\\
&0 \leq x_3 \leq 7\\
&0 \leq x_4 \leq 80\\
&0 \leq x_5 \leq 25.
\end{aligned}$$

The projection of $\widehat{\mathbf{x}}$ onto this set is

$$\mathbf{x}^* = (42.27,\ 0.0,\ 0.0,\ 46.81,\ 17.27).$$

The Euclidean distance between $\widehat{\mathbf{x}}$ and $\mathbf{x}^*$ is

$$dist\ (\mathbf{x}^*, \widehat{\mathbf{x}}) = \|\mathbf{x}^* - \widehat{\mathbf{x}}\| = \left(\sum_{j=1}^{5} (x_j^* - \widehat{x}_j)^2\right)^{\frac{1}{2}} = 46.37691.$$

Example 11.5b Find the projection of $\widehat{\mathbf{x}} = (2, 3, 1, 2)$ onto a feasible region defined as follows:

$$\begin{aligned}
&x_1 + x_2 + x_3 + x_4 = 1\\
&0 \leq x_j \leq 1, \quad j = 1, 2, 3, 4.
\end{aligned}$$

The projection of $\widehat{\mathbf{x}}$ onto this region is

$$\mathbf{x}^* = (0, 1, 0, 0).$$

The Euclidean distance between $\widehat{\mathbf{x}}$ and $\mathbf{x}^*$ is

$$dist\ (\mathbf{x}^*, \widehat{\mathbf{x}}) = \|\mathbf{x}^* - \widehat{\mathbf{x}}\| = \left(\sum_{j=1}^{4} (x_j^* - \widehat{x}_j)^2\right)^{\frac{1}{2}} = 3.60555.$$

Problem 11.6

$$c_j(x_j) = -s_j x_j + m_j x_j^2, \quad m_j > 0, \tag{11.6.0}$$

$$d_j(x_j) = d_j x_j, \quad d_j > 0.$$

Since $c_j''(x_j) = 2m_j > 0$ then $c_j(x_j),\ j \in J$ are strictly convex functions (Theorem 1.32) with a minimum point at $h_j = \frac{s_j}{2m_j}$. Suppose that $a_j < h_j,\ j \in J$ and $\alpha < \sum_{j \in J} \min\{b_j, h_j\}$.

According to Theorem 6.1, problem ($C^=$) with $c_j(x_j) = -s_j x_j + m_j x_j^2$ and $d_j(x_j) = d_j x_j$ has an optimal solution $\mathbf{x}^* \in X^L$ if and only if there exists a $\lambda \in \mathbf{R}^1$ such that

$$x_j^* = a_j, \quad j \in J_a^\lambda = \left\{ j \in J : \lambda \geq \frac{-2m_j a_j + s_j}{d_j} \right\}, \tag{11.6.1}$$

$$x_j^* = b_j, \quad j \in J_b^\lambda = \left\{ j \in J : \lambda \leq \frac{-2m_j b_j + s_j}{d_j} \right\}, \tag{11.6.2}$$

$$x_j^* = \frac{s_j - \lambda d_j}{2m_j},$$

$$j \in J^\lambda = \left\{ j \in J : \frac{-2m_j b_j + s_j}{d_j} \leq \lambda \leq \frac{-2m_j a_j + s_j}{d_j} \right\}. \tag{11.6.3}$$

$$\delta(\lambda) := \sum_{j \in J_a^\lambda} d_j a_j + \sum_{j \in J_b^\lambda} d_j b_j + \frac{1}{2} \sum_{j \in J^\lambda} \frac{d_j s_j}{m_j} - \frac{\lambda}{2} \sum_{j \in J^\lambda} \frac{d_j^2}{m_j} - \alpha, \tag{11.6.4}$$

$$\lambda^{(k)} := \left(\sum_{j \in J^{\lambda(k)}} \frac{d_j^2}{m_j} \right)^{-1} \left(\sum_{j \in J^{\lambda(k)}} \frac{d_j s_j}{m_j} - 2\alpha^{(k)} \right), \tag{11.6.5}$$

$$k = 0, 1, \dots,$$

$$\lambda^* = \left(\sum_{j \in J^\lambda} \frac{d_j^2}{m_j} \right)^{-1} \left(2 \sum_{j \in J_a^\lambda} d_j a_j + 2 \sum_{j \in J_b^\lambda} d_j b_j + \sum_{j \in J^\lambda} \frac{d_j s_j}{m_j} - 2\alpha \right), \tag{11.6.6}$$

$$h_j^{\leq} = h_j^{=} = h_j^{\geq} = \frac{s_j}{2m_j}. \tag{11.6.7}$$

The following two examples are of this type.

Example 11.6a

$$\min \left\{ c(\mathbf{x}) = \sum_{j=1}^{10} x_j^2 + 4x_1 + 2x_2 + 6x_3 + 5x_4 - 12x_5 - x_6 - 9x_7 - x_8 + 2x_9 + 5x_{10} \right\}$$

subject to

$$\sum_{j=1}^{10} x_j = 57.5$$

$$7 \leq x_1 \leq 10$$

$$-1 \leq x_2 \leq 1.5$$

$$-2 \leq x_3 \leq 2.4$$

$$-2.5 \leq x_4 \leq 0.5$$

$$18 \leq x_5 \leq 21$$

$$1.5 \leq x_6 \leq 3.5$$

$$14.5 \leq x_7 \leq 19.5$$

$$3.5 \leq x_8 \leq 6.1$$

$$1 \leq x_9 \leq 2.5$$

$$-1.5 \leq x_{10} \leq 5.5.$$

The optimal solution obtained by Algorithm 2 is

$$\mathbf{x}^* = (7,\ 1.5,\ 2,\ 0.5,\ 18,\ 3.5,\ 14.5,\ 5.5,\ 2.5,\ 2.5)$$

$$c_{\min} = c(\mathbf{x}^*) = 352.25$$

Example 11.6b

$$\min \left\{c(\mathbf{x}) = 4x_1^2 + 10x_2^2 + 4x_3^2 + 3x_4^2 + 7x_5^2 + 3x_6^2 + x_7^2\right\}$$

subject to

$$\sum_{j=1}^{7} x_j = 72$$

$$4 \leq x_1 \leq 7$$

$$4.5 \leq x_2 \leq 10$$

$$8 \leq x_3 \leq 13$$

$$5 \leq x_4 \leq 8$$

$$4 \leq x_5 \leq 7$$

$$30 \leq x_6 \leq 40$$

$$4 \leq x_7 \leq 7.$$

The optimal solution obtained by Algorithm 2 is

$$\mathbf{x}^* = (7,\ 4.5,\ 9.8636,\ 8,\ 5.6363,\ 30,\ 7)$$

$$c_{\min} = c(\mathbf{x}^*) = 3951.0454$$

Problem 11.7

$$c_j(x_j) = \frac{1}{2}\left(\frac{h}{S} - \frac{x_j}{s_j}\right)^2, \quad s_j > 0,\ S \neq 0, \tag{11.7.0}$$

$$d_j(x_j) = d_j x_j, \quad d_j > 0.$$

Since $c_j''(x_j) = 1/s_j^2 > 0$, then $c_j(x_j),\ j \in J$ are always strictly convex functions (Theorem 1.32).

According to Theorem 6.1, components of the optimal solution to problem $(C^=)$ with $c_j(x_j) = \frac{1}{2}\left(\frac{h}{S} - \frac{x_j}{s_j}\right)^2$ and $d_j(x_j) = d_j x_j$ are

$$x_j^* = a_j, \quad j \in J_a^\lambda = \left\{ j \in J : \lambda \geq \frac{hs_j - Sa_j}{Ss_j^2 d_j} \right\}, \tag{11.7.1}$$

$$x_j^* = b_j, \quad j \in J_b^\lambda = \left\{ j \in J : \lambda \leq \frac{hs_j - Sb_j}{Ss_j^2 d_j} \right\}, \tag{11.7.2}$$

$$x_j^* = \frac{hs_j}{S} - \lambda d_j s_j^2,$$

$$j \in J^\lambda = \left\{ j \in J : \frac{hs_j - Sb_j}{Ss_j^2 d_j} \leq \lambda \leq \frac{hs_j - Sa_j}{Ss_j^2 d_j} \right\}. \tag{11.7.3}$$

$$\delta(\lambda) := \sum_{j \in J_a^\lambda} d_j a_j + \sum_{j \in J_b^\lambda} d_j b_j + \frac{h}{S} \sum_{j \in J^\lambda} d_j s_j - \lambda \sum_{j \in J^\lambda} d_j^2 s_j^2 - \alpha, \tag{11.7.4}$$

$$\lambda^{(k)} := \left(\sum_{j \in J^{\lambda(k)}} d_j^2 s_j^2 \right)^{-1} \left(\frac{h}{S} \sum_{j \in J^{\lambda(k)}} d_j s_j - \alpha^{(k)} \right), \tag{11.7.5}$$

$$k = 0, 1, \ldots,$$

$$\lambda^* = \left(\sum_{j \in J^\lambda} d_j^2 s_j^2 \right)^{-1} \left(\sum_{j \in J_a^\lambda} d_j a_j + \sum_{j \in J_b^\lambda} d_j b_j + \frac{h}{S} \sum_{j \in J^\lambda} d_j s_j - \alpha \right), \tag{11.7.6}$$

$$h_j^{\leq} = h_j^{=} = h_j^{\geq} = \frac{hs_j}{S}. \tag{11.7.7}$$

Problem 11.8

$$c_j(x_j) = -s_j \ln(1 + m_j x_j), \quad m_j > 0,\ s_j > 0,\ x_j > -\frac{1}{m_j}, \tag{11.8.0}$$

$$d_j(x_j) = d_j x_j, \quad d_j > 0.$$

Since $c_j''(x_j) = \frac{s_j m_j^2}{(1+m_j x_j)^2} > 0$, then $c_j(x_j),\ j \in J$ are strictly convex functions (Theorem 1.32).

According to Theorem 6.1, components of the optimal solution to problem $(C^=)$ with $c_j(x_j) = -s_j \ln(1 + m_j x_j)$ and $d_j(x_j) = d_j x_j$ are

$$x_j^* = a_j, \quad j \in J_a^\lambda = \left\{ j \in J : \lambda \geq \frac{s_j m_j}{d_j(1 + m_j a_j)} \right\}, \tag{11.8.1}$$

$$x_j^* = b_j, \quad j \in J_b^\lambda = \left\{ j \in J : \lambda \leq \frac{s_j m_j}{d_j(1 + m_j b_j)} \right\}, \tag{11.8.2}$$

$$x_j^* = \frac{s_j}{\lambda d_j} - \frac{1}{m_j},$$

$$j \in J^\lambda = \left\{ j \in J : \frac{s_j m_j}{d_j(1 + m_j b_j)} \leq \lambda \leq \frac{s_j m_j}{d_j(1 + m_j a_j)} \right\}. \tag{11.8.3}$$

$$\delta(\lambda) := \sum_{j \in J_a^\lambda} d_j a_j + \sum_{j \in J_b^\lambda} d_j b_j - \sum_{j \in J^\lambda} \frac{d_j}{m_j} + \frac{1}{\lambda} \sum_{j \in J^\lambda} s_j - \alpha, \tag{11.8.4}$$

$$\lambda^{(k)} := \left(\alpha^{(k)} + \sum_{j \in J^{\lambda(k)}} \frac{d_j}{m_j} \right)^{-1} \left(\sum_{j \in J^{\lambda(k)}} s_j \right), \ k = 0, 1, \dots, \tag{11.8.5}$$

$$\lambda^* = \left(\alpha - \sum_{j \in J_a^\lambda} d_j a_j - \sum_{j \in J_b^\lambda} d_j b_j + \sum_{j \in J^\lambda} \frac{d_j}{m_j} \right)^{-1} \left(\sum_{j \in J^\lambda} s_j \right), \tag{11.8.6}$$

$$h_j^{\leq} := -\infty,\ h_j^{=} := +\infty,\ h_j^{\geq} := +\infty. \tag{11.8.7}$$

Problem 11.9

$$c_j(x_j) = s_j(e^{-m_j x_j} - 1), \quad m_j > 0,\ s_j > 0, \tag{11.9.0}$$

$$d_j(x_j) = d_j x_j, \quad d_j > 0.$$

Since $c''_j(x_j) = s_j m_j^2 e^{-m_j x_j} > 0$, then $c_j(x_j),\ j \in J$ are strictly convex functions (Theorem 1.32).

According to Theorem 6.1, components of the optimal solution to problem $(C^=)$ with $c_j(x_j) = s_j(e^{-m_j x_j} - 1)$ and $d_j(x_j) = d_j x_j$ are

$$x_j^* = a_j, \quad j \in J_a^\lambda = \left\{ j \in J : \lambda \geq \frac{s_j m_j e^{-m_j a_j}}{d_j} \right\}, \tag{11.9.1}$$

$$x_j^* = b_j, \quad j \in J_b^\lambda = \left\{ j \in J : \lambda \leq \frac{s_j m_j e^{-m_j b_j}}{d_j} \right\}, \tag{11.9.2}$$

$$x_j^* = \frac{\ln (s_j m_j) - \ln (\lambda d_j)}{m_j},$$

$$j \in J^\lambda = \left\{ j \in J : \frac{s_j m_j e^{-m_j b_j}}{d_j} \leq \lambda \leq \frac{s_j m_j e^{-m_j a_j}}{d_j} \right\}. \tag{11.9.3}$$

$$\delta(\lambda) := \sum_{j \in J_a^\lambda} d_j a_j + \sum_{j \in J_b^\lambda} d_j b_j + \sum_{j \in J^\lambda} \frac{d_j (\ln (s_j m_j) - \ln d_j)}{m_j}$$

$$- \ln \lambda \sum_{j \in J^\lambda} \frac{d_j}{m_j} - \alpha, \tag{11.9.4}$$

$$\lambda^{(k)} := \exp \left\{ \left[\sum_{j \in J^{\lambda(k)}} \frac{d_j}{m_j} \right]^{-1} \left[\sum_{j \in J^{\lambda(k)}} \frac{d_j}{m_j} \ln \frac{s_j m_j}{d_j} - \alpha^{(k)} \right] \right\}, \tag{11.9.5}$$

$$k = 0, 1, \dots ,$$

$$\lambda^* = \exp \left\{ \left[\sum_{j \in J^\lambda} \frac{d_j}{m_j} \right]^{-1} \left[\sum_{j \in J_a^\lambda} d_j a_j + \sum_{j \in J_b^\lambda} d_j b_j + \sum_{j \in J^\lambda} \frac{d_j}{m_j} \ln \frac{s_j m_j}{d_j} - \alpha \right] \right\}. \tag{11.9.6}$$

Obviously $\lambda^{(k)} > 0,\ \lambda^* > 0$.

$$h_j^{\leq} := -\infty,\ h_j^{=} := +\infty,\ h_j^{\geq} := +\infty. \tag{11.9.7}$$

Problem 11.10

$$c_j(x_j) = e^{k_j x_j}, \quad k_j > 0, \tag{11.10.0}$$

$$d_j(x_j) = d_j x_j, \quad d_j > 0.$$

Since $c_j''(x_j) = k_j^2 e^{k_j x_j} > 0$, then $c_j(x_j)$, $j \in J$ are strictly convex functions (Theorem 1.32) and since $c_j'(x_j) = k_j e^{k_j x_j} > 0$, then $c_j(x_j)$, $j \in J$ are increasing.

According to Theorem 6.1, the optimal solution to problem $(C^=)$ with $c_j(x_j) = e^{k_j x_j}$ and $d_j(x_j) = d_j x_j$ is $\mathbf{x}^* = (x_j^*)_{j \in J}$ where

$$x_j^* = a_j, \quad j \in J_a^\lambda = \left\{ j \in J : \lambda \geq -\frac{k_j e^{k_j a_j}}{d_j} \right\}, \tag{11.10.1}$$

$$x_j^* = b_j, \quad j \in J_b^\lambda = \left\{ j \in J : \lambda \leq -\frac{k_j e^{k_j b_j}}{d_j} \right\}, \tag{11.10.2}$$

$$x_j^* = \frac{1}{k_j} \ln \left(-\frac{\lambda d_j}{k_j} \right),$$

$$j \in J^\lambda = \left\{ j \in J : -\frac{k_j e^{k_j b_j}}{d_j} \leq \lambda \leq -\frac{k_j e^{k_j a_j}}{d_j} \right\}. \tag{11.10.3}$$

$$\begin{aligned} \delta(\lambda) := & \sum_{j \in J_a^\lambda} d_j a_j + \sum_{j \in J_b^\lambda} d_j b_j + \ln(-\lambda) \sum_{j \in J^\lambda} \frac{d_j}{k_j} \\ & + \sum_{j \in J^\lambda} \frac{d_j (\ln d_j - \ln k_j)}{k_j} - \alpha, \end{aligned} \tag{11.10.4}$$

$$\lambda^{(k)} := -\exp \left\{ \left[\sum_{j \in J^{\lambda(k)}} \frac{d_j}{k_j} \right]^{-1} \left[\alpha^{(k)} - \sum_{j \in J^{\lambda(k)}} \frac{d_j}{k_j} \ln \frac{d_j}{k_j} \right] \right\}, \tag{11.10.5}$$

$$k = 0, 1, \ldots,$$

$$\lambda^* = -\exp \left\{ \left[\sum_{j \in J^\lambda} \frac{d_j}{k_j} \right]^{-1} \left[\alpha - \sum_{j \in J_a^\lambda} d_j a_j - \sum_{j \in J_b^\lambda} d_j b_j - \sum_{j \in J^\lambda} \frac{d_j}{k_j} \ln \frac{d_j}{k_j} \right] \right\}. \tag{11.10.6}$$

Obviously $\lambda^{(k)} < 0$, $\lambda^* < 0$.

$$h_j^{\leq} := -\infty, \; h_j^{=} := -\infty, \; h_j^{\geq} := +\infty. \tag{11.10.7}$$

Similarly, we are able to consider other convex objective functions $c(\mathbf{x}) = \sum_{j \in J} c_j(x_j)$ and convex constraint functions $d_j(x_j),\ j \in J$.

11.2 Computational Experiments

The computations have been performed on an Intel Pentium(R) Dual-Core CPU E5800 3.20GHz/2.00GB. Each problem was run 30 times. Parameters and data have been randomly generated in intervals where the functions $c_j(x_j)$ are strictly convex.

Type 11.1.

Number of variables	$n = 1200$	$n = 1500$
(Average) Number of iterations	2.03	2.07
Average run time (in seconds)	0.0010	0.0012

Type 11.2.

Number of variables	$n = 1200$	$n = 1500$
(Average) Number of iterations	2.07	3.03
Average run time (in seconds)	0.0012	0.0018

Type 11.3.

Number of variables	$n = 1200$	$n = 1500$
(Average) Number of iterations	3.03	3.07
Average run time (in seconds)	0.0012	0.0016

Type 11.4.

Number of variables	$n = 1200$	$n = 1500$
(Average) Number of iterations	4.07	7.03
Average run time (in seconds)	0.0001	0.0001

Type 11.5.

Number of variables	$n = 1200$	$n = 1500$
(Average) Number of iterations	2.03	2.10
Average run time (in seconds)	0.00009	0.0001

Type 11.6.

Number of variables	$n = 1200$	$n = 1500$
(Average) Number of iterations	2.07	2.10
Average run time (in seconds)	0.00009	0.0001

Type 11.7.

Number of variables	$n = 1200$	$n = 1500$
(Average) Number of iterations	2.03	7.033
Average run time (in seconds)	0.0001	0.0001

Type 11.8.

Number of variables	$n = 1200$	$n = 1500$
(Average) Number of iterations	3.33	4.03
Average run time (in seconds)	0.00009	0.00011

Type 11.9.

Number of variables	$n = 1200$	$n = 1500$
(Average) Number of iterations	3.03	3.13
Average run time (in seconds)	0.000096	0.0001

Type 11.10.

Number of variables	$n = 1200$	$n = 1500$
(Average) Number of iterations	2.07	5.07
Average run time (in seconds)	0.00001	0.0001

When $n < 1200$, the run time of the algorithms is so small that the timer does not recognize the corresponding value from its computer zero. In such cases, the timer displays 0 seconds.

Effectiveness of Algorithms 1, 2, and 3 for problems (C), $(C^{=})$, and $(C^{\geq})$, respectively, has been tested by many other examples. As it can be observed, the (average) number of iterations is much less than the number of variables n of the corresponding problem for large values of n.

NOTES TO PART TWO

As it was pointed out in the *Preface* and in Chapter Five, problems like (C), $(C^=)$, $(C^\geq)$ arise in many cases: in production planning and scheduling (Bitran and Hax [11]), in the allocation of financial resources and in the allocation of promotional resources among competing activities (Bitran and Hax [11], Luss and Gupta [77, 78], Zipkin [181]), in the theory of search, in subgradient optimization, in facility location (Rockafellar and Wets [109]), in the implementation of the projection methods when the feasible region has the same form as the feasible sets under consideration (Rockafellar and Wets [109], Stefanov [119, 124, 130, 152]), etc.

Problems like (C), $(C^=)$, and $(C^\geq)$, related problems and methods for them are considered in [8], [11]–[12], [14]–[17], [20], [28]–[29], [32]–[33], [35], [40]–[42], [46]–[47], [49], [56]–[64], [68], [73]–[79], [84], [88], [90]–[93], [96], [98]–[99], [101]–[102], [105]–[106], [109], [115]–[116], [118]–[120], [124]–[126], [129]–[134], [137]–[140], [142], [144]–[145], [149], [151], [155], [158], [169], [171], [174], [176], [180]–[181], etc.

Quadratic knapsack problems and related to them are studied in papers of Brucker [17], Pardalos, Ye, and Han [99], Robinson, Jiang, and Lerme [105], etc. An indefinite version of these problems is considered by Vavasis [169]. Algorithms for the convex quadratic knapsack problem are proposed in papers of Brucker [17], Dussault, Ferland, and Lemaire [41], Helgason, Kennington, and Lall [57], Pardalos and Kovoor [98], etc.

The solution of continuous knapsack problems with arbitrary convex or concave objective functions is studied in the works of Bitran and Hax [11], Luss and Gupta [78], Moré and Vavasis [93], Stefanov [120], [125]–[126], [129], [131]–[134], [137]–[140], [144]–[145], [149, 151], etc. A version of this problem with a linear inequality constraint of the form "$\geq$" is considered by Bitran and Hax [11], Stefanov [125]–[126], [129, 131], etc. The knapsack sharing problem is discussed in papers of Brown [14]–[16], etc.

Algorithms for bound constrained quadratic programming problems are proposed in the works of Dembo and Tulowitzki [35], Moré and Toraldo [92], Pardalos and Kovoor [98]. Algorithms for quadratic programs with simple bounds are proposed in the papers of Coleman and Hulbert [28, 29].

Methods for large convex quadratic programs subject to box constraints are suggested in the report of Yang and Tolle [176]. Direct methods for convex quadratic programs with box constraints are suggested in the work of Júdice and Pires [64].

An active set Newton method for large-scale nonlinear programs with box constraints is considered by Facchinei, Júdice, and Soares [46].

Barrier functions and interior-point algorithms for linear programs with bounded variables are suggested in the paper of Freund and Todd [49]. An interior-point algorithm for quadratic programs with box constraints is considered in the work of Han, Pardalos, and Ye [56].

Iterative algorithms for problems (C), $(C^=)$ and $(C^\geq)$ are suggested in the papers of Stefanov [125, 131], etc.

An algorithm for solving a quadratic program with (strictly) convex objective function of the form $c(\mathbf{x}) = \frac{1}{2} \sum_{j \in J} (x_j - y_j)^2$ and constraints like those of problem (C) is suggested in the paper of Rockafellar and Wets [109]. This problem itself is equivalent to projecting a point $\mathbf{y} = (y_1, \ldots, y_n)$ onto the convex feasible set X of problem (C) and has always a unique solution when $X \neq \emptyset$. An algorithm for finding a projection onto the canonical simplex of $\mathbf{R}^n$ is proposed, for example, by Michelot [88].

A Lagrangian relaxation algorithm for minimizing a strictly convex separable quadratic function over a transportation polytope is proposed in the paper of Cottle, Duvall, and Zikan [32].

Numerical solution of optimization problems subject to one constraint is considered by McCord [84], Trevisan [158], etc.

Methods for least squares problems with bounds on the variables are suggested in papers of Björck [12], Lötstedt [74], etc.

Algorithms for the least distance programming problem are proposed by Berman, Kovoor and Pardalos [8], Wolfe [174], etc. In [8], a linear time algorithm is suggested, based on the construction of hyperplane arrangements.

Trust region affine scaling algorithms for linearly constrained convex and concave programs are suggested by Monteiro and Wang [91].

A survey of iterative methods for large convex quadratic programs is given in the paper of Lin and Pang [73].

The resource allocation problem is discussed by Bitran and Hax [11], Dunstan [40], Ibaraki and Katoh [59], Luss and Gupta [77, 78], Luss and Smith [79], Wilkinson and Gupta [171], Zipkin [181], etc. The resource allocation is described by a problem of the form (C) with strictly convex objective function, linear equality/linear inequality constraint of the form "$\leq$", and bounds on the variables. A nonlinear minimax allocation problem with multiple knapsack constraints is considered in the paper of Luss [76], and an algorithm for separable nonlinear minimax problems is suggested in paper [75] by the same author. Various aspects of convex resource allocation problems on directed acyclic graphs are discussed by Monma, Schrijver, Todd, and Wei [90]. A polynomial time algorithm for the resource allocation problem with a convex objective function and nonnegative integer variables is suggested in the paper of Katoh, Ibaraki, and Mine [68].

A stochastic version of convex allocation problems is considered in the work of Edirisinghe, Atkins, and Iyogun [42].

A geometric programming duality is applied to the resource allocation problem in the paper of Scott, Jefferson, and Kerdvonbundit [115], and geometric programming with applications to problems under consideration and related to them is discussed in the papers of Jefferson and Scott [60]–[62]. A convex dual approach to quadratic concave fractional programming is considered by Scott and Jefferson [116].

Exact algorithms for a scheduling problem with unrelated parallel machines and sequence and machine-dependent setup times are proposed in Rocha, Ravetti, Mateus, and Pardalos [106].

Tychonov and Hadamard well-posedness and well-posedness in the generalized sense are considered, for example, in the work of Cavazzuti and Morgan [21], in the book of Dontchev and Zolezzi [38], in works of Hadamard [54, 55], etc. The definition of Tychonov well-posedness is given by Tychonov in [165]. Other topics regarding stability, ill-posed problems, and Tychonov regularization can be found in [159]–[164], [166]–[167], etc. Sometimes Tychonov is written as Tykhonov, Tikhonov or Tychonoff in references. Well-posedness is also discussed in the book of Rockafellar and Wets [110], in the paper of Stefanov [142], etc.

Calmness in the sense of Clarke is proposed and studied in the works of Clarke [24, 25, 27]. For Definitions 9.4, 9.5, and Theorems 9.1–9.4, see the same references. About Theorem 9.1, see also the paper of Roberts and Varberg [104].

Stability of the set of saddle points of Lagrangians is considered in the papers of Gol'shtein [51], Gol'shtein and Tret'iakov [53], etc.

Theorem 6.2 and Corollary 6.1 are due to Berman, Kovoor, and Pardalos [8], and Definitions 9.4, 9.5, and Theorems 9.2–9.4 are due to Clarke [27]. These results are included in this book with permission from World Scientific Publishing and SIAM Classics Series, respectively.

Optimization problems and methods for solving these problems, considered in Part Two of this book, are interesting not only from the viewpoint of optimization theory, optimization methods, and their applications, but also from the viewpoint of other fields of science, especially computer science. Artificial intelligence, machine learning, and other areas of computer science use various optimization methods, for example, convex optimization methods, large-scale optimization methods, discrete (including integer) optimization methods, stochastic gradient descent methods (including stochastic quasigradient methods), etc. In particular, machine learning algorithms can be formulated as optimization algorithms. These applications revived the optimization methods in the past few years.

Part III
Selected Supplementary Topics and Applications

Chapter 12
Applications of Convex Separable Unconstrained Nondifferentiable Optimization to Approximation Theory

In this chapter, the data fitting problem is considered, that is, the problem of approximating a function (of several variables) given by tabulated data, and the analogous problem for inconsistent systems of linear equations. Also, the problem of numerical solution of systems of nonlinear algebraic equations with even powers and systems of nonlinear equations, defined by convex differentiable functions, is studied. A traditional approach for solving these problems is the least squares data fitting, which is based on discrete ℓ_2-norm. An alternative approach is applied in this chapter: with each of these problems, nondifferentiable (nonsmooth) unconstrained minimization problems are associated, with objective functions based on discrete ℓ_1- and ℓ_∞-norms, respectively, that is, these norms are used as proximity criteria. In other words, the problems under consideration are solved by minimizing the residuals using these two norms. Some computational results, obtained by an appropriate iterative method, are presented at the end of the chapter. These results are compared with the results, obtained by the iterative gradient method for the corresponding "differentiable" least squares problems.

12.1 Introduction. Statement of Problems Under Consideration

Problem No. 1

Let $f : \mathbb{R}^p \to \mathbb{R}$ be a real-valued function in p variables and let the following table of data be given

$\mathbf{x} \in \mathbb{R}^p$	$\mathbf{x}_1$	$\mathbf{x}_2$	$\ldots$	$\mathbf{x}_m$
$f(\mathbf{x})$	$f(\mathbf{x}_1)$	$f(\mathbf{x}_2)$	$\ldots$	$f(\mathbf{x}_m)$

. (12.1)

© The Author(s), under exclusive license to Springer Nature Switzerland AG 2021
S. M. Stefanov, *Separable Optimization*, Springer Optimization and Its Applications 177,
https://doi.org/10.1007/978-3-030-78401-0_12

Find a generalized polynomial $P_n(\mathbf{x}) = \sum_{j=0}^{n} a_j\varphi_j(\mathbf{x})$ based on the system of linearly independent functions $\{\varphi_j(\mathbf{x})\}_{j=0}^{n}$, that is, a polynomial of generalized degree n, which approximates function $f(\mathbf{x})$ with respect to some distance/norm. Depending on the distance/norm, $P_n(\mathbf{x})$ is an optimal solution to different problems. We discuss the approximation with respect to weighted discrete ℓ_1-norm

$$\|f\|_{\ell_1} = \sum_{i=1}^{m} w_i|f(\mathbf{x}_i)| \tag{12.2}$$

and weighted discrete ℓ_∞-norm

$$\|f\|_{\ell_\infty} = \sup_{1\le i\le m} w_i|f(\mathbf{x}_i)|, \tag{12.3}$$

and, only for comparison, a weighted discrete ℓ_2-norm

$$\|f\|_{\ell_2} = \left(\sum_{i=1}^{m} w_i f^2(\mathbf{x}_i)\right)^{\frac{1}{2}}, \tag{12.4}$$

where $w_i > 0,\ i = 1, \ldots, m$ are weights.

According to the theory of best approximation, it is necessary that $m \ge n + 1$. This requirement means that we must have at least $n + 1$ values of f, where $n + 1$ is the number of unknown coefficients $a_0, a_1, \ldots, a_n$ of $P_n(\mathbf{x})$.

Thus, the polynomial $P_n(\mathbf{x})$ of best approximation to function $f(\mathbf{x})$ with respect to ℓ_1-norm is an optimal solution to the problem

$$\min_{\mathbf{a}} \left\{\Phi_1(\mathbf{a}) \stackrel{\text{def}}{=} \|P_n - f\|_{\ell_1} = \sum_{i=1}^{m} w_i\left|\sum_{j=0}^{n} a_j\varphi_j(\mathbf{x}_i) - f(\mathbf{x}_i)\right|\right\}, \tag{12.5}$$

and the best approximation to $f(\mathbf{x})$ with respect to ℓ_∞-norm is an optimal solution to the problem

$$\min_{\mathbf{a}} \left\{\Phi_2(\mathbf{a}) \stackrel{\text{def}}{=} \|P_n - f\|_{\ell_\infty} = \max_{1\le i\le m} w_i\left|\sum_{j=0}^{n} a_j\varphi_j(\mathbf{x}_i) - f(\mathbf{x}_i)\right|\right\}. \tag{12.6}$$

The corresponding least squares problem which can be associated with problem (12.1) is

$$\min_{\mathbf{a}} \left\{ \Phi_3(\mathbf{a}) \stackrel{\text{def}}{=} \|P_n - f\|_{\ell_2} \right.$$
$$\left. = \left(\sum_{i=1}^{m} w_i \left(\sum_{j=0}^{n} a_j \varphi_j(\mathbf{x}_i) - f(\mathbf{x}_i) \right)^2 \right)^{\frac{1}{2}} \right\}. \tag{12.7}$$

In (12.5), (12.6), and (12.7), $\mathbf{a} = (a_0, a_1, \ldots, a_n) \in \mathbf{R}^{n+1}$. Unless other values of the weights w_i, $i = 1, \ldots, m$ are given, w_i usually are chosen to be equal to 1 (> 0) for all $i = 1, \ldots, m$.

Recall that the system of functions $\{\varphi_0, \ldots, \varphi_n\}$ is said to be *linearly independent* if whenever

$$c_0\varphi_0(\mathbf{x}) + c_1\varphi_1(\mathbf{x}) + \cdots + c_n\varphi_n(\mathbf{x}) = 0$$

then $c_0 = c_1 = \cdots = c_n = 0$. Otherwise the set of functions is said to be *linearly dependent*.

The system $\{\varphi_j(\mathbf{x})\}_{j=0}^n$ of linearly independent functions for problems (12.5), (12.6), and (12.7) can be chosen as follows:

$$\varphi_j(\mathbf{x}) = \sum_{k=1}^{p} x_k^j, \quad j = 0, 1, \ldots, n,$$

that is

$$\begin{aligned} \varphi_0(\mathbf{x}) &= 1 + 1 + \cdots + 1 = p \\ \varphi_1(\mathbf{x}) &= x_1 + x_2 + \cdots + x_p \\ \varphi_2(\mathbf{x}) &= x_1^2 + x_2^2 + \cdots + x_p^2 \\ &\cdots\cdots\cdots\cdots\cdots \\ \varphi_n(\mathbf{x}) &= x_1^n + x_2^n + \cdots + x_p^n. \end{aligned} \tag{12.8}$$

It is easy to prove that functions $\{\varphi_j(\mathbf{x})\}_{j=0}^n$ in (12.8) are linearly independent.

Theorem 12.1 (Linear independence of a system of multivariate functions)
If $\varphi_j(\mathbf{x})$ is a polynomial of degree j for each $j = 0, 1, \ldots, n$, then the set of functions $\{\varphi_0, \ldots, \varphi_n\}$ is linearly independent.

Proof. Let $c_0, \ldots, c_n$ be real numbers such that

$$P(\mathbf{x}) \stackrel{\text{def}}{=} \sum_{j=0}^{n} c_j\varphi_j(\mathbf{x}) = 0.$$

Since the polynomial $P(\mathbf{x})$ of degree n vanishes, the coefficients of x_k^n, $k = 1, \ldots, p$ are zero. Since $c_n\varphi_n(\mathbf{x})$ is the only term in $P(\mathbf{x})$ containing x_k^n, then c_n must be equal to zero. Therefore,

$$P(\mathbf{x}) = \sum_{j=0}^{n-1} c_j \varphi_j(\mathbf{x}).$$

In this representation of $P(\mathbf{x})$, the only term that contains powers of x_k^{n-1} is $c_{n-1}\varphi_{n-1}(\mathbf{x})$. Hence, we must have $c_{n-1} = 0$, and

$$P(\mathbf{x}) = \sum_{j=0}^{n-2} c_j \varphi_j(\mathbf{x}).$$

Continuing in this way, we obtain that the remaining coefficients $c_{n-2}, \ldots, c_1, c_0$ are also equal to zero. Therefore, the functions $\{\varphi_0, \ldots, \varphi_n\}$ are linearly independent by definition. ■

Problem No. 2

Given an inconsistent (overdetermined) system of linear equations

$$\sum_{j=1}^{n} a_{ij} x_j = b_i, \quad i = 1, \ldots, m; \quad m > n. \tag{12.9}$$

This system does not have a solution in the general case when $m > n$ and all the equations are linearly independent.

The following minimization problems can be associated with (12.9):

$$\min \left\{ \Phi_4(\mathbf{x}) \overset{\text{def}}{=} \max_{1 \le i \le m} w_i \left| \sum_{j=1}^{n} a_{ij} x_j - b_i \right| \right\} \tag{12.10}$$

and

$$\min \left\{ \Phi_5(\mathbf{x}) \overset{\text{def}}{=} \sum_{i=1}^{m} w_i \left| \sum_{j=1}^{n} a_{ij} x_j - b_i \right| \right\}, \tag{12.11}$$

where $\mathbf{x} = (x_1, \ldots, x_n) \in \mathbf{R}^n$. The corresponding least squares problem is

$$\min \left\{ \Phi_6(\mathbf{x}) \overset{\text{def}}{=} \left(\sum_{i=1}^{m} w_i \left(\sum_{j=1}^{n} a_{ij} x_j - b_i \right)^2 \right)^{\frac{1}{2}} \right\}. \tag{12.12}$$

Problem (12.10) is a special case of the problem of best Chebyshev approximation based on ℓ_∞-norm (12.3), problem (12.11) is based on ℓ_1-norm (12.2), and problem (12.12) is based on ℓ_2-norm (12.4).

Approximations with respect to ℓ_1-norm are known as ℓ_1- (or *absolute deviation*) *approximations*, and approximations with respect to ℓ_∞-norm, as *Chebyshev*, or *minimax*, or *uniform*, *approximations*.

Problems like (12.1) and (12.9), connected with measurement of physical quantities, arise in physics, engineering, etc. The weights $w_i, i = 1, \dots, m$ mean the reliability, with which each value (measurement, empirical datum) $f(\mathbf{x}_i)$ at $\mathbf{x}_i$ for Problem No. 1, or each equation for Problem No. 2, can be accepted.

12.2 Some Properties of Objective Functions and Solvability of Problems

Functions $\sum_{j=0}^{n} a_j \varphi_j(\mathbf{x}_i) - f(\mathbf{x}_i),\ i = 1, \dots, m$ are linear functions of $a_j,\ j = 0, \dots, n$, the "absolute value" function $|f| = \max\{f, -f\}$ is convex (Theorem 1.20) when f is a linear function (and therefore, both f and $-f$ are convex, see Example 1.2, Chapter One). Hence, $\Phi_1(\mathbf{a})$ is a convex (and separable) function of $\mathbf{a}$ as a linear combination with nonnegative (even positive) coefficients w_i of convex functions (Theorem 1.17).

Using the same reasoning, we obtain that $\Phi_5(\mathbf{x})$ is a convex separable function of $\mathbf{x}$.

According to Theorem 1.20, $\Phi_2(\mathbf{a})$ is a convex (and separable) function of $\mathbf{a}$ as maximum of the convex functions $w_i \left| \sum_{j=0}^{n} a_j \varphi_j(\mathbf{x}_i) - f(\mathbf{x}_i) \right|$, where $w_i > 0$, because functions $\sum_{j=0}^{n} a_j \varphi_j(\mathbf{x}_i) - f(\mathbf{x}_i),\ i = 1, \dots, m$ are both convex and concave as linear functions of $a_j,\ j = 0, \dots, n$.

Function $\Phi_4(\mathbf{x})$ is a convex separable function of $x_j,\ j = 1, \dots, n$ because of similar reasons.

Function $\Phi_3^2(\mathbf{a})$ is a strictly convex (and separable) function of $\mathbf{a}$ as a linear combination with nonnegative (positive) coefficients w_i of the quadratic functions $\left(\sum_{j=0}^{n} a_j \varphi_j(\mathbf{x}_i) - f(\mathbf{x}_i)\right)^2,\ i = 1, \dots, m$ which, as it is known, are strictly convex (Example 1.3, Chapter One).

Similarly, $\Phi_6^2(\mathbf{x})$ is a strictly convex separable function of $\mathbf{x}$.

Functions $\Phi_1(\mathbf{a})$, $\Phi_2(\mathbf{a})$, $\Phi_4(\mathbf{x})$ and $\Phi_5(\mathbf{x})$ are nondifferentiable (nonsmooth), whereas functions $\Phi_3(\mathbf{a})$ and $\Phi_6(\mathbf{x})$ are differentiable.

On Problems Associated with Problem No. 1 (12.1)

Since (12.5) is a minimization problem, $\Phi_1(\mathbf{a})$ is a continuous (and, therefore, both lower and upper semicontinuous) function, bounded from below from 0 as a sum of nonnegative terms, and $\Phi_1(\mathbf{a}) \to +\infty$ as $\|\mathbf{a}\| \to \infty$, problem (12.5) has an optimal solution according to Corollary C.1 (Appendix C) with $X = \mathbf{R}^n$.

Using the same reasoning, we can conclude that problems (12.6) and (12.7) are also solvable.

Since $\min \Phi_3(\mathbf{a})$ and $\min \Phi_3^2(\mathbf{a})$ are attained at the same point (vector) $\mathbf{a} = (a_0, a_1, \dots, a_n)$, we can consider problem

$$\min_{\mathbf{a}} \left\{ F(\mathbf{a}) \stackrel{\text{def}}{=} \Phi_3^2(\mathbf{a}) = \sum_{i=1}^{m} w_i \left(\sum_{j=0}^{n} a_j \varphi_j(\mathbf{x}_i) - f(\mathbf{x}_i) \right)^2 \right\} \qquad (12.7')$$

instead of problem (12.7). Since $\Phi_3^2(\mathbf{a})$ is a strictly convex function, (12.7′) has a unique solution (Theorem 1.38).

The existence of solutions to these problems can also be proved by using some general results.

As it is known, ℓ_1, ℓ_2 and ℓ_∞ are normed linear spaces; they are Banach spaces with the norms (12.2), (12.4) and (12.3), respectively; ℓ_1, ℓ_2 are separable spaces and ℓ_∞ is not a separable space (see Appendix B; Kantorovich and Akilov [66], Yosida [178], etc.).

Linear independence of $\{\varphi_j(\mathbf{x})\}_{j=0}^{n}$, proved in Theorem 12.1, guarantees the existence of an element of best approximation for problems (12.5), (12.6) and (12.7) (see Theorem D.1, Appendix D).

Furthermore, since ℓ_p, $1 < p < \infty$ are strictly convex spaces (see Appendix B), problem (12.7) (and (12.7′)) has a unique solution (Theorem D.3, Appendix D), and since ℓ_1 and ℓ_∞ are not strictly convex spaces (Appendix B), in the general case, we cannot conclude uniqueness of the optimal solution to problems (12.5) and (12.6).

The $n+1$-tuple $\mathbf{a} = (a_0, a_1, \ldots, a_n) \in \mathbf{R}^{n+1}$, which is obtained as an optimal solution to problem (12.5) (problem (12.6), problem (12.7′), respectively) gives coefficients of the generalized polynomial $P_n(\mathbf{x})$ of best approximation for $f(\mathbf{x})$ (12.1), $\mathbf{x} \in \mathbf{R}^p$ with respect to ℓ_1-norm (ℓ_∞-norm, ℓ_2-norm, respectively).

When $p = 1$, that is, when $f(x)$ is a single-variable function, the generalized polynomial $P_n(\mathbf{x}) = \sum_{j=0}^{n} a_j \varphi_j(\mathbf{x})$ becomes an algebraic polynomial of degree n: $P_n(x) = \sum_{j=0}^{n} a_j x^j$ and problem (12.7) [or equivalently (12.7′)] with $p = 1$ is the well-known least squares data fitting problem.

On Problems Associated with Problem No. 2 (12.9)

Solvability of problems (12.10), (12.11), and (12.12) follows from Corollary C.1 (Appendix C): using $\Phi_4(\mathbf{x}) \geq 0$, $\Phi_5(\mathbf{x}) \geq 0$ and $\Phi_6(\mathbf{x}) \geq 0$ are continuous functions; a_{ij}, $i = 1, \ldots, m$; $j = 1, \ldots, n$; b_i, $i = 1, \ldots, m$ are the fixed coefficients, given by (12.9), it follows that $\Phi_4(\mathbf{x}) \to +\infty$, $\Phi_5(\mathbf{x}) \to +\infty$ and $\Phi_6(\mathbf{x}) \to +\infty$ when $\|\mathbf{x}\| \to \infty$.

In addition, using the same reasoning, problem (12.12′): min $\Phi_6^2(\mathbf{x})$ has also an optimal solution and it is unique (Theorem 1.38) because $\Phi_6^2(\mathbf{x})$ is a strictly convex function. The existence and the uniqueness of the optimal solution to the problem (12.12′) can also be proved by using an approach, analogous to the alternative approach for the problem (12.7′).

From Theorem 1.29 and Theorem 1.39, Chapter One, it follows that $\mathbf{a}^*$ is an optimal solution to problem (12.6) if and only if

$$\mathbf{0}_{n+1} \in \partial\,\Phi_2(\mathbf{a}^*) = \text{co}\,\{\partial F_{i2}(\mathbf{a}) : i \in I(\mathbf{a})\},$$

where

$$F_{i2}(\mathbf{a}) = w_i \left| \sum_{j=0}^{n} a_j \varphi_j(\mathbf{x}_i) - f(\mathbf{x}_i) \right|,$$

$$I(\mathbf{a}) = \{i \in \{1, \ldots, m\} : F_{i2}(\mathbf{a}) = \Phi_2(\mathbf{a})\}.$$

Similarly, $\mathbf{x}^*$ is an optimal solution to problem (12.10) if and only if

$$\mathbf{0}_n \in \partial \Phi_4(\mathbf{x}^*) = \text{co}\,\{\partial F_{i4}(\mathbf{x}) : i \in I(\mathbf{x})\},$$

where

$$F_{i4}(\mathbf{x}) = w_i \left| \sum_{j=1}^{n} a_{ij} x_j - b_i \right|, \quad I(\mathbf{x}) = \{i \in \{1, \ldots, m\} : F_{i4}(\mathbf{x}) = \Phi_4(\mathbf{x})\}.$$

12.3 Methods for Solving Problems Under Consideration

12.3.1 Theoretical Matters. The Subgradient Method

Since $\Phi_1(\mathbf{a})$, $\Phi_2(\mathbf{a})$, $\Phi_4(\mathbf{x})$ and $\Phi_5(\mathbf{x})$ are nondifferentiable convex functions, we use the so-called subgradient (generalized gradient) method for solving problems (12.5), (12.6), (12.10), and (12.11).

Let $f(\mathbf{x})$ be a convex proper function defined on $\mathbf{R}^n$.

The *subgradient method* for solving problem

$$\min f(\mathbf{x}) \tag{12.13}$$

is defined as follows

$$\mathbf{x}_{k+1} = \mathbf{x}_k - \rho_k \gamma_k \widehat{\mathbf{f}}(\mathbf{x}_k), \quad k = 0, 1, \ldots, \tag{12.14}$$

where

- $\mathbf{x}_0 \in \mathbf{R}^n$ is an arbitrary initial guess (initial approximation);
- ρ_k is a step size; $\rho_k \to +0$, $\sum_{k=0}^{\infty} \rho_k = +\infty$, $\sum_{k=0}^{\infty} \rho_k^2 < +\infty$;
- γ_k is a norming multiplier; usually $\gamma_k = \frac{1}{\|\widehat{\mathbf{f}}(\mathbf{x}_k)\|}$ or $\gamma_k = 1$;
- $\widehat{\mathbf{f}}(\mathbf{x}_k)$ is a subgradient of f at $\mathbf{x}_k$.

The following theorem states convergence of the subgradient method (12.14).

Theorem 12.2 (Convergence of the subgradient method)
If $\rho_k \to 0$ *when* $k \to \infty$, $\rho_k \geq 0$, $\sum_{k=0}^{\infty} \rho_k = +\infty$, $\gamma_k = 1$ *for all* k *and* $\|\widehat{\mathbf{f}}(\mathbf{x}_k)\| < C = const.$ *for all* $\mathbf{x}_k$*, then there exists a subsequence* $\{f(\mathbf{x}_{k_s})\}$ *of the sequence* $\{f(\mathbf{x}_k)\}$ *such that* $\lim\limits_{s\to\infty} f(\mathbf{x}_{k_s}) = f(\mathbf{x}^*)$*, where* $\mathbf{x}^* \in M^*$, $M^* \stackrel{\text{def}}{=} \Big\{\mathbf{x}^* \in \mathbf{R}^n : f(\mathbf{x}^*) = \inf\limits_{\mathbf{x}\in\mathbf{R}^n} f(\mathbf{x})\Big\}$.

Proof. By assumptions of Theorem 12.2, we have that

$$\begin{aligned}
&\|\mathbf{x}^* - \mathbf{x}_{k+1}\|^2 = \|\mathbf{x}^* - \mathbf{x}_k + \rho_k \widehat{\mathbf{f}}(\mathbf{x}_k)\|^2 \\
&= \|\mathbf{x}^* - \mathbf{x}_k\|^2 + 2\rho_k \langle \widehat{\mathbf{f}}(\mathbf{x}_k), \mathbf{x}^* - \mathbf{x}_k\rangle + \rho_k^2 \|\widehat{\mathbf{f}}(\mathbf{x}_k)\|^2 \qquad (12.15) \\
&\leq \|\mathbf{x}^* - \mathbf{x}_k\|^2 + 2\rho_k \langle \widehat{\mathbf{f}}(\mathbf{x}_k), \mathbf{x}^* - \mathbf{x}_k\rangle + C^2 \rho_k^2.
\end{aligned}$$

Choose some $\delta > 0$. For every $k = 0, 1, \ldots$, there are two possible cases:

$$1) \quad 2\ \langle \widehat{\mathbf{f}}(\mathbf{x}_k), \mathbf{x}^* - \mathbf{x}_k\rangle + C^2 \rho_k \leq -\delta, \qquad (12.16)$$

and

$$2) \quad 2\ \langle \widehat{\mathbf{f}}(\mathbf{x}_k), \mathbf{x}^* - \mathbf{x}_k\rangle + C^2 \rho_k > -\delta. \qquad (12.17)$$

It turns out that there exists a positive integer N such that (12.17) is satisfied for $k \geq N$. Assume the contrary, that (12.16) is satisfied and $k \geq N$. Then from (12.15) it follows that

$$\begin{aligned}
&\|\mathbf{x}^* - \mathbf{x}_{k+1}\|^2 \leq \|\mathbf{x}^* - \mathbf{x}_k\|^2 - \delta\rho_k \\
&\leq \left(\|\mathbf{x}^* - \mathbf{x}_{k-1}\|^2 - \delta\rho_{k-1}\right) - \delta\rho_k \leq \cdots \leq \|\mathbf{x}^* - \mathbf{x}_N\|^2 - \delta \sum_{s=N}^{k} \rho_s. \qquad (12.18)
\end{aligned}$$

The right-hand side of (12.18) tends to $-\infty$ when $k \to \infty$ because $\sum_{k=0}^{\infty} \rho_k = +\infty$ by the assumption, which contradicts $\|\mathbf{x}^* - \mathbf{x}_{k+1}\|^2 \geq 0$. Therefore, there exist sufficiently large numbers k_s, $s = 1, 2, \ldots$ such that

$$2\ \langle \widehat{\mathbf{f}}(\mathbf{x}_{k_s}), \mathbf{x}^* - \mathbf{x}_{k_s}\rangle + C^2 \rho_{k_s} > -\delta, \qquad (12.19)$$

that is, which satisfy (12.17). Since $\rho_{k_s} \to 0$, for any $\varepsilon > 0$, a sequence $\{k_s\}$ and a number S_ε can be found such that

$$\langle \widehat{\mathbf{f}}(\mathbf{x}_{k_s}), \mathbf{x}^* - \mathbf{x}_{k_s}\rangle > -\varepsilon \qquad (12.20)$$

is satisfied for $s \geq S_\varepsilon$. Moreover, using the property of convex functions (Theorem 1.30), we have $f(\mathbf{x}^*) - f(\mathbf{x}_{k_s}) \geq \langle \widehat{\mathbf{f}}(\mathbf{x}_{k_s}), \mathbf{x}^* - \mathbf{x}_{k_s}\rangle > -\varepsilon$, that is, $f(\mathbf{x}_{k_s}) - f(\mathbf{x}^*) < \varepsilon$. However, $\mathbf{x}^* = \operatorname{argmin} f(\mathbf{x})$, therefore $f(\mathbf{x}_{k_s}) - f(\mathbf{x}^*) > 0$.

From both inequalities it follows that $\lim\limits_{s\to\infty} f(\mathbf{x}_{k_s}) = f(\mathbf{x}^*)$. ■

12.3.2 Calculation of Subgradients and Subdifferentials

Using that $\Phi_1(\mathbf{a})$, $\Phi_2(\mathbf{a})$, $\Phi_4(\mathbf{x})$, $\Phi_5(\mathbf{x})$ are convex separable functions and Theorems 1.27, 1.28, 1.29, 1.34, 1.35, we can calculate corresponding subdifferentials (subgradient sets) at iteration k as follows, respectively.

$$\partial\Phi_1^{(k)}(\mathbf{a}_k) = \sum_{i=1}^{m} \partial\mathbf{g}_i^{(k)}(\mathbf{a}_k) \in \mathbf{R}^{n+1},$$

where

$$\partial\mathbf{g}_i^{(k)}(\mathbf{a}_k) = \begin{cases} w_i.(p, \varphi_1(\mathbf{x}_i), \ldots, \varphi_n(\mathbf{x}_i)), & \sum_{j=0}^{n} a_j^{(k)}\varphi_j(\mathbf{x}_i) - f(\mathbf{x}_i) > 0, \\ w_i.(p, \varphi_1(\mathbf{x}_i), \ldots, \varphi_n(\mathbf{x}_i)).[-1, 1], & \sum_{j=0}^{n} a_j^{(k)}\varphi_j(\mathbf{x}_i) - f(\mathbf{x}_i) = 0, \\ -w_i.(p, \varphi_1(\mathbf{x}_i), \ldots, \varphi_n(\mathbf{x}_i)), & \sum_{j=0}^{n} a_j^{(k)}\varphi_j(\mathbf{x}_i) - f(\mathbf{x}_i) < 0, \end{cases} \quad (12.21)$$

$i = 1, \ldots, m.$

Let $\max_{1\le i\le m} w_i\left|\sum_{j=0}^{n} a_j\varphi_j(\mathbf{x}_i) - f(\mathbf{x}_i)\right|$ be attained for the index(es) $i = i(a) \in I(a) \subset \{1, \ldots, m\}$. Then

$$\partial\Phi_2^{(k)}(\mathbf{a}_k) = \text{co}\left\{(\partial g_{i(a),0}^{(k)}(\mathbf{a}_k), \partial g_{i(a),1}^{(k)}(\mathbf{a}_k), \ldots, \partial g_{i(a),n}^{(k)}(\mathbf{a}_k)) : i(a) \in I(a)\right\},$$

where

$$\partial g_{i(a),j}^{(k)}(\mathbf{a}_k) = \begin{cases} w_{i(a)}\varphi_j(\mathbf{x}_{i(a)}), & \sum_{j=0}^{n} a_j^{(k)}\varphi_j(\mathbf{x}_{i(a)}) - f(\mathbf{x}_{i(a)}) > 0, \\ w_{i(a)}\varphi_j(\mathbf{x}_{i(a)}).[-1, 1], & \sum_{j=0}^{n} a_j^{(k)}\varphi_j(\mathbf{x}_{i(a)}) - f(\mathbf{x}_{i(a)}) = 0, \\ -w_{i(a)}\varphi_j(\mathbf{x}_{i(a)}), & \sum_{j=0}^{n} a_j^{(k)}\varphi_j(\mathbf{x}_{i(a)}) - f(\mathbf{x}_{i(a)}) < 0, \end{cases} \quad (12.22)$$

$j = 0, 1, \ldots, n.$

Let $\max_{1\le i\le m} w_i\left|\sum_{j=1}^{n} a_{ij}x_j - b_i\right|$ be attained for the index(es) $i = i(x) \in I(x) \subset \{1, \ldots, m\}$. Then

$$\partial\Phi_4^{(k)}(\mathbf{x}_k) = \text{co}\left\{(\partial g_{i(x),1}^{(k)}(\mathbf{x}_k), \ldots, \partial g_{i(x),n}^{(k)}(\mathbf{x}_k)) : i(x) \in I(x)\right\},$$

where

$$\partial g^{(k)}_{i(x),j}(\mathbf{x}_k) = \begin{cases} w_{i(x)}a_{i(x),j}, & \sum_{j=1}^{n} a_{i(x),j}x_j^{(k)} - b_{i(x)} > 0, \\ w_{i(x)}a_{i(x),j}.[-1,1], & \sum_{j=1}^{n} a_{i(x),j}x_j^{(k)} - b_{i(x)} = 0, \\ -w_{i(x)}a_{i(x),j}, & \sum_{j=1}^{n} a_{i(x),j}x_j^{(k)} - b_{i(x)} < 0, \end{cases} \tag{12.23}$$

$j = 1, \ldots, n$.

$$\partial \Phi_5^{(k)}(\mathbf{x}_k) = \sum_{i=1}^{m} \partial \mathbf{g}_i^{(k)}(\mathbf{x}_k) \ \in \mathbf{R}^n,$$

where

$$\partial \mathbf{g}_i^{(k)}(\mathbf{x}_k) = \begin{cases} w_i.(a_{i1}, a_{i2}, \ldots, a_{in}), & \sum_{j=1}^{n} a_{ij}x_j^{(k)} - b_i > 0, \\ w_i.(a_{i1}, a_{i2}, \ldots, a_{in}).[-1,1], & \sum_{j=1}^{n} a_{ij}x_j^{(k)} - b_i = 0, \\ -w_i.(a_{i1}, a_{i2}, \ldots, a_{in}), & \sum_{j=1}^{n} a_{ij}x_j^{(k)} - b_i < 0, \end{cases} \tag{12.24}$$

$i = 1, \ldots, m$.

Obviously, elements of $\partial\Phi_1^{(k)}(\mathbf{a}_k)$, $\partial\Phi_2^{(k)}(\mathbf{a}_k)$, $\partial\Phi_4^{(k)}(\mathbf{x}_k)$ and $\partial\Phi_5^{(k)}(\mathbf{x}_k)$ depend on the sign of the corresponding expression in (12.21), (12.22), (12.23), and (12.24), and therefore, $\partial\Phi_1^{(k)}(\mathbf{a}_k)$ and $\partial\Phi_2^{(k)}(\mathbf{a}_k)$ depend on the current values $a_j^{(k)}$, $j = 0, \ldots, n$, and $\partial\Phi_4^{(k)}(\mathbf{x}_k)$ and $\partial\Phi_5^{(k)}(\mathbf{x}_k)$ depend on the current values $x_j^{(k)}$, $j = 1, \ldots, n$, respectively.

We choose $\rho_0 = 1$; $\rho_k = \frac{c}{k}$, $k = 1, 2, \ldots$; $c = const. > 0$. The requirements for the step size are satisfied for this choice of ρ_k.

12.3.3 The Gradient Method for Differentiable Functions

In order to compare the results obtained by the subgradient method of nonsmooth optimization for problems (12.5) [(12.6)] and (12.10) [(12.11)] with the corresponding results, obtained by methods of "differentiable" optimization for problems (12.7′) and (12.12′), respectively, consider a method for solving the "differentiable" unconstrained minimization problem min $f(\mathbf{x})$ where $f : \mathbf{R}^n \to \mathbf{R}$.

The *gradient method* is defined through

$$\mathbf{x}_{k+1} = \mathbf{x}_k - \rho_k \mathbf{f}'(\mathbf{x}_k), \quad k = 0, 1, \ldots, \tag{12.25}$$

where

- $\mathbf{x}_0 \in \mathbf{R}^n$ is an arbitrary initial guess;
- ρ_k (≥ 0) is a step size;
- $\mathbf{f}'(\mathbf{x}_k)$ is the gradient of the differentiable function $f(\mathbf{x})$ at $\mathbf{x}_k$.

We use, for example, a *line search method* for choosing the step size ρ_k. The gradient method with such a choice of step size is known as the *steepest descent method*. The value of ρ_k at iteration k is an optimal solution to the following single-variable problem of ρ:

$$\min \{f(\mathbf{x}_k - \rho\mathbf{f}'(\mathbf{x}_k))\}$$

subject to

$$\rho \geq 0,$$

that is,

$$f(\mathbf{x}_k - \rho_k\mathbf{f}'(\mathbf{x}_k)) = \min_{\rho\geq 0} f(\mathbf{x}_k - \rho\mathbf{f}'(\mathbf{x}_k)).$$

An alternative way of choosing the step length ρ_k is the so-called *doubling method*. Set, for example, $\rho_0 = 1$. Choose $\rho_k = \rho_{k-1}$. If $f(\mathbf{x}_{k+1}) < f(\mathbf{x}_k)$, then $\rho_k := 2\rho_{k-1}$. If $f(\mathbf{x}_{k+1}) < f(\mathbf{x}_k)$ again, then this doubling continues until $f(\mathbf{x})$ stops to decrease. If $f(\mathbf{x}_{k+1}) \geq f(\mathbf{x}_k)$, then $\rho_k := \frac{1}{2}\rho_{k-1}$. If $f(\mathbf{x}_k - \frac{1}{2}\rho_{k-1}\mathbf{f}'(\mathbf{x}_k)) < f(\mathbf{x}_k)$, then $\mathbf{x}_{k+1} := \mathbf{x}_k - \frac{1}{2}\rho_{k-1}\mathbf{f}'(\mathbf{x}_k)$, go to iteration $k+2$. If $f(\mathbf{x}_k - \frac{1}{2}\rho_{k-1}\mathbf{f}'(\mathbf{x}_k)) \geq f(\mathbf{x}_k)$, then $\rho_k := \frac{1}{4}\rho_{k-1}$, and so on.

The gradient method (12.25) can be considered as a special case of subgradient method (12.14) (with $\gamma_k = 1$) when the function f to be minimized is differentiable.

Gradients of $\Phi_3^2(\mathbf{a})$ and $\Phi_6^2(\mathbf{x})$ at iteration k, respectively, are

$$[(\Phi_3^2)']^{(k)}(\mathbf{a}_k) = (g_0^{(k)}(\mathbf{a}_k), g_1^{(k)}(\mathbf{a}_k), \ldots, g_n^{(k)}(\mathbf{a}_k)),$$

where

$$g_l^{(k)}(\mathbf{a}_k) = 2\sum_{i=1}^{m} w_i\left(\sum_{j=0}^{n} a_j^{(k)}\varphi_j(\mathbf{x}_i) - f(\mathbf{x}_i)\right)\varphi_l(\mathbf{x}_i), \quad l = 0, 1, \ldots, n;$$

$$[(\Phi_6^2)']^{(k)}(\mathbf{x}_k) = (g_1^{(k)}(\mathbf{x}_k), \ldots, g_n^{(k)}(\mathbf{x}_k)),$$

where

$$g_l^{(k)}(\mathbf{x}_k) = 2\sum_{i=1}^{m} w_i a_{il}\left(\sum_{j=1}^{n} a_{ij}x_j^{(k)} - b_i\right), \quad l = 1, \ldots, n.$$

The following theorem holds true.

Theorem 12.3 (An upper bound in gradient minimization)
Let $\mathbf{f}'$ be Lipschitz continuous of rank L, that is,

$$\|\mathbf{f}'(\mathbf{x}) - \mathbf{f}'(\mathbf{y})\| \leq L\|\mathbf{x} - \mathbf{y}\|$$

for each $\mathbf{x}, \mathbf{y} \in \mathbf{R}^n$, *and let* $\mathbf{x}_{k+1}$ *be determined by* (12.25). *Then there exists an* ε *such that*

$$f(\mathbf{x}_{k+1}) - f(\mathbf{x}_k) \leq -\varepsilon\rho_k\|\mathbf{f}'(\mathbf{x}_k)\|^2. \quad (12.26)$$

Proof. Using the Mean value theorem (Theorem A.2, Appendix A), Cauchy-Schwarz inequality (Theorem B.1, Appendix B) and (12.25), we get

$$\begin{aligned}
f(\mathbf{x}_{k+1}) - f(\mathbf{x}_k) &= \langle \mathbf{f}'(\xi), \mathbf{x}_{k+1} - \mathbf{x}_k \rangle \\
&\equiv \langle \mathbf{f}'(\mathbf{x}_k), \mathbf{x}_{k+1} - \mathbf{x}_k \rangle + \langle \mathbf{f}'(\xi) - \mathbf{f}'(\mathbf{x}_k), \mathbf{x}_{k+1} - \mathbf{x}_k \rangle \\
&= \langle \mathbf{f}'(\mathbf{x}_k), -\rho_k\mathbf{f}'(\mathbf{x}_k) \rangle + \langle \mathbf{f}'(\xi) - \mathbf{f}'(\mathbf{x}_k), \mathbf{x}_{k+1} - \mathbf{x}_k \rangle \\
&= -\rho_k\|\mathbf{f}'(\mathbf{x}_k)\|^2 + \langle \mathbf{f}'(\xi) - \mathbf{f}'(\mathbf{x}_k), \mathbf{x}_{k+1} - \mathbf{x}_k \rangle \\
&\leq -\rho_k\|\mathbf{f}'(\mathbf{x}_k)\|^2 + \|\mathbf{f}'(\xi) - \mathbf{f}'(\mathbf{x}_k)\|.\|\mathbf{x}_{k+1} - \mathbf{x}_k\| \\
&\leq -\rho_k\|\mathbf{f}'(\mathbf{x}_k)\|^2 + L\|\xi - \mathbf{x}_k\|.\|\mathbf{x}_{k+1} - \mathbf{x}_k\| \\
&\leq -\rho_k\|\mathbf{f}'(\mathbf{x}_k)\|^2 + L\|\mathbf{x}_{k+1} - \mathbf{x}_k\|^2 \\
&= -\rho_k\|\mathbf{f}'(\mathbf{x}_k)\|^2 + L\| - \rho_k\mathbf{f}'(\mathbf{x}_k)\|^2 \\
&= (\rho_k L - 1)\rho_k\|\mathbf{f}'(\mathbf{x}_k)\|^2 \leq -\varepsilon\rho_k\|\mathbf{f}'(\mathbf{x}_k)\|^2,
\end{aligned}$$

where $\rho_k L - 1 \leq -\varepsilon$ for some ε, that is, $\rho_k \leq \frac{1-\varepsilon}{L}$. We can choose, for example, $\rho_k = \frac{1-\varepsilon}{L}$. ■

Theorem 12.3 presents another approach of choosing ρ_k: by using (12.26). The following algorithm is based on this idea.

Algorithm for generating ρ_k by using Theorem 12.3.
0. (Initialization) Choose $\varepsilon \in (0, 1)$, $\lambda \in (0, 1)$ and $\rho > 0$.
   Set $k := 0$. Go to the Main step.
Main step.
1. Calculate $\mathbf{x}_{k+1} = \mathbf{x}_k - \rho\mathbf{f}'(\mathbf{x}_k)$. Go to 2.
2. Calculate $f(\mathbf{x}_{k+1})$. Go to 3.
3. If $f(\mathbf{x}_{k+1}) - f(\mathbf{x}_k) \leq -\varepsilon\rho\|\mathbf{f}'(\mathbf{x}_k)\|^2$ then go to 5 else go to 4.
4. $\rho := \lambda\rho$. Go to 1.
5. $\rho_k := \rho$.
6. End.

Theorem 12.4 (Norm of the gradient vanishes in gradient minimization)
Let $f(\mathbf{x})$ *be a differentiable function,* $f(\mathbf{x}) > -\infty$, $\mathbf{f}'$ *be Lipschitz continuous of rank* L *and sequence* $\{\mathbf{x}_k\}$ *be generated through* (12.25). *Then*

$$\|\mathbf{f}'(\mathbf{x}_k)\| \to 0 \quad \text{when} \quad k \to \infty. \quad (12.27)$$

Proof. According to Theorem 12.3, whose assumptions are satisfied,

$$f(\mathbf{x}_{k+1}) - f(\mathbf{x}_k) \leq -\varepsilon\rho_k \|\mathbf{f}'(\mathbf{x}_k)\|^2.$$

Therefore

$$\|\mathbf{f}'(\mathbf{x}_k)\|^2 \leq \frac{f(\mathbf{x}_k) - f(\mathbf{x}_{k+1})}{\varepsilon\rho_k}. \tag{12.28}$$

Since $\{f(\mathbf{x}_k)\}$ is a monotone decreasing sequence ($\{\mathbf{x}_k\}$ minimizes f) and it is bounded below by the assumption $f(\mathbf{x}) > -\infty$, then $\{f(\mathbf{x}_k)\}$ is a convergent sequence which tends to some η when $k \to \infty$. Hence,

$$f(\mathbf{x}_k) - f(\mathbf{x}_{k+1}) \to 0 \quad \text{as} \quad k \to \infty.$$

From (12.28), using that $\|\mathbf{f}'(\mathbf{x}_k)\|^2 \geq 0$, it follows that $\|\mathbf{f}'(\mathbf{x}_k)\| \to 0$ when $k \to \infty$. ■

Theorem 12.4 states that, under its assumptions, $\frac{\partial f}{\partial x_j}(\mathbf{x}_k) \to 0$ for $j = 1, \ldots, n$ as $k \to \infty$, which is a necessary condition for a minimum of a function of several variables.

The following theorem gives the rate of convergence of the steepest descent method.

Theorem 12.5 (Rate of convergence of the steepest descent method)
Let $f : \mathbf{R}^n \to \mathbf{R}$, $f \in C^2(\mathbf{R}^n)$, let there exist positive constants m and M such that

$$m\|\mathbf{y}\|^2 \leq \langle f''(\mathbf{x})\mathbf{y}, \mathbf{y}\rangle \leq M\|\mathbf{y}\|^2 \tag{12.29}$$

for any $\mathbf{x} \in \mathbf{R}^n$ and $\mathbf{y} \in \mathbf{R}^n$, and let sequence $\{\mathbf{x}_k\}$ be generated by the steepest descent method (method (12.25) with ρ_k determined by a line search method).

Then f has a unique minimum solution $\mathbf{x}^* \in \mathbf{R}^n$ and for each $\mathbf{x} \in \mathbf{R}^n$ the following inequality holds true

$$\|\mathbf{f}'(\mathbf{x})\|^2 \geq m\left(1 + \frac{m}{M}\right)(f(\mathbf{x}) - f(\mathbf{x}^*)). \tag{12.30}$$

Further, there exist constants C and q: $C > 0,\ 0 \leq q \leq 1$ such that

$$f(\mathbf{x}_{k+1}) - f(\mathbf{x}^*) \leq q^k(f(\mathbf{x}_1) - f(\mathbf{x}^*))$$

and

$$\|\mathbf{x}_{k+1} - \mathbf{x}^*\| \leq Cq^{\frac{k}{2}}, \quad k = 0, 1, \ldots .$$

Proof. Let $\mathbf{x}_1, \mathbf{x}_2 \in \mathbf{R}^n$ be terms of sequence $\{\mathbf{x}_k\}$ defined by (12.25). Using Taylor's second-order theorem (Theorem A.3, Appendix A), we get

$$f(\mathbf{x}_2) = f(\mathbf{x}_1) + \langle \mathbf{f}'(\mathbf{x}_1), \mathbf{x}_2 - \mathbf{x}_1 \rangle + \frac{1}{2}\langle f''(\mathbf{x}_\theta)(\mathbf{x}_2 - \mathbf{x}_1), \mathbf{x}_2 - \mathbf{x}_1 \rangle,$$

where $\mathbf{x}_\theta = \mathbf{x}_1 + \theta(\mathbf{x}_2 - \mathbf{x}_1), \ \theta \in (0, 1)$.

Since $\langle f''(\mathbf{x}_\theta)(\mathbf{x}_2 - \mathbf{x}_1), \mathbf{x}_2 - \mathbf{x}_1 \rangle \geq m\|\mathbf{x}_2 - \mathbf{x}_1\|^2$ according to (12.29) with $\mathbf{x} = \mathbf{x}_\theta$ and $\mathbf{y} = \mathbf{x}_2 - \mathbf{x}_1$, then

$$f(\mathbf{x}_2) \geq f(\mathbf{x}_1) - \langle \mathbf{f}'(\mathbf{x}_1), \mathbf{x}_1 - \mathbf{x}_2 \rangle + \frac{1}{2}m\|\mathbf{x}_2 - \mathbf{x}_1\|^2. \tag{12.31}$$

Replacing $\mathbf{x}_2$ by $\mathbf{x}$ in (12.31), from $f(\mathbf{x}_1) \geq f(\mathbf{x}_2)$ ($\{\mathbf{x}_k\}$ minimizes f) and from Cauchy-Schwarz inequality it follows that

$$f(\mathbf{x}_1) \geq f(\mathbf{x}) \geq f(\mathbf{x}_1) - \|\mathbf{f}'(\mathbf{x}_1)\|.\|\mathbf{x} - \mathbf{x}_1\| + \frac{1}{2}m\|\mathbf{x} - \mathbf{x}_1\|^2.$$

Dividing by $\|\mathbf{x} - \mathbf{x}_1\| > 0$, we obtain

$$\|\mathbf{x} - \mathbf{x}_1\| \leq \frac{2}{m}\|\mathbf{f}'(\mathbf{x}_1)\|.$$

Therefore the *lower-level set*

$$L(\mathbf{x}_1) \stackrel{\text{def}}{=} \{\mathbf{x} \in \mathbf{R}^n : f(\mathbf{x}) \leq f(\mathbf{x}_1)\}$$

is bounded. Since f is a continuous function by assumption, $L(\mathbf{x}_1)$ is closed. Hence, $L(\mathbf{x}_1)$ is compact (see Definition A.6, Appendix A).

Therefore, f attains its minimum value on $L(\mathbf{x}_1)$:

$$\min_{\mathbf{R}^n} f(\mathbf{x}) = \min_{L(\mathbf{x}_1)} f(\mathbf{x}).$$

From the left inequality of (12.29) and Theorem 1.32 it follows that f is strictly convex. Hence, the minimum point $\mathbf{x}^*$ of f is unique.

Consider the point $\widehat{\mathbf{x}}_\theta \stackrel{\text{def}}{=} \mathbf{x}^* + \theta(\mathbf{x} - \mathbf{x}^*), \ \theta \in (0, 1)$.

Using Taylor's second-order theorem and the fact that the minimum solution $\mathbf{x}^*$ is a stationary point for f, that is, $\mathbf{f}'(\mathbf{x}^*) = \mathbf{0}$, we get

$$f(\mathbf{x}) - f(\mathbf{x}^*) = \frac{1}{2}\langle f''(\widehat{\mathbf{x}}_\theta)(\mathbf{x} - \mathbf{x}^*), \mathbf{x} - \mathbf{x}^* \rangle.$$

Therefore, this relation and (12.29) with $\mathbf{x} = \widehat{\mathbf{x}}_\theta$ and $\mathbf{y} = \mathbf{x} - \mathbf{x}^*$ imply

$$\frac{m}{2}\|\mathbf{x} - \mathbf{x}^*\|^2 \leq f(\mathbf{x}) - f(\mathbf{x}^*) \leq \frac{M}{2}\|\mathbf{x} - \mathbf{x}^*\|^2. \tag{12.32}$$

From (12.31), replacing $\mathbf{x}_1$ by $\mathbf{x}$ and $\mathbf{x}_2$ by $\mathbf{x}^*$, we get

$$f(\mathbf{x}) - f(\mathbf{x}^*) \le \langle \mathbf{f}'(\mathbf{x}), \mathbf{x} - \mathbf{x}^* \rangle - \frac{1}{2} m \|\mathbf{x} - \mathbf{x}^*\|^2. \tag{12.31'}$$

From (12.31′) and (12.32) it follows that

$$\frac{m}{2}\|\mathbf{x} - \mathbf{x}^*\|^2 \le f(\mathbf{x}) - f(\mathbf{x}^*) \le \langle \mathbf{f}'(\mathbf{x}), \mathbf{x} - \mathbf{x}^* \rangle - \frac{1}{2} m \|\mathbf{x} - \mathbf{x}^*\|^2,$$

that is,

$$m\|\mathbf{x} - \mathbf{x}^*\|^2 \le \langle \mathbf{f}'(\mathbf{x}), \mathbf{x} - \mathbf{x}^* \rangle,$$

and using Cauchy-Schwarz inequality, we get

$$m\|\mathbf{x} - \mathbf{x}^*\|^2 \le \langle \mathbf{f}'(\mathbf{x}), \mathbf{x} - \mathbf{x}^* \rangle \le \|\mathbf{f}'(\mathbf{x})\|.\|\mathbf{x} - \mathbf{x}^*\|.$$

Hence, dividing by $\|\mathbf{x} - \mathbf{x}^*\| > 0$, we obtain

$$\|\mathbf{x} - \mathbf{x}^*\| \le \frac{1}{m}\|\mathbf{f}'(\mathbf{x})\|. \tag{12.33}$$

Then

$$f(\mathbf{x}) - f(\mathbf{x}^*) \le \|\mathbf{f}'(\mathbf{x})\|.\|\mathbf{x} - \mathbf{x}^*\| - \frac{1}{2} m \|\mathbf{x} - \mathbf{x}^*\|^2$$

$$\le \frac{1}{m}\|\mathbf{f}'(\mathbf{x})\|^2 - \frac{1}{2} m \frac{2}{M}(f(\mathbf{x}) - f(\mathbf{x}^*)),$$

where we have used (12.31′) and Cauchy-Schwarz inequality for the first inequality above, and (12.32) (multiplied by $\frac{2}{M}$) and (12.33) to derive the second inequality above.

Therefore

$$f(\mathbf{x}) - f(\mathbf{x}^*) \le \frac{1}{m}\|\mathbf{f}'(\mathbf{x})\|^2 - \frac{m}{M}(f(\mathbf{x}) - f(\mathbf{x}^*)),$$

that is,

$$\|\mathbf{f}'(\mathbf{x})\|^2 \ge m\left(1 + \frac{m}{M}\right)(f(\mathbf{x}) - f(\mathbf{x}^*)).$$

Thus, (12.30) is proved.

Using Taylor's second-order theorem again, we get

$$f(\mathbf{x}_2) = f(\mathbf{x}_1) + \langle \mathbf{f}'(\mathbf{x}_1), \mathbf{x}_2 - \mathbf{x}_1 \rangle + \frac{1}{2}\langle f''(\mathbf{x}_\theta)(\mathbf{x}_2 - \mathbf{x}_1), \mathbf{x}_2 - \mathbf{x}_1 \rangle,$$

where $\mathbf{x}_\theta = \mathbf{x}_1 + \theta(\mathbf{x}_2 - \mathbf{x}_1)$, $\theta \in (0, 1)$.

Because $\langle f''(\mathbf{x}_\theta)(\mathbf{x}_2 - \mathbf{x}_1), \mathbf{x}_2 - \mathbf{x}_1 \rangle \le M\|\mathbf{x}_2 - \mathbf{x}_1\|^2$ according to (12.29) with $\mathbf{x} = \mathbf{x}_\theta$ and $\mathbf{y} = \mathbf{x}_2 - \mathbf{x}_1$, then

$$f(\mathbf{x}_1) - f(\mathbf{x}_2) \geq \langle \mathbf{f}'(\mathbf{x}_1), \mathbf{x}_1 - \mathbf{x}_2 \rangle - \frac{1}{2} M \|\mathbf{x}_1 - \mathbf{x}_2\|^2. \tag{12.34}$$

Since the steepest descent method moves along the direction $-\mathbf{f}'(\mathbf{x}_k)$, then

$$\begin{aligned} f(\mathbf{x}_k) - f(\mathbf{x}_{k+1}) &\geq f(\mathbf{x}_k) - f(\mathbf{x}_k - \rho \mathbf{f}'(\mathbf{x}_k)) \\ &\geq \langle \mathbf{f}'(\mathbf{x}_k), \rho \mathbf{f}'(\mathbf{x}_k) \rangle - \frac{1}{2} M \|\rho \mathbf{f}'(\mathbf{x}_k)\|^2 \\ &= \rho \|\mathbf{f}'(\mathbf{x}_k)\|^2 - \frac{1}{2} M \rho^2 \|\mathbf{f}'(\mathbf{x}_k)\|^2 = \left(\rho - \frac{M\rho^2}{2}\right) \|\mathbf{f}'(\mathbf{x}_k)\|^2, \end{aligned} \tag{12.35}$$

where for the first inequality in (12.35), we have used that

$$f(\mathbf{x}_{k+1}) \equiv f(\mathbf{x}_k - \rho_k \mathbf{f}'(\mathbf{x}_k)) = \min_{\rho \geq 0} f(\mathbf{x}_k - \rho \mathbf{f}'(\mathbf{x}_k)),$$

and for the second inequality in (12.35) we have applied (12.34) with $\mathbf{x}_1 = \mathbf{x}_k$, $\mathbf{x}_2 = \mathbf{x}_k - \rho \mathbf{f}'(\mathbf{x}_k)$.

Consider the function $g(\rho) \stackrel{\text{def}}{=} \rho - \frac{M\rho^2}{2}$. Since

$$g'(\rho) = \left(\rho - \frac{M\rho^2}{2}\right)'_\rho = 1 - M\rho,$$

then $\rho = \frac{1}{M}$ is a stationary point for $g(\rho)$, and since the second derivative of $g(\rho)$ is negative

$$g''(\rho) = \left(\rho - \frac{M\rho^2}{2}\right)''_\rho = -M < 0,$$

then $\rho = \frac{1}{M}$ is a maximum point for $g(\rho)$. Hence, $g_{\max}(\rho) = g(\frac{1}{M}) = \frac{1}{2M}$.

If we take the maximum with respect to ρ on both sides of (12.35), we get

$$f(\mathbf{x}_k) - f(\mathbf{x}_{k+1}) \geq \frac{1}{2M} \|\mathbf{f}'(\mathbf{x}_k)\|^2. \tag{12.36}$$

From (12.36) and (12.30) proved above, applied for $\mathbf{x} = \mathbf{x}_k$, it follows that

$$\begin{aligned} &[f(\mathbf{x}_k) - f(\mathbf{x}^*)] - [f(\mathbf{x}_{k+1}) - f(\mathbf{x}^*)] \equiv f(\mathbf{x}_k) - f(\mathbf{x}_{k+1}) \\ &\geq \frac{1}{2M} \|\mathbf{f}'(\mathbf{x}_k)\|^2 \geq \frac{1}{2M} m \left(1 + \frac{m}{M}\right) (f(\mathbf{x}_k) - f(\mathbf{x}^*)). \end{aligned}$$

Denoting $a_k = f(\mathbf{x}_k) - f(\mathbf{x}^*)$, we get

$$a_k - a_{k+1} \geq \frac{m(M+m)}{2M^2} a_k,$$

that is,

$$a_{k+1} \leq \left[1 - \frac{m(M+m)}{2M^2}\right] a_k,$$

and so on, continuing in the same manner, we obtain

$$a_{k+1} \leq \left[1 - \frac{m(M+m)}{2M^2}\right]^k a_1.$$

Set $q = 1 - \frac{m(M+m)}{2M^2}$. Obviously $q \in [0, 1]$ because $0 \leq m \leq M$. Therefore

$$f(\mathbf{x}_{k+1}) - f(\mathbf{x}^*) \leq q^k (f(\mathbf{x}_1) - f(\mathbf{x}^*)). \tag{12.37}$$

Hence

$$\frac{m}{2}\|\mathbf{x}_{k+1} - \mathbf{x}^*\|^2 \overset{(12.32)}{\leq} f(\mathbf{x}_{k+1}) - f(\mathbf{x}^*) \overset{(12.37)}{\leq} q^k (f(\mathbf{x}_1) - f(\mathbf{x}^*)).$$

Then

$$\|\mathbf{x}_{k+1} - \mathbf{x}^*\| \leq \left[\frac{2}{m}(f(\mathbf{x}_1) - f(\mathbf{x}^*))\right]^{\frac{1}{2}} q^{\frac{k}{2}}.$$

Denote $C = \left[\frac{2}{m}(f(\mathbf{x}_1) - f(\mathbf{x}^*))\right]^{\frac{1}{2}} = const$. Since $\mathbf{x}^*$ is a minimum solution to problem $\min f(\mathbf{x})$, then $f(\mathbf{x}_1) > f(\mathbf{x}^*)$, and since $m > 0$ by assumption, hence $C > 0$. Therefore

$$\|\mathbf{x}_{k+1} - \mathbf{x}^*\| \leq C q^{\frac{k}{2}}, \quad k = 0, 1, \ldots$$

with $C > 0$, $q \in [0, 1]$. The proof of Theorem 12.5 is complete. ■

12.4 Numerical Solution of Some Systems of Nonlinear Algebraic Equations

The problem of numerical solution of systems of nonlinear algebraic equations with even powers is studied in this section. Similarly to the approach of Section 12.1, non-differentiable (nonsmooth) unconstrained minimization problems are associated with the systems under consideration, with the objective functions, based on the weighted discrete ℓ_1- and ℓ_∞-norm, respectively. By using these two norms as proximity criteria and by applying a subgradient method, the considered systems are solved approximately by minimizing the sum of absolute values of residuals, in the case of ℓ_1-norm, and by minimizing the maximal residual, in the case of ℓ_∞-norm, respectively. Also, "differentiable" unconstrained minimization problems (least squares problems) are associated with the considered systems of nonlinear equations, with objective functions, based on the weighted discrete ℓ_2-norm. The considered systems are solved approximately by minimizing the sum of squares of the residuals, using a gradient method. Subgradients of the objective functions of unconstrained minimization problems, as well as gradients of the objective functions of the corresponding "differentiable" unconstrained minimization problems, are calculated, which are necessary for the subgradient and gradient methods, applied for the associated nondifferentiable and differentiable problems, respectively.

12.4.1 Statement of Problems Under Consideration

Consider a system of nonlinear equations of the form

$$f_i(x_1, \dots, x_n) = 0, \quad i = 1, \dots, m, \tag{12.38}$$

where $f_1, \dots, f_m$ are nonlinear functions in n variables $x_1, \dots, x_n$ and $m > n$. These variables can be considered as components of the n−tuple (n−vector) $\mathbf{x} = (x_1, \dots, x_n)$.

When functions $f_1, \dots, f_m$ are algebraic polynomials of n variables with even powers and nonnegative coefficients, system (12.38) becomes

$$\begin{aligned}
&a_{11}^{(2)} x_1^2 + a_{12}^{(2)} x_2^2 + \dots + a_{1n}^{(2)} x_n^2 + \dots + a_{11}^{(2n)} x_1^{2n} + a_{12}^{(2n)} x_2^{2n} + \dots + a_{1n}^{(2n)} x_n^{2n} = a_{10}\\
&a_{21}^{(2)} x_1^2 + a_{22}^{(2)} x_2^2 + \dots + a_{2n}^{(2)} x_n^2 + \dots + a_{21}^{(2n)} x_1^{2n} + a_{22}^{(2n)} x_2^{2n} + \dots + a_{2n}^{(2n)} x_n^{2n} = a_{20}\\
&\dots\dots\dots\dots\dots\dots\dots\dots\dots\dots\dots\dots\dots\\
&a_{m1}^{(2)} x_1^2 + a_{m2}^{(2)} x_2^2 + \dots + a_{mn}^{(2)} x_n^2 + \dots + a_{m1}^{(2n)} x_1^{2n} + a_{m2}^{(2n)} x_2^{2n} + \dots + a_{mn}^{(2n)} x_n^{2n} = a_{m0},
\end{aligned} \tag{12.39}$$

where $a_{ij}^{(2k)} \geq 0, \ i = 1, \dots, m; \ j, k = 1, \dots, n$.

System (12.39) is one of the most general forms of the system of nonlinear algebraic equations of n variables (unknowns) of the considered type. System (12.39) can also be written as follows:

$$\sum_{j=1}^{n} \sum_{k=1}^{n} a_{ij}^{(2k)} x_j^{2k} = a_{i0}, \quad i = 1, \dots, m. \tag{12.39'}$$

A special case of a system of nonlinear algebraic equations of n variables with even powers and nonnegative coefficients is

$$\begin{aligned}
&a_{12}(x_1^2 + x_2^2 + \dots + x_n^2) + \dots + a_{1,2n}(x_1^{2n} + x_2^{2n} + \dots + x_n^{2n}) = a_{10}\\
&a_{22}(x_1^2 + x_2^2 + \dots + x_n^2) + \dots + a_{2,2n}(x_1^{2n} + x_2^{2n} + \dots + x_n^{2n}) = a_{20}\\
&\dots\dots\dots\dots\dots\dots\dots\dots\dots\dots\dots\dots\\
&a_{m2}(x_1^2 + x_2^2 + \dots + x_n^2) + \dots + a_{m,2n}(x_1^{2n} + x_2^{2n} + \dots + x_n^{2n}) = a_{m0},
\end{aligned} \tag{12.40}$$

where $a_{i,2j} \geq 0,\ i = 1, \ldots, m;\ j = 1, \ldots, n.$

System (12.40) can also be written as follows:

$$\sum_{j=1}^{n} \sum_{k=1}^{n} a_{i,2j} x_k^{2j} = a_{i0}, \quad i = 1, \ldots, m. \tag{12.40'}$$

Another special case of a system of nonlinear algebraic equations of n unknowns with even powers and nonnegative coefficients is

$$\begin{aligned} a_{11}x_1^2 + a_{12}x_2^2 + \cdots + a_{1n}x_n^2 &= a_{10} \\ a_{21}x_1^4 + a_{22}x_2^4 + \cdots + a_{2n}x_n^4 &= a_{20} \\ \cdots\cdots\cdots\cdots\cdots\cdots\cdots\cdots\cdots \\ a_{m1}x_1^{2m} + a_{m2}x_2^{2m} + \cdots + a_{mn}x_n^{2m} &= a_{m0}, \end{aligned} \tag{12.41}$$

where $a_{ij} \geq 0,\ i = 1, \ldots, m;\ j = 1, \ldots, n$, which can also be written as follows:

$$\sum_{j=1}^{n} a_{ij} x_j^{2i} = a_{i0}, \quad i = 1, \ldots, m. \tag{12.41'}$$

The following minimization problems can be associated with system (12.39):

$$\min \left\{ F_1(\mathbf{x}) \stackrel{\text{def}}{=} \sum_{i=1}^{m} w_i \left| \sum_{j=1}^{n} \sum_{k=1}^{n} a_{ij}^{(2k)} x_j^{2k} - a_{i0} \right| \right\} \tag{12.42}$$

and

$$\min \left\{ F_2(\mathbf{x}) \stackrel{\text{def}}{=} \max_{1 \leq i \leq m} w_i \left| \sum_{j=1}^{n} \sum_{k=1}^{n} a_{ij}^{(2k)} x_j^{2k} - a_{i0} \right| \right\}. \tag{12.43}$$

Problem (12.42) is based on the weighted discrete ℓ_1-norm (12.2) and problem (12.43) is a special case of the problem of best Chebyshev approximation, based on the weighted discrete ℓ_∞-norm (12.3).

The corresponding discrete least squares data fitting problem, which can be associated with system (12.39), based on the weighted discrete ℓ_2-norm (12.4), is

$$\min \left\{ F_3(\mathbf{x}) \stackrel{\text{def}}{=} \left(\sum_{i=1}^{m} w_i \left(\sum_{j=1}^{n} \sum_{k=1}^{n} a_{ij}^{(2k)} x_j^{2k} - a_{i0} \right)^2 \right)^{\frac{1}{2}} \right\}. \tag{12.44}$$

The weights $w_i > 0,\ i = 1, \ldots, m$, mean the "reliability", with which data of each equation of the system can be accepted. If all equations are equally reliable, then we accept $w_i = 1,\ i = 1, \ldots, m$.

Similarly, following minimization problems can be associated with system (12.40):

$$\min \left\{ F_4(\mathbf{x}) \stackrel{\text{def}}{=} \sum_{i=1}^{m} w_i \left| \sum_{j=1}^{n} \sum_{k=1}^{n} a_{i,2j} x_k^{2j} - a_{i0} \right| \right\}, \tag{12.45}$$

based on the weighted discrete ℓ_1-norm, and

$$\min \left\{ F_5(\mathbf{x}) \stackrel{\text{def}}{=} \max_{1 \le i \le m} w_i \left| \sum_{j=1}^{n} \sum_{k=1}^{n} a_{i,2j} x_k^{2j} - a_{i0} \right| \right\}, \tag{12.46}$$

based on the weighted discrete ℓ_∞-norm.

The corresponding discrete least squares data fitting problem, which can be associated with system (12.40), based on the weighted discrete ℓ_2-norm, is

$$\min \left\{ F_6(\mathbf{x}) \stackrel{\text{def}}{=} \left(\sum_{i=1}^{m} w_i \left(\sum_{j=1}^{n} \sum_{k=1}^{n} a_{i,2j} x_k^{2j} - a_{i0} \right)^2 \right)^{\frac{1}{2}} \right\}. \tag{12.47}$$

Finally, following minimization problems can be associated with system (12.41):

$$\min \left\{ F_7(\mathbf{x}) \stackrel{\text{def}}{=} \sum_{i=1}^{m} w_i \left| \sum_{j=1}^{n} a_{ij} x_j^{2i} - a_{i0} \right| \right\}, \tag{12.48}$$

based on the weighted discrete ℓ_1-norm, and

$$\min \left\{ F_8(\mathbf{x}) \stackrel{\text{def}}{=} \max_{1 \le i \le m} w_i \left| \sum_{j=1}^{n} a_{ij} x_j^{2i} - a_{i0} \right| \right\}, \tag{12.49}$$

based on the weighted discrete ℓ_∞-norm.

The corresponding discrete least squares data fitting problem, which can be associated with system (12.41), based on the weighted discrete ℓ_2-norm, is

$$\min \left\{ F_9(\mathbf{x}) \stackrel{\text{def}}{=} \left(\sum_{i=1}^{m} w_i \left(\sum_{j=1}^{n} a_{ij} x_j^{2i} - a_{i0} \right)^2 \right)^{\frac{1}{2}} \right\}. \tag{12.50}$$

12.4.2 *Properties of the Objective Functions and Solvability of the Considered Problems*

Under the assumptions, $\sum_{j=1}^{n} \sum_{k=1}^{n} a_{ij}^{(2k)} x_j^{2k} \ge 0,\ i = 1, \dots, m$. Functions $\sum_{j=1}^{n} \sum_{k=1}^{n} a_{ij}^{(2k)} x_j^{2k} - a_{i0},\ i=1, \dots, m$, are convex functions of $x_j,\ j = 1, \dots, n$, and the "absolute value" function is convex. Hence, $F_1(\mathbf{x})$ is a convex function of $\mathbf{x}$ as a linear combination with nonnegative (even positive) coefficients $w_i,\ i = 1, \ldots, m$, of convex functions (Theorem 1.17).

Using the same reasoning, we obtain that $F_4(\mathbf{x})$ and $F_7(\mathbf{x})$ are also convex functions of $\mathbf{x}$.

According to Theorem 1.20, $F_2(\mathbf{x})$ is a convex function of $\mathbf{x}$ as maximum of the convex functions $w_i\left|\sum_{j=1}^{n}\sum_{k=1}^{n} a_{ij}^{(2k)} x_j^{2k} - a_{i0}\right|$, where $w_i > 0,\ i = 1, \ldots, m$.

Similarly, $F_5(\mathbf{x})$ and $F_8(\mathbf{x})$ are convex functions of $x_j,\ j = 1, \ldots, n$.

Since $\min F_3(\mathbf{x})$ and $\min F_3^2(\mathbf{x})$ are attained at the same point (vector) $\mathbf{x} = (x_1, \ldots, x_n)$, the problem $\min F_3^2(\mathbf{x})$ can be considered instead of problem $\min F_3(\mathbf{x})$. An analogous remark is valid for problems $\min F_6^2(\mathbf{x})$ and $\min F_6(\mathbf{x})$, and for problems $\min F_9^2(\mathbf{x})$ and $\min F_9(\mathbf{x})$.

Function $F_3^2(\mathbf{x})$ is a strictly convex function of $\mathbf{x}$ as a linear combination with nonnegative (positive) coefficients $w_i,\ i = 1, \ldots, m$, of the quadratic functions $\left(\sum_{j=1}^{n}\sum_{k=1}^{n} a_{ij}^{(2k)} x_j^{2k} - a_{i0}\right)^2,\ i = 1, \ldots, m$, which are strictly convex.

Analogously, $F_6^2(\mathbf{x})$ and $F_9^2(\mathbf{x})$ are strictly convex functions of $\mathbf{x}$.

Functions $F_1(\mathbf{x}),\ F_2(\mathbf{x}),\ F_4(\mathbf{x}),\ F_5(\mathbf{x}),\ F_7(\mathbf{x})$ and $F_8(\mathbf{x})$ are nondifferentiable (nonsmooth) whereas functions $F_3(\mathbf{x}),\ F_6(\mathbf{x})$ and $F_9(\mathbf{x})$ are differentiable.

All functions $F_s(\mathbf{x}),\ s = 1, \ldots, 9$ are *separable functions*, that is, these functions are expressed as the sums of single-variable (univariate) functions (Definition 2.1), which follows from definitions of these nine functions.

Solvability of problems (12.42) and (12.43), as well as of problems (12.45) and (12.46), and of problems (12.48) and (12.49), follows from Corollary C.1 (Appendix C).

For example, solvability of problems (12.42) and (12.43) can be obtained as follows. Since $F_1(\mathbf{x}) \geq 0$, $F_2(\mathbf{x}) \geq 0$ are continuous functions; $a_{ij}^{(2k)},\ i = 1, \ldots, m;\ j, k = 1, \ldots, n$ and $a_{i0},\ i = 1, \ldots, m$, are the fixed coefficients given by system (12.39), it follows that $F_1(\mathbf{x}) \to +\infty$ and $F_2(\mathbf{x}) \to +\infty$ when $\|\mathbf{x}\| \to \infty$. Then Corollary C.1 (Appendix C) implies that problem (12.42) and problem (12.43) are solvable.

Solvability of problems (12.45) and (12.46) and of problems (12.48) and (12.49) can be obtained similarly.

In addition, using the same reasoning, the problem $\min F_3^2(\mathbf{x})$ has also an optimal solution and it is unique (Theorem 1.38) because $F_3^2(\mathbf{x})$ is a strictly convex function.

Theorem 1.29 and Theorem 1.39, Chapter One imply that $\mathbf{x}^*$ is an optimal solution of problem (12.43) if and only if

$$\mathbf{0}_n \in \partial F_2(\mathbf{x}^*) = \text{co}\,\{\partial\Psi_{i2}(\mathbf{x}) : i \in I(\mathbf{x})\},$$

where

$$\Psi_{i2}(\mathbf{x}) = w_i\left|\sum_{j=1}^{n}\sum_{k=1}^{n} a_{ij}^{(2k)} x_j^{2k} - a_{i0}\right|,$$

$$I(\mathbf{x}) = \{i \in \{1, \ldots, m\} : \Psi_{i2}(\mathbf{x}) = F_2(\mathbf{x})\}.$$

Similarly, $\mathbf{x}^*$ is an optimal solution to problem (12.46) if and only if

$$\mathbf{0}_n \in \partial F_5(\mathbf{x}^*) = \text{co}\,\{\partial \Psi_{i5}(\mathbf{x}) : i \in I(\mathbf{x})\},$$

where

$$\Psi_{i5}(\mathbf{x}) = w_i \left| \sum_{j=1}^{n} \sum_{k=1}^{n} a_{i,2j} x_k^{2j} - a_{i0} \right|,$$

$$I(\mathbf{x}) = \{i \in \{1, \ldots, m\} : \Psi_{i5}(\mathbf{x}) = F_5(\mathbf{x})\},$$

and $\mathbf{x}^*$ is an optimal solution to problem (12.49) if and only if

$$\mathbf{0}_n \in \partial F_8(\mathbf{x}^*) = \text{co}\,\{\partial \Psi_{i8}(\mathbf{x}) : i \in I(\mathbf{x})\},$$

where

$$\Psi_{i8}(\mathbf{x}) = w_i \left| \sum_{j=1}^{n} a_{ij} x_j^{2i} - a_{i0} \right|,$$

$$I(\mathbf{x}) = \{i \in \{1, \ldots, m\} : \Psi_{i8}(\mathbf{x}) = F_8(\mathbf{x})\},$$

and "co X" is notation of the convex hull (convex envelope) of X.

12.4.3 Calculation of Subgradients and Subdifferentials

In order to apply the subgradient method (12.14) for solving the nondifferentiable problems under consideration, we have to calculate the corresponding subgradients.

Using that $F_1(\mathbf{x}),\ F_2(\mathbf{x}),\ F_4(\mathbf{x}),\ F_5(\mathbf{x}),\ F_7(\mathbf{x}),\ F_8(\mathbf{x})$ are convex separable functions and statements of Theorems 1.27, 1.28, 1.29, 1.34, and 1.35 of Chapter One, we can calculate corresponding subdifferentials (subgradient sets) at iteration κ as follows, respectively.

$$\partial F_1^{(\kappa)}(\mathbf{x}_\kappa) = \sum_{i=1}^{m} \partial \mathbf{g}_i^{(\kappa)}(\mathbf{x}_\kappa) \in \mathbf{R}^n,$$

where

$$\partial \mathbf{g}_i^{(\kappa)}(\mathbf{x}_\kappa) = \begin{cases} w_i.\Big(2a_{i1}^{(2)}x_1^{(\kappa)} + \cdots + 2na_{i1}^{(2n)}(x_1^{(\kappa)})^{2n-1}, \ldots, \\ \quad 2a_{in}^{(2)}x_n^{(\kappa)} + \cdots + 2na_{in}^{(2n)}(x_n^{(\kappa)})^{2n-1}\Big), \\ \quad \text{if } \sum_{j=1}^{n}\sum_{k=1}^{n} a_{ij}^{(2k)}.(x_j^{(\kappa)})^{2k} - a_{i0} > 0, \\ w_i.\Big(2a_{i1}^{(2)}x_1^{(\kappa)} + \cdots + 2na_{i1}^{(2n)}(x_1^{(\kappa)})^{2n-1}, \ldots, \\ \quad 2a_{in}^{(2)}x_n^{(\kappa)} + \cdots + 2na_{in}^{(2n)}(x_n^{(\kappa)})^{2n-1}\Big).[-1, 1], \\ \quad \text{if } \sum_{j=1}^{n}\sum_{k=1}^{n} a_{ij}^{(2k)}.(x_j^{(\kappa)})^{2k} - a_{i0} = 0, \\ -w_i.\Big(2a_{i1}^{(2)}x_1^{(\kappa)} + \cdots + 2na_{i1}^{(2n)}(x_1^{(\kappa)})^{2n-1}, \ldots, \\ \quad 2a_{in}^{(2)}x_n^{(\kappa)} + \cdots + 2na_{in}^{(2n)}(x_n^{(\kappa)})^{2n-1}\Big), \\ \quad \text{if } \sum_{j=1}^{n}\sum_{k=1}^{n} a_{ij}^{(2k)}.(x_j^{(\kappa)})^{2k} - a_{i0} < 0, \end{cases} \tag{12.51}$$

$i = 1, \ldots, m$, and $\mathbf{x}_\kappa = (x_1^{(\kappa)}, \ldots, x_n^{(\kappa)})$.

Let $\max_{1 \le i \le m} w_i \left| \sum_{j=1}^{n}\sum_{k=1}^{n} a_{ij}^{(2k)} x_j^{2k} - a_{i0} \right|$ be attained for the index(es) $i = i(x) \in I(x) \subset \{1, \ldots, m\}$. Then

$$\partial F_2^{(\kappa)}(\mathbf{x}_\kappa) = \text{co}\left\{(\partial g_{i(x),1}^{(\kappa)}(\mathbf{x}_\kappa), \ldots, \partial g_{i(x),n}^{(\kappa)}(\mathbf{x}_\kappa)) : i(x) \in I(x)\right\},$$

where

$$\partial g_{i(x),j}^{(\kappa)}(\mathbf{x}_\kappa) = \begin{cases} w_{i(x)}.\Big(2a_{i(x),j}^{(2)}x_j^{(\kappa)} + \cdots + 2na_{i(x),j}^{(2n)}(x_j^{(\kappa)})^{2n-1}\Big), \\ \quad \text{if } \sum_{j=1}^{n}\sum_{k=1}^{n} a_{i(x),j}^{(2k)}.(x_j^{(\kappa)})^{2k} - a_{i(x),0} > 0, \\ w_{i(x)}.\Big(2a_{i(x),j}^{(2)}x_j^{(\kappa)} + \cdots + 2na_{i(x),j}^{(2n)}(x_j^{(\kappa)})^{2n-1}\Big).[-1, 1], \\ \quad \text{if } \sum_{j=1}^{n}\sum_{k=1}^{n} a_{i(x),j}^{(2k)}.(x_j^{(\kappa)})^{2k} - a_{i(x),0} = 0, \\ -w_{i(x)}.\Big(2a_{i(x),j}^{(2)}x_j^{(\kappa)} + \cdots + 2na_{i(x),j}^{(2n)}(x_j^{(\kappa)})^{2n-1}\Big), \\ \quad \text{if } \sum_{j=1}^{n}\sum_{k=1}^{n} a_{i(x),j}^{(2k)}.(x_j^{(\kappa)})^{2k} - a_{i(x),0} < 0, \end{cases} \tag{12.52}$$

$j = 1, \ldots, n$, and $\mathbf{x}_\kappa = (x_1^{(\kappa)}, \ldots, x_n^{(\kappa)})$.

$$\partial F_4^{(\kappa)}(\mathbf{x}_\kappa) = \sum_{i=1}^{m} \partial \mathbf{g}_i^{(\kappa)}(\mathbf{x}_\kappa) \ \in \mathbb{R}^n,$$

where

$$\partial \mathbf{g}_i^{(\kappa)}(\mathbf{x}_\kappa) = \begin{cases} w_i.\Big(2a_{i2}x_1^{(\kappa)} + \cdots + 2na_{i,2n}(x_1^{(\kappa)})^{2n-1}, \ldots, \\ \quad 2a_{i2}x_n^{(\kappa)} + \cdots + 2na_{i,2n}(x_n^{(\kappa)})^{2n-1}\Big), \\ \quad \text{if } \sum_{j=1}^{n}\sum_{k=1}^{n} a_{i,2j}.(x_k^{(\kappa)})^{2j} - a_{i0} > 0, \\ w_i.\Big(2a_{i2}x_1^{(\kappa)} + \cdots + 2na_{i,2n}(x_1^{(\kappa)})^{2n-1}, \ldots, \\ \quad 2a_{i2}x_n^{(\kappa)} + \cdots + 2na_{i,2n}(x_n^{(\kappa)})^{2n-1}\Big).[-1, 1], \\ \quad \text{if } \sum_{j=1}^{n}\sum_{k=1}^{n} a_{i,2j}.(x_k^{(\kappa)})^{2j} - a_{i0} = 0, \\ -w_i.\Big(2a_{i2}x_1^{(\kappa)} + \cdots + 2na_{i,2n}(x_1^{(\kappa)})^{2n-1}, \ldots, \\ \quad 2a_{i2}x_n^{(\kappa)} + \cdots + 2na_{i,2n}(x_n^{(\kappa)})^{2n-1}\Big), \\ \quad \text{if } \sum_{j=1}^{n}\sum_{k=1}^{n} a_{i,2j}.(x_k^{(\kappa)})^{2j} - a_{i0} < 0, \end{cases} \tag{12.53}$$

$i = 1, \ldots, m$, and $\mathbf{x}_\kappa = (x_1^{(\kappa)}, \ldots, x_n^{(\kappa)})$.

Let $\max_{1 \le i \le m} w_i \left| \sum_{j=1}^{n}\sum_{k=1}^{n} a_{i,2j}x_k^{2j} - a_{i0} \right|$ be attained for the index(es) $i = i(x) \in I(x) \subset \{1, \ldots, m\}$. Then

$$\partial F_5^{(\kappa)}(\mathbf{x}_\kappa) = \text{co}\left\{(\partial g_{i(x),1}^{(\kappa)}(\mathbf{x}_\kappa), \ldots, \partial g_{i(x),n}^{(\kappa)}(\mathbf{x}_\kappa)) : i(x) \in I(x)\right\},$$

where

$$\partial g_{i(x),j}^{(\kappa)}(\mathbf{x}_\kappa) = \begin{cases} w_{i(x)}.\Big(2a_{i(x),2j}x_j^{(\kappa)} + \cdots + 2na_{i(x),2j}(x_j^{(\kappa)})^{2n-1}\Big), \\ \quad \text{if } \sum_{j=1}^{n}\sum_{k=1}^{n} a_{i(x),2j}.(x_k^{(\kappa)})^{2j} - a_{i(x),0} > 0, \\ w_{i(x)}.\Big(2a_{i(x),2j}x_j^{(\kappa)} + \cdots + 2na_{i(x),2j}(x_j^{(\kappa)})^{2n-1}\Big).[-1, 1], \\ \quad \text{if } \sum_{j=1}^{n}\sum_{k=1}^{n} a_{i(x),2j}.(x_k^{(\kappa)})^{2j} - a_{i(x),0} = 0, \\ -w_{i(x)}.\Big(2a_{i(x),2j}x_j^{(\kappa)} + \cdots + 2na_{i(x),2j}(x_j^{(\kappa)})^{2n-1}\Big), \\ \quad \text{if } \sum_{j=1}^{n}\sum_{k=1}^{n} a_{i(x),2j}.(x_k^{(\kappa)})^{2j} - a_{i(x),0} < 0, \end{cases} \tag{12.54}$$

$j = 1, \ldots, n$, and $\mathbf{x}_\kappa = (x_1^{(\kappa)}, \ldots, x_n^{(\kappa)})$.

$$\partial F_7^{(\kappa)}(\mathbf{x}_\kappa) = \sum_{i=1}^{m} \partial \mathbf{g}_i^{(\kappa)}(\mathbf{x}_\kappa) \in \mathbf{R}^n,$$

where

$$\partial \mathbf{g}_i^{(\kappa)}(\mathbf{x}_\kappa) = \begin{cases} w_i.\Big(2ia_{i1}(x_1^{(\kappa)})^{2i-1}, 2ia_{i2}(x_2^{(\kappa)})^{2i-1}, \ldots, 2ia_{in}(x_n^{(\kappa)})^{2i-1}\Big), \\ \quad \text{if } \sum_{j=1}^{n} a_{ij}.(x_j^{(\kappa)})^{2i} - a_{i0} > 0, \\ w_i.\Big(2ia_{i1}(x_1^{(\kappa)})^{2i-1}, 2ia_{i2}(x_2^{(\kappa)})^{2i-1}, \ldots, 2ia_{in}(x_n^{(\kappa)})^{2i-1}\Big) \times \\ \quad \times[-1, 1], \quad \text{if } \sum_{j=1}^{n} a_{ij}.(x_j^{(\kappa)})^{2i} - a_{i0} = 0, \\ -w_i.\Big(2ia_{i1}(x_1^{(\kappa)})^{2i-1}, 2ia_{i2}(x_2^{(\kappa)})^{2i-1}, \ldots, 2ia_{in}(x_n^{(\kappa)})^{2i-1}\Big), \\ \quad \text{if } \sum_{j=1}^{n} a_{ij}.(x_j^{(\kappa)})^{2i} - a_{i0} < 0, \end{cases} \tag{12.55}$$

$i = 1, \ldots, m$, and $\mathbf{x}_\kappa = (x_1^{(\kappa)}, \ldots, x_n^{(\kappa)})$.

Let $\max_{1\le i\le m} w_i \left| \sum_{j=1}^{n} a_{ij}x_j^{2i} - a_{i0} \right|$ be attained for the index(es) $i = i(x) \in I(x) \subset \{1, \ldots, m\}$. Then

$$\partial F_8^{(\kappa)}(\mathbf{x}_\kappa) = \mathrm{co}\left\{\partial g_{i(x)}^{(\kappa)}(\mathbf{x}_\kappa) : i(x) \in I(x)\right\},$$

where

$$\partial g_{i(x)}^{(\kappa)}(\mathbf{x}_\kappa) = \begin{cases} w_{i(x)}.\Big(2i(x).a_{i(x),1}.(x_1^{(\kappa)})^{2i(x)-1}, \ldots, \\ \quad 2i(x).a_{i(x),n}.(x_n^{(\kappa)})^{2i(x)-1}\Big), \\ \quad \text{if } \sum_{j=1}^{n} a_{i(x),j}.(x_j^{(\kappa)})^{2i(x)} - a_{i(x),0} > 0, \\ w_{i(x)}.\Big(2i(x).a_{i(x),1}.(x_1^{(\kappa)})^{2i(x)-1}, \ldots, \\ \quad 2i(x).a_{i(x),n}.(x_n^{(\kappa)})^{2i(x)-1}\Big).[-1, 1], \\ \quad \text{if } \sum_{j=1}^{n} a_{i(x),j}.(x_j^{(\kappa)})^{2i(x)} - a_{i(x),0} = 0, \\ -w_{i(x)}.\Big(2i(x).a_{i(x),1}.(x_1^{(\kappa)})^{2i(x)-1}, \ldots, \\ \quad 2i(x).a_{i(x),n}.(x_n^{(\kappa)})^{2i(x)-1}\Big), \\ \quad \text{if } \sum_{j=1}^{n} a_{i(x),j}.(x_j^{(\kappa)})^{2i(x)} - a_{i(x),0} < 0, \end{cases} \tag{12.56}$$

$\mathbf{x}_\kappa = (x_1^{(\kappa)}, \dots, x_n^{(\kappa)})$.

It is clear that elements of $\partial F_1^{(\kappa)}(\mathbf{x}_\kappa)$, $\partial F_2^{(\kappa)}(\mathbf{x}_\kappa)$, $\partial F_4^{(\kappa)}(\mathbf{x}_\kappa)$, $\partial F_5^{(\kappa)}(\mathbf{x}_\kappa)$, $\partial F_7^{(\kappa)}(\mathbf{x}_\kappa)$, $\partial F_8^{(\kappa)}(\mathbf{x}_\kappa)$ depend on the sign of the corresponding expression from (12.51), (12.52), (12.53), (12.54), (12.55), (12.56), respectively, and therefore, they depend on the current values $x_j^{(\kappa)}$, $j = 1, 2, \dots, n$, respectively.

As in Subsection 12.3.2, we can choose, for example, $\rho_0 = 1$; $\rho_\kappa = \frac{c}{\kappa}$, $\kappa = 1, 2, \dots$; $c = const. > 0$. The requirements for the step size are satisfied for this choice of ρ_κ.

12.4.4 Calculation of Gradients

Gradients of differentiable functions $F_3^2(\mathbf{x})$, $F_6^2(\mathbf{x})$ and $F_9^2(\mathbf{x})$, respectively, at iteration κ are

$$[(F_3^2)']^{(\kappa)}(\mathbf{x}_\kappa) = (g_1^{(\kappa)}(\mathbf{x}_\kappa), \dots, g_n^{(\kappa)}(\mathbf{x}_\kappa)),$$

where

$$g_l^{(\kappa)}(\mathbf{x}_\kappa) = 2\sum_{i=1}^{m} w_i \left(\sum_{j=1}^{n} \sum_{k=1}^{n} a_{ij}^{(2k)} (x_j^{(\kappa)})^{2k} - a_{i0} \right) .2k.a_{il}^{(2k)}.(x_l^{(\kappa)})^{2k-1},$$

$l = 1, \dots, n$, and $\mathbf{x}_\kappa = (x_1^{(\kappa)}, \dots, x_n^{(\kappa)})$;

$$[(F_6^2)']^{(\kappa)}(\mathbf{x}_\kappa) = (g_1^{(\kappa)}(\mathbf{x}_\kappa), \dots, g_n^{(\kappa)}(\mathbf{x}_\kappa)),$$

where

$$g_l^{(\kappa)}(\mathbf{x}_\kappa) = 2\sum_{i=1}^{m} w_i \left(\sum_{j=1}^{n} \sum_{k=1}^{n} a_{i,2j} (x_k^{(\kappa)})^{2j} - a_{i0} \right) .2j.a_{i,2j}.(x_l^{(\kappa)})^{2j-1},$$

$l = 1, \dots, n$, and $\mathbf{x}_\kappa = (x_1^{(\kappa)}, \dots, x_n^{(\kappa)})$;

$$[(F_9^2)']^{(\kappa)}(\mathbf{x}_\kappa) = (g_1^{(\kappa)}(\mathbf{x}_\kappa), \dots, g_n^{(\kappa)}(\mathbf{x}_\kappa)),$$

where

$$g_l^{(\kappa)}(\mathbf{x}_\kappa) = 2\sum_{i=1}^{m} w_i \left(\sum_{j=1}^{n} a_{ij} (x_j^{(\kappa)})^{2i} - a_{i0} \right) .2i.a_{il}.(x_l^{(\kappa)})^{2i-1},$$

$l = 1, \dots, n$, and $\mathbf{x}_\kappa = (x_1^{(\kappa)}, \dots, x_n^{(\kappa)})$.

12.5 Solution of Systems of Nonlinear Equations Defined by Convex Functions

In this section, the problem of numerical solution of systems of nonlinear equations, defined by convex differentiable functions, is studied. Similarly to the approach of Section 12.1 and Section 12.4, nonsmooth (nondifferentiable) unconstrained minimization problems are associated with the system, with objective functions, which are based on the weighted discrete ℓ_1- and ℓ_∞-norm, respectively, that is, using these two norms as proximity criteria. Also, a "differentiable" unconstrained minimization problem (least squares data fitting problem) is associated with the considered system of nonlinear equations, with objective function, which is based on the weighted discrete ℓ_2-norm. The special case when nonlinear equations are defined by separable convex differentiable functions is also studied. The considered systems of nonlinear equations are solved approximately, in the case of ℓ_1-norm, by minimizing the sum of absolute values of residuals, and, in the case of ℓ_∞-norm, by minimizing the maximal residual, respectively. In the case of ℓ_2-norm, the systems of nonlinear equations are solved by minimizing the sum of squares of the residuals. Subgradients of the objective functions of the associated unconstrained minimization problems, as well as gradients of the objective functions of the corresponding "differentiable" unconstrained minimization problems, are calculated.

The structure of this section is similar to the structure of Section 12.4.

12.5.1 Systems of Nonlinear Equations Defined by Convex Differentiable Functions

Consider the system of nonlinear equations

$$\begin{aligned} f_1(x_1, \ldots, x_n) &= a_1, \\ \ldots&\ldots \\ f_m(x_1, \ldots, x_n) &= a_m, \end{aligned} \tag{12.57}$$

that is

$$f_i(x_1, \ldots, x_n) = a_i, \quad i = 1, \ldots, m, \tag{12.57'}$$

where $f_1, f_2, \ldots, f_m$ are nonlinear convex differentiable functions with nonnegative values of n variables $x_1, \ldots, x_n$, and $m > n$ in general. Denote the $n-$tuple of these variables as the $n-$vector $\mathbf{x} = (x_1, \ldots, x_n)$.

The following unconstrained minimization problems can be associated with system (12.57):

$$\min \left\{ G_1(\mathbf{x}) \stackrel{\text{def}}{=} \sum_{i=1}^{m} w_i |f_i(x_1, \ldots, x_n) - a_i| \right\} \tag{12.58}$$

and

$$\min \left\{ G_2(\mathbf{x}) \stackrel{\text{def}}{=} \max_{1 \le i \le m} w_i \left| f_i(x_1, \dots, x_n) - a_i \right| \right\}. \tag{12.59}$$

Problem (12.58) is based on the weighted discrete ℓ_1-norm (12.2) and problem (12.59) can be considered as a special case of the problem of best Chebyshev approximation, which is based on the weighted discrete ℓ_∞-norm (12.3).

The corresponding "differentiable" discrete least squares data fitting problem, which can be associated with system (12.57) and which is based on the weighted discrete ℓ_2-norm (12.4), is

$$\min \left\{ G_3(\mathbf{x}) \stackrel{\text{def}}{=} \left(\sum_{i=1}^{m} w_i \left(f_i(x_1, \dots, x_n) - a_i \right)^2 \right)^{\frac{1}{2}} \right\}. \tag{12.60}$$

Weights $w_i > 0,\ i = 1, \dots, m$, again mean the "reliability", with which data of each equation of the nonlinear system (12.57) can be accepted. In the case when all equations are equally reliable, we accept, for example, $w_i = 1,\ i = 1, \dots, m$.

12.5.2 Systems of Nonlinear Equations Defined by Separable Convex Differentiable Functions

Consider the following system of nonlinear equations defined by separable convex differentiable functions

$$\begin{aligned} \sum_{j=1}^{n} f_{1j}(x_j) &= a_1, \\ \sum_{j=1}^{n} f_{2j}(x_j) &= a_2, \\ \dots\dots&\dots\dots \\ \sum_{j=1}^{n} f_{mj}(x_j) &= a_m, \end{aligned} \tag{12.61}$$

that is,

$$\sum_{j=1}^{n} f_{ij}(x_j) = a_i, \quad i = 1, \dots, m, \tag{12.61$'$}$$

where $f_{ij}(x_j) \ge 0,\ i = 1, \dots, m;\ j = 1, \dots, n$ are convex differentiable functions of $x_j,\ j = 1, \dots, n$, respectively, and $m > n$ in general.

The following unconstrained minimization problems can be associated with system (12.61):

$$\min \left\{ G_4(\mathbf{x}) \stackrel{\text{def}}{=} \sum_{i=1}^{m} w_i \left| \sum_{j=1}^{n} f_{ij}(x_j) - a_i \right| \right\} \tag{12.62}$$

and

$$\min \left\{ G_5(\mathbf{x}) \stackrel{\text{def}}{=} \max_{1 \le i \le m} w_i \left| \sum_{j=1}^{n} f_{ij}(x_j) - a_i \right| \right\}. \tag{12.63}$$

Problem (12.62) is based on the weighted discrete ℓ_1-norm (12.2), and problem (12.63) is a problem of best Chebyshev approximation, based on the weighted discrete ℓ_∞-norm (12.3).

The corresponding "differentiable" discrete least squares data fitting problem, which can be associated with system (12.61) and which is based on the weighted discrete ℓ_2-norm (12.4), is

$$\min \left\{ G_6(\mathbf{x}) \stackrel{\text{def}}{=} \left(\sum_{i=1}^{m} w_i \left(\sum_{j=1}^{n} f_{ij}(x_j) - a_i \right)^2 \right)^{\frac{1}{2}} \right\}. \tag{12.64}$$

12.5.3 Solvability of the Considered Problems

Under the assumptions, $f_i(x_1, \dots, x_n) \ge 0,\ i = 1, \dots, m$; functions $f_i(x_1, \dots, x_n) - a_i,\ i = 1, \dots, m$, are convex functions of $x_j,\ j = 1, \dots, n$; the "absolute value" function is also convex. Then $G_1(\mathbf{x})$ is a convex function of $\mathbf{x}$ as a linear combination with nonnegative (even positive) coefficients $w_i,\ i = 1, \dots, m$, of convex functions (Theorem 1.17).

According to Theorem 1.20, $G_2(\mathbf{x})$ is a convex function of $\mathbf{x}$ as maximum of the convex functions $w_i |f_i(x_1, \dots, x_n) - a_i|$, where $w_i > 0, i = 1, \dots, m$.

Similarly, functions $G_4(\mathbf{x})$ and $G_5(\mathbf{x})$ are convex functions of $\mathbf{x}$.

Since min $G_3(\mathbf{x})$ and min $G_3^2(\mathbf{x})$ are attained at the same point (vector) $\mathbf{x} = (x_1, \dots, x_n)$, we can consider problem min $G_3^2(\mathbf{x})$ instead of problem min $G_3(\mathbf{x})$.

Analogously, problem min $G_6^2(\mathbf{x})$ instead of problem min $G_6(\mathbf{x})$ can be considered.

Function $G_3^2(\mathbf{x})$ is a strictly convex function of $\mathbf{x}$ as a linear combination with nonnegative (positive) coefficients $w_i,\ i = 1, \dots, m$, of the strictly convex quadratic functions $(f_i(x_1, \dots, x_n) - a_i)^2,\ i = 1, \dots, m$.

Similarly, function $G_6^2(\mathbf{x})$ is a strictly convex function of $\mathbf{x}$ as a linear combination with nonnegative (positive) coefficients $w_i,\ i = 1, \dots, m$, of the strictly convex quadratic functions $\left(\sum_{j=1}^{n} f_{ij}(x_j) - a_i \right)^2,\ i = 1, \dots, m$.

Functions $G_1(\mathbf{x}),\ G_2(\mathbf{x}),\ G_4(\mathbf{x}),\ G_5(\mathbf{x})$ are nondifferentiable (nonsmooth), whereas functions $G_3(\mathbf{x})$ and $G_6(\mathbf{x})$ are differentiable.

All functions $G_s(\mathbf{x}),\ s = 1, 2, 3, 4, 5, 6$ are *separable functions* because they are expressed as the sums of single-variable (univariate) functions, which follows from definitions of these functions.

Since $G_1(\mathbf{x}) \geq 0$, $G_2(\mathbf{x}) \geq 0$ are continuous functions, bounded from below from zero; $f_i(x_1, \ldots, x_n)$ and $a_i,\ i = 1, \ldots, m$, are given by system (12.57), it follows that $G_1(\mathbf{x}) \to +\infty$ and $G_2(\mathbf{x}) \to +\infty$ when $\|\mathbf{x}\| \to \infty$. Then Corollary C.1 (Appendix C) implies that problems (12.58) and (12.59) are solvable.

Similarly, problems (12.62) and (12.63) are solvable.

In addition, each of problems min $G_3^2(\mathbf{x})$ obtained from (12.60) and min $G_6^2(\mathbf{x})$ obtained from (12.64) has also an optimal solution and it is unique according to Theorem 1.38 because $G_3^2(\mathbf{x}) \geq 0$ and $G_6^2(\mathbf{x}) \geq 0$ are bounded from below from zero and they are strictly convex functions.

Theorem 1.29 implies that $\mathbf{x}^*$ is an optimal solution of problem (12.59) if and only if

$$\mathbf{0}_n \in \partial G_2(\mathbf{x}^*) = \text{co}\, \{\partial \Omega_{i2}(\mathbf{x}) : i \in I(\mathbf{x})\},$$

where

$$\Omega_{i2}(\mathbf{x}) = w_i |f_i(x_1, \ldots, x_n) - a_i|,$$

$$I(\mathbf{x}) = \{i \in \{1, \ldots, m\} : \Omega_{i2}(\mathbf{x}) = G_2(\mathbf{x})\},$$

and "co X" denotes the convex hull (convex envelope) of X.

A similar remark is also valid for problem (12.63).

Theorem D.1 and Theorem D.3 of Appendix D state existence and uniqueness of element of best approximation, respectively.

12.5.4 Calculation of Subgradients and Subdifferentials

Implementation of the subgradient method to the considered nondifferentiable problems requires the corresponding subgradients.

Since $G_1(\mathbf{x})$, $G_2(\mathbf{x})$, $G_4(\mathbf{x})$ and $G_5(\mathbf{x})$ are convex separable functions, using statements of Theorems 1.27, 1.28, 1.29, 1.34, and 1.35 of Chapter One, corresponding subdifferentials (subgradient sets) at iteration No. k can be calculated as follows, respectively.

$$\partial G_1^{(k)}(\mathbf{x}_k) = \sum_{i=1}^{m} \partial \mathbf{f}_i^{(k)}(\mathbf{x}_k) \in \mathbf{R}^n,$$

where

$$\partial \mathbf{f}_i^{(k)}(\mathbf{x}_k) = \begin{cases} w_i.\Big(\dfrac{\partial f_i}{\partial x_1}(\mathbf{x}_k), \ldots, \dfrac{\partial f_i}{\partial x_n}(\mathbf{x}_k)\Big), \\ \quad \text{if} \quad f_i(x_1^{(k)}, \ldots, x_n^{(k)}) - a_i > 0, \\ w_i.\Big(\dfrac{\partial f_i}{\partial x_1}(\mathbf{x}_k), \ldots, \dfrac{\partial f_i}{\partial x_n}(\mathbf{x}_k)\Big).[-1, 1], \\ \quad \text{if} \quad f_i(x_1^{(k)}, \ldots, x_n^{(k)}) - a_i = 0, \\ -w_i.\Big(\dfrac{\partial f_i}{\partial x_1}(\mathbf{x}_k), \ldots, \dfrac{\partial f_i}{\partial x_n}(\mathbf{x}_k)\Big), \\ \quad \text{if} \quad f_i(x_1^{(k)}, \ldots, x_n^{(k)}) - a_i < 0, \end{cases} \tag{12.65}$$

$i = 1, \ldots, m$, and $\mathbf{x}_k = (x_1^{(k)}, \ldots, x_n^{(k)})$.

Let $\max\limits_{1 \le i \le m} w_i |f_i(x_1, \ldots, x_n) - a_i|$ be attained for the index(es) $i = i(x) \in I(x) \subset \{1, \ldots, m\}$. Then

$$\partial G_2^{(k)}(\mathbf{x}_k) = \text{co}\left\{\partial f_{i(x)}^{(k)}(\mathbf{x}_k) : i(x) \in I(x)\right\},$$

where

$$\partial f_{i(x)}^{(k)}(\mathbf{x}_k) = \begin{cases} w_{i(x)}.\Big(\dfrac{\partial f_{i(x)}}{\partial x_1}(\mathbf{x}_k), \ldots, \dfrac{\partial f_{i(x)}}{\partial x_n}(\mathbf{x}_k)\Big), \\ \quad \text{if} \quad f_{i(x)}(x_1^{(k)}, \ldots, x_n^{(k)}) - a_{i(x)} > 0, \\ w_{i(x)}.\Big(\dfrac{\partial f_{i(x)}}{\partial x_1}(\mathbf{x}_k), \ldots, \dfrac{\partial f_{i(x)}}{\partial x_n}(\mathbf{x}_k)\Big).[-1, 1], \\ \quad \text{if} \quad f_{i(x)}(x_1^{(k)}, \ldots, x_n^{(k)}) - a_{i(x)} = 0, \\ -w_{i(x)}.\Big(\dfrac{\partial f_{i(x)}}{\partial x_1}(\mathbf{x}_k), \ldots, \dfrac{\partial f_{i(x)}}{\partial x_n}(\mathbf{x}_k)\Big), \\ \quad \text{if} \quad f_{i(x)}(x_1^{(k)}, \ldots, x_n^{(k)}) - a_{i(x)} < 0, \end{cases} \tag{12.66}$$

$\mathbf{x}_k = (x_1^{(k)}, \ldots, x_n^{(k)})$.

$$\partial G_4^{(k)}(\mathbf{x}_k) = \sum_{i=1}^{m} \partial \mathbf{f}_i^{(k)}(\mathbf{x}_k) \in \mathbf{R}^n,$$

where

$$\partial \mathbf{f}_i^{(k)}(\mathbf{x}_k) = \begin{cases} w_i.\Big(\dfrac{\partial f_{i1}}{\partial x_1}(x_1^{(k)}), \dots, \dfrac{\partial f_{in}}{\partial x_n}(x_n^{(k)})\Big), \\ \quad \text{if } \displaystyle\sum_{j=1}^{n} f_{ij}(x_j^{(k)}) - a_i > 0, \\ w_i.\Big(\dfrac{\partial f_{i1}}{\partial x_1}(x_1^{(k)}), \dots, \dfrac{\partial f_{in}}{\partial x_n}(x_n^{(k)})\Big).[-1, 1], \\ \quad \text{if } \displaystyle\sum_{j=1}^{n} f_{ij}(x_j^{(k)}) - a_i = 0, \\ -w_i.\Big(\dfrac{\partial f_{i1}}{\partial x_1}(x_1^{(k)}), \dots, \dfrac{\partial f_{in}}{\partial x_n}(x_n^{(k)})\Big), \\ \quad \text{if } \displaystyle\sum_{j=1}^{n} f_{ij}(x_j^{(k)}) - a_i < 0, \end{cases} \tag{12.67}$$

$i = 1, \dots, m$, and $\mathbf{x}_k = (x_1^{(k)}, \dots, x_n^{(k)})$.

Let $\max\limits_{1 \le i \le m} w_i \left| \sum\limits_{j=1}^{n} f_{ij}(x_j) - a_i \right|$ be attained for the index(es) $i = i(x) \in I(x) \subset \{1, \dots, m\}$. Then

$$\partial G_5^{(k)}(\mathbf{x}_k) = \text{co}\left\{\partial f_{i(x)}^{(k)}(\mathbf{x}_k) : i(x) \in I(x)\right\},$$

where

$$\partial f_{i(x)}^{(k)}(\mathbf{x}_k) = \begin{cases} w_{i(x)}.\Big(\dfrac{\partial f_{i(x),1}}{\partial x_1}(x_1^{(k)}), \dots, \dfrac{\partial f_{i(x),n}}{\partial x_n}(x_n^{(k)})\Big), \\ \quad \text{if } \displaystyle\sum_{j=1}^{n} f_{i(x),j}(x_j^{(k)}) - a_{i(x)} > 0, \\ w_{i(x)}.\Big(\dfrac{\partial f_{i(x),1}}{\partial x_1}(x_1^{(k)}), \dots, \dfrac{\partial f_{i(x),n}}{\partial x_n}(x_n^{(k)})\Big).[-1, 1], \\ \quad \text{if } \displaystyle\sum_{j=1}^{n} f_{i(x),j}(x_j^{(k)}) - a_{i(x)} = 0, \\ -w_{i(x)}.\Big(\dfrac{\partial f_{i(x),1}}{\partial x_1}(x_1^{(k)}), \dots, \dfrac{\partial f_{i(x),n}}{\partial x_n}(x_n^{(k)})\Big), \\ \quad \text{if } \displaystyle\sum_{j=1}^{n} f_{i(x),j}(x_j^{(k)}) - a_{i(x)} < 0, \end{cases} \tag{12.68}$$

$\mathbf{x}_k = (x_1^{(k)}, \dots, x_n^{(k)})$.

Evidently elements of $\partial G_1^{(k)}(\mathbf{x}_k)$, $\partial G_2^{(k)}(\mathbf{x}_k)$, $\partial G_4^{(k)}(\mathbf{x}_k)$ and $\partial G_5^{(k)}(\mathbf{x}_k)$ depend on the sign of the corresponding expression from (12.65), (12.66), (12.67), (12.68), respectively, and therefore, they depend on the current values $x_j^{(k)}$, $j = 1, \dots, n$, respectively.

The step size ρ_k is chosen as in Subsections 12.3.2 and 12.4.3: $\rho_0 = 1$; $\rho_k = \frac{c}{k}$, $k = 1, \dots$; $c = const. > 0$.

12.5.5 Calculation of Gradients

Gradients of the differentiable functions $G_3^2(\mathbf{x})$ and $G_6^2(\mathbf{x})$ at iteration No. k can be calculated as follows, respectively

$$[(G_3^2)']^{(k)}(\mathbf{x}_k) = (g_1^{(k)}(\mathbf{x}_k), \dots, g_n^{(k)}(\mathbf{x}_k)),$$

where

$$g_l^{(k)}(\mathbf{x}_k) = 2\sum_{i=1}^{m} w_i \Big(f_i(x_1^{(k)}, \dots, x_n^{(k)}) - a_i \Big) . \frac{\partial f_l}{\partial x_l}(\mathbf{x}_k), \quad l = 1, \dots, n,$$

$\mathbf{x}_k = (x_1^{(k)}, \dots, x_n^{(k)})$;

$$[(G_6^2)']^{(k)}(\mathbf{x}_k) = (g_1^{(k)}(\mathbf{x}_k), \dots, g_n^{(k)}(\mathbf{x}_k)),$$

where

$$g_l^{(k)}(\mathbf{x}_k) = 2\sum_{i=1}^{m} w_i \Big(\sum_{j=1}^{n} f_{ij}(x_j^{(k)}) - a_i \Big) . \frac{\partial f_{il}}{\partial x_l}(x_l^{(k)}), \quad l = 1, \dots, n,$$

$\mathbf{x}_k = (x_1^{(k)}, \dots, x_n^{(k)})$.

12.6 Computational Experiments and Conclusions

In this section, some computational results, obtained by the subgradient method for problems (12.5), (12.6), (12.10), and (12.11), are presented. As it was pointed out, only for comparison, the results obtained by the gradient method for solving the least squares problems (12.7) and (12.12) are also presented. Parameters and data have been randomly generated. The computations have been performed on an Intel Pentium(R) Dual-Core CPU E5800 3.20 GHz/2.00GB. Each problem was run 30 times.

For both methods (12.14) and (12.25), two termination tests are used: an "accuracy" stopping criterion

$$\|\mathbf{x}_{k+1} - \mathbf{x}_k\| \equiv \left\{ \begin{array}{l} \|\rho_k \gamma_k \widehat{\mathbf{f}}^{(k)}(\mathbf{x}_k)\| \\ \|\rho_k \mathbf{f}'(\mathbf{x}_k)\| \end{array} \right\} < \varepsilon,$$

where $\varepsilon > 0$ is some tolerance value, and an upper limit criterion on the number of iterations.

Example 12.1 (Problem (12.1))

$$m = 26, \ n = 5, \ p = 10, \ \varepsilon = 0.001.$$

Results

By method (12.14) for problem (12.5)	By method (12.14) for problem (12.6)	By method (12.25) for problem (12.7)
Iterations 101 Run time 0.00046 s	Iterations 98 Run time 0.00054 s	Iterations 97 Run time 0.00037 s

Example 12.2 (Problem (12.1))

$$m = 20, \ n = 5, \ p = 10, \ \varepsilon = 0.001.$$

Results

By method (12.14) for problem (12.5)	By method (12.14) for problem (12.6)	By method (12.25) for problem (12.7)
Iterations 103 Run time 0.00374 s	Iterations 103 Run time 0.00391 s	Iterations 96 Run time 0.00373

Example 12.3 (Problem (12.1))

$$m = 18, \ n = 4, \ p = 15, \ \varepsilon = 0.001.$$

Results

By method (12.14) for problem (12.5)	By method (12.14) for problem (12.6)	By method (12.25) for problem (12.7)
Iterations 100 Run time 0.00134 s	Iterations 105 Run time 0.00191 s	Iterations 82 Run time 0.00165 s

Example 12.4 (Problem (12.9))

$$m = 26,\ n = 20,\ \varepsilon = 0.001.$$

Results

By method (12.14) for problem (12.10)	By method (12.14) for problem (12.11)	By method (12.25) for problem (12.12)
Iterations 100 Run time 0.0046 s	Iterations 104 Run time 0.0054 s	Iterations 108 Run time 0.0037 s

Example 12.5 (Problem (12.9))

$$m = 28,\ n = 22,\ \varepsilon = 0.001.$$

Results

By method (12.14) for problem (12.10)	By method (12.14) for problem (12.11)	By method (12.25) for problem (12.12)
Iterations 108 Run time 0.0372 s	Iterations 118 Run time 0.0319 s	Iterations 111 Run time 0.0073 s

Example 12.6 (Problem (12.9))

$$m = 30,\ n = 28,\ \varepsilon = 0.001.$$

Results

By method (12.14) for problem (12.10)	By method (12.14) for problem (12.11)	By method (12.25) for problem (12.12)
Iterations 102 Run time 0.0011 s	Iterations 119 Run time 0.0008 s	Iterations 101 Run time 0.0010 s

Example 12.7 (Problem (12.1))

$x \in \mathbf{R}^1$	-2	-1	0	1	2	3
$f(x)$	-4	15	1	10	7	6

$$w_i = 1,\ \ i = 1, \ldots, 6; \quad n = 2.$$

Results

By method (12.14) for problem (12.5)	By method (12.25) for problem (12.7)
$a_0 = 1.27$	$a_0 = 7.904$
$a_1 = 2.21$	$a_1 = 1.986$
$a_2 = -0.22$	$a_2 = -0.975$
$\Phi_1(\mathbf{a}) = 25.4635$	$\Phi_3(\mathbf{a}) = 12.9625$
Iterations 101	Iterations 106
Run time 0.00134 s	Run time 0.00191 s

Algebraic polynomials obtained by the two methods are

$$P_2^{(1)}(x) = -0.22x^2 + 2.21x + 1.27$$

and

$$P_2^{(2)}(x) = -0.975x^2 + 1.986x + 7.904,$$

respectively.

Example 12.8 (Problem (12.9))

$$\begin{aligned} x_1 + x_2 \phantom{{}+ x_3} &= 1 \\ x_2 + x_3 &= 1 \\ x_1 \phantom{{}+ x_2} + x_3 &= 1 \\ x_1 + x_2 + x_3 &= 1. \end{aligned}$$

Results

By method (12.14) for problem (12.10)	By method (12.14) for problem (12.11)	By method (12.25) for problem (12.12)
$x_1 = 0.3945$	$x_1 = 0.4999$	$x_1 = 0.4261$
$x_2 = 0.4016$	$x_2 = 0.4999$	$x_2 = 0.4261$
$x_3 = 0.3946$	$x_3 = 0.4999$	$x_3 = 0.4261$
$\Phi_4(\mathbf{x}) = 0.2107$	$\Phi_5(\mathbf{x}) = 0.500$	$\Phi_6(\mathbf{x}) = 0.3780$
Iterations 101	Iterations 84	Iterations 18
Run time 0.0011 s	Run time 0.0008 s	Run time 0.0105 s

Computational experiments presented above, as well as many other experiments, allow us to conclude that the subgradient method (12.14), applied for nondifferentiable functions $\Phi_1(\mathbf{a})$, $\Phi_2(\mathbf{a})$, $\Phi_4(\mathbf{x})$, $\Phi_5(\mathbf{x})$, is computationally comparable with the gradient method (12.25), applied to "differentiable" problems (12.7′) and (12.12′), respectively. In some examples, the gradient method gives better results with respect to number of iterations, and therefore, run time. However, in many cases, it is preferable to approximate with respect to either ℓ_1-norm (12.2) or ℓ_∞-norm (12.3) instead of using the ℓ_2-approximation.

Chapter 13
About Projections in the Implementation of Stochastic Quasigradient Methods to Some Probabilistic Inventory Control Problems

In this chapter, probabilistic inventory control problems are considered. After a brief description of stochastic quasigradient methods (SQMs) for solving stochastic programming problems, the algorithms suggested in Part Two of this book are implemented for projecting the current approximation, generated by the SQM, onto feasible sets of two important inventory models. Some examples and results of computational experiments are also presented. A stochastic version of the problem of best Chebyshev approximation is formulated and the corresponding stochastic quasigradient is calculated.

13.1 Introduction

Consider the following probabilistic inventory control model. Determine the quantities x_j of inventory units (facilities) at locations (points) j in order to meet the demand $\omega_j,\ j = 1, \ldots, n$, respectively. Since the demand $\boldsymbol{\omega} = (\omega_1, \ldots, \omega_n)$ is random, we know only its distribution function $H(\overline{\boldsymbol{\omega}}) = \mathbf{P}\{\omega_1 \leq \overline{\omega_1}, \ldots, \omega_n \leq \overline{\omega_n}\}$. At the moment of decision-making concerning $\mathbf{x} = (x_1, \ldots, x_n)$, the actual value of the demand $\boldsymbol{\omega} = (\omega_1, \ldots, \omega_n)$ is not known.

Suppose that we have made a decision $\mathbf{x}$ about the quantities of materials, facilities, etc., and that the actual demand turned out to be $\boldsymbol{\omega}$. We have to pay for both oversupply and shortfalls. The penalty charged at location j is $\psi_1^j(\omega_j - x_j)$ if $\omega_j \geq x_j$, and $\psi_2^j(x_j - \omega_j)$ if $\omega_j < x_j$, where the functions ψ_1^j and ψ_2^j are nondecreasing. In the simplest case, these functions are linear and the total penalty is

$$f(\mathbf{x}, \boldsymbol{\omega}) = \sum_{j=1}^{n} \max \{p_j(x_j - \omega_j), q_j(\omega_j - x_j)\}, \tag{13.1}$$

© The Author(s), under exclusive license to Springer Nature Switzerland AG 2021
S. M. Stefanov, *Separable Optimization*, Springer Optimization and Its Applications 177,
https://doi.org/10.1007/978-3-030-78401-0_13

where $p_j \geq 0$ and $q_j \geq 0$, $j = 1, \ldots, n$ are the unit holding cost (the cost of maintaining unit of inventory j in stock) and unit shortage cost (the penalty for unit of inventory j when we run out of stock), respectively.

In most cases, $\mathbf{x}$ must be determined such that the average penalty is minimal, that is, we have to minimize the function

$$F(\mathbf{x}) = \mathbf{E}_\omega f(\mathbf{x}, \boldsymbol{\omega})$$

$$= \mathbf{E}_\omega \left\{ \sum_{j=1}^{n} \left(p_j \int_0^{x_j} (x_j - \omega_j)\mathbf{P}(d\omega_j) + q_j \int_{x_j}^{\infty} (\omega_j - x_j)\mathbf{P}(d\omega_j) \right) \right\},$$

where $\mathbf{E}_\omega$ denotes the mathematical expectation with respect to $\boldsymbol{\omega}$.

Often there are some constraints on $\mathbf{x}$.

If the warehouse capacity we can use to store facilities is α and we have to order at least a_j and at most b_j units of product j, $j = 1, \ldots, n$, we obtain the following minimization problem:

Find $\mathbf{x} = (x_1, \ldots, x_n)$ satisfying

$$\min_{\mathbf{x}} \left\{ F(\mathbf{x}) = \mathbf{E}_\omega f(\mathbf{x}, \boldsymbol{\omega}) \right\} \tag{13.2}$$

subject to

$$\mathbf{x} \in X \tag{13.3}$$

where X is defined through

$$\sum_{j=1}^{n} d_j x_j \leq \alpha, \quad d_j > 0, \ j = 1, \ldots, n \tag{13.4}$$

$$a_j \leq x_j \leq b_j, \quad j = 1, \ldots, n \tag{13.5}$$

or

$$\sum_{j=1}^{n} d_j x_j = \alpha, \quad d_j > 0, \ j = 1, \ldots, n \tag{13.6}$$

$$a_j \leq x_j \leq b_j, \quad j = 1, \ldots, n. \tag{13.7}$$

Relations "≤" and "=" in (13.4) and (13.6) mean "warehouse may not be completely filled" and "capacity of warehouse must be completely used", respectively.

Here ω_j are random variables in closed segments $[R1_j, R2_j]$, $j = 1, \ldots, n$, respectively.

Function $F(\mathbf{x})$, defined by (13.2), can be presented as follows:

$$F(\mathbf{x}) = \mathbf{E}_\omega f(\mathbf{x}, \boldsymbol{\omega}) = \mathbf{E}_\omega \sum_{j=1}^{n} f_j(x_j, \omega_j), \tag{13.2$'$}$$

where

$$f_j(x_j, \omega_j) = \begin{cases} p_j(x_j - \omega_j), & \text{if } x_j \geq \omega_j \\ q_j(\omega_j - x_j), & \text{if } x_j < \omega_j. \end{cases}$$

Since $f(\mathbf{x}, \boldsymbol{\omega})$ is a nondifferentiable (nonsmooth) function at $\mathbf{x} = \boldsymbol{\omega}$, then $F(\mathbf{x})$ is also a nondifferentiable function. Moreover, function $f(\mathbf{x}, \boldsymbol{\omega})$ is separable (Definition 2.1, Chapter Two) because it is expressed as the sum of univariate functions $f_j(x_j, \omega_j)$, $j = 1, \ldots, n$, depending on a single pair of arguments (x_j, ω_j), $j = 1, \ldots, n$. Therefore, problem (13.2)–(13.3) is a *stochastic separable nondifferentiable optimization problem.*

Problem (13.2)–(13.3) is known as the *inventory control problem* or as the *multi-item (multy-commodity, multi-) facility location problem.* It is a special case of a *stochastic programming problem*, that is, a problem of the form

Find $\mathbf{x} = (x_1, \ldots, x_n)$ such that

$$\min_{\mathbf{x}} \left\{ F^0(\mathbf{x}) \equiv \mathbf{E}_\omega \, f^0(\mathbf{x}, \boldsymbol{\omega}) = \int f^0(\mathbf{x}, \boldsymbol{\omega}) \mathbf{P}(d\boldsymbol{\omega}) \right\}$$

subject to

$$F^i(\mathbf{x}) \equiv \mathbf{E}_\omega \, f^i(\mathbf{x}, \boldsymbol{\omega}) = \int f^i(\mathbf{x}, \boldsymbol{\omega}) \mathbf{P}(d\boldsymbol{\omega}) \leq 0, \; i = 1, \ldots, m$$

$$\mathbf{x} \in X \subseteq \mathbf{R}^n.$$

Functions $F^i(\mathbf{x})$, $i = 0, 1, \ldots, m$ are called *regression functions*.

13.2 Stochastic Quasigradient Methods

Stochastic quasigradient methods (SQMs) for solving stochastic optimization problems have been suggested by Yu. Ermoliev (see, for example, Ermoliev [44], Ermoliev and Gaivoronski [45], Gaivoronski [50]).

The main idea of SQMs is outlined below.

Given the problem

$$\min F(\mathbf{x})$$

subject to

$$\mathbf{x} \in X$$

where F is a convex function and X is a "deterministic" set, that is, a set defined by "deterministic" functions.

SQMs are defined through

$$\mathbf{x}_{k+1} = \Pi_X \left(\mathbf{x}_k - \rho_k \boldsymbol{\xi}^k \right), \quad k = 0, 1, \ldots, \tag{13.8}$$

where

- $\mathbf{x}_0$ is an arbitrary initial guess (initial approximation);
- $\Pi_X(\mathbf{y})$ is a projection operation of $\mathbf{y}$ onto the feasible region X;
- ρ_k is a step size;
- $\boldsymbol{\xi}^k = \boldsymbol{\xi}^k(\boldsymbol{\omega})$ is a step direction, $\boldsymbol{\xi}^k(\boldsymbol{\omega})$ is a random vector such that

$$\mathbf{E}_\omega(\boldsymbol{\xi}^k / \mathbf{x}_0, \mathbf{x}_1, \ldots, \mathbf{x}_k) = a_k \widehat{\mathbf{F}}(\mathbf{x}_k) + \mathbf{b}_k, \; k = 0, 1, \ldots, \tag{13.9}$$

where $a_k > 0$ is a random variable; $\mathbf{b}_k = (b_1^k, \ldots, b_n^k)$ is a random vector, measurable with respect to the σ−algebra $\mathcal{B}^k$ induced by the family of random variables $(\mathbf{x}_0, \ldots, \mathbf{x}_k)$; $\widehat{\mathbf{F}}(\mathbf{x}_k)$ is a subgradient (generalized gradient) of $F(\mathbf{x})$ at $\mathbf{x}_k$; $\mathbf{E}_\omega(\boldsymbol{\xi}^k / \mathbf{x}_0, \ldots, \mathbf{x}_k)$ is the conditional expectation of $\boldsymbol{\xi}^k$ subject to $\mathbf{x}_0, \ldots, \mathbf{x}_k$; ρ_k's are also measurable with respect to $\mathcal{B}^k$.

If $a_k \equiv 1$, $\mathbf{b}_k \equiv \mathbf{0}$, then $\boldsymbol{\xi}^k$ is said to be *a stochastic quasigradient* (or *a stochastic generalized gradient*) of $F(\mathbf{x})$ at $\mathbf{x}_k$. Method (13.8)–(13.9) is called the *stochastic quasigradient method.*

SQMs are direct methods, that is, they use information about functions $f^i(\mathbf{x}, \boldsymbol{\omega})$, $i = 0, 1, \ldots, m$ (but not about $F^i(\mathbf{x})$, $i = 0, 1, \ldots, m$) and their generalized gradients/quasigradients.

Convergence theorems are proved under certain requirements for $\boldsymbol{\xi}^k$, ρ_k.

Theorem 13.1 (Convergence of stochastic quasigradient methods)
Let ρ_k do not depend on $\boldsymbol{\omega}$ for each $k = 0, 1, \ldots$,

$$\|\boldsymbol{\xi}^k\|^2 \leq C = const., \; k = 0, 1, \ldots; \tag{13.10}$$

$$a_k(\boldsymbol{\omega}) \geq c_k, \quad \|\mathbf{b}_k(\boldsymbol{\omega})\| \leq b_k; \tag{13.11}$$

$$\rho_k \geq 0, \; \sum_{k=0}^{\infty} \rho_k^2 < \infty, \; \sum_{k=0}^{\infty} \rho_k \mathbf{E} b_k < \infty; \tag{13.12}$$

$$\sum_{k=0}^{\infty} \rho_k c_k = \infty \text{ almost surely;} \tag{13.13}$$

let the set X be bounded and $\mathbf{E}\|\mathbf{x}_0\|^2 < \infty$.

Then the sequence $\{\mathbf{x}_k(\omega)\}$ defined by (13.8)–(13.9) converges to X^* with probability one (almost surely) with $\lim\limits_{k\to\infty} \mathbf{x}_k(\omega) \in X^*$, where X^* is the set of optimal solutions to problem $\min\limits_{\mathbf{x}\in X} F(\mathbf{x})$.

Proof. Let $\mathbf{x}^* \in X^*$. Using definition of projection operation $\Pi_X(.)$ (Section 1.2, Chapter One) and (13.8), we get

$$\begin{aligned}\|\mathbf{x}^* - \mathbf{x}_{k+1}\|^2 &= \left\|\mathbf{x}^* - \Pi_X\left(\mathbf{x}_k - \rho_k \boldsymbol{\xi}^k\right)\right\| \le \|\mathbf{x}^* - \mathbf{x}_k + \rho_k \boldsymbol{\xi}^k\|^2 \\ &= \|\mathbf{x}^* - \mathbf{x}_k\|^2 + 2\rho_k \langle \boldsymbol{\xi}^k, \mathbf{x}^* - \mathbf{x}_k \rangle + \rho_k^2 \|\boldsymbol{\xi}^k\|^2. \end{aligned} \tag{13.14}$$

Take conditional expectation on both sides of (13.14) provided that $\mathbf{x}_0, \mathbf{x}_1, \ldots, \mathbf{x}_k$ have occurred. Using (13.9), we obtain

$$\mathbf{E}\{\|\mathbf{x}^* - \mathbf{x}_{k+1}\|^2 / \mathbf{x}_0, \mathbf{x}_1, \ldots, \mathbf{x}_k\} \tag{13.15}$$

$$\le \|\mathbf{x}^* - \mathbf{x}_k\|^2 + 2\rho_k a_k \langle \widehat{\mathbf{F}}(\mathbf{x}_k), \mathbf{x}^* - \mathbf{x}_k \rangle + 2\rho_k \langle \mathbf{b}_k, \mathbf{x}^* - \mathbf{x}_k \rangle + \rho_k^2 \|\boldsymbol{\xi}^k\|^2.$$

Since $F(\mathbf{x})$ is a convex function and $\mathbf{x}^* \in X^*$ by assumption, then

$$0 \ge F(\mathbf{x}^*) - F(\mathbf{x}_k) \ge \langle \widehat{\mathbf{F}}(\mathbf{x}_k), \mathbf{x}^* - \mathbf{x}_k \rangle,$$

that is

$$\langle \widehat{\mathbf{F}}(\mathbf{x}_k), \mathbf{x}^* - \mathbf{x}_k \rangle \le 0. \tag{13.16}$$

By using the Cauchy-Schwarz inequality $\langle \mathbf{b}_k, \mathbf{x}^* - \mathbf{x}_k \rangle \le \|\mathbf{b}_k\| \|\mathbf{x}^* - \mathbf{x}_k\|$ and (13.10), (13.11), (13.16), $\rho_k \ge 0$, $a_k > 0$, from (13.15) it follows that

$$\begin{aligned} &\mathbf{E}\{\|\mathbf{x}^* - \mathbf{x}_{k+1}\|^2 / \mathbf{x}_0, \mathbf{x}_1, \ldots, \mathbf{x}_k\} \\ &\le \|\mathbf{x}^* - \mathbf{x}_k\|^2 + 2\rho_k \|\mathbf{b}_k\| \|\mathbf{x}^* - \mathbf{x}_k\| + C\rho_k^2 \\ &\le \|\mathbf{x}^* - \mathbf{x}_k\|^2 + 2\rho_k b_k \gamma + C\rho_k^2, \end{aligned} \tag{13.17}$$

where γ is a constant. Denote

$$z_k = \|\mathbf{x}^* - \mathbf{x}_k\|^2 + 2\gamma \sum_{s=k}^{\infty} \rho_s b_s + C \sum_{s=k}^{\infty} \rho_s^2. \tag{13.18}$$

From (13.17) it follows that

$$\mathbf{E}\{z_{k+1}/\mathbf{x}_0, \mathbf{x}_1, \ldots, \mathbf{x}_k\} \leq z_k,$$

and since z_k depends only on $\mathbf{x}_k$, we get

$$\mathbf{E}\{z_{k+1}/z_0, z_1, \ldots, z_k\} \leq z_k. \tag{13.19}$$

A sequence $\{z_k\}$ of random variables with this property is called a *supermartingale*. Since the terms of the supermartingale are nonnegative ($z_k \geq 0$), it is convergent with probability 1 (almost surely). By using this fact and (13.12) it follows that the sequence $\|\mathbf{x}^* - \mathbf{x}_k\|^2$ is convergent with probability 1, therefore, the sequence $\{\mathbf{x}_k\}$ is bounded and the set X^* of its limit points is nonempty.

Let $\mathbf{x}'(\omega)$ and $\mathbf{x}''(\omega)$ be two arbitrary limit points of the sequence $\{\mathbf{x}_k(\omega)\}$. Then for any $\mathbf{x}^* \in X^*$, we have

$$\|\mathbf{x}^* - \mathbf{x}'(\omega)\|^2 = \|\mathbf{x}^* - \mathbf{x}''(\omega)\|^2. \tag{13.20}$$

We want to prove that one limit point of the sequence $\{\mathbf{x}_k(\omega)\}$ belongs to the set X^* with probability 1.

Applying the inequality (13.14) k times, we get

$$\begin{aligned}\|\mathbf{x}^* - \mathbf{x}_{k+1}\|^2 &\leq \|\mathbf{x}^* - \mathbf{x}_k\|^2 + 2\rho_k\langle \boldsymbol{\xi}^k, \mathbf{x}^* - \mathbf{x}_k\rangle + C\rho_k^2 \\ &\leq \left(\|\mathbf{x}^* - \mathbf{x}_{k-1}\|^2 + 2\rho_{k-1}\langle \boldsymbol{\xi}^{k-1}, \mathbf{x}^* - \mathbf{x}_{k-1}\rangle + C\rho_{k-1}^2\right) \\ &\quad + 2\rho_k\langle \boldsymbol{\xi}^k, \mathbf{x}^* - \mathbf{x}_k\rangle + C\rho_k^2 \leq \cdots \\ &\leq \|\mathbf{x}^* - \mathbf{x}_0\|^2 + 2\sum_{s=0}^{k} \rho_s\langle \boldsymbol{\xi}^s, \mathbf{x}^* - \mathbf{x}_s\rangle + C\sum_{s=0}^{k} \rho_s^2.\end{aligned}$$

Then by using (13.9) and Cauchy-Schwarz inequality (see also (13.17)), we obtain

$$\mathbf{E}\{\|\mathbf{x}^* - \mathbf{x}_{k+1}\|^2\} \tag{13.21}$$

$$\leq \mathbf{E}\{\|\mathbf{x}^* - \mathbf{x}_0\|^2\} + 2\sum_{s=0}^{k} \rho_s\mathbf{E}(a_s\langle\widehat{\mathbf{F}}(\mathbf{x}_s), \mathbf{x}^* - \mathbf{x}_s\rangle) + 2\gamma\sum_{s=0}^{k} \rho_s\mathbf{E}b_s + C\sum_{s=0}^{k} \rho_s^2.$$

Since the left-hand side of (13.21) is nonnegative ($\mathbf{E}\{\|\mathbf{x}^* - \mathbf{x}_{k+1}\|^2\} \geq 0$), the following inequality holds

$$\mathbf{E}\{\|\mathbf{x}^* - \mathbf{x}_0\|^2\} + 2\sum_{s=0}^{k} \rho_s \mathbf{E}(a_s \langle \widehat{\mathbf{F}}(\mathbf{x}_s), \mathbf{x}^* - \mathbf{x}_s\rangle) + 2\gamma \sum_{s=0}^{k} \rho_s \mathbf{E} b_s + C \sum_{s=0}^{k} \rho_s^2 \geq 0.$$

When $k \to \infty$, taking into account (13.12), we obtain

$$2\sum_{s=0}^{\infty} \rho_s \mathbf{E}(a_s \langle \widehat{\mathbf{F}}(\mathbf{x}_s), \mathbf{x}^* - \mathbf{x}_s\rangle)$$

$$\geq -\mathbf{E}\{\|\mathbf{x}^* - \mathbf{x}_0\|^2\} - 2\gamma \sum_{s=0}^{\infty} \rho_s \mathbf{E} b_s - C \sum_{s=0}^{\infty} \rho_s^2 > -\infty$$

with probability 1. By using (13.11), we get

$$\sum_{s=0}^{\infty} \rho_s c_s \mathbf{E}\langle \widehat{\mathbf{F}}(\mathbf{x}_s), \mathbf{x}^* - \mathbf{x}_s\rangle > -\infty. \tag{13.22}$$

Since $\sum_{s=0}^{\infty} \rho_s c_s = \infty$ with probability 1 according to the assumption (13.13), there exists a subsequence $\{k_l\}$ such that

$$\langle \widehat{\mathbf{F}}(\mathbf{x}_{k_l}), \mathbf{x}^* - \mathbf{x}_{k_l}\rangle \to 0$$

with probability 1 when $l \to \infty$. However, $\widehat{\mathbf{F}}(\mathbf{x})$ is a subgradient of the convex function $F(\mathbf{x})$, that is, it satisfies

$$F(\mathbf{x}^*) - F(\mathbf{x}_{k_l}) \geq \langle \widehat{\mathbf{F}}(\mathbf{x}_{k_l}), \mathbf{x}^* - \mathbf{x}_{k_l}\rangle \to 0.$$

Hence, for any limit point $\mathbf{x}^b(\omega)$ of the subsequence $\{\mathbf{x}_{k_l}\}$ we have that $F(\mathbf{x}^*) = F(\mathbf{x}^b(\omega))$, that is, $\mathbf{x}^b(\omega) \in X^*$ with probability 1.

Then (13.20) implies that this limit point $\mathbf{x}^b(\omega)$ is the unique limit point of the sequence $\{\mathbf{x}_k(\omega)\}$ and $\lim_{k\to\infty} \mathbf{x}_k(\omega) = \mathbf{x}^b(\omega) \in X^*$ with probability 1. ■

SQMs are slow methods. That is why one of the main problems concerning their implementation is the choice of the step-size sequence $\{\rho_k\}$. The simplest way is to do this before the iterative process. Convergence theorems (for example, Theorem 13.1) state that any sequence with the properties

$$\rho_k \geq 0; \quad \rho_k \to 0,\ k \to \infty; \quad \sum_{k=0}^{\infty} \rho_k = \infty; \quad \sum_{k=0}^{\infty} \rho_k^2 < \infty$$

can be used as a step-size sequence. However, this approach does not use information obtained during the iterative process.

A modern method for choosing the step size ρ_k is the so-called *adaptive step-size regulation* (see, for example, the paper of Urasiev [168]).

Due to the specificity of stochastic programming problems and stochastic quasi-gradient methods (slow convergence, nonmonotonicity, and sometimes oscillatory behavior), it is advisable to average the values of variables and of the objective function during a certain number of last iterations and take these average values as the final approximation to the solution.

The second main problem regarding the implementation of SQMs is finding the projection of a current point $\mathbf{y}_k \equiv \mathbf{x}_k - \rho_k \boldsymbol{\xi}^k$ onto the feasible set X. As it is known (Section 1.2, Chapter One), this is equivalent to solving the quadratic optimization problem

$$\min \frac{1}{2} \|\mathbf{y}_k - \mathbf{x}\|^2$$

subject to

$$\mathbf{x} \in X.$$

This problem is to be solved at *each* iteration of the algorithm performance. That is why projection is the most onerous and time-consuming part of the SQM (and of any gradient-type projection method for constrained optimization) and we need efficient algorithms for solving this problem.

The third important problem concerning the implementation of SQMs is calculation of a stochastic quasigradient of the function to be minimized.

For example, components of the stochastic quasigradient of $F(\mathbf{x})$ (13.2) [(13.2′)] at iteration k are

$$\xi_j^k = \widehat{f}_j(x_j^k, \omega_j^k) = \begin{cases} p_j, \text{ if } x_j^k \geq \omega_j^k, \\ -q_j, \text{ if } x_j^k < \omega_j^k, \end{cases} \quad j = 1, \ldots, n, \tag{13.23}$$

where x_j^k is the jth component of $\mathbf{x}$ at iteration k, ω_j^k is the jth component of the observation of $\boldsymbol{\omega}$ at iteration k.

13.3 On the Projection in the Implementation of Stochastic Quasigradient Methods to Some Probabilistic Inventory Control Problems. Computational Experiments

As pointed out in Chapter Five and in Section 13.2, we need efficient algorithms for projecting a point onto certain feasible regions.

Consider the problem of projecting an arbitrary point $\widehat{\mathbf{x}} = (\widehat{x}_1, \ldots, \widehat{x}_n) \in \mathbf{R}^n$ onto the set X defined by (13.4)–(13.5) or (13.6)–(13.7). This problem is equivalent to

$$\min \left\{ c(\mathbf{x}) \equiv \sum_{j=1}^{n} c_j(x_j) \equiv \frac{1}{2} \sum_{j=1}^{n} (x_j - \widehat{x}_j)^2 \right\}$$

subject to

$$\mathbf{x} \in X.$$

Denote this problem by $(P^{\leq})$ in the first case (feasible set X defined by (13.4)–(13.5)) and by $(P^{=})$ in the second case (feasible set X defined by (13.6)–(13.7)).

It was pointed out that problems $(P^{\leq})$ and $(P^{=})$ are special cases of problems (C) (with linear inequality constraint) and $(C^{=})$, respectively, with $c_j(x_j) = \frac{1}{2}(x_j - \widehat{x}_j)^2$. For the respective explicit expressions, see Chapter Eleven, Problem 11.5.

Results of some numerical experiments, obtained by applying a SQM with adaptive step-size regulation to multi-commodity facility location problems, are presented below. Projection of the current approximation onto the feasible region has been found by using the polynomial algorithms, suggested in Part Two of this book.

Example 13.1

$$\min \left\{ F(\mathbf{x}) = E_\omega \sum_{j=1}^{5} \max \{p_j(x_j - \omega_j), q_j(\omega_j - x_j)\} \right\}$$

subject to

$$\begin{aligned} &x_1 + x_2 + 2x_3 + 3x_4 + x_5 = 200 \\ &0 \leq x_1 \leq 50 \\ &0 \leq x_2 \leq 7 \\ &0 \leq x_3 \leq 7 \\ &0 \leq x_4 \leq 80 \\ &0 \leq x_5 \leq 25. \end{aligned}$$

Here

$$\mathbf{p} = (p_1, p_2, p_3, p_4, p_5) = (1, 0, 3, 1, 2),$$

$$\mathbf{q} = (q_1, q_2, q_3, q_4, q_5) = (3, 4, 1, 2, 3),$$

and $\omega_j,\ j = 1, 2, 3, 4, 5$ are random variables, uniformly distributed on the closed segments

$$[0, 60],\ [0, 15],\ [0, 17],\ [0, 90],\ [0, 40],$$

respectively.

Obviously this is a problem of the form (13.2)–(13.3) with $n = 5$ and feasible set X of the form (13.6)–(13.7).

Optimal Solution (by using SQM)
(Values of the last 10 iterations have been averaged):

$$\mathbf{x}^* = (42.08259,\ 6.98305,\ 3.76966,\ 41.86273,\ 17.80680).$$

Minimum value of $F(\mathbf{x}):\ F(\mathbf{x}^*) = 96.63854.$
The equality constraint is satisfied with tolerance: 0.00000000419.

This special test example can also be solved analytically by using a nonlinear programming approach.

The analytical expression of the objective function is

$$F(\mathbf{x}) = \frac{1}{30}x_1^2 + \frac{2}{15}x_2^2 + \frac{2}{17}x_3^2 + \frac{1}{60}x_4^2 + \frac{1}{16}x_5^2 - 3x_1 - 4x_2 - x_3 - 2x_4 - 3x_5 + 278.5$$

Analytic Solution
(by using the nonlinear programming approach):

$$\mathbf{x}_a^* = (41.88057,\ 7.00000,\ 2.48092,\ 41.27456,\ 22.33456).$$

Minimum value of $F(\mathbf{x}):\ F(\mathbf{x}_a^*) = 98.10089.$

Example 13.2

$$\min\left\{F(\mathbf{x}) = E_\omega \sum_{j=1}^{6} \max\left\{p_j(x_j - \omega_j), q_j(\omega_j - x_j)\right\}\right\}$$

subject to

$$\begin{aligned}
&3x_1 + 4x_2 + 7x_3 + 5x_5 + x_6 \le 872\\
&0 \le x_1 \le 45\\
&0 \le x_2 \le 56\\
&0 \le x_3 \le 32\\
&0 \le x_4 \le 27\\
&0 \le x_5 \le 456\\
&0 \le x_6 \le 45.
\end{aligned}$$

Here

$$\mathbf{p} = (p_1, p_2, p_3, p_4, p_5, p_6) = (5, 7, 8, 3, 5, 1),$$

$$\mathbf{q} = (q_1, q_2, q_3, q_4, q_5, q_6) = (4, 56, 7, 8, 4, 4),$$

and $\omega_j,\ j = 1, 2, 3, 4, 5, 6$ are uniformly distributed random variables on closed segments

$$[0, 56],\ [0, 57],\ [0, 36],\ [0, 34],\ [0, 468],\ [0, 65],$$

respectively.

Obviously this is a problem of the form (13.2)–(13.3) with $n = 6$ and feasible set X of the form (13.4)–(13.5) (with $d_4 = 0$, see Chapter Ten).

Optimal Solution (by SQM)
(Values of the last 10 iterations have been averaged):

$\mathbf{x}^* = (25.13417,\ 48.01999,\ 15.62868,\ 25.32763,\ 49.51728,\ 33.14936)$.

Minimum value of $F(\mathbf{x})$: $F(\mathbf{x}^*) = 972.59309$.

Results of some computational experiments are presented below, obtained by applying a stochastic quasigradient method with adaptive step-size regulation to inventory control problems of the form $(P^{\leq})$ and $(P^{=})$, respectively. The projections of the current approximations, generated by SQM, on the feasible regions have been determined by the polynomial algorithms, proposed in Part Two of this book.

The computations have been performed on an Intel Pentium(R) Dual-Core CPU E5800 3.20 GHz/2.00GB using RZTools interactive system. Each problem was run 30 times.

Problem type	$(P^{\leq})$		$(P^{=})$	
Number of the variables	$n = 1220$	$n = 1550$	$n = 1220$	$n = 1550$
Average number of iterations	3.033	3.166	2.166	4.33
Average run time (seconds)	0.000097	0.00012	0.00001	0.000102

Effectiveness of the algorithms for solving problems $(P^{\leq})$ and $(P^{=})$ has been tested by many other numerical examples, also for greater values of n. As we can observe, the average number of iterations is much less than the number of variables n of the problems under consideration for large values of n.

13.4 The Stochastic Problem of Best Chebyshev Approximation

Problem of the form

$$\min \left\{ f(\mathbf{x}, \mathbf{y}, \omega) = \left| \sum_{j=1}^{n} a_j(\mathbf{y}, \omega) x_j - b(\mathbf{y}, \omega) \right| \right\} \tag{13.24}$$

is known as the *stochastic problem of best Chebyshev approximation* and it is analogous to problems (12.6), (12.10), etc., considered in Chapter Twelve.

In this case, components of the stochastic quasigradient $\boldsymbol{\xi} = (\xi_1, \xi_2, ..., \xi_n)$, defined as $\boldsymbol{\xi} = \widehat{\mathbf{f}_{\mathbf{x}}}(\mathbf{x}, \mathbf{y}(\mathbf{x}, \omega), \omega)$, are

$$\xi_j = a_j(\mathbf{y}(\mathbf{x}, \omega), \omega).\text{sign}\left(\sum_{j=1}^{n} a_j(\mathbf{y}(\mathbf{x}, \omega), \omega)x_j - b(\mathbf{y}(\mathbf{x}, \omega), \omega)\right),$$

$j = 1, \ldots, n$, that is

$$\xi_j = \begin{cases} a_j(\mathbf{y}(\mathbf{x}, \omega), \omega), \\ \quad \text{if } \sum_{j=1}^{n} a_j(\mathbf{y}(\mathbf{x}, \omega), \omega)x_j - b(\mathbf{y}(\mathbf{x}, \omega), \omega) > 0, \\ a_j(\mathbf{y}(\mathbf{x}, \omega), \omega).[-1, 1], \\ \quad \text{if } \sum_{j=1}^{n} a_j(\mathbf{y}(\mathbf{x}, \omega), \omega)x_j - b(\mathbf{y}(\mathbf{x}, \omega), \omega) = 0, \\ -a_j(\mathbf{y}(\mathbf{x}, \omega), \omega), \\ \quad \text{if } \sum_{j=1}^{n} a_j(\mathbf{y}(\mathbf{x}, \omega), \omega)x_j - b(\mathbf{y}(\mathbf{x}, \omega), \omega) < 0, \end{cases} \tag{13.25}$$

$j = 1, \ldots, n.$

Since problem (13.24) is an unconstrained minimization problem ($X = \mathbf{R}^n$, and therefore, $\Pi_X(\mathbf{z}) = \mathbf{z}$), the following version of the SQM can be applied to this problem

$$\mathbf{x}_{k+1} = \mathbf{x}_k - \rho_k \boldsymbol{\xi}^k, \quad k = 0, 1, \ldots, \tag{13.8'}$$

where $\mathbf{x}_0$ is an arbitrary initial guess, ρ_k is a step size, and $\boldsymbol{\xi}^k = \boldsymbol{\xi}^k(\omega)$ is a stochastic quasigradient of the objective function at iteration k.

Then components of the stochastic quasigradient $\boldsymbol{\xi}^k = (\xi_1^k, \xi_2^k, ..., \xi_n^k)$ at iteration k of implementation of SQM (13.8′) are

$$\xi_j^k = a_j(\mathbf{y}(\mathbf{x}_k, \omega_k), \omega_k).\text{sign}\left(\sum_{j=1}^{n} a_j(\mathbf{y}(\mathbf{x}_k, \omega_k), \omega_k)x_j^k - b(\mathbf{y}(\mathbf{x}_k, \omega_k), \omega_k)\right),$$

$j = 1, \ldots, n$, that is

$$\xi_j^k = \begin{cases} a_j(\mathbf{y}(\mathbf{x}_k, \boldsymbol{\omega}_k), \boldsymbol{\omega}_k), \\ \quad \text{if } \sum_{j=1}^{n} a_j(\mathbf{y}(\mathbf{x}_k, \boldsymbol{\omega}_k), \boldsymbol{\omega}_k) x_j^k - b(\mathbf{y}(\mathbf{x}_k, \boldsymbol{\omega}_k), \boldsymbol{\omega}_k) > 0, \\ a_j(\mathbf{y}(\mathbf{x}_k, \boldsymbol{\omega}_k), \boldsymbol{\omega}_k).[-1, 1], \\ \quad \text{if } \sum_{j=1}^{n} a_j(\mathbf{y}(\mathbf{x}_k, \boldsymbol{\omega}_k), \boldsymbol{\omega}_k) x_j^k - b(\mathbf{y}(\mathbf{x}_k, \boldsymbol{\omega}_k), \boldsymbol{\omega}_k) = 0, \\ -a_j(\mathbf{y}(\mathbf{x}_k, \boldsymbol{\omega}_k), \boldsymbol{\omega}_k), \\ \quad \text{if } \sum_{j=1}^{n} a_j(\mathbf{y}(\mathbf{x}_k, \boldsymbol{\omega}_k), \boldsymbol{\omega}_k) x_j^k - b(\mathbf{y}(\mathbf{x}_k, \boldsymbol{\omega}_k), \boldsymbol{\omega}_k) < 0, \end{cases} \quad (13.25')$$

$j = 1, \ldots, n; \;\; k = 0, 1, \ldots$

Chapter 14
Valid Inequalities, Cutting Planes, and Integrality of the Knapsack Polytope

In this chapter, the so-called knapsack problem and knapsack polytope, defined by linear inequality/inequalities and bounds on the variables, are considered. Some important concepts and preliminaries are given at the beginning. As we have observed, knapsack polytope is a feasible region of some problems considered in Part One and Part Two of this book. Some results connected with the generation of valid and dominating inequalities (including the modular arithmetic approach) and cutting planes are presented. Necessary and sufficient conditions are also proved for the knapsack polytope to be integral. Such integrality criteria are useful because the optimal solution to a linear programming problem, when it is solvable, is attained at a vertex of the feasible polytope/polyhedron, and in the case of integrality of the knapsack polytope, the integer problem and the corresponding continuous problem have the same solution(s).

14.1 Basic Formulations, Definitions, and Results

14.1.1 Formulations

As usual, $\mathbf{R}$ is the set of real numbers, $\mathbf{R}^n$ is the totality of vectors with n real components, $\mathbf{R}_+$ is the set of all real nonnegative numbers, and $\mathbf{R}^n_+$ is the totality of vectors with n real nonnegative components. Analogous notation applies to the set of integers $\mathbf{Z}$.

© The Author(s), under exclusive license to Springer Nature Switzerland AG 2021

S. M. Stefanov, *Separable Optimization*, Springer Optimization and Its Applications 177,
https://doi.org/10.1007/978-3-030-78401-0_14

Consider the knapsack problem

$$\max \left\{\mathbf{cx} = \sum_{j=1}^{n} c_j x_j\right\} \tag{14.1}$$

subject to

$$\sum_{j=1}^{n} a_j x_j \leq a_0 \tag{14.2}$$

$$0 \leq x_j \leq d_j, \quad j = 1, \ldots, n \tag{14.3}$$

$$x_j \in \mathbf{Z}, \quad j = 1, \ldots, n, \tag{14.4}$$

where $a_0 > 0,\ a_j \geq 0,\ c_j > 0,\ d_j > 0,\ j = 1, \ldots, n.$

Remark 14.1 *We can assume* $a_j > 0,\ c_j > 0,\ j = 1, \ldots, n$, *because if* $c_j < 0$ *and* $a_j < 0$ *for some* j *among* $1, \ldots, n$ *then by substituting* $\overline{x_j} = d_j - x_j$ *for such* j *in* (14.1)–(14.4), *we transform this case to the case* $a_j > 0,\ c_j > 0$. *If* $c_j > 0,\ a_j < 0$, *then the corresponding component of the optimal solution to* (14.1)–(14.4) *is* $x_j^* = d_j$, *and if* $c_j < 0,\ a_j > 0$, *then* $x_j^* = 0,\ j = 1, \ldots, n$. *We also assume that* $a_0 < \sum_{j=1}^{n} a_j d_j$.

When $x_j \in \{0, 1\},\ j = 1, \ldots, n$, we obtain the *boolean* (*zero-one*, $0 - 1$) *knapsack problem*. Zero-one knapsack problem belongs to the class of the so-called *packing problems*.

14.1.2 Definitions

In this subsection, definitions of some important concepts are introduced, which are used in the next sections of this chapter.

Definition 14.1 A *polyhedron* $P \subset \mathbf{R}^n$ is the set of points that satisfy a finite number of linear inequalities, that is, $P = \{\mathbf{x} \in \mathbf{R}^n : A\mathbf{x} \leq \mathbf{b},\ A_{m\times n},\ \mathbf{b} \in \mathbf{R}^m\}$. A bounded polyhedron is called a *polytope*.

Definition 14.2 Let $S \subset \mathbf{R}^n$, $\boldsymbol{\pi}, \mathbf{x} \in \mathbf{R}^n$, $\pi_0 \in \mathbf{R}$. The inequality $\boldsymbol{\pi}\mathbf{x} \leq \pi_0$ $\left(\text{that is, } \sum_{j=1}^{n} \pi_j x_j \leq \pi_0\right)$ is said to be a *valid inequality* for S if it is satisfied by all $\mathbf{x} \in S$.

Every inequality is valid for the empty set $\emptyset$ by definition.

Definition 14.3 The valid inequalities $\boldsymbol{\pi}\mathbf{x} \leq \pi_0$ and $\boldsymbol{\gamma}\mathbf{x} \leq \gamma_0$ are said to be *equivalent* if $\boldsymbol{\gamma} = \mu\boldsymbol{\pi}$ and $\gamma_0 = \mu\pi_0$ for some $\mu > 0$.

Definition 14.4 If the valid inequalities $\boldsymbol{\pi}\mathbf{x} \leq \pi_0$ and $\boldsymbol{\gamma}\mathbf{x} \leq \gamma_0$ are not equivalent and there exists $\lambda > 0$ such that $\lambda\boldsymbol{\gamma} \geq \boldsymbol{\pi}$ and $\lambda\gamma_0 \leq \pi_0$ then we say that $\boldsymbol{\gamma}\mathbf{x} \leq \gamma_0$ *dominates* or is *stronger than* $\boldsymbol{\pi}\mathbf{x} \leq \pi_0$ or that $\boldsymbol{\pi}\mathbf{x} \leq \pi_0$ is *dominated by* or is *weaker than* $\boldsymbol{\gamma}\mathbf{x} \leq \gamma_0$.

A *maximal* valid inequality is one that is not dominated by any other valid inequality.

Let C be a closed convex set.

Definition 14.5 A convex subset F of C is called a *face* of C if for any two points $y, z \in C,\ y \neq z$ so that $(y, z) \cap F \neq \emptyset$, we have $[y, z] \subset F$.

Here (y, z) and $[y, z]$ denote open and closed line segments, respectively. Since F is a convex set, then $[y, z] \subset F$ is equivalent to $y, z \in F$.

Definition 14.6 If $\boldsymbol{\pi}\mathbf{x} \leq \pi_0$ is a valid inequality for polyhedron P and $F = \{\mathbf{x} \in P : \boldsymbol{\pi}\mathbf{x} = \pi_0\}$, F is called a *face* of P and $\boldsymbol{\pi}\mathbf{x} = \pi_0$ is said to *represent* F. A face F is said to be *proper* for P if $F \neq \emptyset$ and $F \neq P$.

The empty set $\emptyset$ and polyhedron P are faces of P by definition and they are called *improper faces*.

Definition 14.7 A face F of C such that $0 \leq \dim F = \dim C - 1$ is called a *facet* of C, where $\dim C$ denotes the dimension of C.

An equivalent definition is

Definition 14.8 Let $P \subset \mathbf{R}^n$. The face defined by the inequality $\boldsymbol{\pi}\mathbf{x} \leq \pi_0$ is called a *facet* of P if the following conditions are satisfied:

(i) $\mathbf{x} \in P$ implies $\boldsymbol{\pi}\mathbf{x} \leq \pi_0$ (that is, $\boldsymbol{\pi}\mathbf{x} \leq \pi_0$ is a valid inequality for P);

(ii) there exist exactly n affinely independent vertices $\mathbf{x}^i$ of P satisfying $\boldsymbol{\pi}\mathbf{x}^i \leq \pi_0,\ i = 1, \ldots, n$, with an equality, while all $\mathbf{x} \in P$ satisfy $\boldsymbol{\pi}\mathbf{x} \leq \pi_0$.

We use the term *facet* for the face from Definition 14.7 and Definition 14.8 as well as for the inequality $\boldsymbol{\pi}\mathbf{x} \leq \pi_0$ which produces this facet.

There are also other equivalent definitions of *face* and *facet*.

Definition 14.9 A nonempty polyhedron $P \subset \mathbf{R}^n$ is said to be *integral* if each of its nonempty faces contains an integral point.

Proposition 14.1 *A nonempty polyhedron* $P = \{\mathbf{x} \in \mathbf{R}^n : A\mathbf{x} \leq \mathbf{b}\}$ *with* $\operatorname{rank}(A) = n$ *is integral if and only if all of its extreme points are integral.*

Definition 14.10 An additional linear constraint to the linear programming problem is said to define a *regular cutting plane* (a *regular cut*) if it is:

a) not satisfied by the "nonintegral" solution(s) of the continuous linear problem;

b) satisfied by all integral points of the polyhedron of this problem.

Therefore, cutting plane cuts off points from the polyhedron of the continuous problem but does not cut off "integral" points.

According to Definition 14.7 (or 14.8) and Definition 14.10, facets of the integral polytope

$$P_I \stackrel{\text{def}}{=} \text{co}\,\{\mathbf{x} \in \mathbf{Z}^n : \mathbf{ax} \le a_0,\ \mathbf{d}' \le \mathbf{x} \le \mathbf{d}\}$$

of the "integral" problem (14.1)–(14.4) are regular cutting planes of corresponding "continuous" polytope.

M. W. Padberg has noticed that while cutting planes generally constitute valid inequalities, facets of P_I generate "deepest" cutting planes, and facets belong to the class of inequalities that uniquely determines P_I.

Remark 14.2 *The inequality which defines a facet of a polyhedron is the "strongest" inequality for this polyhedron.*

Any maximal valid inequality for P defines a nonempty face of P_I and the set of maximal valid inequalities contains all of the facet-defining inequalities for P_I, where $P \subset \mathbf{Z}^n$.

If $S \subseteq P$ and $\boldsymbol{\pi}\mathbf{x} \le \pi_0$ is a valid inequality for P, then obviously $\boldsymbol{\pi}\mathbf{x} \le \pi_0$ is a valid inequality for S as well.

Definition 14.11 An $m \times n$ integral matrix A is *totally unimodular* (TU) if the determinant of each square submatrix of A is equal to 0, 1, or $-$ 1.

It is evident that if A is TU, then $a_{ij} = 0, 1, \text{or } -1$, that is, A is a $(0, 1, -1)$ matrix.

In addition, the intersection of an arbitrary family $\mathcal{A}$ of faces of set C is a face of C as well.

Denote by $\mathcal{F}(C)$ the set of all faces of C. Then partially ordered set $(\mathcal{F}(C), \subset)$ is a complete lattice with the following lattice operations

$$\inf \mathcal{A} \stackrel{\text{def}}{=} \bigcap \{F \in \mathcal{F}(C) : F \in \mathcal{A}\},$$

$$\sup \mathcal{A} \stackrel{\text{def}}{=} \bigcap \{G \in \mathcal{F}(C) : \forall F \in \mathcal{A} : F \subset G\}.$$

We will call $(\mathcal{F}(C), \subset)$ a *face-lattice* for the set C.

14.1.3 Preliminaries

In this subsection, some preliminary results, called lemmas, are recalled without their proofs.

Lemma 14.1 *Let variables $x_1, x_2, \ldots, x_n$ be ordered such that*

$$\frac{c_1}{a_1} \ge \frac{c_2}{a_2} \ge \cdots \ge \frac{c_n}{a_n}. \tag{14.5}$$

Then the solution to the continuous knapsack problem defined by (14.1), (14.2), $x_j \in \mathbb{R}_+$, $j = 1, \ldots, n$, *is*

$$x_1^* = \frac{a_0}{a_1}, \quad x_j^* = 0, \ j = 2, \ldots, n. \tag{14.6}$$

Lemma 14.2 *Let variables* $x_1, x_2, \ldots, x_n$ *be ordered so that* (14.5) *holds. Then the solution to continuous problem* (14.1), (14.2), (14.3), $x_j \in \mathbb{R}$, $j = 1, \ldots, n$, *is* $\mathbf{x}^* = (x_1^*, \ldots, x_n^*)$ *with*

$$x_j^* = d_j, \quad j = 1, \ldots, s_0,$$

$$x_j^* = 0, \quad j = s_0 + 2, \ldots, n,$$

$$x_{s_0+1}^* = \frac{1}{a_{s_0+1}} \left\{ a_0 - \sum_{j=1}^{s_0} a_j d_j \right\},$$

where $s_0 = \max \left\{ s : \sum_{j=1}^{s} a_j d_j \leq a_0 \right\}$.

When $x_j \in \mathbb{R}_+$, $j = 1, \ldots, n$, that is, $d_j = +\infty$, $j = 1, \ldots, n$, then $s_0 = 0$.

Remark 14.3 a) *If* $\sum_{j=1}^{n} a_j d_j \leq a_0$ *then* $x_j^* = d_j$, $j = 1, \ldots, n$.

b) *If* $\sum_{j=1}^{n} a_j d_j > a_0$ *then the solution to the continuous problem* (14.1), (14.2), (14.3), $x_j \in \mathbb{R}$, $j = 1, \ldots, n$, *belongs to the set*

$$P' \stackrel{\text{def}}{=} \left\{ \mathbf{x} : \sum_{j=1}^{n} a_j x_j = a_0, \quad 0 \leq x_j \leq d_j, \ j = 1, \ldots, n \right\}.$$

Lemma 14.3 *If A is TU, then* $P(\mathbf{b}) = \{\mathbf{x} \in \mathbb{R}_+^n : A\mathbf{x} \leq \mathbf{b}\}$ *is integral for all* $\mathbf{b} \in \mathbb{Z}^m$ *for which it is not empty.*

Lemma 14.4 *If* $P(\mathbf{b})$ *is integral for all* $\mathbf{b} \in \mathbb{Z}^m$ *for which it is not empty, then A is TU.*

Lemma 14.5 *Let A be a* (0, 1, −1) *matrix with no more than two nonzero elements in each column. Then A is TU if and only if the rows of A can be partitioned into two subsets* Q_1 *and* Q_2 *such that if a column contains two nonzero elements, the following statements are true:*

a. *If both nonzero elements have the same sign, then one is in a row contained in* Q_1, *and the other is in a row contained in* Q_2.

b. *If the two nonzero elements have opposite signs, then both are in rows contained in the same subset.*

14.2 Valid Inequalities for Knapsack Polytope

14.2.1 Valid, Equivalent, and Dominating Inequalities

The following Theorem 14.1 refers to the so-called multidimensional knapsack problem and to its polytope, denoted by $P(A, \boldsymbol{\alpha}, \mathbf{d}', \mathbf{d})$ below.

Theorem 14.1 (Valid, equivalent, and dominating inequalities for the knapsack polytope)
Let A *be an* $m \times n$ *matrix with columns* $\mathbf{a}_j,\ j = 1, \ldots, n$, $\boldsymbol{\alpha} = (\alpha_1, \ldots, \alpha_m) \in \mathbf{R}^m$, $\mathbf{d}' = (d_1', \ldots, d_n')$, $\mathbf{d} = (d_1, \ldots, d_n) \in \mathbf{R}^n$,

$$P(A, \boldsymbol{\alpha}, \mathbf{d}', \mathbf{d}) \stackrel{\text{def}}{=} \{\mathbf{x} \in \mathbf{R}^n : A\mathbf{x} \leq \boldsymbol{\alpha},\quad d_j' \leq x_j \leq d_j,\ j = 1, \ldots, n\},$$

and $\boldsymbol{\pi}\mathbf{x} \leq \pi_0$ *be a valid inequality for* $P(A, \boldsymbol{\alpha}, \mathbf{d}', \mathbf{d})$, $\boldsymbol{\pi} \in \mathbf{R}^n,\ \pi_0 \in \mathbf{R}$.
Then $\boldsymbol{\pi}\mathbf{x} \leq \pi_0$ *is either equivalent to or is dominated by an inequality of the form*

$$(\boldsymbol{\lambda}A - \mathbf{u} + \mathbf{v})\mathbf{x} \leq \boldsymbol{\lambda}\boldsymbol{\alpha} - \mathbf{u}\mathbf{d}' + \mathbf{v}\mathbf{d},\quad \boldsymbol{\lambda} \in \mathbf{R}_+^m,\ \mathbf{u}, \mathbf{v} \in \mathbf{R}_+^n$$

if any of the following conditions holds:
i) $P(A, \boldsymbol{\alpha}, \mathbf{d}', \mathbf{d}) \neq \emptyset$;
ii) $D \stackrel{\text{def}}{=} \{\boldsymbol{\lambda} \in \mathbf{R}_+^m,\ \mathbf{u}, \mathbf{v} \in \mathbf{R}_+^n : \boldsymbol{\lambda}A - \mathbf{u} + \mathbf{v} = \boldsymbol{\pi}\} \neq \emptyset$;
iii) $A = \begin{pmatrix} A' \\ I \end{pmatrix}$, *where* I *is an* $n \times n$ *identity matrix.*

Proof. i) Since $\boldsymbol{\pi}\mathbf{x} \leq \pi_0$ is a valid inequality for $P(A, \boldsymbol{\alpha}, \mathbf{d}', \mathbf{d})$ and $P(A, \boldsymbol{\alpha}, \mathbf{d}', \mathbf{d}) \neq \emptyset$ then the (primal) linear programming problem

$$\max\ \boldsymbol{\pi}\mathbf{x}$$

subject to

$$\mathbf{x} \in P(A, \boldsymbol{\alpha}, \mathbf{d}', \mathbf{d})$$

has an optimal solution $\mathbf{x}_0$, and $\boldsymbol{\pi}\mathbf{x} \leq \pi_0$ for all $\mathbf{x} \in P(A, \boldsymbol{\alpha}, \mathbf{d}', \mathbf{d})$. According to a Duality theorem, the dual programming problem

$$\min\ (\boldsymbol{\lambda}\boldsymbol{\alpha} - \mathbf{u}\mathbf{d}' + \mathbf{v}\mathbf{d})$$

where

$$\mathbf{y} \equiv (\lambda_1, \ldots, \lambda_m, u_1, \ldots, u_n, v_1, \ldots, v_n) \in D$$

also has an optimal solution $(\boldsymbol{\lambda}^0, \mathbf{u}^0, \mathbf{v}^0) \in \mathbf{R}_+^{m+2n}$ such that

$$\boldsymbol{\lambda}^0 A - \mathbf{u}^0 + \mathbf{v}^0 = \boldsymbol{\pi} \text{ and } \boldsymbol{\lambda}^0\boldsymbol{\alpha} - \mathbf{u}^0\mathbf{d}' + \mathbf{v}^0\mathbf{d} = \boldsymbol{\pi}\mathbf{x}_0.$$

Therefore,

$$\boldsymbol{\pi}\mathbf{x} = (\boldsymbol{\lambda}^0 A - \mathbf{u}^0 + \mathbf{v}^0)\mathbf{x} \equiv \boldsymbol{\lambda}^0(A\mathbf{x}) - \mathbf{u}^0\mathbf{x} + \mathbf{v}^0\mathbf{x}$$

$$\leq \boldsymbol{\lambda}^0\boldsymbol{\alpha} - \mathbf{u}^0\mathbf{d}' + \mathbf{v}^0\mathbf{d} = \boldsymbol{\pi}\mathbf{x}_0 \leq \pi_0.$$

Here we have used the inequalities that define $P(A, \boldsymbol{\alpha}, \mathbf{d}', \mathbf{d})$.

ii) Assume that $P(A, \boldsymbol{\alpha}, \mathbf{d}', \mathbf{d}) = \emptyset$, otherwise we are in case i). Since $D \neq \emptyset$ by assumption, then there exist $\widehat{\boldsymbol{\lambda}} \in \mathbf{R}^m_+$ and $\widehat{\mathbf{u}}, \widehat{\mathbf{v}} \in \mathbf{R}^n_+$ such that

$$\widehat{\boldsymbol{\lambda}}A - \widehat{\mathbf{u}} + \widehat{\mathbf{v}} = \boldsymbol{\pi}.$$

If $\widehat{\boldsymbol{\lambda}}\boldsymbol{\alpha} - \widehat{\mathbf{u}}\mathbf{d}' + \widehat{\mathbf{v}}\mathbf{d} \leq \pi_0$, the statement is proved. Otherwise, since $P(A, \boldsymbol{\alpha}, \mathbf{d}', \mathbf{d}) = \emptyset$ (that is, $\mathbf{x} \in \mathbf{R}^n$, $\mathbf{d}' \leq \mathbf{x} \leq \mathbf{d}$ imply $(A\mathbf{x})_i > \alpha_i$ for at least one i) it follows that there exist $\overline{\boldsymbol{\lambda}} \in \mathbf{R}^m_+$ and $\overline{\mathbf{u}}, \overline{\mathbf{v}} \in \mathbf{R}^n_+$ such that

$$\overline{\boldsymbol{\lambda}}A - \overline{\mathbf{u}} + \overline{\mathbf{v}} = \mathbf{0}, \quad \overline{\boldsymbol{\lambda}}\boldsymbol{\alpha} - \overline{\mathbf{u}}\mathbf{d}' + \overline{\mathbf{v}}\mathbf{d} < 0.$$

Hence, for some $\mu > 0$, we have

$$(\widehat{\boldsymbol{\lambda}} + \mu\overline{\boldsymbol{\lambda}})A - (\widehat{\mathbf{u}} + \mu\overline{\mathbf{u}}) + (\widehat{\mathbf{v}} + \mu\overline{\mathbf{v}}) \equiv (\widehat{\boldsymbol{\lambda}}A - \widehat{\mathbf{u}} + \widehat{\mathbf{v}}) + \mu(\overline{\boldsymbol{\lambda}}A - \overline{\mathbf{u}} + \overline{\mathbf{v}}) = \boldsymbol{\pi},$$

$$(\widehat{\boldsymbol{\lambda}} + \mu\overline{\boldsymbol{\lambda}})\boldsymbol{\alpha} - (\widehat{\mathbf{u}} + \mu\overline{\mathbf{u}})\mathbf{d}' + (\widehat{\mathbf{v}} + \mu\overline{\mathbf{v}})\mathbf{d}$$

$$\equiv (\widehat{\boldsymbol{\lambda}}\boldsymbol{\alpha} - \widehat{\mathbf{u}}\mathbf{d}' + \widehat{\mathbf{v}}\mathbf{d}) + \mu(\overline{\boldsymbol{\lambda}}\boldsymbol{\alpha} - \overline{\mathbf{u}}\mathbf{d}' + \overline{\mathbf{v}}\mathbf{d}) \leq \pi_0.$$

For $\widehat{\boldsymbol{\lambda}}, \overline{\boldsymbol{\lambda}} \in \mathbf{R}^m_+$; $\widehat{\mathbf{u}}, \overline{\mathbf{u}}, \widehat{\mathbf{v}}, \overline{\mathbf{v}} \in \mathbf{R}^n_+$; $\mu > 0$ denote

$$\boldsymbol{\lambda}^0 \equiv \widehat{\boldsymbol{\lambda}} + \mu\overline{\boldsymbol{\lambda}} \in \mathbf{R}^m_+; \quad \mathbf{u}^0 \equiv \widehat{\mathbf{u}} + \mu\overline{\mathbf{u}} \in \mathbf{R}^n_+; \quad \mathbf{v}^0 \equiv \widehat{\mathbf{v}} + \mu\overline{\mathbf{v}} \in \mathbf{R}^n_+.$$

Then $\boldsymbol{\lambda}^0 A - \mathbf{u}^0 + \mathbf{v}^0 = \boldsymbol{\pi}$, that is, $(\boldsymbol{\lambda}^0, \mathbf{u}^0, \mathbf{v}^0) \in D$, $\boldsymbol{\lambda}^0\boldsymbol{\alpha} - \mathbf{u}^0\mathbf{d}' + \mathbf{v}^0\mathbf{d} \leq \pi_0$ and since we have $\boldsymbol{\pi}\mathbf{x} \leq \pi_0$, the statement of ii) is proved.

iii) Let $A = \begin{pmatrix} A' \\ I \end{pmatrix}$. Therefore, $A' = (a'_{ij})$ is an $(m - n) \times n$ matrix. Choose an arbitrary vector $\boldsymbol{\pi} \in \mathbf{R}^n$. From the vector equation $\boldsymbol{\lambda}A - \mathbf{u} + \mathbf{v} = \boldsymbol{\pi}$, where $\boldsymbol{\lambda} \in \mathbf{R}^m_+$; $\mathbf{u}, \mathbf{v} \in \mathbf{R}^n_+$, which in this case has the form

$$\left(\sum_{i=1}^{m-n} \lambda_i a'_{i1} + \lambda_{m-n+1} - u_1 + v_1, \; \dots, \; \sum_{i=1}^{m-n} \lambda_i a'_{in} + \lambda_m - u_n + v_n\right)^T = (\pi_1, \dots, \pi_n)^T, \tag{14.7}$$

if we fix arbitrary nonnegative values $\lambda_1 \geq 0, \dots, \lambda_{m-n} \geq 0$, we determine the values

$$l_j := \lambda_{m-n+j} - u_j + v_j, \quad j = 1, \dots, n \tag{14.8}$$

uniquely in terms of a'_{ij}, π_j, λ_i, $i = 1, \ldots, m - n$; $j = 1, \ldots, n$. Since we have these values, we can choose $\lambda_{m-n+j} \geq 0$, $u_j \geq 0$, $v_j \geq 0$, $j = 1, \ldots, n$ so that (14.8) holds. We are able to do this due to the structure of the system (14.7) (of the matrix A, respectively).

Hence, $D \neq \emptyset$, and with the use of case ii), we obtain the statement of iii).

Therefore, in the three cases considered, the inequality

$$(\boldsymbol{\lambda}^0 A - \mathbf{u}^0 + \mathbf{v}^0)\mathbf{x} \leq \boldsymbol{\lambda}^0 \boldsymbol{\alpha} - \mathbf{u}^0 \mathbf{d}' + \mathbf{v}^0 \mathbf{d}$$

dominates or is equivalent to $\boldsymbol{\pi}\mathbf{x} \leq \pi_0$ by definition. ■

When $d_j = +\infty$, $j = 1, \ldots, n$, then the dual variables v_j are equal to zero: $v_j = 0$, $j = 1, \ldots, n$, because the right (upper) inequalities do not exist in this case.

Corollary 14.1 *Let* $P_0 \overset{\text{def}}{=} \{\mathbf{x} \in \mathbf{R}^n : A\mathbf{x} \leq \boldsymbol{\alpha},\ 0 \leq x_j \leq d_j,\ j = 1, \ldots, n\}$, *$A$ be an $m \times n$ matrix,* $\boldsymbol{\alpha} \in \mathbf{R}^m$, $\mathbf{d} = (d_1, \ldots, d_n) \in \mathbf{R}^n$, $\boldsymbol{\pi}\mathbf{x} \leq \pi_0$ *be a valid inequality for* P_0, $\boldsymbol{\pi} \in \mathbf{R}^n$, $\pi_0 \in \mathbf{R}$.

Then $\boldsymbol{\pi}\mathbf{x} \leq \pi_0$ *is either equivalent to or is dominated by the inequality of the form* $(\boldsymbol{\lambda} A + \mathbf{v})\mathbf{x} \leq \boldsymbol{\lambda}\boldsymbol{\alpha} + \mathbf{v}\mathbf{d}$, *where* $\boldsymbol{\lambda} \in \mathbf{R}^m_+$, $\mathbf{v} \in \mathbf{R}^n_+$, *if any of the following conditions holds:*

i) $P_0 \neq \emptyset$;

ii) $L_0 \overset{\text{def}}{=} \{\boldsymbol{\lambda} \in \mathbf{R}^m_+, \mathbf{v} \in \mathbf{R}^n_+ : \boldsymbol{\lambda} A + \mathbf{v} \geq \boldsymbol{\pi}\} \neq \emptyset$;

iii) $A = \begin{pmatrix} A' \\ I \end{pmatrix}$, *where I is an $n \times n$ identity matrix.*

Corollary 14.1 can also be proved directly using a technique similar to that of the proof of Theorem 14.1.

Proof. i) Since $\boldsymbol{\pi}\mathbf{x} \leq \pi_0$ is a valid inequality for P_0 and $P_0 \neq \emptyset$ then the (primal) linear programming problem

$$\max\ \boldsymbol{\pi}\mathbf{x} \quad \text{subject to} \quad \mathbf{x} \in P_0$$

has an optimal solution $\mathbf{x}_0$. According to Duality theorem, the dual programming problem

$$\min\ (\boldsymbol{\lambda}\boldsymbol{\alpha} + \mathbf{v}\mathbf{d})$$

where

$$\mathbf{y} \equiv (\lambda_1, \ldots, \lambda_m, v_1, \ldots, v_n) \in L_0$$

also has an optimal solution $(\boldsymbol{\lambda}^0, \mathbf{v}^0) \in \mathbf{R}^{m+n}_+$ such that $\boldsymbol{\lambda}^0 A + \mathbf{v}^0 \geq \boldsymbol{\pi}$ and $\boldsymbol{\lambda}^0 \boldsymbol{\alpha} + \mathbf{v}^0 \mathbf{d} = \boldsymbol{\pi}\mathbf{x}_0$. Therefore,

$$\boldsymbol{\pi}\mathbf{x} \leq (\boldsymbol{\lambda}^0 A + \mathbf{v}^0)\mathbf{x} \equiv \boldsymbol{\lambda}^0 (A\mathbf{x}) + \mathbf{v}^0 \mathbf{x} \leq \boldsymbol{\lambda}^0 \boldsymbol{\alpha} + \mathbf{v}^0 \mathbf{d} = \boldsymbol{\pi}\mathbf{x}_0 \leq \pi_0.$$

Here we have used $\boldsymbol{\lambda}^0 \in \mathbf{R}^m_+$, $\mathbf{v} \in \mathbf{R}^n_+$ and the inequalities that define P_0.

ii) Assume that $P_0 = \emptyset$, otherwise we are in case i). Since $L_0 \neq \emptyset$ then there exist $\widehat{\boldsymbol{\lambda}} \in \mathbf{R}^m_+$, $\widehat{\mathbf{v}} \in \mathbf{R}^n_+$ such that

$$\widehat{\boldsymbol{\lambda}} A + \widehat{\mathbf{v}} \geq \boldsymbol{\pi}.$$

If $\widehat{\boldsymbol{\lambda}} \boldsymbol{\alpha} + \widehat{\mathbf{v}} \mathbf{d} \leq \pi_0$, the statement is proved because

$$\boldsymbol{\pi}\mathbf{x} \leq (\widehat{\boldsymbol{\lambda}} A + \widehat{\mathbf{v}})\mathbf{x} \equiv \widehat{\boldsymbol{\lambda}}(A\mathbf{x}) + \widehat{\mathbf{v}}\mathbf{x} \leq \widehat{\boldsymbol{\lambda}} \boldsymbol{\alpha} + \widehat{\mathbf{v}} \mathbf{d} \leq \pi^0.$$

Otherwise, when $\widehat{\boldsymbol{\lambda}} \boldsymbol{\alpha} + \widehat{\mathbf{v}} \mathbf{d} > \pi_0$, since $P_0 = \emptyset$, that is, $\mathbf{x} \in \mathbf{R}^n$ and $\mathbf{0} \leq \mathbf{x} \leq \mathbf{d}$ imply $(A\mathbf{x})_i > \alpha_i$ for at least one i, it follows that there exist $\overline{\boldsymbol{\lambda}} \in \mathbf{R}^m_+$ and $\overline{\mathbf{v}} \in \mathbf{R}^n_+$ such that

$$\overline{\boldsymbol{\lambda}} A + \overline{\mathbf{v}} = \mathbf{0}, \quad \overline{\boldsymbol{\lambda}} \boldsymbol{\alpha} + \overline{\mathbf{v}} \mathbf{d} < 0.$$

Hence, for some $\mu > 0$ we have

$$(\widehat{\boldsymbol{\lambda}} + \mu \overline{\boldsymbol{\lambda}})A + (\widehat{\mathbf{v}} + \mu \overline{\mathbf{v}}) \equiv (\widehat{\boldsymbol{\lambda}} A + \widehat{\mathbf{v}}) + \mu(\overline{\boldsymbol{\lambda}} A + \overline{\mathbf{v}}) \geq \boldsymbol{\pi},$$

$$(\widehat{\boldsymbol{\lambda}} + \mu \overline{\boldsymbol{\lambda}})\boldsymbol{\alpha} + (\widehat{\mathbf{v}} + \mu \overline{\mathbf{v}})\mathbf{d} \equiv (\widehat{\boldsymbol{\lambda}} \boldsymbol{\alpha} + \widehat{\mathbf{v}} \mathbf{d}) + \mu(\overline{\boldsymbol{\lambda}} \boldsymbol{\alpha} + \overline{\mathbf{v}} \mathbf{d}) \leq \pi_0.$$

For $\widehat{\boldsymbol{\lambda}}, \overline{\boldsymbol{\lambda}} \in \mathbf{R}^m_+$; $\widehat{\mathbf{v}}, \overline{\mathbf{v}} \in \mathbf{R}^n_+$; $\mu > 0$ denote

$$\boldsymbol{\lambda}^0 \equiv \widehat{\boldsymbol{\lambda}} + \mu \overline{\boldsymbol{\lambda}} \in \mathbf{R}^m_+; \quad \mathbf{v}^0 \equiv \widehat{\mathbf{v}} + \mu \overline{\mathbf{v}} \in \mathbf{R}^n_+.$$

Then $\boldsymbol{\lambda}^0 A + \mathbf{v}^0 \geq \boldsymbol{\pi}$, that is, $(\boldsymbol{\lambda}^0, \mathbf{v}^0) \in L_0$, $\boldsymbol{\lambda}^0 \boldsymbol{\alpha} + \mathbf{v}^0 \mathbf{d} \leq \pi_0$, and since $\boldsymbol{\pi}\mathbf{x} \leq \pi_0$, the statement of ii) is proved.

iii) Let $A = \begin{pmatrix} A' \\ I \end{pmatrix}$. Therefore, $A' = (a'_{ij})$ is an $(m - n) \times n$ matrix. Choose an arbitrary vector $\boldsymbol{\pi} \in \mathbf{R}^n$. Consider the vector inequality $\boldsymbol{\lambda} A + \mathbf{v} \geq \boldsymbol{\pi}$, where $\boldsymbol{\lambda} \in \mathbf{R}^m_+$, $\mathbf{v} \in \mathbf{R}^n_+$. In this case, it has the form

$$\begin{aligned} &\left(\sum_{i=1}^{m-n} \lambda_i a'_{i1} + \lambda_{m-n+1} + v_1, \sum_{i=1}^{m-n} \lambda_i a'_{i2} + \lambda_{m-n+2} + v_2, \ldots, \right. \\ &\qquad \left. \sum_{i=1}^{m-n} \lambda_i a'_{in} + \lambda_m + v_n \right)^T \geq (\pi_1, \pi_2, \ldots, \pi_n)^T. \end{aligned} \tag{14.9}$$

We can choose arbitrary nonnegative values $\lambda_1 \geq 0, \ldots, \lambda_{m-n} \geq 0$ such that the values

$$l_j^0 := \lambda_{m-n+j} + v_j, \quad j = 1, \ldots, n, \tag{14.10}$$

determined in terms of a'_{ij}, π_j, λ_i, $i = 1, \ldots, m - n$; $j = 1, \ldots, n$, be nonnegative. Since we have these values, we can choose $\lambda_{m-n+j} \geq 0$, $v_j \geq 0$, $j = 1, \ldots, n$ so that (14.10) holds. The structure of the system (14.9) (of matrix A, respectively) allows us to do this.

Hence, $L_0 \neq \emptyset$, and with the use of case ii), we get the statement of iii).

Therefore, in the three cases: i), ii), and iii), the inequality of the form $(\boldsymbol{\lambda}^0 A + \mathbf{v}^0)\mathbf{x} \leq \boldsymbol{\lambda}^0 \boldsymbol{\alpha} + \mathbf{v}^0 \mathbf{d}$ dominates or is equivalent to the inequality $\boldsymbol{\pi}\mathbf{x} \leq \pi_0$ by definition. ■

14.2.2 Valid Inequalities Generation—Modular Arithmetic Approach

Let

$$S = \left\{ \mathbf{x} \in \mathbf{Z}_+^n : \sum_{j=1}^{n} a_j x_j = a_0 \right\},\ a_j \in \mathbf{R},\ j = 0, 1, \ldots, n. \tag{14.11}$$

Assume $q \in \mathbf{Z},\ q > 0$. Present every $a_j,\ j = 0, 1, \ldots, n$, as follows:

$$a_j = \alpha_j q + b_j, \tag{14.12}$$

where $0 \leq b_j < q,\ \alpha_j \in \mathbf{Z},\ j = 0, 1, \ldots, n$, that is, $a_j \equiv b_j (\text{mod}\ q),\ b_j$ are the remainders when a_j is divided by $q,\ j = 0, 1, \ldots, n$.

Consider the (continuous) linear programming problem

$$\max\ \mathbf{cx} \tag{14.13}$$

subject to

$$\mathbf{x} \in S_0 \stackrel{\text{def}}{=} \left\{ \mathbf{x} \in \mathbf{R}_+^n : \sum_{j=1}^{n} a_j x_j = a_0 \right\}, \tag{14.14}$$

where $\mathbf{c}$ is an arbitrary n-vector. The set S_0 is obtained from S (14.11) by relaxing integrality requirements $\mathbf{x} \in \mathbf{Z}^n$. Without loss of generality we can consider (14.13)–(14.14) as a canonical form (with equality constraint) of the one-dimensional linear knapsack problem.

Hence, $\sum_{j=1}^{n} (\alpha_j q + b_j) x_j = \alpha_0 q + b_0$ according to (14.12) and (14.14), that is

$$\sum_{j=1}^{n} b_j x_j - b_0 = \alpha_0 q - q \sum_{j=1}^{n} \alpha_j x_j. \tag{14.15}$$

Let $\mathbf{x}$ be an integral feasible solution to problem (14.13)–(14.14), that is, $\mathbf{x} \in S$ (14.11). Therefore, the right-hand side of (14.15) is integer. Hence

$$\sum_{j=1}^{n} b_j x_j - b_0 \equiv \gamma \in \mathbf{Z}. \tag{14.16}$$

Assume that $\gamma \leq -q$. Hence, $\sum_{j=1}^{n} b_j x_j \leq b_0 - q$, and $\sum_{j=1}^{n} b_j x_j + q \leq b_0$. For $x_j \in \mathbf{Z}_+$, $0 \leq b_j < q$, $j = 1, \ldots, n$, we have $\sum_{j=1}^{n} b_j x_j \geq 0$. Hence, $b_0 \geq q$. However, $0 \leq b_0 < q$, a contradiction. It follows that the assumption was wrong. Therefore, $\gamma > -q$, and since $\gamma \in \mathbf{Z}$, it follows that $\gamma \geq -q + 1$. From (14.16) it follows that

$$\sum_{j=1}^{n} b_j x_j \geq b_0 + 1 - q \tag{14.17}$$

for each integral $\mathbf{x} \in S_0$ (14.14). This means that (14.17) is a valid inequality for S_0 for each integral $\mathbf{x} \in S_0$, that is, for each point of S (14.11), and this is *a class* of valid inequalities for S (14.11) with different $q \in \mathbf{Z}$, $q > 0$.

When $q = 1$ we have $\alpha_j = [a_j]$, $0 \leq b_j < 1$, $j = 0, 1, \ldots, n$, where $[x]$ denotes the greatest integer, not greater than x. From (14.17), with $q = 1$, we have that $\sum_{j=1}^{n} b_j x_j \geq b_0$ is a valid inequality, and it is the strongest inequality among inequalities (14.17). In this case, it has the form

$$\sum_{j=1}^{n} (a_j - [a_j]) x_j \geq a_0 - [a_0] \tag{14.18}$$

and it is called a *Gomory cutting plane*. The inequality (14.18) can also be written in the following equivalent form

$$x_{n+1} \equiv \sum_{j=1}^{n} (a_j - [a_j]) x_j - (a_0 - [a_0]), \quad x_{n+1} \geq 0, \tag{14.18$'$}$$

where x_{n+1} is an additional variable to linear programming problem (14.13)–(14.14).

The inequality (14.18) is well-known. The contribution of the approach suggested here consists in the fact that we have obtained a *class* of valid inequalities and (14.18) is only *an element* of this class.

Without loss of generality let x_1 be the (single) basic variable to the linear programming problem (14.13)–(14.14). Therefore, $x_j = 0$, $j = 2, \ldots, n$. Moreover, according to Lemma 14.1, the optimal solution to the (continuous) linear programming problem (14.13)–(14.14) when (14.5) holds (we can assume that these requirements are satisfied) is namely (14.6). From (14.14) it follows that

$$x_1 = \frac{a_0}{a_1} - \sum_{j=2}^{n} \frac{a_j}{a_1} x_j.$$

Therefore,

$$x_1 = \left[\frac{a_0}{a_1}\right] + \gamma_0 - \sum_{j=2}^{n} \left(\left[\frac{a_j}{a_1}\right] + \gamma_j\right) x_j,$$

where

$$\gamma_j \equiv \frac{a_j}{a_1} - \left[\frac{a_j}{a_1}\right], \quad 0 \leq \gamma_j < 1, \quad j = 0, 2, \ldots, n. \tag{14.19}$$

Then

$$x_1 - \left[\frac{a_0}{a_1}\right] + \sum_{j=2}^{n} \left[\frac{a_j}{a_1}\right] x_j = \gamma_0 - \sum_{j=2}^{n} \gamma_j x_j. \tag{14.20}$$

A) If $\mathbf{x}$ is an integral feasible solution to problem (14.13)–(14.14) then left-hand side of (14.20) is integer. Therefore, the same is fulfilled for the right-hand side

$$\gamma_0 - \sum_{j=2}^{n} \gamma_j x_j \equiv \xi \in \mathbf{Z}.$$

Hence, $\gamma_0 - \sum_{j=2}^{n} \gamma_j x_j \leq \gamma_0 < 1$ because $\gamma_0 \geq 0,\ x_j \geq 0,\ j = 1, \ldots, n$, by assumption and according to (14.19). However, an integer which is less than 1, is less than or equal to 0. Therefore, the necessary and sufficient condition for integrality of the right-hand side of (14.20) is

$$\gamma_0 - \sum_{j=2}^{n} \gamma_j x_j \leq 0,$$

that is

$$x_{n+1} \equiv \sum_{j=2}^{n} \gamma_j x_j - \gamma_0 \geq 0, \tag{14.21}$$

where x_1 is the single basic variable for problem (14.13)–(14.14), x_{n+1} is an additional variable for (14.13)–(14.14). The inequality (14.21) is a valid inequality for S (14.11) and it is called a *Gomory cutting plane*.

Each integral feasible solution to problem (14.13)–(14.14) satisfies (14.21) with equality because $\gamma_0 = 0$ in this case and $x_j = 0,\ j = 2, \ldots, n$.

B) If we write equality $\sum_{j=1}^{n} a_j x_j = a_0$ from (14.11) in the equivalent form $\sum_{j=1}^{n} \frac{a_j}{a_1} x_j = \frac{a_0}{a_1}$ (assuming that $a_1 \neq 0$) then for $\mathbf{x} \notin \mathbf{Z}^n$ we have $\frac{a_0}{a_1} \notin \mathbf{Z}$ in general. Therefore, $\gamma_0 \equiv \frac{a_0}{a_1} - \left[\frac{a_0}{a_1}\right] > 0$ strictly, and from (14.21) we obtain that $x_{n+1} \equiv -\gamma_0 < 0$ for an arbitrary "nonintegral" solution to problem (14.13)–(14.14). Hence, this "nonintegral" solution does not satisfy inequality (14.21).

From cases A) and B) it follows that valid inequality (14.21) for S is a regular cutting plane for S_0 by definition.

14.3 Integrality of the Knapsack Polytope

Integrality of the knapsack polytope is an important topic because the solution to a linear programming problem, when it is solvable, is attained at the vertex (vertices) of the feasible polytope. If we do not fix the linear objective function, the optimal solution is attainable at any vertex of the polytope. Thus, the integrality of the feasible polytope is equivalent to the fact that corresponding maximization linear programming problem has an integral optimal solution for an arbitrary objective function for which it has optimal solution(s).

The following theorem gives a necessary and sufficient condition for integrality of the knapsack polytope.

Theorem 14.2 (Integrality of the knapsack polytope)
Let $a_j,\ j = 0, \dots, n$, be positive integers, $\mathbf{a} = (a_1, \dots, a_n)$, $\mathbf{d} = (d_1, \dots, d_n)$, $\mathbf{d}' = (d'_1, \dots, d'_n) \in \mathbf{Z}^n$, $d'_j \leq d_j$, $a_j d_j \leq a_0$, $j = 1, \dots, n$, and $\sum_{j=1}^{n} a_j d'_j < a_0 < \sum_{j=1}^{n} a_j d_j$.

Polytope

$$P(\mathbf{a}, a_0, \mathbf{d}', \mathbf{d}) = \left\{\mathbf{x} \in \mathbf{R}^n : \sum_{j=1}^{n} a_j x_j \leq a_0,\ d'_j \leq x_j \leq d_j,\ j = 1, \dots, n\right\} \tag{14.22}$$

is an integral polytope for all positive integers $a_j,\ j = 0, 1, \dots, n$, and for all $\mathbf{d}', \mathbf{d} \in \mathbf{Z}^n$, for which it is not empty, if and only if $a_j = \lambda$, $j = 1, \dots, n$ and $a_0 = \lambda k$, where λ and k are positive integers.

Proof. i) Let $a_j = \lambda,\ j = 1, \dots, n;\ a_0 = \lambda k$, where λ and k are positive integers. Consider linear programming problem

$$\max \quad \mathbf{cx} \tag{14.23}$$

subject to

$$\mathbf{x} \in P(\mathbf{a}, a_0, \mathbf{d}', \mathbf{d}), \tag{14.24}$$

where $\mathbf{c}$ is an arbitrary n-vector. Components of any vertex of polytope $P(\mathbf{a}, a_0, \mathbf{d}', \mathbf{d})$ are of the form

$$x_j = d_j, \quad j \in J_1 \tag{14.25}$$

$$x_j = d'_j, \quad j \in J_2 \tag{14.26}$$

$$x_r = \frac{1}{a_r}\Big(a_0 - \sum_{j \in J_1} a_j d_j - \sum_{j \in J_2} a_j d'_j\Big), \quad r \notin J_1 \cup J_2, \tag{14.27}$$

where $J_1 \cup J_2 \cup \{r\} = \{1, \dots, n\}$, or of the form

$$x_i = d_i, \quad i \in I_1 \tag{14.25'}$$

$$x_i = d'_i, \quad i \in I_2, \tag{14.26'}$$

where $I_1 \cup I_2 = \{1, \dots, n\}$. Obviously, components (14.25), (14.26) [(14.25′), (14.26′), respectively] and

$$x_r = \frac{1}{\lambda}\Big(\lambda k - \sum_{j \in J_1} \lambda d_j - \sum_{j \in J_2} \lambda d'_j\Big) = k - \sum_{j \in J_1} d_j - \sum_{j \in J_2} d'_j$$

are integer for $d'_j,\ d_j \in \mathbf{Z},\ j = 1, \dots, n$ and k positive integer. Therefore, each vertex of $P(\mathbf{a}, a_0, \mathbf{d}', \mathbf{d})$ is integral under the above assumptions. Hence, $P(\mathbf{a}, a_0, \mathbf{d}', \mathbf{d})$ is an integral polytope.

ii) Conversely, let $P(\mathbf{a}, a_0, \mathbf{d}', \mathbf{d})$ be an integral polytope for all $\mathbf{d}', \mathbf{d} \in \mathbf{Z}^n$ and for all positive integers a_j, that is, let all vertices of $P(\mathbf{a}, a_0, \mathbf{d}', \mathbf{d})$ be integral points. Polytope $P(\mathbf{a}, a_0, \mathbf{d}', \mathbf{d})$ is a convex combination of these integral vertices because it is a polyhedral set. Therefore, the linear programming problem (14.23)–(14.24) has integral optimal solution(s) for each linear objective function $\mathbf{cx}$.

Components (14.25), (14.26) [(14.25′), (14.26′), respectively] are integer because $\mathbf{d}', \mathbf{d} \in \mathbf{Z}^n$. In order that x_r (14.27) be integer under the assumptions of Theorem 14.2, it is necessary $a_j = \lambda,\ j = 1, \dots, n;\ a_0 = \lambda k$, that is, $(\mathbf{a}, a_0) = \lambda(\mathbf{e}, k)$, where λ and k are positive integers, and $\mathbf{e} = (1, \dots, 1)_n$. ■

Theorem 14.2 gives a criterion for integrality of the knapsack polytope $P(\mathbf{a}, a_0, \mathbf{d}', \mathbf{d})$. When $P(\mathbf{a}, a_0, \mathbf{d}', \mathbf{d})$ is integral, then the integer linear programming problem (14.23), (14.24), $x_j \in \mathbf{Z},\ j = 1, \dots, n$, which belongs to the class of NP-problems according to the classification of S. A. Cook and R. M. Karp, and the continuous linear programming problem (14.23)–(14.24), which belongs to the class of P-problems for which there exist polynomial algorithms, have the same solution(s).

When $d'_j = 0,\ j = 1, \ldots, n$, in the definition of $P(\mathbf{a}, a_0, \mathbf{d}', \mathbf{d})$, then Theorem 14.2 can also be proved with the use of the properties of totally unimodular matrices, considered in Subsections 14.1.2 and 14.1.3.

Corollary 14.2 (Integrality of the knapsack polytope: a modified version)
Let a_0 be a positive integer, $\mathbf{a} = (a_1, \ldots, a_n)$ be a vector of positive integers, $\mathbf{d} = (d_1, \ldots, d_n) \in \mathbf{Z}^n$, $d_j > 0,\ a_j d_j \le a_0,\ j = 1, \ldots, n$, and $a_0 < \sum_{j=1}^{n} a_j d_j$.
Polytope

$$P(\mathbf{a}, a_0, \mathbf{d}) \stackrel{\text{def}}{=} \left\{\mathbf{x} \in \mathbf{R}^n : \sum_{j=1}^{n} a_j x_j \le a_0, \quad 0 \le x_j \le d_j,\ j = 1, \ldots, n\right\}$$

is an integral polytope for all positive integers $a_j,\ j = 0, 1, \ldots, n$, and for all $\mathbf{d} \in \mathbf{Z}^n$, for which it is not empty, if and only if $a_j = \lambda,\ j = 1, \ldots, n$ and $a_0 = \lambda k$, where λ and k are positive integers.

Proof. i) Let $a_j = \lambda,\ j = 1, \ldots, n$ and $a_0 = \lambda k$, where λ and k are positive integers. Using these assumptions, we get

$$P(\mathbf{a}, a_0, \mathbf{d}) = \left\{\mathbf{x} \in \mathbf{R}^n_+ : \sum_{j=1}^{n} x_j \le k, \quad x_j \le d_j,\ j = 1, \ldots, n\right\}.$$

The $(n+1) \times n$ matrix A of coefficients of the constraints of $P(\mathbf{a}, a_0, \mathbf{d})$ is

$$A = \begin{pmatrix} 1\ 1 \ldots 1 \\ 1\ 0 \ldots 0 \\ 0\ 1 \ldots 0 \\ .\ \ . \ \ldots . \\ 0\ 0 \ldots 1 \end{pmatrix}.$$

A is a (0, 1) matrix with no more than two nonzero elements in each column and the rows of A can be partitioned into two subsets: Q_1 containing row No. 1, and Q_2 containing rows No. 2 to No. $n+1$, such that Q_1 and Q_2 satisfy conditions a. and b. of Lemma 14.5, Subsection 14.1.3. Therefore, A is TU, and $P(\mathbf{a}, a_0, \mathbf{d})$ is an integral polytope for all positive integers a_0 and for all $\mathbf{d} \in \mathbf{Z}^n$ according to Lemma 14.3.

ii) Conversely, let $P(\mathbf{a}, a_0, \mathbf{d})$ be an integral polytope for all positive integers $a_j,\ j = 0, 1, \ldots, n$ and for all $\mathbf{d} \in \mathbf{Z}^n$ for which it is not empty, that is, all its vertices are points with integer components. Therefore, the $(n+1) \times n$ matrix of coefficients of the constraints of $P(\mathbf{a}, a_0, \mathbf{d})$

$$A = \begin{pmatrix} a_1\ a_2 \ldots a_n \\ 1\ \ 0\ \ \ldots 0 \\ 0\ \ 1\ \ \ldots 0 \\ .\ \ \ .\ \ \ \ldots . \\ 0\ \ 0\ \ \ldots 1 \end{pmatrix}$$

is TU according to Lemma 14.4. Hence, A is a $(0, 1, -1)$ matrix. Under assumptions of Corollary 14.2, this is possible if and only if $a_j = 1, \ j = 1, \ldots, n$, or equivalently, if and only if $a_j = \lambda, \ j = 1, \ldots, n$, and $a_0 = \lambda k$, where λ and k are positive integers. ■

Proposition 1 ([97]) is equivalent to Corollary 14.2, and therefore, it follows from Theorem 14.2 with $d'_j = 0, \ j = 1, \ldots, n$.

Chapter 15
Relaxation of the Equality Constrained Convex Continuous Knapsack Problem

In this chapter, problem $(C^{=})$, defined by (6.1)–(6.3) in Chapter Six of Part Two, is considered again. Necessary and sufficient condition (characterization theorem) for problem $(C^{=})$ is stated and proved in terms of a relaxed version (6.1)–(6.2) of this problem.

15.1 Introduction

Consider the following separable continuous convex knapsack problem with bounded variables

$(C^{=})$

$$\min \left\{ c(\mathbf{x}) = \sum_{j \in J} c_j(x_j) \right\} \tag{15.1}$$

subject to

$$\sum_{j \in J} d_j x_j = \alpha \tag{15.2}$$

$$a_j \le x_j \le b_j, \quad j \in J, \tag{15.3}$$

where $c_j(x_j),\ j \in J \stackrel{\text{def}}{=} \{1, ..., n\}$, are convex differentiable functions, $d_j > 0$ for each $j \in J, \mathbf{x} = (x_j)_{j \in J}$, and $x_j,\ j \in J$ are the variables to be optimally determined.

As pointed out above, problem (15.1)–(15.3) is the problem $(C^{=})$, defined by (6.1)–(6.3) in Chapter Six.

Some assumptions and results of Chapter Six are recalled below.

Suppose that the following assumptions are satisfied.

© The Author(s), under exclusive license to Springer Nature Switzerland AG 2021

S. M. Stefanov, *Separable Optimization*, Springer Optimization and Its Applications 177,
https://doi.org/10.1007/978-3-030-78401-0_15

15.a) $a_j \le b_j$ for every $j \in J$. If $a_k = b_k$ for some index $k \in J$, then the value $x_k := a_k = b_k$ is known in advance.

15.b) $\sum_{j\in J} d_j a_j \le \alpha \le \sum_{j\in J} d_j b_j$. Otherwise, the constraints (15.2)–(15.3) are inconsistent and the feasible set of problem $(C^=)$ is empty.

The Lagrangian function for problem $(C^=)$ is defined as follows:

$$L(\mathbf{x}, \mathbf{u}, \mathbf{v}, \lambda)$$

$$= \sum_{j\in J} c_j(x_j) + \lambda\Big(\sum_{j\in J} d_j x_j - \alpha\Big) + \sum_{j\in J} u_j(a_j - x_j) + \sum_{j\in J} v_j(x_j - b_j), \tag{15.4}$$

where $\lambda \in \mathbf{R}$; $\mathbf{u}$, $\mathbf{v} \in \mathbf{R}^n_+$, and $\mathbf{R}^n_+$ is the totality of all $n-$tuples with n real nonnegative components.

The Karush-Kuhn-Tucker (KKT) necessary and sufficient optimality conditions for $\mathbf{x}^* \in \mathbf{R}^n$ to be minimum solution of problem $(C^=)$ are

$$c'_j(x^*_j) + \lambda d_j - u_j + v_j = 0, \quad j \in J, \qquad (15.5) \equiv (6.7)$$

$$u_j(a_j - x^*_j) = 0, \quad j \in J, \qquad (15.6) \equiv (6.8)$$

$$v_j(x^*_j - b_j) = 0, \quad j \in J, \qquad (15.7) \equiv (6.9)$$

$$\lambda \in \mathbf{R};\ u_j \in \mathbf{R}_+,\ v_j \in \mathbf{R}_+, \quad j \in J, \qquad (15.8) \equiv (6.10)$$

$$\sum_{j\in J} d_j x^*_j = \alpha, \qquad (15.9) \equiv (6.11)$$

$$a_j \le x^*_j \le b_j, \quad j \in J, \qquad (15.10) \equiv (6.12)$$

where $\lambda \in \mathbf{R}$; $\mathbf{u}, \mathbf{v} \in \mathbf{R}^n_+$ are the Lagrange multipliers (dual variables), associated with constraint (15.2) and the two inequalities of constraint (15.3), respectively.

The following Theorem 6.1 is stated and proved in Chapter Six.

Theorem 15.1 (Characterization of the optimal solution to problem $(C^=)$) *Under assumptions* 15.a)–15.b), *a feasible solution* $\mathbf{x}^* = (x^*_j)_{j\in J}$ *is an optimal solution to problem* $(C^=)$ *if and only if there exists a real number* $\lambda \in \mathbf{R}^1$ *such that*

$$x^*_j = a_j, \quad j \in J^\lambda_a \stackrel{\text{def}}{=} \left\{ j \in J : \lambda \ge -\frac{c'_j(a_j)}{d_j} \right\}, \qquad (15.11) \equiv (6.4)$$

$$x^*_j = b_j, \quad j \in J^\lambda_b \stackrel{\text{def}}{=} \left\{ j \in J : \lambda \le -\frac{c'_j(b_j)}{d_j} \right\}, \qquad (15.12) \equiv (6.5)$$

$$x_j^* : \lambda d_j = -c_j'(x_j^*), \quad j \in J^\lambda \stackrel{\text{def}}{=} \left\{ j \in J : -\frac{c_j'(b_j)}{d_j} \le \lambda \le -\frac{c_j'(a_j)}{d_j} \right\}. \qquad (15.13) \equiv (6.6)$$

When $c(\mathbf{x})$ is strictly convex, then inequalities, defining J^λ in (15.13), *are strict, and the optimal solution to problem $(C^=)$ is unique.*

15.2 The Relaxed Problem

The optimal solution of the original problem $(C^=)$ is characterized below in terms of the optimal solution of a relaxed problem.

Denote by $(C_r^=)$ the following relaxation of problem $(C^=)$:
$(C_r^=)$

$$\min \left\{ c(\mathbf{x}) = \sum_{j \in J} c_j(x_j) \right\} \qquad (15.14) \equiv (15.1)$$

subject to

$$\sum_{j \in J} d_j x_j = \alpha. \qquad (15.15) \equiv (15.2)$$

The KKT conditions for the relaxed problem $(C_r^=)$ are

$$c_j'(x_j) + \lambda d_j = 0, \quad j \in J, \quad \lambda \in \mathbf{R}, \qquad (15.16)$$

$$\sum_{j \in J} d_j x_j = \alpha. \qquad (15.17)$$

The following theorem gives a characterization of the optimal solution to problem $(C^=)$ in terms of the optimal solution to the relaxed problem $(C_r^=)$ and the Lagrange multiplier (dual variable) λ, associated with the equality constraint (15.2).

Theorem 15.2 (Characterization of the optimal solution to problem $(C^=)$ in terms of the optimal solution to a relaxed problem)
The optimal solution $\mathbf{x}^* = (x_j^*)_{j \in J}$ to problem $(C^=)$, defined by (15.1)–(15.3), *can be presented as follows*

$$x_j^* = a_j, \quad j \in J_a \stackrel{\text{def}}{=} \{j \in J : x_j \le a_j\} = J_a^\lambda, \qquad (15.18)$$

$$x_j^* = b_j, \quad j \in J_b \stackrel{\text{def}}{=} \{j \in J : x_j \ge b_j\} = J_b^\lambda, \qquad (15.19)$$

$$x_j^* : \lambda d_j = -c_j'(x_j^*), \quad j \in J_{ab} \stackrel{\text{def}}{=} \{j \in J : a_j \le x_j \le b_j\} = J^\lambda, \qquad (15.20)$$

where $\mathbf{x} = (x_j)_{j \in J}$ is an optimal solution to the relaxed problem $(C_r^=)$.

Proof. Let $\mathbf{x} = (x_j)_{j \in J}$ be an optimal solution to the relaxed problem $(C_r^=)$.

1) Let $k \in J_a$. From $x_k \leq a_k$ for $k \in J_a$ and monotonicity of the derivatives of the differentiable convex functions $c_j(x_j),\ j \in J$ (Theorem 1.31) it follows that $c_k'(a_k) \geq c_k'(x_k)$. Therefore

$$-\frac{c_k'(a_k)}{d_k} \leq -\frac{c_k'(x_k)}{d_k} = \lambda,$$

where we have used that $d_j > 0$ for all $j \in J$ by the assumption. Hence $k \in J_a$ implies $\lambda \geq -\frac{c_k'(a_k)}{d_k}$, that is, $k \in J_a^\lambda$.

2) Conversely, if $j \in J_a^\lambda$ (15.11), then

$$-\frac{c_j'(a_j)}{d_j} \leq \lambda \overset{(15.16)}{=} -\frac{c_j'(x_j)}{d_j} = -\frac{c_i'(x_i)}{d_i} \leq -\frac{c_i'(a_i)}{d_i}$$

for all $i \in J \setminus J_a^\lambda$ because $\lambda \leq -\frac{c_i'(a_i)}{d_i}$ for such indices i according to (15.11), (15.12), and (15.13). Using that $d_j > 0,\ j \in J$, by assumption, it follows that $-c_j'(a_j) \leq -c_j'(x_j)$, hence $a_j \geq x_j$ according to monotonicity of the derivatives of the convex differentiable functions $c_j(x_j),\ j \in J$. Therefore $j \in J_a^\lambda$ implies $j \in J_a$.

From cases 1) and 2) it follows that $J_a^\lambda = J_a$.

3)–4) The proof of equivalence of index sets J_b^λ and J_b is omitted because it is similar to the proof of equivalence of index sets J_a^λ and J_a.

5) Let $k \in J_{ab}$. Therefore $a_k \leq x_k \leq b_k$. According to monotonicity of the derivatives of the convex differentiable functions $c_j(x_j),\ j \in J$ and using $d_j > 0,\ j \in J$, it follows that

$$-\frac{c_k'(a_k)}{d_k} \geq -\frac{c_k'(x_k)}{d_k} \geq -\frac{c_k'(b_k)}{d_k}.$$

Denote $\lambda = -\frac{c_k'(x_k)}{d_k}$ (see (15.16) for $j = k$). Then $\lambda \in \left[-\frac{c_k'(b_k)}{d_k}, -\frac{c_k'(a_k)}{d_k}\right]$. Using monotonicity of $-\frac{c_j'(x_j)}{d_j}$ (because $c_j(x_j)$ are convex differentiable functions and $d_j > 0,\ j \in J$) and $k \notin J_a = J_a^\lambda,\ k \notin J_b = J_b^\lambda$, then $k \in J^\lambda$.

6) Conversely, let $j \in J^\lambda$ (15.13). Therefore, using monotonicity of the derivatives of the convex differentiable functions $c_j(x_j),\ j \in J; d_j > 0,\ j \in J$ by assumption, $a_j \leq b_j$ for all $j \in J$, and $j \notin J_a^\lambda = J_a,\ j \notin J_b^\lambda = J_b$, it follows that $a_j \leq x_j \leq b_j$, that is, $j \in J_{ab}$.

From cases 5) and 6) it follows that $J^\lambda = J_{ab}$. ∎

Chapter 16
On the Solution of Multidimensional Convex Separable Continuous Knapsack Problem with Bounded Variables

A minimization problem with a convex separable objective function subject to linear equality constraints and box constraints (bounds on the variables), known as the multidimensional convex separable continuous knapsack problem with bounded variables, is considered in this chapter. A necessary and sufficient optimality condition (characterization theorem) is proved for a feasible solution to be an optimal solution to this problem. Primal-dual analysis for the considered problem is also included.

16.1 Statement of the Problem

Consider the following convex separable knapsack problem

$(K_m^=)$

$$\min \left\{ c(\mathbf{x}) = \sum_{j \in J} c_j(x_j) \right\} \tag{16.1}$$

subject to

$$\sum_{j \in J} d_{ij} x_j = \alpha_i, \quad i = 1, \ldots, m \tag{16.2}$$

$$a_j \le x_j \le b_j, \quad j \in J, \tag{16.3}$$

where $c_j(x_j),\ j \in J$, are differentiable convex functions, defined on the open convex sets $X_j,\ j \in J$, respectively; $d_{ij} > 0$ for every $i = 1, \ldots, m$ and $j \in J$; $\alpha_i \in \mathbb{R},\ i = 1, \ldots, m;\ a_j,\ b_j \in \mathbb{R},\ j \in J$; $\mathbf{x} = (x_j)_{j \in J}$, and $J \equiv \{1, ..., n\}$.

The feasible region X, defined by constraints (16.2)–(16.3), is a convex set because it is an intersection of the convex set (16.2), defined by m hyperplanes, and the box (16.3) of dimension $|J| = n$.

© The Author(s), under exclusive license to Springer Nature Switzerland AG 2021
S. M. Stefanov, *Separable Optimization*, Springer Optimization and Its Applications 177,
https://doi.org/10.1007/978-3-030-78401-0_16

Feasible region of the form (16.2)–(16.3) is a polytope and it is known as the (multiple) *knapsack polytope*.

Suppose that the following assumptions are satisfied.

16.a) $a_j \le b_j$ for all $j \in J$. If $a_k = b_k$ for some $k \in J$, then the value $x_k := a_k = b_k$ is determined a priori.

16.b) $\sum_{j \in J} d_{ij} a_j \le \alpha_i \le \sum_{j \in J} d_{ij} b_j$, $i = 1, \ldots, m$. Otherwise the constraints (16.2)–(16.3) are inconsistent and the feasible region X, defined by (16.2)–(16.3), is empty.

16.2 Characterization Theorem

The Lagrangian for problem $(K_m^=)$ is

$$L(\mathbf{x}, \mathbf{u}, \mathbf{v}, \boldsymbol{\lambda})$$

$$= \sum_{j \in J} c_j(x_j) + \sum_{i=1}^{m} \lambda_i \Big(\sum_{j \in J} d_{ij} x_j - \alpha_i \Big) + \sum_{j \in J} u_j (a_j - x_j) + \sum_{j \in J} v_j (x_j - b_j), \tag{16.4}$$

where $\boldsymbol{\lambda} \in \mathbb{R}^m$; $\mathbf{u}, \ \mathbf{v} \in \mathbb{R}^n_+$, and $\mathbb{R}^n_+$ consists of all vectors with n real nonnegative components.

The Karush-Kuhn-Tucker (KKT) necessary and sufficient optimality conditions for the minimum solution $\mathbf{x}^* = (x_j^*)_{j \in J}$ of problem $(K_m^=)$ are

$$c_j'(x_j^*) + \sum_{i=1}^{m} \lambda_i d_{ij} - u_j + v_j = 0, \quad j \in J, \tag{16.5}$$

$$u_j (a_j - x_j^*) = 0, \quad j \in J, \tag{16.6}$$

$$v_j (x_j^* - b_j) = 0, \quad j \in J, \tag{16.7}$$

$$\sum_{j \in J} d_{ij} x_j^* = \alpha_i, \quad i = 1, \ldots, m, \tag{16.8}$$

$$a_j \le x_j^* \le b_j, \quad j \in J, \tag{16.9}$$

$$\lambda_i \in \mathbb{R}^1, \ i = 1, \ldots, m, \quad u_j \in \mathbb{R}^1_+, \ v_j \in \mathbb{R}^1_+, \quad j \in J, \tag{16.10}$$

where λ_i $(i = 1, \ldots, m)$, u_j, v_j $(j \in J)$ are the Lagrange multipliers, associated with the constraints (16.2), $a_j \le x_j$, $x_j \le b_j$ $(j \in J)$, respectively. If $a_j = -\infty$ or $b_j = +\infty$ for some $j \in J$, we do not consider the corresponding KKT condition

(16.6) (KKT condition (16.7), respectively) and multiplier u_j (multiplier v_j, respectively).

The following theorem gives necessary and sufficient optimality condition (characterization) of the optimal solution to problem $(K_m^=)$.

Theorem 16.1 (Characterization of the optimal solution to problem $(K_m^=)$)
Let assumptions 16.a)–16.b) *be satisfied. A feasible solution* $\mathbf{x}^* = (x_j^*)_{j \in J} \in X$ *is an optimal solution to problem* $(K_m^=)$ *if and only if there exist* $\lambda_i \in \mathbb{R},\ i = 1, \ldots, m$, *such that*

$$x_j^* = a_j, \quad j \in J_a^\lambda \stackrel{\text{def}}{=} \left\{ j \in J : -c_j'(a_j) \le \sum_{i=1}^m \lambda_i d_{ij} \right\} \tag{16.11}$$

$$x_j^* = b_j, \quad j \in J_b^\lambda \stackrel{\text{def}}{=} \left\{ j \in J : -c_j'(b_j) \ge \sum_{i=1}^m \lambda_i d_{ij} \right\} \tag{16.12}$$

$$x_j^* : -c_j'(x_j^*) = \sum_{i=1}^m \lambda_i d_{ij},$$

$$j \in J^\lambda \stackrel{\text{def}}{=} \left\{ j \in J : -c_j'(b_j) \le \sum_{i=1}^m \lambda_i d_{ij} \le -c_j'(a_j) \right\}. \tag{16.13}$$

When convex functions $c_j(x_j),\ j \in J$, *are strictly convex, then inequalities, defining* J^λ *in* (16.13), *are strict.*

Proof. *Necessity.* Let $\mathbf{x}^* = (x_j^*)_{j \in J}$ be an optimal solution to problem $(K_m^=)$. According to KKT theorem, there exist constants $\lambda_i,\ i = 1, \ldots, m$ and $u_j, v_j,\ j \in J$, such that KKT conditions (16.5) – (16.10) are satisfied.

(a) If $x_j^* = a_j$, then $u_j \ge 0,\ v_j = 0$ according to KKT conditions (16.6), (16.7), and (16.10). Therefore (16.5) and $u_j \ge 0,\ j \in J$ imply

$$c_j'(x_j^*) \equiv c_j'(a_j) = u_j - \sum_{i=1}^m \lambda_i d_{ij} \ge -\sum_{i=1}^m \lambda_i d_{ij}.$$

Multiplying this inequality by (-1), we obtain

$$-c_j'(a_j) \le \sum_{i=1}^m \lambda_i d_{ij}.$$

(b) If $x_j^* = b_j$, then $u_j = 0,\ v_j \ge 0$ according to KKT conditions (16.6), (16.7), and (16.10). Therefore (16.5) and $v_j \ge 0,\ j \in J$ imply

$$c'_j(x^*_j) \equiv c'_j(b_j) = -v_j - \sum_{i=1}^{m} \lambda_i d_{ij} \le -\sum_{i=1}^{m} \lambda_i d_{ij}.$$

Multiplying this inequality by (-1), we get

$$-c'_j(b_j) \ge \sum_{i=1}^{m} \lambda_i d_{ij}.$$

(c) If $a_j < x^*_j < b_j$, then $u_j = v_j = 0$ according to KKT conditions (16.6) and (16.7), respectively. Therefore (16.5) implies

$$-c'_j(x^*_j) = \sum_{i=1}^{m} \lambda_i d_{ij}.$$

Using that $a_j < x^*_j < b_j$ and $c_j(x_j),\ j \in J$, are convex functions, it follows that $c'_j(a_j) \le c'_j(x^*_j) \le c'_j(b_j)$, that is, in case (c) we have

$$-c'_j(b_j) \le -c'_j(x^*_j) = \sum_{i=1}^{m} \lambda_i d_{ij} \le -c'_j(a_j).$$

When convex functions $c_j(x_j),\ j \in J$, are strictly convex, these inequalities are strict.

In order to describe cases (a), (b), (c), we introduce the index sets J_a^λ, J_b^λ, J^λ, defined by (16.11), (16.12) and (16.13), respectively. It is obvious that $J_a^\lambda \cup J_b^\lambda \cup J^\lambda = J$. The "necessity" part of Theorem 16.1 is proved.

Sufficiency. Conversely, let $\mathbf{x}^* \in X$ and components of $\mathbf{x}^*$ satisfy (16.11), (16.12), (16.13).

Set:

$$\sum_{i=1}^{m} \lambda_i d_{ij} = -c'_j(x^*_j),\ \ u_j = v_j = 0 \ \text{ for } j \in J^\lambda;$$

$$u_j = c'_j(a_j) + \sum_{i=1}^{m} \lambda_i d_{ij} \ \ (\ge 0 \text{ according to (16.11)}),\ \ v_j = 0 \ \text{ for } j \in J_a^\lambda;$$

$$u_j = 0,\ \ v_j = -c'_j(b_j) - \sum_{i=1}^{m} \lambda_i d_{ij} \ \ (\ge 0 \text{ according to (16.12)}) \ \text{ for } j \in J_b^\lambda.$$

By the use of these expressions, it is easy to check that KKT conditions (16.5), (16.6), (16.7), (16.10) are satisfied. KKT conditions (16.8), (16.9) are also satisfied according to the assumption that $\mathbf{x}^* \in X$.

Hence, in all possible cases, considered above, $\lambda_i,\ i = 1, \ldots, m;\ x^*_j, u_j, v_j,\ j \in J$, satisfy KKT conditions (16.5)–(16.10) for problem $(K_m^=)$, which are necessary and sufficient conditions for a feasible solution to be an optimal solution to the convex minimization problem. Therefore $\mathbf{x}^*$ is an optimal solution to problem $(K_m^=)$. When $c_j(x_j),\ j \in J$, are strictly convex functions, the optimal solution is unique. ■

Since the optimal solution $\mathbf{x}^*$ to problem $(K_m^=)$ depends on $\boldsymbol{\lambda} = (\lambda_1, \ldots, \lambda_m)$ (expressions (16.11)–(16.13)), consider components of $\mathbf{x}^*$ as functions of $\boldsymbol{\lambda}$ for different $\boldsymbol{\lambda} \in \mathbb{R}^m$:

$$x_j(\boldsymbol{\lambda}) = \begin{cases} a_j, & j \in J_a^{\lambda}, \\ b_j, & j \in J_b^{\lambda}, \\ x_j^* : \quad -c_j'(x_j^*) = \sum_{i=1}^{m} \lambda_i d_{ij}, & j \in J^{\lambda}. \end{cases} \tag{16.14}$$

Functions $x_j(\boldsymbol{\lambda}),\ j \in J$, are piecewise, monotone, piecewise differentiable functions of $\boldsymbol{\lambda}$, with a pair of break hyperplanes

$$A_j = \left\{ \boldsymbol{\lambda} \in \mathbb{R}^m : -c_j'(a_j) = \sum_{i=1}^{m} \lambda_i d_{ij} \right\}, \quad j \in J$$

and

$$B_j = \left\{ \boldsymbol{\lambda} \in \mathbb{R}^m : -c_j'(b_j) = \sum_{i=1}^{m} \lambda_i d_{ij} \right\}, \quad j \in J.$$

Complexity of problem $(K_m^=)$.

Since $u_j \geq 0,\ v_j \geq 0,\ j \in J$, and since the complementary KKT conditions (16.6), (16.7) must be satisfied, in order to find $x_j^*,\ j \in J$, from the KKT system (16.5)–(16.10), we have to consider all possible cases for u_j and $v_j,\ j \in J$: all u_j, v_j equal to 0; all u_j, v_j different from 0; some of u_j, v_j equal to 0 and some of u_j, v_j different from 0. The number of these cases is $2^{2|J|} = 2^{2n}$, where $2n$ is the number of Lagrange multipliers $u_j, v_j,\ j \in J$, where $|J| = n$. Obviously this is an enormous number of cases, especially for large-scale problems. For example, when $n = 1500$, we have to consider $2^{3000} \approx 10^{900}$ cases. (Only for comparison, recall that astrophysicists have determined that the age of the Universe is approximately $1,5.10^{10}$ years, that is, approximately "only" 10^{18} seconds.) Furthermore, in each case, we have to solve a large-scale system of nonlinear equations in $\lambda_i,\ i = 1, \ldots, m;\ x_j^*, u_j, v_j,\ j \in J$. Therefore, the direct application of the KKT theorem, using explicit enumeration of all possible cases, for solving large-scale problems of the considered form would not give a result in real time, and efficient characterization theorems and methods for solving the considered problem will be very useful.

Importance of Theorem 16.1, similarly to Theorem 5.1, Theorem 6.1., Theorem 6.2, etc., consists in the fact that it describes components of the optimal solution to problem $(K_m^=)$ only through the Lagrange multipliers $\lambda_i,\ i = 1, \ldots, m$, associated with the equality constraints (16.2).

16.3 Primal-dual Analysis

When the optimal Lagrange multipliers λ_i^*, $i = 1, \ldots, m$, associated with equality constraints (16.2), are known, then problem ($K_m^=$), defined by (16.1)–(16.3), can be replaced by the following convex separable optimization problem

$$\min \left\{ \sum_{j \in J} c_j(x_j) + \sum_{i=1}^{m} \lambda_i^* \left(\sum_{j \in J} d_{ij} x_j \right) - \sum_{i=1}^{m} \lambda_i^* \alpha_i \right\}$$

subject to

$$\mathbf{x} \equiv (x_1, \ldots, x_n) \in A \stackrel{\text{def}}{=} \{\mathbf{x} \in \mathbb{R}^n : a_j \le x_j \le b_j, \ j \in J\}.$$

The problem, dual to problem ($K_m^=$), is

$$\max \ \Psi(\boldsymbol{\lambda}) \tag{16.15}$$

subject to

$$\boldsymbol{\lambda} \equiv (\lambda_1, \ldots, \lambda_m) \in \mathbb{R}^m, \tag{16.16}$$

where

$$\Psi(\boldsymbol{\lambda}) = \min_{\mathbf{x} \in A} \left\{ \sum_{j \in J} c_j(x_j) + \sum_{i=1}^{m} \lambda_i \left(\sum_{j \in J} d_{ij} x_j \right) - \sum_{i=1}^{m} \lambda_i \alpha_i \right\}.$$

Thus, using the Lagrangian duality and Theorem 16.1, we can replace the multivariate problem ($K_m^=$) of $\mathbf{x} \in \mathbb{R}^n$ by another multivariate problem (16.15)–(16.16) of $\boldsymbol{\lambda} \in \mathbb{R}^m$. Since usually the number m of "general" linear equality constraints (16.2) is less than the number n of variables (of box constraints (16.3)) of the original problem ($K_m^=$), it turns out that in many cases the dual problem (16.15)–(16.16) is easier to be solved than the original problem ($K_m^=$).

Chapter 17
Characterization of the Optimal Solution of the Convex Generalized Nonlinear Transportation Problem

Convex generalized nonlinear transportation problem and convex generalized nonlinear transportation problem with bounds on the variables are considered in this chapter. A necessary and sufficient optimality condition (characterization theorem) for a feasible solution to be an optimal solution to each of these problems is formulated and proved.

17.1 Statement of the Problems

The transportation problem (TP) consists in determining the minimum cost plan for distribution of a product from m sources (supply points, factories) to n destinations (markets, consumers), subject to availability (capacity) of each source and necessity (demand) of each destination. In the case when the transportation costs depend nonlinearly on the transported units, we obtain the *nonlinear transportation problem* (NTP). The version of the NTP, in which the amount of the transported product from sources to destinations is changed (usually it is decreased) during the transportation process, is called the *generalized nonlinear transportation problem* (GNTP).

Let for each $i = 1, \dots, m; \quad j = 1, \dots n$,

- x_{ij} be the amount of product, sent from source i to destination j $(x_{ij} \geq 0)$,
- c_{ij} be the unit transportation cost from source i to destination j,
- a_i be the available amount of product at source i (capacity of source i),
- b_j be the demand at destination j (necessity of destination j).

Denote by

$$\mathbf{x} = (x_{ij})_{m\times n} \equiv (x_{11}, \dots, x_{1n} \mid x_{21}, \dots, x_{2n} \mid \dots \mid x_{m1}, \dots, x_{mn})$$

© The Author(s), under exclusive license to Springer Nature Switzerland AG 2021
S. M. Stefanov, *Separable Optimization*, Springer Optimization and Its Applications 177,
https://doi.org/10.1007/978-3-030-78401-0_17

the $m \times n$ matrix ($m \times n$ vector) of transported amounts x_{ij}, $i = 1, \ldots, m;\ j = 1, \ldots, n$.

Then the classical *linear transportation problem* (LTP) is formulated as follows: (LTP)

$$\min \left\{ f(\mathbf{x}) = \sum_{i=1}^{m} \sum_{j=1}^{n} c_{ij} x_{ij} \right\} \tag{17.1}$$

subject to

$$\sum_{j=1}^{n} x_{ij} = a_i, \quad i = 1, \ldots, m, \tag{17.2}$$

$$\sum_{i=1}^{m} x_{ij} = b_j, \quad j = 1, \ldots, n, \tag{17.3}$$

$$x_{ij} \geq 0, \quad i = 1, \ldots, m;\ j = 1, \ldots, n. \tag{17.4}$$

Constraints (17.2) ensure that for each source, the sum of all shipped amounts is equal to its available amount, and conditions (17.3) require that for each destination, the sum of all transported amounts to this destination satisfies its demand.

If

$$\sum_{i=1}^{m} a_i = \sum_{j=1}^{n} b_j, \tag{17.5}$$

that is, if all available amounts of the product are equal to all demand amounts of this product, the transportation model is called *balanced*. However, in many real situations, condition (17.5) is not satisfied, and the transportation model is said to be *unbalanced* in this case. The unbalanced model can be reduced to a balanced model by introducing a dummy source or a dummy destination, depending on what unbalanced situation occurs: $\sum_{i=1}^{m} a_i < \sum_{j=1}^{n} b_j$ or $\sum_{i=1}^{m} a_i > \sum_{j=1}^{n} b_j$.

As mentioned above, usually in real situations transportation costs depend nonlinearly on the transported units, and the amounts of the transported product from sources to destinations are changed. This is the motivation to consider the so-called *generalized nonlinear transportation problem* (GNTP), which is formulated as follows:

(GNTP)

$$\min \left\{ c(\mathbf{x}) = \sum_{i=1}^{m} \sum_{j=1}^{n} c_{ij}(x_{ij}) \right\} \tag{17.6}$$

subject to

$$\sum_{j=1}^{n} x_{ij} = a_i, \quad i = 1, \dots, m, \qquad (17.7) \equiv (17.2)$$

$$\sum_{i=1}^{m} x_{ij} = b_j, \quad j = 1, \dots, n, \qquad (17.8) \equiv (17.3)$$

$$x_{ij} \geq 0, \quad i = 1, \dots, m; \ j = 1, \dots, n. \qquad (17.9) \equiv (17.4)$$

Here, $c_{ij}(x_{ij}),\ i = 1, \dots, m;\ j = 1, \dots, n$, are convex continuously differentiable functions. Furthermore, the objective function and the constraint functions of problem (GNTP) are *separable*, that is, they are expressed as the sums of single-variable functions.

Therefore, the (GNTP) is a convex separable optimization problem. In the case of *strictly* convex functions $c_{ij}(x_{ij}),\ i = 1, \dots, m;\ j = 1, \dots, n$, the optimal solution of problem (GNTP), if there exists such a solution, is *unique*.

17.2 Characterization Theorem for Problem (GNTP)

The Lagrangian function for problem (GNTP) is defined as follows:

$$L(\mathbf{x}, \boldsymbol{\lambda}, \boldsymbol{\mu}) = \sum_{i=1}^{m} \sum_{j=1}^{n} c_{ij}(x_{ij}) + \sum_{i=1}^{m} \lambda_i \left(a_i - \sum_{j=1}^{n} x_{ij} \right) + \sum_{j=1}^{n} \mu_j \left(b_j - \sum_{i=1}^{m} x_{ij} \right), \qquad (17.10)$$

where $\boldsymbol{\lambda} = (\lambda_1, \dots, \lambda_m) \in \mathbf{R}^m$, $\boldsymbol{\mu} = (\mu_1, \dots, \mu_n) \in \mathbf{R}^n$ consist of Lagrange multipliers (dual variables), associated with constraints (17.7) and (17.8), respectively.

Denote by T the feasible set of problem (GNTP), defined by constraints (17.7)–(17.9).

The following theorem gives necessary and sufficient condition (characterization) of the optimal solution of problem (GNTP).

Theorem 17.1 (Characterization of the optimal solution of problem (GNTP)) *A feasible solution* $\mathbf{x}^* = (x^*_{ij})_{m \times n} \in T$ *is an optimal solution to problem* (GNTP) *if and only if there exist constants* $\lambda_i,\ i = 1, \dots, m$, *and* $\mu_j,\ j = 1, \dots, n$, *such that*

$$c'_{ij}(x^*_{ij}) - \lambda_i - \mu_j = 0 \quad \text{for } x^*_{ij} > 0, \ 1 \leq i \leq m, \ 1 \leq j \leq n, \qquad (17.11)$$

$$c'_{ij}(x^*_{ij}) - \lambda_i - \mu_j \geq 0 \quad \text{for } x^*_{ij} = 0, \ 1 \leq i \leq m, \ 1 \leq j \leq n. \qquad (17.12)$$

Proof. The Karush-Kuhn-Tucker (KKT) necessary and sufficient optimality conditions for the convex problem (GNTP) are

$$c'_{ij}(x^*_{ij}) - \lambda_i - \mu_j \geq 0, \quad i = 1, \ldots, m; \quad j = 1, \ldots, n, \tag{17.13}$$

$$x^*_{ij}.[c'_{ij}(x^*_{ij}) - \lambda_i - \mu_j] = 0, \quad i = 1, \ldots, m; \quad j = 1, \ldots, n, \tag{17.14}$$

$$\sum_{j=1}^{n} x^*_{ij} = a_i, \quad i = 1, \ldots, m, \tag{17.15}$$

$$\sum_{i=1}^{m} x^*_{ij} = b_j, \quad j = 1, \ldots, n, \tag{17.16}$$

$$x^*_{ij} \geq 0, \quad i = 1, \ldots, m; \quad j = 1, \ldots, n. \tag{17.17}$$

In fact, KKT conditions (17.15) – (17.17) are constraints (17.7)–(17.9) of problem (GNTP), respectively, written for the optimal solution $\mathbf{x}^*$ of this problem.

Necessity. Let $(x^*_{ij})_{m\times n} \in T$ be optimal solution of problem (GNTP). In accordance with KKT theorem, there exist constants λ_i, $i = 1, \ldots, m$; μ_j, $j = 1, \ldots, n$, such that KKT necessary and sufficient optimality conditions (17.13)–(17.17) for problem (GNTP) are satisfied.

Since $x^*_{ij} \geq 0$, $i = 1, \ldots, m$; $j = 1, \ldots, n$, according to (17.17), then considering both possible cases: x^*_{ij} be strictly positive or equal to zero, from (17.13) and (17.14) it follows that

$$c'_{ij}(x^*_{ij}) - \lambda_i - \mu_j = 0 \qquad \text{for } x^*_{ij} > 0, \ 1 \leq i \leq m, \ 1 \leq j \leq n, \tag{17.18}$$

$$c'_{ij}(x^*_{ij}) - \lambda_i - \mu_j \geq 0 \qquad \text{for } x^*_{ij} = 0, \ 1 \leq i \leq m; \ 1 \leq j \leq n, \tag{17.19}$$

respectively, and these are namely conditions (17.11) and (17.12) of the statement of Theorem 17.1.

The "necessity" part of Theorem 17.1 is proved.

Sufficiency. Conversely, let $(x^*_{ij})_{m\times n} \in T$ and conditions (17.11)–(17.12) be satisfied.

Set

$$\lambda_i + \mu_j = c'_{ij}(x^*_{ij}).$$

From (17.11) and (17.12) it follows that KKT condition (17.13) is satisfied for both $x^*_{ij} > 0$ and $x^*_{ij} = 0$, that is, for all feasible variables $(x^*_{ij})_{m\times n} \in T$.

Condition (17.11) implies that KKT condition (17.14) is trivially satisfied for $x_{ij}^* > 0$, and condition (17.12) implies that KKT condition (17.14) is trivially satisfied for $x_{ij}^* = 0$. Therefore, KKT condition (17.14) is also satisfied for all feasible $(x_{ij}^*)_{m\times n} \in T$.

Since $(x_{ij}^*)_{m\times n} \in T$ by assumption, then KKT conditions (17.15)–(17.17) are also fulfilled.

Therefore, all KKT conditions (17.13)–(17.17) for problem (GNTP) are satisfied.

From "sufficiency" part of KKT theorem it follows that $(x_{ij}^*)_{m\times n} \in T$ is an optimal solution to problem (GNTP).

As pointed out in Section 17.1, when the objective function of problem (GNTP) is *strictly* convex, then optimal solution to this problem, if it exists, is *unique*. ■

17.3 Generalized Nonlinear Transportation Problem with Bounded Variables (GNTPBV)

Consider problem (GNTP) with two-sided bounds on the variables instead of non-negativity constraints (17.9):

(GNTPBV)

$$\min \left\{ c(\mathbf{x}) = \sum_{i=1}^{m} \sum_{j=1}^{n} c_{ij}(x_{ij}) \right\} \qquad (17.20) \equiv (17.6)$$

subject to

$$\sum_{j=1}^{n} x_{ij} = a_i, \quad i = 1, \dots, m, \qquad (17.21) \equiv (17.7)$$

$$\sum_{i=1}^{m} x_{ij} = b_j, \quad j = 1, \dots, n, \qquad (17.22) \equiv (17.8)$$

$$l_{ij} \le x_{ij} \le u_{ij}, \quad i = 1, \dots, m; \; j = 1, \dots, n. \qquad (17.23)$$

According to economic interpretation of x_{ij}'s, we have $l_{ij} \ge 0, \; i = 1, \dots, m;$ $j = 1, \dots, n$. The values of $l_{ij} \ge 0$ and u_{ij} $(\ge l_{ij})$, $i = 1, \dots, m; \; j = 1, \dots, n$, mean the minimal nonnegative and the maximal (finite) values of amounts x_{ij}, if any, to be transported from source i to destination j, where $i = 1, \dots, m; \; j = 1, \dots, n$. When $l_{ij} = 0$ and $u_{ij} = +\infty$ for all $i = 1, \dots, m; \; j = 1, \dots, n$, constraints (17.23) are reduced to constraints (17.9).

Denote by TBV the feasible set of problem (GNTPBV), defined by constraints (17.21)–(17.23).

Suppose that the following assumptions are satisfied.

17.a) $l_{ij} \leq u_{ij}$ for all $i = 1, \dots, m$; $j = 1, \dots, n$.

If $l_{ij} = u_{ij}$ for some pair of indices (p, q), then the value $x_{pq} := l_{pq} = u_{pq}$ is known a priori.

17.b) $$\sum_{j=1}^{n} l_{ij} \leq a_i \leq \sum_{j=1}^{n} u_{ij}, \quad i = 1, \dots, m,$$

$$\sum_{i=1}^{m} l_{ij} \leq b_j \leq \sum_{i=1}^{m} u_{ij}, \quad j = 1, \dots, n.$$

Otherwise the constraints (17.21), (17.22), and (17.23) are inconsistent, the feasible region TBV is empty, and problem (GNTPBV) has neither feasible nor optimal solution.

Theorem 17.2 (Characterization of the optimal solution of problem (GNTPBV)) *A feasible solution* $\mathbf{x}^* = (x_{ij}^*)_{m \times n} \in TBV$ *is an optimal solution to problem* (GNTPBV) *if and only if there exist constants* λ_i, $i = 1, \dots, m$, *and* μ_j, $j = 1, \dots, n$, *such that*

$$x_{ij}^* = l_{ij}, \quad (i, j) \in K_l \stackrel{\text{def}}{=} \{(i, j) : -c'_{ij}(l_{ij}) \leq \lambda_i + \mu_j\}, \tag{17.24}$$

$$x_{ij}^* = u_{ij}, \quad (i, j) \in K_u \stackrel{\text{def}}{=} \{(i, j) : -c'_{ij}(u_{ij}) \geq \lambda_i + \mu_j\}, \tag{17.25}$$

$$x_{ij}^* : \; - c'_{ij}(x_{ij}^*) = \lambda_i + \mu_j,$$
$$(i, j) \in K \stackrel{\text{def}}{=} \{(i, j) : -c'_{ij}(u_{ij}) \leq \lambda_i + \mu_j \leq -c'_{ij}(l_{ij})\}. \tag{17.26}$$

When convex functions $c_{ij}(x_{ij})$, $i = 1, \dots, m$; $j = 1, \dots, n$, *are strictly convex, then inequalities, defining* K *in* (17.26), *are strict.*

Proof. The Lagrangian function for problem (GNTPBV) is

$$L(\mathbf{x}, \boldsymbol{\lambda}, \boldsymbol{\mu}, \mathbf{s}, \mathbf{t}) = \sum_{i=1}^{m} \sum_{j=1}^{n} c_{ij}(x_{ij}) + \sum_{i=1}^{m} \lambda_i \Bigl(\sum_{j=1}^{n} x_{ij} - a_i \Bigr) + \sum_{j=1}^{n} \mu_j \Bigl(\sum_{i=1}^{m} x_{ij} - b_j \Bigr)$$

$$+ \sum_{i=1}^{m} \sum_{j=1}^{n} s_{ij}(l_{ij} - x_{ij}) + \sum_{i=1}^{m} \sum_{j=1}^{n} t_{ij}(x_{ij} - u_{ij}), \tag{17.27}$$

where $\boldsymbol{\lambda} = (\lambda_1, \dots, \lambda_m) \in \mathbf{R}^m$, $\boldsymbol{\mu} = (\mu_1, \dots, \mu_n) \in \mathbf{R}^n$, $\mathbf{s} = (s_{ij}) \in \mathbf{R}_+^{m \times n}$, $\mathbf{t} = (t_{ij}) \in \mathbf{R}_+^{m \times n}$ consist of Lagrange multipliers (dual variables), associated with constraints (17.21), (17.22), and both inequality constraints in (17.23), respectively.

The KKT necessary and sufficient optimality conditions for the minimum solution $\mathbf{x}^* = (x_{ij}^*)_{m \times n}$ for problem (GNTPBV) are

$$c'_{ij}(x^*_{ij}) + \lambda_i + \mu_j - s_{ij} + t_{ij} \geq 0, \quad i = 1, \ldots, m; \quad j = 1, \ldots, n, \tag{17.28}$$

$$s_{ij}(l_{ij} - x^*_{ij}) = 0, \quad s_{ij} \geq 0, \quad i = 1, \ldots, m; \quad j = 1, \ldots, n, \tag{17.29}$$

$$t_{ij}(x^*_{ij} - u_{ij}) = 0, \quad t_{ij} \geq 0, \quad i = 1, \ldots, m; \quad j = 1, \ldots, n, \tag{17.30}$$

$$\sum_{j=1}^{n} x^*_{ij} = a_i, \quad i = 1, \ldots, m, \tag{17.31}$$

$$\sum_{i=1}^{m} x^*_{ij} = b_j, \quad j = 1, \ldots, n, \tag{17.32}$$

$$l_{ij} \leq x^*_{ij} \leq u_{ij}, \quad i = 1, \ldots, m; \quad j = 1, \ldots, n. \tag{17.33}$$

If $l_{ij} = -\infty$ or $u_{ij} = +\infty$ for some pair of indices (i, j), $1 \leq i \leq m$; $1 \leq j \leq n$, we do not consider the corresponding KKT condition (17.29) (KKT condition (17.30)) and multiplier s_{ij} (multiplier t_{ij}), respectively.

Necessity. Let $\mathbf{x}^* = (x^*_{ij})_{m\times n} \in TBV$ be an optimal solution to problem (GNTPBV). According to KKT theorem, there exist constants λ_i, $i = 1, \ldots, m$; μ_j, $j = 1, \ldots, n$; $s_{ij} \geq 0$, $t_{ij} \geq 0$, $i = 1, \ldots, m$; $j = 1, \ldots, n$, such that KKT conditions (17.28) –(17.33) are satisfied.

(a) If $x^*_{ij} = l_{ij}$, then $s_{ij} \geq 0$, $t_{ij} = 0$ according to KKT conditions (17.29) and (17.30), respectively. Therefore, (17.28) implies

$$c'_{ij}(x^*_{ij}) \equiv c'_{ij}(l_{ij}) = s_{ij} - \lambda_i - \mu_j \geq -\lambda_i - \mu_j.$$

Multiplying this inequality by (-1), we get

$$-c'_{ij}(l_{ij}) \leq \lambda_i + \mu_j.$$

(b) If $x^*_{ij} = u_{ij}$, then $s_{ij} = 0$, $t_{ij} \geq 0$ according to KKT conditions (17.29) and (17.30), respectively. Hence, (17.28) implies

$$c'_{ij}(x^*_{ij}) \equiv c'_{ij}(u_{ij}) = -t_{ij} - \lambda_i - \mu_j \leq -\lambda_i - \mu_j.$$

If we multiply this inequality by (-1), we obtain

$$-c'_{ij}(u_{ij}) \geq \lambda_i + \mu_j.$$

(c) If $l_{ij} < x^*_{ij} < u_{ij}$, then $s_{ij} = t_{ij} = 0$ according to KKT conditions (17.29) and (17.30), respectively. Therefore, (17.28) implies

$$-c'_{ij}(x^*_{ij}) = \lambda_i + \mu_j. \tag{17.34}$$

Using that $l_{ij} < x^*_{ij} < u_{ij}$ and $c_{ij}(x_{ij}),\ i = 1, \dots, m;\ j = 1, \dots, n$, are convex functions, we get $c'_{ij}(l_{ij}) \le c'_{ij}(x^*_{ij}) \le c'_{ij}(u_{ij})$ (Theorem 1.31), that is, in case (c) we have

$$-c'_{ij}(u_{ij}) \le -c'_{ij}(x^*_{ij}) \le -c'_{ij}(l_{ij}),$$

whence from (17.34) it follows that

$$-c'_{ij}(u_{ij}) \le \lambda_i + \mu_j \le -c'_{ij}(l_{ij}).$$

When convex functions $c_{ij}(x_{ij}),\ i = 1, \dots, m;\ j = 1, \dots, n$, are *strictly* convex, these inequalities are strict.

To describe cases (a), (b), (c), considered above, we introduce the index sets K_l, K_u, K, defined by (17.24), (17.25), (17.26), respectively. Obviously $K_l \cup K_u \cup K = \{(i, j) : i = 1, \dots, m;\ j = 1, \dots, n\}$. The "necessity" part of Theorem 17.2 is proved.

Sufficiency. Conversely, let $\mathbf{x}^* = (x^*_{ij})_{m\times n} \in TBV$ and components of $\mathbf{x}^* = (x^*_{ij})_{m\times n}$ satisfy (17.24), (17.25), (17.26).

Set:

$$\lambda_i + \mu_j = -c'_{ij}(x^*_{ij}), \quad s_{ij} = t_{ij} = 0 \quad \text{for } (i, j) \in K;$$

$$s_{ij} = c'_{ij}(l_{ij}) + \lambda_i + \mu_j \quad (\ge 0 \text{ according to (17.24)}), \quad t_{ij} = 0$$
$$\text{for } (i, j) \in K_l;$$

$$s_{ij} = 0, \quad t_{ij} = -c'_{ij}(u_{ij}) - \lambda_i - \mu_j \quad (\ge 0 \text{ according to (17.25)})$$
$$\text{for } (i, j) \in K_u.$$

By the use of these expressions, it is easy to check that KKT conditions (17.28), (17.29), and (17.30) are satisfied. KKT conditions (17.31), (17.32), and (17.33) are also satisfied according to the assumption that $\mathbf{x}^* = (x^*_{ij})_{m\times n} \in TBV$.

Hence, in all possible cases, considered above, $\lambda_i, \mu_j, x^*_{ij}, s_{ij}, t_{ij},\ i = 1, \dots, m;\ j = 1, \dots, n$, satisfy KKT conditions (17.28)–(17.33) for problem (GNTPBV), which are necessary and sufficient conditions for a feasible solution to be an optimal solution to the convex minimization problem. Therefore, $\mathbf{x}^* = (x^*_{ij})_{m\times n}$ is an optimal solution to problem (GNTPBV).

When the objective function $c(\mathbf{x})$ (17.20) is *strictly* convex, the optimal solution of problem (GNTPBV) is *unique*. ■

Theorem 17.2 describes components of the optimal solution to problem (GNTPBV) only by the Lagrange multipliers $\lambda_i,\ i = 1, \dots, m$ and $\mu_j,\ j=1, \dots, n$, associated with the equality constraints (17.21) and (17.22), respectively.

NOTES TO PART THREE

Problems discussed in Chapter Twelve and related to them are considered in [1]–[4], [10], [13], [18]–[19], [22, 27], [30]–[31], [48], [71]–[72], [85, 95, 100, 122, 136, 143, 147], [153]–[154], [170, 172, 175], etc.

The ℓ_1-approximation is considered, for example, in papers of Barrodale and Roberts [2, 3], Coleman and Li [31], Parhomenko [100], Stefanov [122, 136, 143, 147, 153, 154], and the ℓ_1-solution to overdetermined linear systems is discussed in Bartels, Conn and Sinclair [4], Stefanov [143], etc. ℓ_p-approximations are considered, for example, in papers of Calamai and Conn [19], Fisher [48], Li [72], Merle and Späth [85], Stefanov [147], Watson [170], Wolfe [172], etc. A global quadratically convergent method for linear ℓ_∞ problems is suggested in the paper of Coleman and Li [30].

Least squares approach is discussed by Bertsekas [10], Björck [13], Lawson and Hanson [71], etc. Papers of Andersen [1], Calamai and Conn [18], Overton [95], Xue and Ye [175] consider minimization of a sum of Euclidean norms. Books of Clarke [27] and Demyanov and Vassiliev [36] are devoted to nondifferentiable optimization, and the book of Korneichuk [70] to extremum problems of the approximation theory. A subgradient algorithm for certain minimax and minisum problems is suggested in the paper of Chaleton, Hearn and Lowe [22].

Stochastic programming is discussed, for example, in the works of Ermoliev [44], Ermoliev and Gaivoronski [45], Gaivoronski [50], Kall and Wallace [65], etc. Adaptive stochastic quasigradient step-size regulation method is suggested, for example, by Urasiev [168].

Algorithms for projecting a point onto a set defined by an inequality or equality constraint and bounds on the variables are suggested, for example, by Rockafellar and Wets [109], Stefanov [118, 119, 124, 130, 152], etc.

The basic concept of integrality of polytopes and the relationship between integrality of polytopes and integer programming are discussed, for example, in the works of Gol'shtein and Yudin [52], Nemhauser and Wolsey [94], Padberg [97], Stefanov [121, 141], etc. Integer programming is also considered in the books of Schrijver [111]–[114], Taha [155], Yemelichev, Kovalev and Kravtsov [177], etc.

Valid inequalities are studied, for example, in the book of Nemhauser and Wolsey [94], in the papers of Stefanov [121, 123, 141], and in many other works. Concerning properties of totally unimodular matrices see, for example, the book of Nemhauser and Wolsey [[94], Section III.1.2, Proposition 2.2, Theorem 2.5, Corollary 2.8, etc.].

In the paper of Stefanov [144], the solution of the multidimensional convex separable continuous knapsack problem with bounded variables is studied.

Characterization theorems for the optimal solution of convex generalized nonlinear transportation problems are formulated and proved, e.g., in Stefanov [149].

Characterization of the optimal solution of problem $(C^=)$ via the characterization of the optimal solution of a relaxed problem is discussed, for example, in the paper of Stefanov [151].

Appendix A
Some Definitions and Theorems of Calculus

Definition A.1 An *ε-neighborhood around* $\mathbf{x}$ is the set

$$N(\mathbf{x}, \varepsilon) = \{\mathbf{y} \in \mathbf{R}^n : \|\mathbf{y} - \mathbf{x}\| \leq \varepsilon\}$$

where $\|\mathbf{y} - \mathbf{x}\|$ is the norm of $\mathbf{y} - \mathbf{x}$ (see Definition B.4, Appendix B).

Note that in fact $N(\mathbf{x}, \varepsilon)$ is a closed ball with radius ε around $\mathbf{x}$ in $\mathbf{R}^n$.

Definition A.2 Let X be a set in $\mathbf{R}^n$. A point $\mathbf{x}$ is said to be in the *closure* cl X of X if $X \cap N(\mathbf{x}, \varepsilon) \neq \emptyset$ for each $\varepsilon > 0$.

If $X = \text{cl } X$, the set X is called *closed.*

An equivalent definition is the following. The set X is said to be *closed* if it contains all its limit points.

Recall that a point $\overline{\mathbf{x}}$ is said to be a *limit point* (*cluster point*, *accumulation point*) of the sequence $\{\mathbf{x}_n\}$ if $\|\mathbf{x}_n - \overline{\mathbf{x}}\| < \varepsilon$ for some $\varepsilon > 0$ and all $n \geq n_0$ where n_0 is some positive integer.

Definition A.3 A point $\mathbf{x}$ is said to be in the *interior* int X of X if $N(\mathbf{x}, \varepsilon) \subset X$ for some $\varepsilon > 0$.

Definition A.4 A point $\mathbf{x}$ is said to be in the *boundary* ∂X of X if for each $\varepsilon > 0$, $N(\mathbf{x}, \varepsilon)$ contains at least one point in X and at least one point not in X.

Definition A.5 A set X is called *bounded* if it can be contained in a ball of finite radius, that is, if there exists some $r \in \mathbf{R}, r < +\infty$ such that $\|\mathbf{x}\| \leq r$ for each $\mathbf{x} \in X$.

Definition A.6 A set X in $\mathbf{R}^n$ which is both closed and bounded is called a *compact set.*

Definition A.7 A function $f : X \to \mathbf{R}$ is called *continuous at* $\mathbf{x}_0 \in X$ if for any sequence $\{\mathbf{x}_n\} \to \mathbf{x}_0$ such that $\{f(\mathbf{x}_n)\} \to f_0$ we have $f(\mathbf{x}_0) = f_0$.

Equivalently, a function f is said to be *continuous at* $\mathbf{x}_0 \in X$ if for any $\varepsilon > 0$ there exists a $\delta > 0$ such that $\mathbf{x} \in X$ and $\|\mathbf{x} - \mathbf{x}_0\| \leq \delta$ imply that $|f(\mathbf{x}) - f(\mathbf{x}_0)| < \varepsilon$.

© The Editor(s) (if applicable) and The Author(s), under exclusive license to Springer Nature Switzerland AG 2021

S. M. Stefanov, *Separable Optimization*, Springer Optimization and Its Applications 177,
https://doi.org/10.1007/978-3-030-78401-0

Function f is said to be *continuous on* X if it is continuous at each point of X.

Theorem A.1 (Bolzano-Weierstrass theorem)
Let X *be a compact set in* $\mathbf{R}^n$. *Then every sequence of points in* X *has a limit point in* X.

Sometimes the Bolzano-Weierstrass theorem is formulated as follows.
Every bounded sequence in X *has a convergent subsequence.*

Theorem A.2 (Mean value theorem)
Let f *be a differentiable numerical function on the open convex set* X *in* $\mathbf{R}^n$ *and let* $\overline{\mathbf{x}}, \mathbf{x} \in X$. *Then*

$$f(\mathbf{x}) = f(\overline{\mathbf{x}}) + \langle \mathbf{f}'(\overline{\mathbf{x}} + \lambda(\mathbf{x} - \overline{\mathbf{x}})), \mathbf{x} - \overline{\mathbf{x}} \rangle$$

for some $\lambda \in (0, 1)$, *where* $\langle ., . \rangle$ *denotes the inner (scalar) product (see Definition B.8).*

Theorem A.3 (Taylor's second-order theorem)
Let f *be a twice differentiable numerical function on the open convex set* X *in* $\mathbf{R}^n$ *and let* $\overline{\mathbf{x}} \in X$. *Then*

$$f(\mathbf{x}) = f(\overline{\mathbf{x}}) + \langle \mathbf{f}'(\overline{\mathbf{x}}), \mathbf{x} - \overline{\mathbf{x}} \rangle + \frac{1}{2} \langle f''(\overline{\mathbf{x}} + \lambda(\mathbf{x} - \overline{\mathbf{x}}))(\mathbf{x} - \overline{\mathbf{x}}), \mathbf{x} - \overline{\mathbf{x}} \rangle$$

for each $\mathbf{x} \in X$ *and some* $\lambda \in (0, 1)$, *where* $f''(\overline{\mathbf{x}} + \lambda(\mathbf{x} - \overline{\mathbf{x}}))$ *is the Hessian of* f *at* $\overline{\mathbf{x}} + \lambda(\mathbf{x} - \overline{\mathbf{x}})$.

Theorem A.4 (Darboux theorem)
Let $f(x)$ *have a finite derivative in the closed line segment* $[\alpha, \beta]$ *(that is, let* $f(x)$ *have a derivative at each point of the open segment* (α, β), *a derivative on the right at* α *and a derivative on the left at* β*), and* $f'(\alpha + 0) = A$, $f'(\beta - 0) = B$. *Then for any* $C \in [A, B]$ *there is a* $\xi \in [\alpha, \beta]$ *such that* $f'(\xi) = C$.

Note that the Darboux theorem does not assume continuity of $f'(x)$.

Theorem A.5 (Interior, closure, and boundary of a convex set)
Let X *be a convex set in* $\mathbf{R}^n$ *and* int $X \neq \emptyset$. *Then* int $X =$ int (cl X), cl $X =$ cl (int X), $\partial X = \partial(\text{cl } X)$.

Appendix B
Metric, Banach and Hilbert Spaces

Definition B.1 Let X and Y be sets. Denote by $X \times Y$ the set of all ordered pairs $(\mathbf{x}, \mathbf{y})$, where $\mathbf{x} \in X$ and $\mathbf{y} \in Y$. The set $X \times Y$ is said to be the *Cartesian product* of X and Y.

Definition B.2 X is called a *metric space* if there is defined a function $dist : X \times X \to \mathbb{R}$ such that

1. $dist(\mathbf{x}_1, \mathbf{x}_2) \geq 0$
   and $dist(\mathbf{x}_1, \mathbf{x}_2) = 0$ if and only if $\mathbf{x}_1 = \mathbf{x}_2$ — nonnegativity,
2. $dist(\mathbf{x}_1, \mathbf{x}_2) = dist(\mathbf{x}_2, \mathbf{x}_1)$ — symmetry,
3. $dist(\mathbf{x}_1, \mathbf{x}_2) \leq dist(\mathbf{x}_1, \mathbf{x}_3) + dist(\mathbf{x}_3, \mathbf{x}_2)$ — triangle inequality.

dist (. , .) is called the *distance function* on X.

The n-dimensional Euclidean space $\mathbb{R}^n$ is a metric space by defining

$$dist(\mathbf{x}, \mathbf{y}) = \left(\sum_{j=1}^{n} (x_j - y_j)^2 \right)^{\frac{1}{2}},$$

where $\mathbf{x} = (x_1, \ldots, x_n),\ \mathbf{y} = (y_1, \ldots, y_n) \in \mathbb{R}^n$.

Definition B.3 A real-valued function $s(\mathbf{x})$, defined on a linear space X, is called a *semi-norm* on X if the following conditions are satisfied:

1. $s(\mathbf{x} + \mathbf{y}) \leq s(\mathbf{x}) + s(\mathbf{y})$ — subadditivity,
2. $s(\alpha\mathbf{x}) = |\alpha| s(\mathbf{x})$.

For example, the Euclidean space $\mathbb{R}^n$ is an n-dimensional linear space by the following operations:

© The Editor(s) (if applicable) and The Author(s), under exclusive license to Springer Nature Switzerland AG 2021

S. M. Stefanov, *Separable Optimization*, Springer Optimization and Its Applications 177,
https://doi.org/10.1007/978-3-030-78401-0

$$\mathbf{x} + \mathbf{y} = (x_1 + y_1, \ldots, x_n + y_n)$$
$$\alpha\mathbf{x} = (\alpha x_1, \ldots, \alpha x_n).$$

Then $s(\mathbf{x}) = \max_{1 \le j \le n} |x_j|$ is a semi-norm. Also, $s(\mathbf{x}) = \left(\sum_{j=1}^{n} |x_j|^p \right)^{\frac{1}{p}}$, where $p \ge 1$, is a semi-norm on $\mathbf{R}^n$.

Definition B.4 A linear space X is called a *normed linear space* if for every $\mathbf{x} \in X$ there is associated a real number $\|\mathbf{x}\|$, called the *norm* of $\mathbf{x}$, such that

1. $\|\mathbf{x}\| \ge 0$
   and $\|\mathbf{x}\| = 0$ if and only if $\mathbf{x} = \mathbf{0}$ — nonnegativity,
2. $\|\alpha\mathbf{x}\| = |\alpha| \|\mathbf{x}\|$
3. $\|\mathbf{x} + \mathbf{y}\| \le \|\mathbf{x}\| + \|\mathbf{y}\|$ — triangle inequality.

Then $dist(\mathbf{x}, \mathbf{y}) = \|\mathbf{x} - \mathbf{y}\|$.

The set of all vectors $\mathbf{x} = (x_1, \ldots, x_n)$ such that $\sum_{j=1}^{n} |x_j|^p < \infty$ constitutes a normed linear space ℓ_p by the norm $\|\mathbf{x}\| = \left(\sum_{j=1}^{n} |x_j|^p\right)^{\frac{1}{p}}$.

The linear space ℓ_∞ normed by $\|\mathbf{x}\| = \sup_j |x_j|$ is another example of a normed space.

Definition B.5 A sequence $\{\mathbf{x}_k\}$ in a metric (normed) space X is said to satisfy *Cauchy's convergence condition* if

$$\lim_{l,m\to\infty} dist(\mathbf{x}_l, \mathbf{x}_m) = 0 \qquad \left(\lim_{l,m\to\infty} \|\mathbf{x}_l - \mathbf{x}_m\| = 0 \right).$$

Any sequence $\{\mathbf{x}_k\}$ in a metric (normed) space satisfying this condition is called a *Cauchy sequence*.

Definition B.6 A metric (normed) space X is said to be *complete* if every Cauchy sequence in it converges to some limit point $\mathbf{x}_0 \in X$.
A complete normed space is called a *Banach space*.

For example, $\mathbf{R}^n$ is a Banach space with each norm defined in it.

Definition B.7 A normed linear space X is called a *pre-Hilbert space* if its norm satisfies the condition

$$\|\mathbf{x} + \mathbf{y}\|^2 + \|\mathbf{x} - \mathbf{y}\|^2 = 2(\|\mathbf{x}\|^2 + \|\mathbf{y}\|^2).$$

Sometimes the pre-Hilbert space is defined as follows.

Definition B.8 A real (complex) linear space X is called a *pre-Hilbert space* if to every pair of elements $\mathbf{x}, \mathbf{y} \in X$ there is associated a number $\langle \mathbf{x}, \mathbf{y} \rangle$, called *inner (scalar) product*, satisfying the following conditions:

1. $\langle \mathbf{x}, \mathbf{x} \rangle \geq 0$
   and $\langle \mathbf{x}, \mathbf{x} \rangle = 0$ if and only if $\mathbf{x} = \mathbf{0}$, nonnegativity,
2. $\langle \mathbf{x}, \mathbf{y} \rangle = \langle \mathbf{y}, \mathbf{x} \rangle$ $\big(\langle \mathbf{x}, \mathbf{y} \rangle = \overline{\langle \mathbf{y}, \mathbf{x} \rangle}$ in the complex case$\big)$, symmetry,
3. $\langle \alpha \mathbf{x} + \beta \mathbf{y}, \mathbf{z} \rangle = \alpha \langle \mathbf{x}, \mathbf{z} \rangle + \beta \langle \mathbf{y}, \mathbf{z} \rangle$ linearity.

Each pre-Hilbert space in the sense of Definition B.8 is a normed linear space with a norm defined by $\|\mathbf{x}\| = \langle \mathbf{x}, \mathbf{x} \rangle^{\frac{1}{2}}$.

Definition B.9 A pre-Hilbert space which is complete with respect to the norm $\|\mathbf{x}\| = \langle \mathbf{x}, \mathbf{x} \rangle^{\frac{1}{2}}$ is called a *Hilbert space*.

For example, $\mathbb{R}^n$ is a Hilbert space with the inner product of $\mathbf{x} = (x_1, \ldots, x_n)$, $\mathbf{y} = (y_1, \ldots, y_n) \in \mathbb{R}^n$ defined by

$$\langle \mathbf{x}, \mathbf{y} \rangle = \sum_{j=1}^{n} x_j y_j.$$

Theorem B.1 (Cauchy-Schwarz inequality)
For $\mathbf{x}, \mathbf{y}$ *in a pre-Hilbert space, the following inequality holds true*

$$|\langle \mathbf{x}, \mathbf{y} \rangle| \leq \|\mathbf{x}\|.\|\mathbf{y}\|,$$

where $|\langle ., . \rangle|$ *denotes the absolute value of the inner product.*

The equality is satisfied if and only if $\mathbf{x}$ *and* $\mathbf{y}$ *are linearly dependent.*

Definition B.10 A (Banach) space X is called *strictly convex* when the triangle inequality of Definition B.4 is satisfied as an equality if and only if $\mathbf{x}$ and $\mathbf{y}$ are linearly dependent with a positive constant, that is, $\|\mathbf{x} + \mathbf{y}\| = \|\mathbf{x}\| + \|\mathbf{y}\|$ if and only if $\mathbf{x} = \lambda \mathbf{y}$ for some $\lambda > 0$.

If X is a strictly convex space, $\mathbf{x}, \mathbf{y} \in X$ and $\|\mathbf{x}\| = \|\mathbf{y}\| = \frac{1}{2}\|\mathbf{x} + \mathbf{y}\| = 1$, then $\mathbf{x} = \mathbf{y}$.

It is known that ℓ_p are strictly convex spaces for $1 < p < \infty$.

Definition B.11 Let X be a Banach space, and $S_1 \subset S_2 \subset X$. The set S_1 is said to be *dense* in S_2 if cl $S_1 = S_2$.

Definition B.12 A Banach space X is called *separable* if it has a dense subset which is at most countable.

For example, $\mathbb{R}$ is separable because the set of rational numbers is a countable dense set in $\mathbb{R}$.

ℓ_p is separable for each $1 \leq p < \infty$ and ℓ_∞ is not separable.

Each finite-dimensional space is separable.

Appendix C
Existence of Solutions to Optimization Problems—A General Approach

Definition C.1 Let $\inf\limits_{\mathbf{x} \in X} f(\mathbf{x}) = \alpha, \ \alpha \geq -\infty$. A sequence $\{\mathbf{x}_k\}_{k=1}^{\infty} \in X$ such that

$$\lim_{k \to \infty} f(\mathbf{x}_k) = \alpha$$

is called a *minimizing sequence* for f in X.

Definition C.2 A function $f : X \to \mathbb{R}, \ X \subseteq \mathbb{R}^n$ is said to be *lower semicontinuous* at $\mathbf{x}_0 \in X$ (with respect to X) if either of the following two equivalent conditions is satisfied:

i) For each $\varepsilon > 0$ there exists a $\delta > 0$ such that

$$\|\mathbf{x} - \mathbf{x}_0\| < \delta, \ \mathbf{x} \in X \quad \text{imply} \quad -\varepsilon < f(\mathbf{x}) - f(\mathbf{x}_0).$$

ii) For each sequence $\mathbf{x}_1, \mathbf{x}_2, \ldots$ in X, converging to $\mathbf{x}_0$,

$$\liminf_{k \to \infty} f(\mathbf{x}_k) \geq f(\lim_{k \to \infty} \mathbf{x}_k) = f(\mathbf{x}_0),$$

where $\liminf\limits_{k \to \infty} f(\mathbf{x}_k)$ denotes the infimum of the limit points of the sequence of real numbers $f(\mathbf{x}_1), f(\mathbf{x}_2), \ldots$.

f is said to be *lower semicontinuous* on X if it is lower semicontinuous (with respect to X) at each point $\mathbf{x} \in X$.

Definition C.3 A function $f : X \to \mathbb{R}, \ X \subseteq \mathbb{R}^n$ is said to be *upper semicontinuous* at $\mathbf{x}_0 \in X$ (with respect to X) if either of the following two equivalent conditions is satisfied:

i) For each $\varepsilon > 0$ there exists a $\delta > 0$ such that

$$\|\mathbf{x} - \mathbf{x}_0\| < \delta, \ \mathbf{x} \in X \quad \text{imply} \quad f(\mathbf{x}) - f(\mathbf{x}_0) < \varepsilon.$$

© The Editor(s) (if applicable) and The Author(s), under exclusive license to Springer Nature Switzerland AG 2021

S. M. Stefanov, *Separable Optimization*, Springer Optimization and Its Applications 177,
https://doi.org/10.1007/978-3-030-78401-0

ii) For each sequence $\mathbf{x}_1, \mathbf{x}_2, \dots$ in X, converging to $\mathbf{x}_0$,

$$\limsup_{k\to\infty} f(\mathbf{x}_k) \le f(\lim_{k\to\infty} \mathbf{x}_k) = f(\mathbf{x}_0),$$

where $\limsup\limits_{k\to\infty} f(\mathbf{x}_k)$ denotes the supremum of the limit points of the sequence of real numbers $f(\mathbf{x}_1), f(\mathbf{x}_2), \dots$.

f is said to be *upper semicontinuous* on X if it is upper semicontinuous (with respect to X) at each point $\mathbf{x} \in X$.

From these two definitions it follows that f is lower semicontinuous at $\mathbf{x}_0 \in X$ (with respect to X) if and only if $-f$ is upper semicontinuous at $\mathbf{x}_0 \in X$ (with respect to X), and vice versa, f is upper semicontinuous at $\mathbf{x}_0 \in X$ (with respect to X) if and only if $-f$ is lower semicontinuous at $\mathbf{x}_0 \in X$ (with respect to X).

f is *continuous* at $\mathbf{x}_0 \in X$ (with respect to X) if and only if it is both lower semicontinuous and upper semicontinuous at $\mathbf{x}_0 \in X$ (with respect to X).

Establish the equivalence of Definitions C.2 i) and C.2 ii).

Let $f : X \to \mathbf{R}$ be lower semicontinuous at $\mathbf{x}_0$ according to Definition C.2 i). Choose an arbitrary sequence $\{\mathbf{x}_k\}_{k=1}^{\infty} \to \mathbf{x}_0$, $\{\mathbf{x}_k\}_{k=1}^{\infty} \subset X$. Then for each $\varepsilon > 0$ there exists an integer N_ε and $\delta_\varepsilon > 0$ such that $\|\mathbf{x}_k - \mathbf{x}_0\| < \delta_\varepsilon$ for $k > N_\varepsilon$. Therefore

$$f(\mathbf{x}_k) - f(\mathbf{x}_0) > -\varepsilon \quad \text{for } k > N_\varepsilon,$$

$$\liminf_{k\to\infty} f(\mathbf{x}_k) \ge f(\mathbf{x}_0) - \varepsilon \quad \text{for each } \varepsilon > 0,$$

$$\liminf_{k\to\infty} f(\mathbf{x}_k) \ge f(\mathbf{x}_0),$$

that is, f is lower semicontinuous at $\mathbf{x}_0 \in X$ according to Definition C.2 ii).

Let f be lower semicontinuous at $\mathbf{x}_0 \in X$ according to Definition C.2 ii). Assume that conditions of Definition C.2 i) are violated. Then there exists an $\varepsilon > 0$ such that for each positive integer k there is at least one $\mathbf{x}_k \in X$ with

$$\|\mathbf{x}_k - \mathbf{x}_0\| < \frac{1}{k}, \quad f(\mathbf{x}_k) - f(\mathbf{x}_0) \le -\varepsilon.$$

However, $\{\mathbf{x}_k\}_{k=1}^{\infty} \to \mathbf{x}_0$, hence $\liminf\limits_{k\to\infty} f(\mathbf{x}_k) \le f(\mathbf{x}_0) - \varepsilon < f(\mathbf{x}_0)$, a contradiction with Definition C.2 ii). Therefore the assumption is incorrect and f is lower semicontinuous at $\mathbf{x}_0 \in X$ according to Definition C.2 i) as well. ■

Equivalence of Definitions C.3 i) and C.3 ii) is established similarly.

The following Weierstrass theorem and the corollary are useful for determining solvability of many optimization problems.

Theorem C.1 (Weierstrass theorem)
A lower (upper) semicontinuous function f defined on a compact set X in $\mathbf{R}^n$ is bounded from below (above) and attains in X the value

$$\overline{\alpha} = \inf_{\mathbf{x}\in X} f(\mathbf{x}) \quad \left(\overline{\beta} = \sup_{\mathbf{x}\in X} f(\mathbf{x})\right).$$

Proof. Let $\inf_{\mathbf{x}\in X} f(\mathbf{x}) = \overline{\alpha}$ and let $\{\mathbf{x}_k\}_{k=1}^{\infty} \in X$ be a minimizing sequence for f in X. Since X is a compact set (and therefore it is bounded) then $\{\mathbf{x}_k\}_{k=1}^{\infty}$ is a bounded sequence. Hence, a convergent subsequence $\{\mathbf{x}_{k_i}\}_{i=1}^{\infty} \to \mathbf{x}_0$ can be chosen according to the Bolzano-Weierstrass theorem. Since X is a closed set (X is compact by assumption), then $\mathbf{x}_0 \in X$. From the lower semicontinuity of f on X it follows that

$$f(\mathbf{x}_0) \leq \liminf_{i\to\infty} f(\mathbf{x}_{k_i}) = \overline{\alpha}.$$

Therefore $\overline{\alpha} > -\infty$. On the other hand, $f(\mathbf{x}_0) \geq \inf_{\mathbf{x}\in X} f(\mathbf{x}) = \overline{\alpha}$ by the definition of infimum of a numerical function. From both opposite inequalities it follows that $f(\mathbf{x}_0) = \overline{\alpha} \overset{\text{def}}{=} \inf_{\mathbf{x}\in X} f(\mathbf{x})$, that is, $\overline{\alpha}$ is attained at $\mathbf{x}_0 \in X$.

The upper semicontinuous case can be proved analogously. ■

In the proof of Theorem C.1, we have used semicontinuity of f to conclude that $f(\mathbf{x}_0) \leq \overline{\alpha}$, we have used that the set X is closed in order to conclude that $\mathbf{x}_0 \in X$, and boundedness of X has been used in order to derive that $\{\mathbf{x}_k\}$ is a bounded sequence and therefore the Bolzano-Weierstrass theorem is applicable to it.

Hence, each of the assumptions of Theorem C.1 is essential.

Corollary C.1 (Of the Weierstrass theorem)
Let $f : X \to \mathbb{R}$ be a lower (upper) semicontinuous function on the nonempty closed set X in $\mathbb{R}^n$ and

$$\lim_{\substack{\|\mathbf{x}_k\| \to \infty \\ k \to \infty}} f(\mathbf{x}_k) = +\infty \quad \left(\lim_{\substack{\|\mathbf{x}_k\| \to \infty \\ k \to \infty}} f(\mathbf{x}_k) = -\infty\right), \quad \{\mathbf{x}_k\} \subset X. \tag{C.1}$$

Then $f(\mathbf{x})$ attains in X the value

$$\overline{\alpha} = \inf_{\mathbf{x}\in X} f(\mathbf{x}) \quad \left(\overline{\beta} = \sup_{\mathbf{x}\in X} f(\mathbf{x})\right).$$

Proof. Consider the case $\lim_{\|\mathbf{x}_k\|\to\infty} f(\mathbf{x}_k) = +\infty$ when $k \to \infty$. The second case is proved similarly.

From the assumptions of Corollary C.1 $\left(\{\mathbf{x}_k\}_{k=1}^{\infty} \subset X \text{ and } \lim_{k\to\infty} \|\mathbf{x}_k\| = +\infty\right.$ imply $\left.\lim_{k\to\infty} f(\mathbf{x}_k) = +\infty\right)$ it follows that each minimizing sequence for f is bounded in X.

According to the Bolzano-Weierstrass theorem, we can choose a subsequence $\{\mathbf{x}_{k_i}\}_{i=1}^{\infty}$ of the minimizing sequence $\{\mathbf{x}_k\}_{k=1}^{\infty}$ which converges to some $\mathbf{x}_0 \in X$ because

X is a closed set: $\mathbf{x}_{k_i} \to \mathbf{x}_0, \quad i \to \infty$. However, f is a lower semicontinuous function on X. Hence

$$f(\mathbf{x}_0) \leq \liminf_{i \to \infty} f(\mathbf{x}_{k_i}) \equiv \overline{\alpha}.$$

Therefore $\overline{\alpha} > -\infty$ strictly. However, $f(\mathbf{x}_0) \geq \overline{\alpha}$ since $\overline{\alpha}$ is the infimum of f in X. Hence $f(\mathbf{x}_0) = \overline{\alpha} \stackrel{\text{def}}{=} \inf_{\mathbf{x} \in X} f(\mathbf{x})$. ■

Remark C.1 *Corollary C.1 does not assume boundedness of the set X, whereas Theorem C.1 does. However, Corollary C.1 assumes that one of conditions* (C.1) *is satisfied.*

Theorem C.1 and Corollary C.1 mean that under their assumptions, the problem

$$\min_{\mathbf{x} \in X} f(\mathbf{x}) \qquad \left(\max_{\mathbf{x} \in X} f(\mathbf{x})\right)$$

has a minimim (maximum) solution.

In some cases it is possible to apply Theorem C.1 to check whether particular optimization problems are solvable and in other cases Corollary C.1 is applicable. The fact that sometimes Theorem C.1 or Corollary C.1 does not answer the question whether a particular problem is solvable does not imply *directly* that this problem does not have an optimal solution. The explanation lies in the fact that Theorem C.1 and Corollary C.1 are not both necessary and sufficient conditions, but sufficient conditions only.

Theorem C.2 (Ekeland's theorem, 1974)
Let V be a complete metric space with associated metric dist $(.,.)$, *and let* $F : V \to \mathbf{R} \cup \{+\infty\}$ *be a lower semicontinuous function which is bounded from below. If* $\mathbf{u}$ *is a point in V satisfying*

$$F(\mathbf{u}) \leq \inf F + \varepsilon$$

for some $\varepsilon > 0$, *then for every* $\lambda > 0$ *there exists a point* $\mathbf{v}$ *in V such that*

1. $F(\mathbf{v}) \leq F(\mathbf{u})$,
2. *dist* $(\mathbf{u}, \mathbf{v}) \leq \lambda$,
3. *for all* $\mathbf{w} \neq \mathbf{v}$ *in* V, *one has*

$$F(\mathbf{w}) + \frac{\varepsilon}{\lambda}\, dist\,(\mathbf{w}, \mathbf{v}) > F(\mathbf{v}).$$

Ekeland's theorem states that there is a "nearby point" $\mathbf{v}$ which actually minimizes a slightly perturbed functional F.

A proof of Ekeland's theorem may be found, for example, in the paper of Ekeland [43] and in the book of Clarke [27].

Appendix D
Best Approximation: Existence and Uniqueness

Let X be a normed linear space and $\{\varphi_j\}_{j=0}^n$ be $n + 1$ linearly independent elements of X. Consider the linear subspace L of X generated by $\{\varphi_j\}_{j=0}^n$:

$$L = \{\chi \in X : \chi = c_0\varphi_0 + c_1\varphi_1 + \cdots + c_n\varphi_n\},$$

where $c_j, \ j = 0, 1, \ldots, n$ are real numbers.

Definition D.1 For each element f of X, the *best approximation* $E_L(f)$ of f by elements of L is defined as

$$E_L(f) = \inf_{\varphi \in L} \|f - \varphi\| \equiv \delta(f, L),$$

where $\delta(f, L)$ is the distance from f to L.

Consider the question about existence and uniqueness of an element of best approximation.

Theorem D.1 (Existence of element of best approximation)
Let L be a linear subspace of the normed linear space X and let L be generated by the linearly independent elements $\{\varphi_j\}_{j=0}^n$ of X. Then for each $f \in X$ there exists an element of best approximation in L.

Proof. Consider $\Upsilon(\mathbf{c}) = \|\chi\| = \|c_0\varphi_0 + c_1\varphi_1 + \cdots + c_n\varphi_n\|$. It is a continuous function of $c_0, c_1, \ldots, c_n$ because

$$\left| \left\| \sum_{j=0}^{n} c_j\varphi_j \right\| - \left\| \sum_{j=0}^{n} a_j\varphi_j \right\| \right| \le \left\| \sum_{j=0}^{n} (c_j - a_j)\varphi_j \right\| \le \sum_{j=0}^{n} |c_j - a_j| . \|\varphi_j\| .$$

© The Editor(s) (if applicable) and The Author(s), under exclusive license to Springer Nature Switzerland AG 2021

S. M. Stefanov, *Separable Optimization*, Springer Optimization and Its Applications 177,
https://doi.org/10.1007/978-3-030-78401-0

Therefore $\Upsilon(\mathbf{c})$ attains its minimum value m on the $n+1$-dimensional unit sphere $S_1 = \left\{(c_0, c_1, \ldots, c_n) : \sum_{j=0}^{n} c_j^2 = 1\right\}$:

$$m = \min_{S_1} \|c_0\varphi_0 + c_1\varphi_1 + \cdots + c_n\varphi_n\|$$

according to the Weierstrass theorem because S_1 is a compact set. Since $\Upsilon(\mathbf{c}) \geq 0$ by definition and $\{\varphi_j\}_{j=0}^{n}$ are linearly independent, then $m > 0$.

Function $\|f - \chi\| = \|f - (c_0\varphi_0 + c_1\varphi_1 + \cdots + c_n\varphi_n)\|$ is also a continuous function of $c_0, c_1, \ldots, c_n$. Evaluate this function over the sphere $S_r = \{(c_0, c_1, \ldots, c_n) : \sum_{j=0}^{n} c_j^2 = r^2\}$ of radius r:

$$\left\| f - \sum_{j=0}^{n} c_j\varphi_j \right\| \geq \|c_0\varphi_0 + c_1\varphi_1 + \cdots + c_n\varphi_n)\| - \|f\|$$

$$= |r| \left\| \sum_{j=0}^{n} \frac{c_j}{|r|}\varphi_j \right\| - \|f\| \geq rm - \|f\|.$$

Here $\sum_{j=0}^{n} \left(\frac{c_j}{|r|}\right)^2 = 1$. If $|r| > \frac{2\|f\|}{m} = r_0$, then

$$\left\| f - \sum_{j=0}^{n} c_j\varphi_j \right\| \geq \frac{2\|f\|}{m} m - \|f\| = \|f\| \geq E_L(f).$$

Therefore

$$E_L(f) \stackrel{\text{def}}{=} \inf_{\varphi \in L} \|f - \varphi\| = \inf_{c_0^2 + \cdots + c_n^2 \leq r_0^2} \left\| f - \sum_{j=0}^{n} c_j\varphi_j \right\|.$$

However, $S_{r_0} = \{(c_0, c_1, \ldots, c_n) : c_0^2 + c_1^2 + \cdots + c_n^2 \leq r_0^2\}$ is a compact set in the finite-dimensional normed linear space L. Therefore the continuous function $\|f - (c_0\varphi_0 + c_1\varphi_1 + \cdots + c_n\varphi_n)\|$ attains its minimum value in L. Hence there exists some $\psi = c_0^*\varphi_0 + c_1^*\varphi_1 + \cdots + c_n^*\varphi_n \in L$ such that

$$\|f - \psi\| = \min_{\varphi \in L} \|f - \varphi\| = \inf_{\varphi \in L} \|f - \varphi\| \stackrel{\text{def}}{=} E_L(f).$$

■

Denote by $P(f)$ the collection of elements of best approximation for $f \in X$. According to Theorem D.1, $P(f) \neq \emptyset$.

Theorem D.2 (The set of elements of best approximation is closed and convex) *$P(f)$ is a closed and convex set.*

Proof. Let $\{\varphi_j\}_{j=0}^{\infty} \subset P(f)$ (hence $\|f - \varphi_j\| = E_L(f)$) and $\|\varphi_j - \varphi\| \to 0$ when $j \to \infty$. Then

$$\|f - \varphi\| \leq \|f - \varphi_j\| + \|\varphi_j - \varphi\| \to E_L(f) \quad \text{when } j \to \infty.$$

Therefore $\varphi \in P(f)$ as well, that is, $P(f)$ is closed.

Let $\varphi \in P(f)$ and $\psi \in P(f)$, that is, $\|f - \varphi\| = \|f - \psi\| = E_L(f)$. Then for $\lambda \in [0, 1]$ we have $\lambda\varphi + (1 - \lambda)\psi \in L$ by definition of L, and using definitions of $E_L(f)$ and of the norm, we get

$$E_L(f) \leq \|f - (\lambda\varphi + (1 - \lambda)\psi)\| \leq \lambda\|f - \varphi\| + (1 - \lambda)\|f - \psi\| = E_L(f).$$

Hence $E_L(f) = \|f - (\lambda\varphi + (1 - \lambda)\psi)\|$, that is, $\lambda\varphi + (1 - \lambda)\psi \in P(f)$. Therefore $P(f)$ is convex by definition. ∎

An element of best approximation is not unique in the general case. However, if the normed linear space has some special properties, the element of best approximation is unique.

Theorem D.3 (Uniqueness of the element of best approximation)
If X is a strictly convex space, then the element of best approximation is unique.

Proof. Assume that there are two different elements of best approximation for $f \in X$: $\varphi \in L$, $\psi \in L$, $\varphi \neq \psi$. By Theorem D.2, $\frac{1}{2}\varphi + \frac{1}{2}\psi$ is an element of best approximation for f as well. Therefore

$$\begin{aligned} E_L(f) &= \left\| f - \frac{\varphi + \psi}{2} \right\| \leq \frac{1}{2}\|f - \varphi\| + \frac{1}{2}\|f - \psi\| = \frac{1}{2}E_L(f) + \frac{1}{2}E_L(f) \\ &= E_L(f), \end{aligned}$$

that is,

$$\left\| f - \frac{\varphi + \psi}{2} \right\| = \frac{1}{2}\|f - \varphi\| + \frac{1}{2}\|f - \psi\|.$$

Strict convexity of X (Definition B.10, Appendix B) implies that there exists an $\alpha > 0$ such that $f - \varphi = \alpha(f - \psi)$. If $\alpha = 1$ then $\varphi = \psi$. If $\alpha \neq 1$ then f is a linear combination of $\varphi_0, \ldots, \varphi_n$, that is, $f \in L$, hence $f = \varphi = \psi$. Therefore the element of best approximation is unique under the assumptions of Theorem D.3. ∎

Consider the space $C(K)$ of all continuous real-valued functions f on the compact set K. $C(K)$ becomes a normed space by the norm

$$\|f\|_{C(K)} = \sup_{\mathbf{x} \in K} |f(\mathbf{x})|.$$

The space $C(K)$ is not strictly convex. This can be seen by the following example.

Let $K = [0, 1]$ and $\varphi_1 = 1,\ \varphi_2 = x$. Obviously φ_1 and φ_2 are linearly independent. However, $\|\varphi_1\|_{C[0,1]} = \|\varphi_2\|_{C[0,1]} = 1$, $\|\varphi_1 + \varphi_2\|_{C[0,1]} = 2$, and $\|\varphi_1 + \varphi_2\| = \|\varphi_1\| + \|\varphi_2\|$.

Hence, if $X = C(K)$, then Theorem D.3 does not directly imply uniqueness of the element of best approximation for $f \in C(K)$. This property can be derived from other results which are formulated below.

Definition D.2 The set G of points $\mathbf{x} \in K$ for which

$$|f(\mathbf{x})| = \|f\|_{C(K)},$$

that is, for which a supremum in the definition of sup-norm is attained, is called the *critical set* for $f \in C(K)$.

Theorem D.4 (Kolmogorov's characterization criterion)
Let $f \in C(K)$, $\{\varphi_j\}_{j=0}^n \subset C(K)$ and L be the subspace of $X = C(K)$ defined above. The generalized polynomial $\varphi = \sum_{j=0}^n c_j \varphi_j \in L$ is an element of best approximation for $f \in C(K)$ in L if and only if for each $\psi \in L$ there is at least one $\tilde{\mathbf{x}} \in G$ with

$$\psi(\tilde{\mathbf{x}})[f(\tilde{\mathbf{x}}) - \varphi(\tilde{\mathbf{x}})] \le 0.$$

Theorem D.5 (Haar's theorem)
Let the compact set K contain at least $n + 2$ points. A necessary and sufficient condition for $f \in C(K)$ to have a unique best approximation $\varphi = \sum_{j=0}^n c_j \varphi_j \in L$, $\varphi_j \in C(K)$ is that no element of the finite-dimensional subspace L (except 0) has n or more zeroes in K.

A subspace L with this property is sometimes called the *Haar subspace*.

Appendix E
On the Solvability of a Quadratic Optimization Problem with a Feasible Region Defined as a Minkowski Sum of a Compact Set and Finitely Generated Convex Closed Cone

Consider the following quadratic optimization problem

$$\min\ (\max) \left\{ Q(\mathbf{x}) = \frac{1}{2}\langle A\mathbf{x}, \mathbf{x}\rangle + \langle \mathbf{q}, \mathbf{x}\rangle \right\} \tag{E.1}$$

subject to

$$\mathbf{x} \in X \stackrel{\text{def}}{=} C + K, \tag{E.2}$$

where A is an $n \times n$ symmetric matrix, $\mathbf{q},\ \mathbf{x} \in \mathbf{R}^n$, $\langle ., . \rangle$ denotes the scalar (inner, dot) product in $\mathbf{R}^n$, C is a compact set, K is a finitely generated convex closed cone, and "+" denotes the Minkowski sum of C and K (Definition 1.5).

Recall that

- *Minkowski sum* of the sets X and Y, denoted by $X + Y$, is defined as follows

$$X + Y = \{\mathbf{z} : \ \mathbf{z} = \mathbf{x} + \mathbf{y},\ \mathbf{x} \in X, \mathbf{y} \in Y\},$$

- K is a *(pointed) cone* if $\mathbf{x} \in K$ and $t \geq 0$ imply $t\mathbf{x} \in K$, that is, if $\mathbf{x} \in K$, then the ray $\Lambda = \{t\mathbf{x} : \ t \geq 0\}$ also belongs to the cone K,
- K is a *finitely generated cone* if there exist vectors $\mathbf{y}_1, \ldots, \mathbf{y}_k \in K$, called *generating vectors*, such that

$$K = \left\{\mathbf{y} : \ \mathbf{y} = \sum_{i=1}^{k} \alpha_i \mathbf{y}_i, \quad \alpha_i \geq 0,\ i = 1, \ldots, k\right\},$$

- *conical hull* of a set S is the intersection of all convex cones, containing S:

© The Editor(s) (if applicable) and The Author(s), under exclusive license to Springer Nature Switzerland AG 2021

S. M. Stefanov, *Separable Optimization*, Springer Optimization and Its Applications 177,
https://doi.org/10.1007/978-3-030-78401-0

$$\text{con } S = \left\{ \mathbf{z} : \mathbf{z} = \sum_{i=1}^{k} \alpha_i \mathbf{z}_i, \ \alpha_i \geq 0, \ \mathbf{z}_i \in S, \ i = 1, \ldots, k \right\}.$$

When A is a *positive semidefinite matrix*, that is, if $\langle A\mathbf{x}, \mathbf{x} \rangle \geq 0$ for each $\mathbf{x} \in \mathbf{R}^n$, then quadratic function $Q(\mathbf{x})$ is convex, and if A is a *positive definite matrix*, that is, if $\langle A\mathbf{x}, \mathbf{x} \rangle > 0$ for each $\mathbf{x} \in \mathbf{R}^n$, $\mathbf{x} \neq \mathbf{0}$, then quadratic function $Q(\mathbf{x})$ is strictly convex (Theorem 1.32).

Since the Minkowski sum of a compact set and finitely generated convex closed cone is unbounded in the general case, the question about existence of solution of the considered optimization problem is open.

Problems like (E.1)–(E.2) arise in optimization theory and its applications.

Theorem E.1 (Solvability of problem (E.1)–(E.2))
Let the quadratic function $Q(\mathbf{x}) = \frac{1}{2}\langle A\mathbf{x}, \mathbf{x} \rangle + \langle \mathbf{q}, \mathbf{x} \rangle$ *be bounded below (bounded above, respectively) in the set* $X = C + K$, *where* A *is an* $n \times n$ *symmetric matrix,* $\mathbf{q}, \mathbf{x} \in \mathbf{R}^n$, $\langle ., . \rangle$ *denotes the scalar (inner, dot) product in* $\mathbf{R}^n$, C *is a compact set,* K *is a finitely generated convex closed cone, and "+" denotes the Minkowski sum of* C *and* K.

Then $Q(\mathbf{x})$ *attains its infimum (its supremum, respectively) on* X, *that is, optimization problem* (E.1)–(E.2) *has optimal solution under the assumption of Theorem E.1.*

Proof. Let $Q(\mathbf{x})$ be bounded above function on the set X; the case of bounded below objective function is considered similarly.

For all $\mathbf{y} \in C$, $\mathbf{z} \in K$ and $\lambda \geq 0$ we have $\mathbf{x} \equiv \mathbf{y} + \lambda \mathbf{z} \in X$ (because $X = C + K$ by definition, where K is a cone) and

$$\begin{aligned} Q(\mathbf{x}) &\equiv Q(\mathbf{y} + \lambda \mathbf{z}) = \tfrac{1}{2}\langle A(\mathbf{y} + \lambda \mathbf{z}), \mathbf{y} + \lambda \mathbf{z} \rangle + \langle \mathbf{q}, \mathbf{y} + \lambda \mathbf{z} \rangle \\ &= \tfrac{1}{2}\langle A\mathbf{y}, \mathbf{y} \rangle + \tfrac{1}{2}\lambda \langle A\mathbf{y}, \mathbf{z} \rangle + \tfrac{1}{2}\lambda \langle A\mathbf{z}, \mathbf{y} \rangle + \tfrac{1}{2}\lambda^2 \langle A\mathbf{z}, \mathbf{z} \rangle + \langle \mathbf{q}, \mathbf{y} \rangle + \lambda \langle \mathbf{q}, \mathbf{z} \rangle \\ &= \tfrac{1}{2}\langle A\mathbf{y}, \mathbf{y} \rangle + \langle \mathbf{q}, \mathbf{y} \rangle + \lambda \langle A\mathbf{y} + \mathbf{q}, \mathbf{z} \rangle + \tfrac{1}{2}\lambda^2 \langle A\mathbf{z}, \mathbf{z} \rangle \\ &= Q(\mathbf{y}) + \lambda \langle A\mathbf{y} + \mathbf{q}, \mathbf{z} \rangle + \tfrac{1}{2}\lambda^2 \langle A\mathbf{z}, \mathbf{z} \rangle, \end{aligned} \tag{E.3}$$

where we have used that $\langle A\mathbf{y}, \mathbf{z} \rangle = \langle A\mathbf{z}, \mathbf{y} \rangle$ for all $\mathbf{y}, \mathbf{z} \in \mathbf{R}^n$. Then

$$\langle A\mathbf{z}, \mathbf{z} \rangle \geq 0 \quad \text{for every } \mathbf{z} \in K,$$

and if for some $\overline{\mathbf{z}} \in K$ we have $\langle A\overline{\mathbf{z}}, \overline{\mathbf{z}} \rangle = 0$, then $\langle A\mathbf{y} + \mathbf{q}, \overline{\mathbf{z}} \rangle \geq 0$ for every $\mathbf{y} \in C$.

Denote $L = \{\mathbf{z} \in K : \|\mathbf{z}\| = 1\}$.

Consider both possible cases concerning the value of $\langle A\mathbf{z}, \mathbf{z} \rangle \geq 0$.

Case 1. Let $\langle A\mathbf{z}, \mathbf{z} \rangle > 0$ for every $\mathbf{z} \in L$.

Since the scalar product $\langle A\mathbf{z}, \mathbf{z} \rangle$ is a continuous function of $\mathbf{z}$ and L is a compact set by definition, then $\langle A\mathbf{z}, \mathbf{z} \rangle > 0$ for $\mathbf{z} \in L$ implies that there exists a constant $\delta > 0$, such that $\langle A\mathbf{z}, \mathbf{z} \rangle > \delta$ for every $\mathbf{z} \in L$.

Therefore, if $\mathbf{x} = \mathbf{y} + \lambda \mathbf{z}$, $\mathbf{y} \in C$, $\mathbf{z} \in L$, $\lambda \geq 0$, then

$$Q(\mathbf{x}) \equiv Q(\mathbf{y} + \lambda\mathbf{z}) = \frac{1}{2}\langle A\mathbf{y}, \mathbf{y}\rangle + \langle \mathbf{q}, \mathbf{y}\rangle + \lambda\langle A\mathbf{y} + \mathbf{q}, \mathbf{z}\rangle + \frac{1}{2}\lambda^2\langle A\mathbf{z}, \mathbf{z}\rangle$$

$$\stackrel{\text{def}}{=} \Phi(\mathbf{y}, \mathbf{z}, \lambda) \leq \max_{\lambda \geq 0} \Phi(\mathbf{y}, \mathbf{z}, \lambda) \stackrel{\text{def}}{=} \Psi(\mathbf{y}, \mathbf{z}),$$

so that

$$\sup_X Q(\mathbf{x}) = \sup_{(\mathbf{y},\mathbf{z}) \in C \times L} \Psi(\mathbf{y}, \mathbf{z}).$$

Function $\Psi(\mathbf{y}, \mathbf{z})$ is continuous in the compact set $C \times L$, the Cartesian product of C and L. Therefore there exists a point $(\mathbf{y}^0, \mathbf{z}^0) \in C \times L$ such that

$$\Psi(\mathbf{y}^0, \mathbf{z}^0) = \max\,\{\Psi(\mathbf{y}, \mathbf{z}) : (\mathbf{y}, \mathbf{z}) \in C \times L\}.$$

For $\mathbf{x}^0 = \mathbf{y}^0 + \lambda\mathbf{z}^0$, where $\mathbf{y}^0 \in C,\ \mathbf{z}^0 \in L,\ \lambda \geq 0$, we have

$$Q(\mathbf{x}^0) = \Psi(\mathbf{y}^0, \mathbf{z}^0).$$

Hence

$$Q(\mathbf{x}^0) = \max\,\{Q(\mathbf{x}) : \mathbf{x} \in X\}.$$

Case 2. Let there exist $\bar{\mathbf{z}} \in K \setminus \{\mathbf{0}\}$ such that $\langle A\bar{\mathbf{z}}, \bar{\mathbf{z}}\rangle = 0$.

Statement of Theorem E.1 will be proved by induction on the number of elements, which generate the finitely generated cone K. If $K = \text{con}\,\{\mathbf{z}_1, \ldots, \mathbf{z}_s\}$, the conical hull of generating vectors $\mathbf{z}_1, \ldots, \mathbf{z}_s$, then denote

$$K_p = \text{con}\,\{\mathbf{z}_1, \ldots, \mathbf{z}_{p-1}, \mathbf{z}_{p+1}, \ldots, \mathbf{z}_s\} \qquad \text{(E.4)}$$

and

$$X_p = C + K_p, \quad p = 1, \ldots, s.$$

Let $s = 1$. Then we have

$$K = \{\lambda\mathbf{z}_1 : \lambda \geq 0\}.$$

From the assumption $\langle A\mathbf{z}_1, \mathbf{z}_1\rangle = 0$ it follows that $\langle A\mathbf{y} + \mathbf{q}, \mathbf{z}_1\rangle \geq 0$ for every $\mathbf{y} \in C$.

Take an arbitrary point $\mathbf{x} \in X$. Since $\mathbf{x} = \mathbf{y} + \lambda\mathbf{z}_1,\ \mathbf{y} \in C,\ \mathbf{z}_1 \in K,\ \lambda \geq 0$ and

$$Q(\mathbf{x}) \equiv Q(\mathbf{y} + \lambda\mathbf{z}_1) \stackrel{\text{(E.3)}}{=} Q(\mathbf{y}) + \lambda\langle A\mathbf{y} + \mathbf{q}, \mathbf{z}_1\rangle + \frac{1}{2}\lambda^2\langle A\mathbf{z}_1, \mathbf{z}_1\rangle \geq Q(\mathbf{y}),$$

then we have

$$\sup\,\{Q(\mathbf{x}) : \mathbf{x} \in X\} = \sup\,\{Q(\mathbf{y}) : \mathbf{y} \in C\}.$$

The set C is compact by assumption, therefore the continuous function $Q(\mathbf{y})$ attains its supremum on C. Hence $Q(\mathbf{x})$ also attains its supremum on the set X.

Induction hypothesis. Let statement of Theorem E.1 hold true for $s < k$.

Concerning $s = k$:

From either the Induction hypothesis or Case 1 it follows that $Q(\mathbf{x})$ attains its supremum on each set $X_p,\ p = 1, \ldots, k$ since $X_p = C + K_p$ by definition, where K_p is generated by $s - 1$ elements according to (E.4) for each $p = 1, \ldots, k$.

Let $\overline{\mathbf{z}} \in K \setminus \{\mathbf{0}\}$ and $\langle A\overline{\mathbf{z}}, \overline{\mathbf{z}}\rangle = 0$, that is, $\langle A\mathbf{x} + \mathbf{q}, \overline{\mathbf{z}}\rangle \geq 0$ for each $\mathbf{x} \in X$.

Take an arbitrary point $\mathbf{x} \in X$. If $\mathbf{x} = \mathbf{y} + \mathbf{z},\ \mathbf{y} \in C,\ \mathbf{z} \in K$,

$$\mathbf{z} = \sum_{i=1}^{k} \alpha_i \mathbf{z}_i, \quad \alpha_i \geq 0,\ i = 1, \ldots, k,$$

$$\overline{\mathbf{z}} = \sum_{i=1}^{k} \overline{\alpha}_i \mathbf{z}_i, \quad \overline{\alpha}_i \geq 0,\ i = 1, \ldots, k$$

and if

$$\lambda_p = \max_{\overline{\alpha}_i > 0} \frac{\alpha_i}{\overline{\alpha}_i} = \frac{\alpha_p}{\overline{\alpha}_p} \quad (\geq 0), \quad i = 1, \ldots, k,$$

then $\widehat{\mathbf{z}} = \mathbf{z} + \lambda_p \overline{\mathbf{z}} \in K_p \subset K$ and $\widehat{\mathbf{x}} = \mathbf{y} + \widehat{\mathbf{z}} \in X_p$.

However,

$$\begin{aligned}
Q(\widehat{\mathbf{x}}) &= Q(\mathbf{y} + \widehat{\mathbf{z}}) = Q(\mathbf{y} + \mathbf{z} + \lambda_p \overline{\mathbf{z}}) \\
&= \tfrac{1}{2}\langle A(\mathbf{y} + \mathbf{z} + \lambda_p \overline{\mathbf{z}}), \mathbf{y} + \mathbf{z} + \lambda_p \overline{\mathbf{z}}\rangle + \langle \mathbf{q}, \mathbf{y} + \mathbf{z} + \lambda_p \overline{\mathbf{z}}\rangle \\
&= \tfrac{1}{2}\langle A(\mathbf{y} + \mathbf{z}), \mathbf{y} + \mathbf{z}\rangle + \tfrac{1}{2}\lambda_p \langle A(\mathbf{y} + \mathbf{z}), \overline{\mathbf{z}}\rangle + \tfrac{1}{2}\lambda_p \langle A\overline{\mathbf{z}}, \mathbf{y} + \mathbf{z}\rangle \\
&\quad + \tfrac{1}{2}\lambda_p^2 \langle A\overline{\mathbf{z}}, \overline{\mathbf{z}}\rangle + \langle \mathbf{q}, \mathbf{y} + \mathbf{z}\rangle + \lambda_p \langle \mathbf{q}, \overline{\mathbf{z}}\rangle \\
&= Q(\mathbf{y} + \mathbf{z}) + \lambda_p \langle A(\mathbf{y} + \mathbf{z}), \overline{\mathbf{z}}\rangle + \lambda_p \langle \mathbf{q}, \overline{\mathbf{z}}\rangle + \tfrac{1}{2}\lambda_p^2 \langle A\overline{\mathbf{z}}, \overline{\mathbf{z}}\rangle \\
&= Q(\mathbf{y} + \mathbf{z}) + \lambda_p \langle A(\mathbf{y} + \mathbf{z}) + \mathbf{q}, \overline{\mathbf{z}}\rangle + \tfrac{1}{2}\lambda_p^2 \langle A\overline{\mathbf{z}}, \overline{\mathbf{z}}\rangle \\
&\equiv Q(\mathbf{y} + \mathbf{z}) + \lambda_p \langle A\mathbf{x} + \mathbf{q}, \overline{\mathbf{z}}\rangle + \tfrac{1}{2}\lambda_p^2 \langle A\overline{\mathbf{z}}, \overline{\mathbf{z}}\rangle \\
&\geq Q(\mathbf{y} + \mathbf{z}) \equiv Q(\mathbf{x}),
\end{aligned} \tag{E.5}$$

where, in order to obtain the inequality in (E.5), we have used that $\lambda_p \geq 0$, $\langle A\mathbf{x} + \mathbf{q}, \overline{\mathbf{z}}\rangle \geq 0$ for each $\mathbf{x} \in X$ and $\langle A\overline{\mathbf{z}}, \overline{\mathbf{z}}\rangle = 0$ in the Case 2 under consideration. From (E.5) it follows that for each point $\mathbf{x} \in X$ there exists a point $\widehat{\mathbf{x}} \in \bigcup_{p=1}^{k} X_p$, such that $Q(\widehat{\mathbf{x}}) \geq Q(\mathbf{x})$.

Therefore

$$\sup_{X} Q(\mathbf{x}) = \sup_{\bigcup_{p=1}^{k} X_p} Q(\mathbf{x}) = \max_{1 \le p \le k} \left\{ \sup_{\mathbf{x} \in X_p} Q(\mathbf{x}) \right\}.$$

However, $Q(\mathbf{x})$ attains its supremum on each set X_p either according to Induction hypothesis or according to Case 1. Therefore $Q(\mathbf{x})$ attains its supremum on the set X.

The case of bounded below quadratic objective function is considered similarly. ■

Appendix F
On the Cauchy-Schwarz Inequality Approach for Solving a Quadratic Optimization Problem

Consider the following quadratic optimization problem

$$\min\ (\max) \left\{ Q(\mathbf{x}) = \langle C\mathbf{x}, \mathbf{x} \rangle \equiv \mathbf{x}^T C\mathbf{x} \right\} \tag{F.1}$$

subject to

$$\langle \mathbf{x}, \mathbf{x} \rangle \equiv \mathbf{x}^T \mathbf{x} \equiv \|\mathbf{x}\|^2 = 1, \tag{F.2}$$

where C is a $n \times n$ symmetric matrix with real entries, $\mathbf{x} \in \mathbf{R}^n$, and $\langle ., . \rangle$ is notation of the scalar (inner, dot) product, defined in the Hilbert space $\mathbf{R}^n$.

Here,

$$\langle \mathbf{x}, \mathbf{y} \rangle = \sum_{i=1}^{n} x_i y_i \tag{F.3}$$

for each $\mathbf{x}, \mathbf{y} \in \mathbf{R}^n$ by definition, and

$$\|\mathbf{x}\| = \sqrt{\langle \mathbf{x}, \mathbf{x} \rangle} \tag{F.4}$$

denotes the Euclidean norm of vector $\mathbf{x} = (x_1, \dots, x_n) \in \mathbf{R}^n$.

Since the objective function $Q(\mathbf{x})$ is continuous as a quadratic form (defined by a scalar product) and the feasible region is the unit sphere, which is a compact set in the finite-dimensional space $\mathbf{R}^n$, then problem under consideration (F.1)–(F.2) has an optimal solution according to Weierstrass theorem.

In this appendix, a method for solving problem (F.1)–(F.2) is suggested, which uses specific characteristics of this problem.

Recall some concepts and propositions which are used in the subsequent part of Appendix F.

© The Editor(s) (if applicable) and The Author(s), under exclusive license to Springer Nature Switzerland AG 2021

S. M. Stefanov, *Separable Optimization*, Springer Optimization and Its Applications 177,
https://doi.org/10.1007/978-3-030-78401-0

Let W be a vector space over the field F ($\mathbb{R}$ or $\mathbb{C}$).

Definition F.1 A numerical function $\|.\| : W \to \mathbb{R}$ is said to be a *vector norm* if for all $\mathbf{x}, \mathbf{y} \in W$ it satisfies the following properties:

1) $\|\mathbf{x}\| \geq 0$	nonnegativity,
$\|\mathbf{x}\| = 0$ iff $\mathbf{x} = \mathbf{0}$	positiveness, nonsingularity,
2) $\|c\mathbf{x}\| = \|c\|\|\mathbf{x}\|$ for all constants $c \in F$	homogeneity,
3) $\|\mathbf{x} + \mathbf{y}\| \leq \|\mathbf{x}\| + \|\mathbf{y}\|$	triangle inequality.

Denote by $\mathcal{M}_n$ the set of all $n \times n$ square matrices over the field F, and let L denote the space of all continuous linear operators in F^n. The vector norms on F^n (on F^{n^2} for $\mathcal{M}_n$), introduced in Definition F.1, can be used as norms on $\mathcal{M}_n$ because $\mathcal{M}_n$ is a vector space of dimension n^2. However, on the basis of the multiplication operation of matrices (of linear continuous operators, respectively), a new concept of *matrix norm* can be introduced.

Definition F.2 A numerical function $\|.\| : \mathcal{M}_n \to \mathbb{R}$ ($\|.\| : L \to \mathbb{R}$) is said to be a *matrix norm* (a *linear operator norm*, respectively) if for all matrices $A, B \in \mathcal{M}_n$ (if for all linear operators $\mathcal{A}, \mathcal{B} \in L$, respectively) it satisfies the following properties:

1) $\|A\| \geq 0$	nonnegativity,
$\|A\| = 0$ iff $A = 0$	positiveness, nonsingularity,
2) $\|cA\| = \|c\|\|A\|$ for all constants $c \in F$	homogeneity,
3) $\|A + B\| \leq \|A\| + \|B\|$	triangle inequality,
4) $\|AB\| \leq \|A\|\|B\|$	submultiplicativeness.

It is known that any square symmetric matrix represents a linear continuous self-conjugate operator and vice versa; that a linear operator is continuous if and only if it is bounded; and that for bounded operators, the concepts of *symmetry* and *self-conjugacy* are equivalent.

Because linear operators in $\mathbb{R}^n$ can be identified with matrices, then both cases—for matrices and for operators—can be considered simultaneously, and it is sufficient to use only matrix norms. Furthermore, a linear operator $\mathcal{A}$ in $\mathbb{R}^n$ is self-conjugate if and only if the matrix A, associated with operator $\mathcal{A}$, is symmetric.

It turns out that some vector norms are matrix norms in the vector space $\mathcal{M}_n$ according to Definition F.2, and other vector norms are not matrix norms.

Definition F.3 Let $\|.\|$ be a vector norm on $\mathbb{C}^n$. A *matrix norm* $\|.\|$ on $\mathcal{M}_n$ is defined by

$$\|C\| := \max_{\|\mathbf{x}\|=1} \|C\mathbf{x}\|, \tag{F.5}$$

or, equivalently, by

$$\|C\| := \max_{\mathbf{x} \neq \mathbf{0}} \frac{\|C\mathbf{x}\|}{\|\mathbf{x}\|}.$$

According to Weierstrass theorem, "max" can be used instead of "sup" in Definition F.3, because the unit sphere $S_{\|.\|} \equiv \{\mathbf{x} \in W : \|\mathbf{x}\| = 1\}$ of the vector norm is a compact set in the finite-dimensional vector space W and $c(\mathbf{x}) = \|C\mathbf{x}\|$ is a continuous function of $\mathbf{x}$.

It turns out that following theorem holds true.

Theorem F.1 (Properties of matrix norms)
The function $\|.\|$ *of* Definition F.3 *is a matrix norm on* $\mathcal{M}_n$*, inequality* $\|C\mathbf{x}\| \leq \|C\|\|\mathbf{x}\|$ *is satisfied for all matrices* $C \in \mathcal{M}_n$ *and for all vectors* $\mathbf{x} \in \mathbb{C}^n$*, and* $\|I\| = 1$*, where* I *is the identity matrix* (*identity operator*).

Definition F.4 The matrix norm of Definition F.3 is called *induced* by the vector norm $\|.\|$, or *operator norm*, or the *least upper bound norm* associated with the vector norm $\|.\|$.

Inequality $\|C\mathbf{x}\| \leq \|C\|\|\mathbf{x}\|$, involved in the statement of Theorem F.1, means that the induced matrix norm is compatible with the associated vector norm $\|.\|$.

It turns out that induced matrix norms have another interesting property, called *minimality*, which means that each induced matrix norm $\|.\|_m$ is minimal, that is, $\|C\|_m \leq \|C\|$ is satisfied for each matrix $C \in \mathcal{M}_n$ and for any other matrix norm $\|.\|$ on $\mathcal{M}_n$.

The matrix norm, induced by the Euclidean (Hilbert, ℓ_2-) vector norm

$$\|\mathbf{x}\| = \sqrt{\langle \mathbf{x}, \mathbf{x} \rangle}, \qquad (F.6) \equiv (F.4)$$

is the so-called *spectral norm*, defined by $\|C\|_2 = \sqrt{\mu_M^*}$, where μ_M^* is the largest eigenvalue of the matrix C^*C, and C^* is the matrix (operator), conjugate to C.

Because C^*C is a symmetric and positive definite matrix (a symmetric and positive definite operator, respectively), then all eigenvalues of C^*C are real and nonnegative.

There is a unique minimal norm in the set of *unitarily invariant matrix norms* (these are matrix norms $\|.\|$ with $\|C\| = \|UCV\|$ for all matrices $C \in \mathcal{M}_n$ and for all *unitary* matrices $U, V \in \mathcal{M}_n$), and this norm is the spectral norm.

The space, conjugate to the Hilbert space $\mathbf{R}^n$ with ℓ_2- norm, is the same space $\mathbf{R}^n$ with ℓ_2- norm. More generally, the space H^*, conjugate to a Hilbert space H, is the same Hilbert space H, that is, $H^* = H$.

The following proposition states the main result in this appendix.

Proposition F.1 (Optimal solution of problem (F.1)–(F.2))
The optimal solution of problem (F.1)–(F.2) *in the case of minimization is the eigenvector* $\mathbf{x}_m$*, associated with the least eigenvalue* μ_m *of matrix* C (*the eigenvector* $\mathbf{x}_M$*, associated with the largest eigenvalue* μ_M *of matrix* C *in the case of maximization, respectively*)*, and the minimal and maximal values of the objective function* $Q(\mathbf{x})$*, defined by* (F.1)*, subject to constraint* (F.2) *is equal to* μ_m *and to* μ_M *in the minimization and the maximization case, respectively.*

Proof. Applying the Cauchy-Schwarz inequality $|\langle C\mathbf{x}, \mathbf{x}\rangle| \leq \|C\mathbf{x}\| \|\mathbf{x}\|$, we get

$$-\|C\mathbf{x}\| \|\mathbf{x}\| \leq \langle C\mathbf{x}, \mathbf{x}\rangle \leq \|C\mathbf{x}\| \|\mathbf{x}\|,$$

where

a) the left inequality is fulfilled as equality iff $C\mathbf{x} = \alpha\mathbf{x}$,
b) the right inequality is fulfilled as equality iff $C\mathbf{x} = \beta\mathbf{x}$

for some constants $\alpha, \beta \in F$.

We have $\|\mathbf{x}\| \stackrel{\text{def}}{=} \sqrt{\langle \mathbf{x}, \mathbf{x}\rangle} = 1$ by assumption. Hence, taking minimum in both sides of the first inequality and taking maximum in both sides of the second inequality of the Cauchy-Shwarz inequality, we obtain

$$\min_{\langle \mathbf{x},\mathbf{x}\rangle=1} (-\|C\mathbf{x}\| \|\mathbf{x}\|) \leq \min_{\langle \mathbf{x},\mathbf{x}\rangle=1} \langle C\mathbf{x}, \mathbf{x}\rangle,$$

$$\max_{\langle \mathbf{x},\mathbf{x}\rangle=1} \langle C\mathbf{x}, \mathbf{x}\rangle \leq \max_{\langle \mathbf{x},\mathbf{x}\rangle=1} \|C\mathbf{x}\| \|\mathbf{x}\|,$$

respectively. Therefore

$$-\max_{\langle \mathbf{x},\mathbf{x}\rangle=1} \|C\mathbf{x}\| \leq \min_{\langle \mathbf{x},\mathbf{x}\rangle=1} \langle C\mathbf{x}, \mathbf{x}\rangle,$$

$$\max_{\langle \mathbf{x},\mathbf{x}\rangle=1} \langle C\mathbf{x}, \mathbf{x}\rangle \leq \max_{\langle \mathbf{x},\mathbf{x}\rangle=1} \|C\mathbf{x}\|,$$

respectively, where it is used that $\min\{-f(\mathbf{x}), \mathbf{x} \in X\} = -\max\{f(\mathbf{x}), \mathbf{x} \in X\}$ for every objective function f and for every feasible set X, and $\langle \mathbf{x}, \mathbf{x}\rangle = \|\mathbf{x}\|^2 = 1$ according to (F.2). Using Definition F.3, we obtain

$$-\|C\| \leq \min_{\langle \mathbf{x},\mathbf{x}\rangle=1} \langle C\mathbf{x}, \mathbf{x}\rangle,$$

$$\max_{\langle \mathbf{x},\mathbf{x}\rangle=1} \langle C\mathbf{x}, \mathbf{x}\rangle \leq \|C\|,$$

respectively. About Euclidean vector norm of vector $\mathbf{x}$ we have

$$-\sqrt{\mu_M^*} \leq \min_{\langle \mathbf{x},\mathbf{x}\rangle=1} \langle C\mathbf{x}, \mathbf{x}\rangle,$$

$$\max_{\langle \mathbf{x},\mathbf{x}\rangle=1} \langle C\mathbf{x}, \mathbf{x}\rangle \leq \sqrt{\mu_M^*},$$

respectively, where μ_M^* is the largest eigenvalue of matrix $C^T C$. However, C is a symmetric matrix (symmetric operator) by assumption. Therefore, $C^T = C$, hence $C^T C = C^2$, and therefore, $\mu_j^* = \mu_j^2$, $j = 1, \ldots, n$, where μ_j, $j = 1, \ldots, n$, are the eigenvalues of matrix C and μ_j^*, $j = 1, \ldots, n$, are the eigenvalues of matrix $C^T C$.

In particular, we have $\mu_M^* = \mu_M^2$. It follows that

$$-|\mu_M| \leq \min_{\langle \mathbf{x},\mathbf{x}\rangle=1} \langle C\mathbf{x}, \mathbf{x}\rangle, \tag{F.7}$$

$$\max_{\langle \mathbf{x},\mathbf{x}\rangle=1} \langle C\mathbf{x}, \mathbf{x}\rangle \leq |\mu_M|, \tag{F.8}$$

respectively, where μ_M is the largest eigenvalue of matrix C.

In accordance with cases a) and b), the left and the right inequalities of the Cauchy-Schwarz inequality are fulfilled as equalities for $C\mathbf{x} = \mu_m\mathbf{x}$ and $C\mathbf{x} = \mu_M\mathbf{x}$, respectively, that is, for the eigenvectors $\mathbf{x}_m$ and $\mathbf{x}_M$, associated with the eigenvalues μ_m and μ_M of matrix C, respectively, and the minimal value and the maximal value of the objective function $Q(\mathbf{x}) = \langle C\mathbf{x}, \mathbf{x}\rangle$ subject to $\|\mathbf{x}\|^2 = \langle \mathbf{x}, \mathbf{x}\rangle = 1$ are μ_m and μ_M, respectively. Therefore we have

$$\min_{\langle \mathbf{x},\mathbf{x}\rangle=1} \langle C\mathbf{x}, \mathbf{x}\rangle = \langle \mu_m\mathbf{x}_m, \mathbf{x}_m\rangle = \mu_m\langle \mathbf{x}_m, \mathbf{x}_m\rangle = \mu_m, \tag{F.9}$$

and

$$\max_{\langle \mathbf{x},\mathbf{x}\rangle=1} \langle C\mathbf{x}, \mathbf{x}\rangle = \langle \mu_M\mathbf{x}_M, \mathbf{x}_M\rangle = \mu_M\langle \mathbf{x}_M, \mathbf{x}_M\rangle = \mu_M. \tag{F.10}$$

Because μ_m and μ_M are eigenvalues of matrix C and since (F.9) and (F.10) hold true, then μ_m is the least eigenvalue of matrix C and μ_M is the largest eigenvalue of matrix C. Assuming the contrary, that there exists eigenvalue ξ of matrix C, smaller than μ_m, and that there exists eigenvalue η of matrix C, larger than μ_M, we would obtain

$$\min_{\langle \mathbf{x},\mathbf{x}\rangle=1} \langle C\mathbf{x}, \mathbf{x}\rangle = \langle \xi\mathbf{x}, \mathbf{x}\rangle = \xi\langle \mathbf{x}, \mathbf{x}\rangle = \xi < \mu_m,$$

and

$$\max_{\langle \mathbf{x},\mathbf{x}\rangle=1} \langle C\mathbf{x}, \mathbf{x}\rangle = \langle \eta\mathbf{x}, \mathbf{x}\rangle = \eta\langle \mathbf{x}, \mathbf{x}\rangle = \eta > \mu_M,$$

respectively, a contradiction with (F.9) and (F.10).

Therefore,

$$\min_{\langle \mathbf{x},\mathbf{x}\rangle=1} \langle C\mathbf{x}, \mathbf{x}\rangle = \mu_m \tag{F.11}$$

and the solution of minimization problem (F.1)–(F.2) is the eigenvector $\mathbf{x}_m$ of matrix C, associated with the least eigenvalue μ_m of matrix C. Similarly, we have obtained that

$$\max_{\langle \mathbf{x},\mathbf{x}\rangle=1} \langle C\mathbf{x}, \mathbf{x}\rangle = \mu_M \tag{F.12}$$

and the solution of maximization problem (F.1)–(F.2) is the eigenvector $\mathbf{x}_M$ of matrix C, associated with μ_M, where μ_M is defined above. ■

The Lagrange multiplier method is a common approach for solving equality constrained optimization problems, including the considered problem (F.1)–(F.2).

The Lagrangian function for problem (F.1)–(F.2) is

$$L(\mathbf{x}, \boldsymbol{\mu}) = \mu_0 \langle C\mathbf{x}, \mathbf{x} \rangle + \mu_1 \langle \mathbf{x}, \mathbf{x} \rangle.$$

In accordance with the Lagrange multiplier theorem, the necessary condition for the minimum solution to this problem is

$$\mu_0 C\mathbf{x} + \mu_1 \mathbf{x} = \mathbf{0},$$

where $\langle \mathbf{x}, \mathbf{x} \rangle = 1$ according to (F.2), and the Lagrange multipliers (dual variables) μ_0 and μ_1 cannot simultaneously be equal to zero. Assuming that $\mu_0 = 0$, we would obtain that $\mu_1 \mathbf{x} = \mathbf{0}$, where μ_1 must be different from zero, because $\mu_0 = 0$ by the assumption. Therefore $\mathbf{x} = \mathbf{0}$. However, $\mathbf{x} = \mathbf{0}$ is not a feasible solution for the considered problem (F.1)–(F.2), because $\langle \mathbf{0}, \mathbf{0} \rangle \neq 1$, a contradiction. Hence $\mu_0 \neq 0$, and $\mu_0 C\mathbf{x} + \mu_1 \mathbf{x} = \mathbf{0}$. If we divide both sides of this vector equality by μ_0 ($\neq 0$), we obtain $C\mathbf{x} + \frac{\mu_1}{\mu_0}\mathbf{x} = \mathbf{0}$, or equivalently, $C\mathbf{x} = -\frac{\mu_1}{\mu_0}\mathbf{x}$. Therefore

$$C\mathbf{x} = \mu_1^* \mathbf{x}, \tag{F.13}$$

where $\mu_1^* \stackrel{\text{def}}{=} -\frac{\mu_1}{\mu_0}$, and this means that the optimal solution $\mathbf{x}^*$ of the considered problem (F.1)–(F.2) is an eigenvector of matrix C, and μ_1^* is the associated eigenvalue. If we multiply both sides of (F.13) by $\mathbf{x}$, we get

$$\min_{\langle \mathbf{x}, \mathbf{x} \rangle = 1} \langle C\mathbf{x}, \mathbf{x} \rangle = \mu_1^* \langle \mathbf{x}, \mathbf{x} \rangle = \mu_1^*,$$

where the constraint (F.2) has been used. This result means that the optimal solution to the minimization problem (F.1)–(F.2) is the eigenvector $\mathbf{x}^*$, associated with the least eigenvalue μ_1^* of matrix C, and μ_1^* is the minimum value of the quadratic objective function $Q(\mathbf{x}) = \langle C\mathbf{x}, \mathbf{x} \rangle$ subject to $\langle \mathbf{x}, \mathbf{x} \rangle = 1$, that is,

$$\min_{\langle \mathbf{x}, \mathbf{x} \rangle = 1} \langle C\mathbf{x}, \mathbf{x} \rangle = \mu_1^*. \tag{F.14}$$

The maximization case is considered similarly.

As we have expected, the method proposed in Proposition F.1 and the Lagrange multiplier method, applied to the considered problem (F.1)–(F.2), give the same result.

Appendix G
Theorems of the Alternative

In this appendix, some theorems of the alternative for linear algebraic equations and linear algebraic inequalities are considered. Such theorems are important for both linear algebra and mathematical optimization, especially for mathematical optimization problems with linear equality and/or inequality constraints.

Some classical theorems of the alternative, such as Gordan's theorem, Farkas' theorem, Stiemke's theorem, Motzkin's theorem, Slater's theorem, Tucker's theorem, Duffin's theorem, and Gale's theorems are recalled below.

Gordan's theorem of the alternative

For a given $m \times n$ matrix A either system

I. $A\mathbf{x} > \mathbf{0}$

has a solution $\mathbf{x} \in \mathbb{R}^n$

or system

II. $A^T\mathbf{y} = \mathbf{0},\ \mathbf{y} \geq \mathbf{0}$

has a solution $\mathbf{y} \in \mathbb{R}^m$,

but never both.

Farkas' theorem of the alternative

For a given $m \times n$ matrix A and each given vector $\mathbf{c} \in \mathbb{R}^n$ either system

I. $A\mathbf{x} \leq \mathbf{0},\ \langle \mathbf{c}, \mathbf{x} \rangle > 0$

has a solution $\mathbf{x} \in \mathbb{R}^n$

or system

II. $A^T\mathbf{y} = \mathbf{c},\ \mathbf{y} \geq \mathbf{0}$

has a solution $\mathbf{y} \in \mathbb{R}^m$,

but never both,

where $\langle ., . \rangle$ denotes the scalar (inner, dot) product, and A^T is the transpose matrix of A.

© The Editor(s) (if applicable) and The Author(s), under exclusive license
to Springer Nature Switzerland AG 2021

S. M. Stefanov, *Separable Optimization*, Springer Optimization and Its Applications 177,
https://doi.org/10.1007/978-3-030-78401-0

Stiemke's theorem of the alternative

For a given $m \times n$ matrix A either system
I. $A\mathbf{x} \geq \mathbf{0}$
has a solution $\mathbf{x} \in \mathbf{R}^n$
or system
II. $A^T\mathbf{y} = \mathbf{0},\ \mathbf{y} > \mathbf{0}$
has a solution $\mathbf{y} \in \mathbf{R}^m$,
but never both.

Motzkin's theorem of the alternative

Let A, B, and C be given $m \times n$ matrices. Then either system
I. $A\mathbf{x} > \mathbf{0},\ B\mathbf{x} \geq \mathbf{0},\ C\mathbf{x} = \mathbf{0}$
has a solution $\mathbf{x} \in \mathbf{R}^n$
or system
II. $A^T\mathbf{y}_1 + B^T\mathbf{y}_2 + C^T\mathbf{y}_3 = \mathbf{0},\ \ \mathbf{y}_1 \geq \mathbf{0},\ \mathbf{y}_2 \geq \mathbf{0}$
has a solution $\mathbf{y}_1, \mathbf{y}_2, \mathbf{y}_3 \in \mathbf{R}^m$,
but never both.

Slater's theorem of the alternative

Let A, B, C, and D be given $m \times n$ matrices. Then either system
I. $A\mathbf{x} > \mathbf{0},\ B\mathbf{x} \geq \mathbf{0},\ C\mathbf{x} \geq \mathbf{0},\ D\mathbf{x} = \mathbf{0}$
has a solution $\mathbf{x} \in \mathbf{R}^n$
or system
II. $A^T\mathbf{y}_1 + B^T\mathbf{y}_2 + C^T\mathbf{y}_3 + D^T\mathbf{y}_4 = \mathbf{0}$
with $\mathbf{y}_1 \geq \mathbf{0},\ \mathbf{y}_2 \geq \mathbf{0},\ \mathbf{y}_3 \geq \mathbf{0}$ or $\mathbf{y}_1 \geq \mathbf{0},\ \mathbf{y}_2 > \mathbf{0},\ \mathbf{y}_3 \geq \mathbf{0}$
has a solution $\mathbf{y}_1, \mathbf{y}_2, \mathbf{y}_3, \mathbf{y}_4 \in \mathbf{R}^m$,
but never both.

Tucker's theorem of the alternative

Let A, B, and C be given $m \times n$ matrices. Then either system
I. $A\mathbf{x} \geq \mathbf{0},\ B\mathbf{x} \geq \mathbf{0},\ C\mathbf{x} = \mathbf{0}$
has a solution $\mathbf{x} \in \mathbf{R}^n$
or system
II. $A^T\mathbf{y}_1 + B^T\mathbf{y}_2 + C^T\mathbf{y}_3 = \mathbf{0},\ \ \mathbf{y}_1 > \mathbf{0},\ \mathbf{y}_2 \geq \mathbf{0}$
has a solution $\mathbf{y}_1, \mathbf{y}_2, \mathbf{y}_3 \in \mathbf{R}^m$,
but never both.

Duffin's theorem of the alternative
(Nonhomogeneous Farkas' theorem of the alternative)

For a given $m \times n$ matrix A, given vectors $\mathbf{c} \in \mathbf{R}^n$, $\mathbf{d} \in \mathbf{R}^m$, and given scalar γ either system
I. $A\mathbf{x} \leq \mathbf{d},\ \langle \mathbf{c}, \mathbf{x} \rangle > \gamma$
has a solution $\mathbf{x} \in \mathbf{R}^n$
or system

II. $A^T\mathbf{y} = \mathbf{c},\ \langle \mathbf{d}, \mathbf{y}\rangle \le \gamma,\ \mathbf{y} \ge \mathbf{0}$ or $A^T\mathbf{y} = \mathbf{0},\ \langle \mathbf{d}, \mathbf{y}\rangle < 0,\ \mathbf{y} \ge \mathbf{0}$
has a solution $\mathbf{y} \in \mathbf{R}^m$,
but never both.

Gale's theorem of the alternative for linear equations

For a given $m \times n$ matrix A and a given vector $\mathbf{c} \in \mathbf{R}^m$ either system
I. $A\mathbf{x} = \mathbf{c}$
has a solution $\mathbf{x} \in \mathbf{R}^n$
or system
II. $A^T\mathbf{y} = \mathbf{0},\ \langle \mathbf{c}, \mathbf{y}\rangle = 1$
has a solution $\mathbf{y} \in \mathbf{R}^m$,
but never both.

Gale's theorem of the alternative for linear inequalities "$\le$"

For a given $m \times n$ matrix A and a given vector $\mathbf{c} \in \mathbf{R}^m$ either system
I. $A\mathbf{x} \le \mathbf{c}$
has a solution $\mathbf{x} \in \mathbf{R}^n$
or system
II. $A^T\mathbf{y} = \mathbf{0},\ \langle \mathbf{c}, \mathbf{y}\rangle = -1,\ \mathbf{y} \ge \mathbf{0}$
has a solution $\mathbf{y} \in \mathbf{R}^m$,
but never both.

Theorem of the alternative (Mangasarian)

Let A, B, C, and D be given $m \times n$ matrices. Then either system
I. $A\mathbf{x} \ge \mathbf{0},\ B\mathbf{x} \ge \mathbf{0},\ C\mathbf{x} \ge \mathbf{0},\ D\mathbf{x} = \mathbf{0}$ or
$A\mathbf{x} \ge \mathbf{0},\ B\mathbf{x} > \mathbf{0},\ C\mathbf{x} \ge \mathbf{0},\ D\mathbf{x} = \mathbf{0}$
has a solution $\mathbf{x} \in \mathbf{R}^n$
or system
II. $A^T\mathbf{y}_1 + B^T\mathbf{y}_2 + C^T\mathbf{y}_3 + D^T\mathbf{y}_4 = \mathbf{0},\ \mathbf{y}_1 > \mathbf{0},\ \mathbf{y}_2 \ge \mathbf{0},\ \mathbf{y}_3 \ge \mathbf{0}$
has a solution $\mathbf{y}_1, \mathbf{y}_2, \mathbf{y}_3, \mathbf{y}_4 \in \mathbf{R}^m$,
but never both.

Theorem of the alternative for linear inequalities "$\le$"
(Mangasarian)

For a given $m \times n$ matrix A and a given vector $\mathbf{c} \in \mathbf{R}^m$ either system
I. $A\mathbf{x} \le \mathbf{c}$
has a solution $\mathbf{x} \in \mathbf{R}^n$
or system
II. $A^T\mathbf{y} = \mathbf{0},\ \langle \mathbf{c}, \mathbf{y}\rangle = -1,\ \mathbf{y} \ge \mathbf{0}$ or $A^T\mathbf{y} = \mathbf{0},\ \langle \mathbf{c}, \mathbf{y}\rangle \le 0,\ \mathbf{y} > \mathbf{0}$
has a solution $\mathbf{y} \in \mathbf{R}^m$,
but never both.

A statement of the form "Either linear system I. has a solution or system II. has a solution but never both", involved in the statement of theorems of the alternative, means that systems I. and II. cannot be solvable simultaneously or unsolvable simultaneously, but always when system I. has a solution then system II. has no solution, and always when system I. has no solution then system II. has a solution. Symbolically, this can be written as follows

$$\text{I.} \Longleftrightarrow \overline{\text{II.}} \quad \text{or} \quad \overline{\text{I.}} \Longleftrightarrow \text{II.},$$

where $\overline{\text{I.}}$ denotes nonexistence of solution of system I.

Typical structure of the proof of a theorem of the alternative is

$$\text{I.} \Rightarrow \overline{\text{II.}} \quad \text{and} \quad \overline{\text{I.}} \Rightarrow \text{II.}$$

$$\text{(or, equivalently,} \quad \overline{\text{I.}} \Rightarrow \text{II.} \quad \text{and} \quad \text{I.} \Rightarrow \overline{\text{II.}},$$

$$\text{or, equivalently,} \quad \text{II.} \Rightarrow \overline{\text{I.}} \quad \text{and} \quad \overline{\text{II.}} \Rightarrow \text{I.).}$$

Implication I. $\Rightarrow \overline{\text{II.}}$ can be proved either directly or indirectly (by proving that systems I. and II. cannot be solved simultaneously, whence it follows that I. $\Rightarrow \overline{\text{II.}}$ (or II. $\Rightarrow \overline{\text{I.}}$), see, for example, proof of Theorem G.1 below).

Proof structures of the form

$$\text{I.} \Rightarrow \overline{\text{II.}} \quad \text{and} \quad \text{II.} \Rightarrow \overline{\text{I.}},$$

as well as of the form

$$\overline{\text{I.}} \Rightarrow \text{II.} \quad \text{and} \quad \overline{\text{II.}} \Rightarrow \text{I.}$$

are not complete because both implications in the first structure prove the same—that both systems I. and II. cannot be solvable simultaneously, and both implications in the second structure prove the same—that both systems I. and II. cannot be unsolvable simultaneously.

The following two propositions will be used below.

Proposition G.1 (Farkas' theorem: an equivalent formulation)
Let $\mathbf{a}_0, \mathbf{a}_1, \ldots, \mathbf{a}_m \in \mathbf{R}^n$, $\mathbf{x} \in \mathbf{R}^n$ *and* $\langle \mathbf{a}_i, \mathbf{x} \rangle \leq 0$, $i = 1, \ldots, m$ *imply* $\langle \mathbf{a}_0, \mathbf{x} \rangle \leq 0$. *Then there exist scalars* $\lambda_i \geq 0$, $i = 1, \ldots, m$, *such that*

$$\mathbf{a}_0 = \sum_{i=1}^{m} \lambda_i \mathbf{a}_i.$$

Proposition G.2 (Corollary of Proposition G.1 for systems of linear equations and linear inequalities)
Let the set X be defined as follows

$\langle \mathbf{a}_i, \mathbf{x} \rangle = 0$, $i = 1, \ldots, l$,

$\langle \mathbf{b}_j, \mathbf{x} \rangle \leq 0$, $j = 1, \ldots, m$.

If the inequality $\langle \mathbf{c}, \mathbf{x} \rangle \leq 0$ *is satisfied for each* $\mathbf{x} \in X$*, then there exist scalars* $\lambda_i,\ i = 1, \ldots, l;\ \ \mu_j \geq 0,\ j = 1, \ldots, m$*, such that*

$$\mathbf{c} = \sum_{i=1}^{l} \lambda_i \mathbf{a}_i + \sum_{j=1}^{m} \mu_i \mathbf{b}_j.$$

Some theorems of the alternative are formulated and proved below.

Theorem G.1 *Let A be an* $m \times n$ *matrix. Then either system*
I. $A\mathbf{x} < \mathbf{0}$
has a solution $\mathbf{x} \in \mathbf{R}^n$
or system
II. $A^T\mathbf{y} = \mathbf{0},\ \mathbf{y} \geq \mathbf{0},\ \mathbf{y} \neq \mathbf{0}$
has a solution $\mathbf{y} \in \mathbf{R}^m$*,*
but never both.

Proof. a) Let system II. have no solution. We will prove that system I. has a solution.
Denote by $\mathbf{a}_i,\ i = 1, \ldots, n$ columns of matrix A, that is, rows of the transpose matrix A^T, and by $\mathbf{e}_j = (0, \ldots, 0, 1, 0, \ldots, 0) \in \mathbf{R}^m,\ j = 1, \ldots, m$, denote the m-dimensional unit vectors with one in the j−th position and zeroes elsewhere. Since system II. has no solution by the assumption, then

$$A^T\mathbf{y} = \mathbf{0},\ \ \mathbf{y} \geq \mathbf{0},\ \ \mathbf{y} \in \mathbf{R}^m \ \text{ imply } \ \mathbf{y} = \mathbf{0},$$

that is,

$$\langle \mathbf{a}_i, \mathbf{y} \rangle = 0,\ i = 1, \ldots, n;\ \ \langle -\mathbf{e}_j, \mathbf{y} \rangle \leq 0,\ \ \mathbf{y} \in \mathbf{R}^m \ \text{ imply } \ \langle \mathbf{e}_{j_0}, \mathbf{y} \rangle = 0,$$

where $1 \leq j_0 \leq m$. From Proposition G.2 it follows that there exist scalars $\lambda_i,\ i = 1, \ldots, n$ and $\mu_j \geq 0,\ j = 1, \ldots, m$, such that

$$\mathbf{e}_{j_0} = \sum_{i=1}^{n} \lambda_i \mathbf{a}_i + \sum_{j=1}^{m} \mu_j(-\mathbf{e}_j),$$

that is,

$$\mathbf{e}_{j_0} + \sum_{j=1}^{m} \mu_j \mathbf{e}_j = \sum_{i=1}^{n} \lambda_i \mathbf{a}_i.$$

Denote $\mathbf{x}_{j_0} \equiv (\lambda_1, \ldots, \lambda_n)$. Taking into account that $\mu_j \geq 0,\ j = 1, \ldots, m$ and that components of vectors $\mathbf{e}_j,\ j = 1, \ldots, m$, and $\mathbf{e}_{j_0}$ are nonnegative, we obtain

$$\mathbf{0} \leq \mathbf{e}_{j_0} + \sum_{j=1}^{m} \mu_j \mathbf{e}_j = A\mathbf{x}_{j_0} \equiv \begin{pmatrix} a_{11} & \ldots & a_{1n} \\ \ldots & \ldots & \ldots \\ a_{m1} & \ldots & a_{mn} \end{pmatrix} \begin{pmatrix} \lambda_1 \\ \ldots \\ \lambda_n \end{pmatrix}.$$

Then for each $j_0 = 1, \ldots, m$, there exists a vector $A\mathbf{x}_{j_0}$, whose j_0–th component is (strictly) positive. Hence

$$A\mathbf{x}^0 \equiv \sum_{j_0=1}^{m} A\mathbf{x}_{j_0} > \mathbf{0}$$

as a sum of positive addends. Therefore $\mathbf{x}_0 \in \mathbf{R}^n$ is a solution of the system $A\mathbf{x} > \mathbf{0}$, that is, $(-\mathbf{x}^0) \in \mathbf{R}^n$ is a solution of the system $A\mathbf{x} < \mathbf{0}$. This means that system I. has a solution.

b) Assume that system I. has a solution $\mathbf{x}^0$ and system II. has a solution $\mathbf{y}^0$ simultaneously. Then $A\mathbf{x}^0 < \mathbf{0},\ \mathbf{x}^0 \in \mathbf{R}^n$ according to I.; $A^T\mathbf{y}^0 = \mathbf{0},\ \mathbf{y}^0 \geq \mathbf{0},\ \mathbf{y}^0 \neq \mathbf{0}$ (whence $\mathbf{y}^0 > \mathbf{0}$) according to II. Therefore

$$0 \overset{\text{I.,II.}}{>} \langle A\mathbf{x}^0, \mathbf{y}^0\rangle = \langle \mathbf{x}^0, A^T\mathbf{y}^0\rangle \overset{\text{II.}}{=} \langle \mathbf{x}^0, \mathbf{0}\rangle = 0,$$

that is, $0 > 0$ strictly, which is a contradiction. Therefore the assumption is wrong, which implies that systems I. and II. cannot be solvable simultaneously. Hence, if system II. has a solution, then system I. has no solution, and vice versa. ■

The following Proposition G.3 will be used in the proof of the next theorem of the alternative.

Proposition G.3 (Existence of solution of a system of linear algebraic inequalities) *Let A be an $m \times n$ matrix and $\mathbf{b} \in \mathbf{R}^m$. The system*

$$\text{I.} \quad A\mathbf{x} \leq \mathbf{b},\ \mathbf{x} \geq \mathbf{0}$$

has a solution $\mathbf{x} \in \mathbf{R}^n$ if and only if $\langle \mathbf{b}, \mathbf{y}\rangle \geq 0$ for each $\mathbf{y} \in \mathbf{R}^m$, such that

$$\text{II.} \quad A^T\mathbf{y} \geq \mathbf{0},\ \mathbf{y} \geq \mathbf{0}.$$

Proof. a) Let system I. have a solution. Therefore there exists a vector $\mathbf{x}^0 \in \mathbf{R}^n$ such that $A\mathbf{x}^0 \leq \mathbf{b},\ \mathbf{x}^0 \geq \mathbf{0}$. Let $\mathbf{y} \in \mathbf{R}^m$ satisfy II. Then

$$\langle \mathbf{b}, \mathbf{y}\rangle \geq \langle A\mathbf{x}^0, \mathbf{y}\rangle = \langle \mathbf{x}^0, A^T\mathbf{y}\rangle \geq 0$$

for each $\mathbf{y} \in \mathbf{R}^m,\ A^T\mathbf{y} \geq \mathbf{0},\ \mathbf{y} \geq \mathbf{0}$, where for the first inequality we have used that $\mathbf{y} \geq \mathbf{0}$ according to II. and $\mathbf{b} \geq A\mathbf{x}^0$ according to I., and for the last inequality we have used that $\mathbf{x}^0 \geq \mathbf{0}$ according to I. and $A^T\mathbf{y} \geq \mathbf{0}$ according to II. We have proved that $\langle \mathbf{b}, \mathbf{y}\rangle \geq 0$ for each $\mathbf{y} \in \mathbf{R}^m$, such that $A^T\mathbf{y} \geq \mathbf{0},\ \mathbf{y} \geq \mathbf{0}$, that is, when system II. is satisfied.

b) Let $\mathbf{y} \in \mathbf{R}^m$ and $A^T\mathbf{y} \geq \mathbf{0},\ \mathbf{y} \geq \mathbf{0}$ (that is, system II. is satisfied) imply $\langle \mathbf{b}, \mathbf{y}\rangle \geq 0$. This is equivalent to the implication

$$-A^T\mathbf{y} \leq \mathbf{0},\ -\mathbf{y} \leq \mathbf{0},\ \mathbf{y} \in \mathbf{R}^m \quad \text{imply} \quad \langle -\mathbf{b}, \mathbf{y}\rangle \leq 0.$$

Let $\mathbf{a}_j,\ j = 1, \ldots, n$ be columns of matrix A, that is, rows of matrix A^T, and let $\mathbf{e}_i$ be the i–th unit m–vector. Then the above implication can be presented in the form

$$\langle -\mathbf{a}_j, \mathbf{y}\rangle \le 0,\ j = 1, \ldots, n;\ \ \langle -\mathbf{e}_i, \mathbf{y}\rangle \le 0,\ i = 1, \ldots, m;\ \ \mathbf{y} \in \mathbf{R}^m$$

$$\text{imply}\ \ \langle -\mathbf{b}, \mathbf{y}\rangle \le 0.$$

From Farkas' theorem (equivalent formulation, Proposition G.1) it follows that there exist scalars $\lambda_j \ge 0,\ j = 1, \ldots, n$ and $\mu_i \ge 0,\ i = 1, \ldots, m$, such that

$$-\mathbf{b} = \sum_{j=1}^{n} \lambda_j(-\mathbf{a}_j) + \sum_{i=1}^{m} \mu_i(-\mathbf{e}_i).$$

Multiplying both sides of this equality by (-1), denoting $\mathbf{x} \equiv (\lambda_1, \ldots, \lambda_n) \in \mathbf{R}^n_+$, $\mathbf{v} \equiv (\mu_1, \ldots, \mu_m) \in \mathbf{R}^m_+$ and taking into account that $\mathbf{a}_j$ are columns of matrix A, that is, rows of matrix A^T, we get

$$\mathbf{b} = A\mathbf{x} + \mathbf{v}.$$

Since $\mathbf{v} \ge \mathbf{0}$, then $A\mathbf{x} \le \mathbf{b}$ for each $\mathbf{x} \ge \mathbf{0},\ \mathbf{x} \in \mathbf{R}^n$, that is, system I. has a solution. ∎

Theorem G.2 *For a given* $m \times n$ *matrix* A *and given vector* $\mathbf{c} \in \mathbf{R}^n$ *either system*

I. $A\mathbf{x} \ge \mathbf{0},\ \langle \mathbf{c}, \mathbf{x}\rangle > 0,\ \mathbf{x} \ge \mathbf{0}$

has a solution $\mathbf{x} \in \mathbf{R}^n$

or system

II. $A^T\mathbf{y} \ge \mathbf{c},\ \mathbf{y} \le \mathbf{0}$

has a solution $\mathbf{y} \in \mathbf{R}^m$,

but never both.

Proof. a) Let system I. have no solution. Then

$$A\mathbf{x} \ge \mathbf{0},\ \mathbf{x} \ge \mathbf{0},\ \mathbf{x} \in \mathbf{R}^n\ \ \text{imply}\ \ \langle \mathbf{c}, \mathbf{x}\rangle \le 0.$$

Interchanging n by m, A by A^T, $\mathbf{b}$ by $-\mathbf{c}$, $\mathbf{x}$ by $\mathbf{y}$ and vice versa (which is correct with respect to the dimensions of the new structures because n and m have been interchanged), then from sufficiency part of Proposition G.3 we get

$$A^T\mathbf{y} \le -\mathbf{c},\ \mathbf{y} \ge \mathbf{0},\ \mathbf{y} \in \mathbf{R}^m,$$

that is,

$$A^T(-\mathbf{y}) \ge \mathbf{c},\ (-\mathbf{y}) \le \mathbf{0},\ -\mathbf{y} \in \mathbf{R}^m.$$

Therefore system II. has a solution $(-\mathbf{y}) \in \mathbf{R}^m$.

b) Assume that both systems I. and II. are solvable simultaneously and their solutions are $\mathbf{x}^0$ and $\mathbf{y}^0$, respectively. Then for $\mathbf{c} \leq A^T\mathbf{y}^0$, $\mathbf{x}^0 \geq \mathbf{0}$, $\mathbf{y}^0 \leq \mathbf{0}$, $A\mathbf{x}^0 \geq \mathbf{0}$ we have

$$0 \overset{\text{I.}}{<} \langle \mathbf{c}, \mathbf{x}^0 \rangle \leq \langle A^T\mathbf{y}^0, \mathbf{x}^0 \rangle = \langle \mathbf{y}^0, \ A\mathbf{x}^0 \rangle \leq 0.$$

The strict inequality $0 < 0$, obtained above, is a contradiction. Therefore the assumption is wrong, which means that systems I. and II. cannot be solvable simultaneously. Hence, if system I. has a solution, then system II. has no solution, and vice versa. ■

The following Theorem G.3 can be proved with an application of Farkas' theorem of the alternative.

Theorem G.3 *Let A be an $m \times n$ matrix and $\mathbf{c} \in \mathbf{R}^n$. Then either system*

I. $A\mathbf{x} \leq \mathbf{0}, \ \mathbf{x} \geq \mathbf{0}, \ \langle \mathbf{c}, \mathbf{x} \rangle < 0$

has a solution $\mathbf{x} \in \mathbf{R}^n$

or system

II. $A^T\mathbf{y} = \mathbf{c}, \ \mathbf{y} \geq \mathbf{0}$

has a solution $\mathbf{y} \in \mathbf{R}^m$,

but never both.

Proof. Rewrite system I. in the following equivalent form

$$\overline{A}\mathbf{x} \leq \mathbf{0}, \ \langle \mathbf{c}, \mathbf{x} \rangle < 0,$$

where

$$\overline{A} = \begin{pmatrix} A \\ -I \end{pmatrix},$$

I is the unit matrix of the corresponding dimension, where we have taken into account that the inequality $\mathbf{x} \geq \mathbf{0}$ is equivalent to the inequality $-I\mathbf{x} \leq \mathbf{0}$. Applying Farkas' theorem of the alternative with $A := \overline{A}$ and $\langle \mathbf{c}, \mathbf{x} \rangle < 0$ instead of $\langle \mathbf{c}, \mathbf{x} \rangle \geq 0$ in system I. (with $\mathbf{c} := -\mathbf{c}$), we get the statement of Theorem G.3. ■

Bibliography

1. K.D. Andersen, An efficient Newton barrier method for minimizing a sum of Euclidean norms. SIAM J. Optim. **6**(1), 74–95 (1996)
2. I. Barrodale, F. Roberts, An improved algorithm for discrete ℓ_1 linear approximation. SIAM J. Numer. Anal. **10**(5), 839–848 (1973)
3. I. Barrodale, F. Roberts, An efficient algorithm for discrete ℓ_1 linear approximation with constraints. SIAM J. Numer. Anal. **15**(3), 603–611 (1978)
4. R.H. Bartels, A.R. Conn, J.W. Sinclair, Minimization techniques for piecewise differentiable functions: the ℓ_1 solution to an overdetermined linear system. SIAM J. Numer. Anal. **15**(2), 224–240 (1978)
5. M.S. Bazaraa, H.D. Sherali, C.M. Shetty, *Nonlinear Programming: Theory and Algorithms*, 3rd edn. (Wiley, New York, 2006)
6. R.E. Bellman, *Dynamic Programming* (Princeton University Press, Princeton, 1957)
7. R.E. Bellman, S.E. Dreyfus, *Applied Dynamic Programming* (Princeton University Press, Princeton, 1962)
8. P. Berman, N. Kovoor, P.M. Pardalos, Algorithms for the least distance problem, in *Complexity in Numerical Optimization*, ed. by P.M. Pardalos (World Scientific Publishing, Co., Singapore, 1993), pp. 33–56
9. D.P. Bertsekas, *Dynamic Programming. Deterministic and Stochastic Models* (Prentice Hall, Inc., Englewood Cliffs, 1987)
10. D.P. Bertsekas, A new class of incremental gradient methods for least squares problems. SIAM J. Optim. **7**(4), 913–926 (1997)
11. G.R. Bitran, A.C. Hax, Disaggregation and resource allocation using convex knapsack problems with bounded variables. Manag. Sci. **27**(4), 431–441 (1981)
12. Å. Björck, A direct method for sparse least squares problems with lower and upper bounds. Numer. Math. **54**(1), 19–32 (1988)
13. Å. Björck, *Numerical Methods for Least Squares Problems* (SIAM, Philadelphia, 1996)
14. J.R. Brown, The knapsack sharing problem. Oper. Res. **27**(2), 341–355 (1979)
15. J.R. Brown, Solving knapsack sharing with general tradeoff functions. Math. Program. **51**(1), 55–73 (1991)
16. J.R. Brown, Bounded knapsack sharing. Math. Program. **67**(3), 343–382 (1994)
17. P. Brucker, An $\mathcal{O}(n)$ algorithm for quadratic knapsack problems. Oper. Res. Lett. **3**(3), 163–166 (1984)
18. P. Calamai, A. Conn, A stable algorithm for solving the multifacility location problem involving Euclidean distances. SIAM J. Sci. Stat. Comput. **1**(4), 512–526 (1980)

© The Editor(s) (if applicable) and The Author(s), under exclusive license to Springer Nature Switzerland AG 2021

S. M. Stefanov, *Separable Optimization*, Springer Optimization and Its Applications 177,
https://doi.org/10.1007/978-3-030-78401-0

19. P. Calamai, A. Conn, A projected Newton method for ℓ_p norm location problems. Math. Program. **38**(1), 75–109 (1987)
20. P.H. Calamai, J.J. Moré, Quasi-Newton updates with bounds. SIAM J. Numer. Anal. **24**(6), 1434–1441 (1987)
21. E. Cavazzuti, J. Morgan, Well-posed saddle point problems, in *Optimization: Theory and Algorithms*. Lecture Notes in Pure and Applied Mathematics, vol. 86, ed. by J.-B. Hiriart-Urruty, W. Oettli, J. Stoer (Marcel Dekker, New York, 1983), pp. 61–76
22. J.A. Chaleton, D.W. Hearn, T.J. Lowe, A subgradient algorithm for certain minimax and minisum problems. Math. Program. **15**(1), 130–145 (1978)
23. A. Charnes, W.W. Cooper, Nonlinear power of adjacent extreme point methods in linear programming. Econometrica **25**(1), 132–153 (1957)
24. F.H. Clarke, Necessary conditions for nonsmooth problems in optimal control and the calculus of variations. Ph.D. thesis, University of Washington, Seattle (1973)
25. F.H. Clarke, A new approach to Lagrange multipliers. Math. Oper. Res. **1**(2), 165–174 (1976)
26. F.H. Clarke, *Methods of Dynamic and Nonsmooth Optimization*. CBMS - NSF Regional Conference Series in Applied Mathematics, vol. 57 (SIAM, Philadelphia, 1989)
27. F.H. Clarke, *Optimization and Nonsmooth Analysis*. Classics in Applied Mathematics, vol. 5 (SIAM, Philadelphia, 1990)
28. T.F. Coleman, L.A. Hulbert, A direct active set algorithm for large sparse quadratic programs with simple bounds. Math. Program. **45**(3), 373–406 (1989)
29. T.F. Coleman, L.A. Hulbert, A globally and superlinearly convergent algorithm for convex quadratic programs with simple bounds. SIAM J. Optim. **3**(2), 298–321 (1993)
30. T.F. Coleman, Y. Li, A global and quadratically-convergent method for linear ℓ_∞ problems. SIAM J. Numer. Anal. **29**(4), 1166–1186 (1992)
31. T.F. Coleman, Y. Li, A globally and quadratically convergent affine scaling method for linear ℓ_1 problems. Math. Program. **56**(1–3), 189–222 (1992)
32. R.W. Cottle, S.G. Duvall, K. Zikan, A Lagrangean relaxation algorithm for the constrained matrix problem. Nav. Res. Logist. Q. **33**(1), 55–76 (1986)
33. K. Dahiya, S.K. Suneja, V. Verma, Convex programming with single separable constraint and bounded variables. Comput. Optim. Appl. Int. J. **36**(1), 67–82 (2007)
34. G.B. Dantzig, S.M. Johnson, W.B. White, A linear programming approach to the chemical equilibrium problem. Manag. Sci. **5**(1), 38–43 (1958)
35. R.S. Dembo, U. Tulowitzki, On the minimization of quadratic functions subject to box constraints. Working Paper Series B # 71, School of Organization and Management, Yale University, New Haven (1983)
36. V.F. Demyanov, L.V. Vassiliev, *Nondifferentiable Optimization* (Nauka, Moscow, 1981) (in Russian)
37. E. Denardo, *Dynamic Programming. Models and Applications*, 2nd edn. (Dover Publications, Inc., New York, 2003)
38. A.L. Dontchev, T. Zolezzi, *Well-Posed Optimization Problems*. Lecture Notes in Mathematics, vol. 1543 (Springer, Berlin, 1993)
39. S.E. Dreyfus, A.M. Law, *The Art and Theory of Dynamic Programming* (Academic Press, New York, 1977)
40. F.D.J. Dunstan, An algorithm for solving a resource allocation problem. Oper. Res. Q. **28**(4), Part 1, 839–851 (1977)
41. J. Dussault, J. Ferland, B. Lemaire, Convex quadratic programming with one constraint and bounded variables. Math. Program. **36**(1), 90–104 (1986)
42. C. Edirisinghe, D. Atkins, P. Iyogun, Bounds on stochastic convex allocation problems. Appl. Stoch. Models Data Anal. **10**(4), 123–140 (1994)
43. I. Ekeland, On the variational principle. J. Math. Anal. Appl. **47**(2), 324–353 (1974)
44. Yu. Ermoliev, *Methods of Stochastic Programming* (Nauka, Moscow, 1976) (in Russian)
45. Yu. Ermoliev, A. Gaivoronski, Stochastic quasigradient methods and their implementation. Working Paper A-2361, IIASA, Laxenburg (1984)

46. F. Facchinei, J. Júdice, J. Soares, An active set Newton algorithm for large-scale nonlinear programs with box constraints. SIAM J. Optim. **8**(1), 158–186 (1998)
47. J.A. Ferland, B. Lemaire, P. Robert, Analytic solutions for nonlinear programs with one or two equality constraints. Publication # 285, Departement d'informatique et de recherche operationnelle, Université de Montréal, Montréal (1978)
48. J. Fisher, An algorithm for discrete linear ℓ_p approximation. Numer. Math. **38**(1), 129–139 (1981)
49. R.M. Freund, M.J. Todd, Barrier functions and interior-point algorithms for linear programming with zero-, one-, or two-sided bounds on the variables. Math. Oper. Res. **20**(2), 415–440 (1995)
50. A. Gaivoronski, Stochastic quasigradient methods and their implementation, in *Numerical Techniques for Stochastic Optimization*. ed. by Yu. Ermoliev, R.J.-B. Wets (Springer, Berlin, 1988), pp. 313–351
51. E.G. Gol'shtein, Generalized gradient method for finding saddle points. Econ. Math. Methods **VIII**(4), 569–579 (1972) (in Russian)
52. E.G. Gol'shtein, D.B. Yudin, *New Trends in Linear Programming* (Soviet Radio, Moscow, 1966) (in Russian)
53. E.G. Gol'shtein, N.V. Tret'iakov, Modified Lagrangians. Econ. Math. Methods **X**(3), 568–591 (1974) (in Russian)
54. J. Hadamard, Sur les problemes aux dérivees partielles et leur signification physique. Princet. Univ. Bull. **13**, 49–52 (1902)
55. J. Hadamard, *Lectures on Cauchy's Problem in Linear Partial Differential Equations* (Dover Publications, New York, 1953)
56. C.-G. Han, P.M. Pardalos, Y. Ye, Computational aspects of an interior point algorithm for quadratic programming problems with box constraints, in *Large-Scale Numerical Optimization*. ed. by T.F. Coleman, Y. Ye (SIAM, Philadelphia, 1990), pp. 92–112
57. R. Helgason, J. Kennington, H. Lall, A polynomially bounded algorithm for a singly constrained quadratic program. Math. Program. **18**(1), 338–343 (1980)
58. G.T. Herman, A. Lent, A family of iterative quadratic optimization algorithms for pairs of inequalities, with application in diagnostic radiology. Math. Program. Study **9**, 15–29 (1978)
59. T. Ibaraki, N. Katoh, *Resource Allocation Problems. Algorithmic Approaches* (The MIT Press, Cambridge, 1988)
60. T.R. Jefferson, C.H. Scott, Avenues of geometric programming. NZOR **6**(2), 109–136 (1978)
61. T.R. Jefferson, C.H. Scott, Avenues of geometric programming - Applications. NZOR **7**(1), 51–68 (1979)
62. T.R. Jefferson, C.H. Scott, The analysis of entropy models with equality and inequality constraints. Transp. Res., Pergamon Press Ltd. **13B**, 123–132 (1979)
63. V. Jeyakumar, G.Y. Li, New strong duality results for convex programs with separable constraints. Eur. J. Oper. Res. **207**(3), 1203–1209 (2010)
64. J.J. Júdice, F.M. Pires, Direct methods for convex quadratic programs subject to box constraints. Technical report, Departamento de Matemática, Universidade de Coimbra, Coimbra (1989)
65. P. Kall, S.W. Wallace, *Stochastic Programming* (Wiley, Chichester, 1994)
66. L.V. Kantorovich, G.P. Akilov, *Functional Analysis*, 3rd edn. (Nauka, Moscow, 1983) (in Russian)
67. V.G. Karmanov, *Mathematical Programming*, 3rd edn. (Nauka, Moscow, 1986) (in Russian)
68. N. Katoh, T. Ibaraki, H. Mine, A polynomial time algorithm for the resource allocation problem with a convex objective function. J. Oper. Res. Soc. **30**(5), 449–455 (1979)
69. P. Kenderov, G. Hristov, A. Dontchev, *Mathematical Programming* (Kliment Ohridski Sofia University Press, Sofia, 1989) (in Bulgarian)
70. N.P. Korneichuk, *Extremum Problems of Approximation Theory* (Nauka, Moscow, 1976) (in Russian)
71. C.L. Lawson, R.J. Hanson, *Solving Least Squares Problems* (SIAM, Philadelphia, 1995)
72. Y. Li, A globally convergent method for ℓ_p problems. SIAM J. Optim. **3**(3), 609–629 (1993)

73. Y.Y. Lin, J.-S. Pang, Iterative methods for large convex quadratic programs: a survey. SIAM J. Control Optim. **25**(2), 383–411 (1987)
74. P. Lötstedt, Solving the minimal least squares problem subject to bounds on the variables. BIT **24**(2), 206–224 (1984)
75. H. Luss, An algorithm for separable non-linear minimax problems. Oper. Res. Lett. **6**(4), 159–162 (1987)
76. H. Luss, A nonlinear minimax allocation problem with multiple knapsack constraints. Oper. Res. Lett. **10**(4), 183–187 (1991)
77. H. Luss, S.K. Gupta, Allocation of marketing effort among P substitutional products in N territories. Oper. Res. Q. **25**(1), 77–88 (1974)
78. H. Luss, S.K. Gupta, Allocation of effort resources among competing activities. Oper. Res. **23**(2), 360–366 (1975)
79. H. Luss, D.R. Smith, Resource allocation among competing activities: a lexicographic minimax approach. Oper. Res. Lett. **5**(5), 227–231 (1986)
80. R. Luus, *Iterative Dynamic Programming* (Chapman & Hall/CRC, Boca Raton, 2000)
81. O.L. Mangasarian, *Nonlinear Programming*. Classics in Applied Mathematics, vol. 10 (SIAM, Philadelphia, 1994)
82. H.M. Markowitz, A.S. Manne, On the solution of discrete programming problems. Econometrica **25**(1), 84–110 (1957)
83. B. Martos, *Nonlinear Programming: Theory and Methods* (North-Holland Publishing Company; American Elsevier Publishing Co., Inc., Amsterdam; New York, 1975)
84. R.K. McCord, Minimization with one linear equality constraint and bounds on the variables. Technical report SOL 79-20, System Optimization Laboratory, Department of Operations Research, Stanford University, Stanford (1979)
85. G. Merle, H. Späth, Computational experience with discrete ℓ_p-approximation. Computing **12**(4), 315–321 (1974)
86. R.R. Meyer, Two-segment separable programming. Manag. Sci. **25**(4), 385–395 (1979)
87. R.R. Meyer, Computational aspects of two-segment separable programming. Computer Sciences Technical report # 382, University of Wisconsin – Madison (1980)
88. C. Michelot, A finite algorithm for finding the projection of a point onto the canonical simplex of $\mathbb{R}^n$. J. Optim. Theory Appl. **50**(1), 195–200 (1986)
89. C.E. Miller, The simplex method for local separable programming, in *Recent Advances in Mathematical Programming*. ed. by R.L. Graves, P. Wolfe (McGraw-Hill, New York, 1963), pp. 89–100
90. C.L. Monma, A. Schrijver, M.J. Todd, V.K. Wei, Convex resource allocation problems on directed acyclic graphs: duality, complexity, special cases, and extensions. Math. Oper. Res. **15**(4), 736–748 (1990)
91. R.D.C. Monteiro, Y. Wang, Trust region affine scaling algorithms for linearly constrained convex and concave programs. Math. Program. **80**(3), 283–313 (1998)
92. J.J. Moré, G. Toraldo, Algorithms for bound constrained quadratic programming problems. Numer. Math. **55**(4), 377–400 (1989)
93. J.J. Moré, S.A. Vavasis, On the solution of concave knapsack problems. Math. Program. **49**(3), 397–411 (1991)
94. G.L. Nemhauser, L.A. Wolsey, *Integer and Combinatorial Optimization* (Wiley, New York, 1999)
95. M.L. Overton, A quadratically convergent method for minimizing a sum of Euclidean norms. Math. Program. **27**(1), 34–63 (1983)
96. A. Padakandla, R. Sundaresan, Separable convex optimization problems with linear ascending constraints. SIAM J. Optim. **20**(3), 1185–1204 (2010)
97. M.W. Padberg, $(1, k)-$ configurations and facets for packing problems. Math. Program. **18**(1), 94–99 (1980)
98. P.M. Pardalos, N. Kovoor, An algorithm for a singly constrained class of quadratic programs subject to upper and lower bounds. Math. Program. **46**(3), 321–328 (1990)

99. P.M. Pardalos, Y. Ye, C.-G. Han, Algorithms for the solution of quadratic knapsack problems. Linear Algebra Appl. **152**, 69–91 (1991)
100. T. Parhomenko, Minimization of the sum of absolute values of linear functions under linear constraints, in *Mathematics and Education in Mathematics* (Bulgarian Academy of Sciences Press, Sofia, 1990), pp. 378–381 (in Russian)
101. M. Patriksson, A survey on the continuous nonlinear resource allocation problem. Eur. J. Oper. Res. **185**(1), 1–46 (2008)
102. M. Patriksson, C. Strömberg, Algorithms for the continuous nonlinear resource allocation problem - new implementations and numerical studies. Eur. J. Oper. Res. **243**(3), 703–722 (2015)
103. B.N. Pshenichnii, *Convex Analysis and Extremum Problems* (Nauka, Moscow, 1980) (in Russian)
104. A.W. Roberts, D.E. Varberg, Another proof that convex functions are locally Lipschitz. Am. Math. Mon. **81**(9), 1014–1016 (1974)
105. A.G. Robinson, N. Jiang, C.S. Lerme, On the continuous quadratic knapsack problem. Math. Program. **55**(1), 99–108 (1992)
106. P.L. Rocha, M.G. Ravetti, G.R. Mateus, P.M. Pardalos, Exact algorithms for a scheduling problem with unrelated parallel machines and sequence and machine-dependent setup times. Comput. Oper. Res. **35**(4), 1250–1264 (2008)
107. R.T. Rockafellar, *Convex Analysis* (Princeton University Press, Princeton, 1970)
108. R.T. Rockafellar, *Conjugate Duality and Optimization* (SIAM, Philadelphia, 1974)
109. R.T. Rockafellar, R.J.-B. Wets, A note about projections in the implementation of stochastic quasigradient methods, in *Numerical Techniques for Stochastic Optimization*. ed. by Yu. Ermoliev, R.J.-B. Wets (Springer, Berlin, 1988), pp. 385–392
110. R.T. Rockafellar, R.J.-B. Wets, *Variational Analysis*, 2nd edn. (Springer, Berlin, 2004)
111. A. Schrijver, *Theory of Linear and Integer Programming* (Wiley, Chichester, 1998)
112. A. Schrijver, *Combinatorial Optimization. Polyhedra and Efficiency, vol. A: Paths, Flows, Matchings* (Springer, Berlin, 2003)
113. A. Schrijver, *Combinatorial Optimization. Polyhedra and Efficiency, vol. B: Matroids, Trees, Stable Sets* (Springer, Berlin, 2003)
114. A. Schrijver, *Combinatorial Optimization. Polyhedra and Efficiency, vol. C: Disjoint Paths, Hypergraphs* (Springer, Berlin, 2003)
115. C.H. Scott, T.R. Jefferson, P. Kerdvonbundit, Geometric programming duality applied to the resource allocation problem. INFOR **17**(2), 133–137 (1979)
116. C.H. Scott, T.R. Jefferson, Convex dual for quadratic concave fractional programming. J. Optim. Theory Appl. **91**(1), 115–122 (1996)
117. M. Sniedovich, *Dynamic Programming* (Marcel Dekker, New York, 1991)
118. S.M. Stefanov, Facility location problems and methods for solving them. Thesis, Sofia University, Sofia (1989) (in Bulgarian)
119. S.M. Stefanov, Stochastic quasigradient methods - some theoretical and numerical aspects, *Mathematics and Its Applications* (SWU, 1993), pp. 121–135
120. S.M. Stefanov, Convex knapsack problem with bounded variables, in *Applications of Mathematics in Engineering*, vol. 19, ed. by D. Ivanchev (Sofia, 1994), pp. 189–194
121. S.M. Stefanov, Valid inequalities and integrality of the knapsack polytope, *Discrete Mathematics and Applications*. Research in Mathematics, vol. 5 (1995), pp. 220–228
122. S.M. Stefanov, Some applications of nonsmooth optimization to approximation theory, in *Applications of Mathematics in Engineering and Business*, vol. 22, ed. by N. Shopolov (Sofia, 1997), pp. 137–142
123. S.M. Stefanov, Valid inequalities and cutting planes for some polytopes. Math. Inequalities Appl. **1**(2), 285–294 (1998)
124. S.M. Stefanov, On the implementation of stochastic quasigradient methods to some facility location problems. YUJOR **10**(2), 235–256 (2000)
125. S.M. Stefanov, Convex separable minimization subject to bounded variables. Comput. Optim. Appl. Int. J. **18**(1), 27–48 (2001)

126. S.M. Stefanov, Convex separable minimization problem with a linear constraint and bounds on the variables, in *Applications of Mathematics in Engineering and Economics*, vol. 27 (Heron Press, Sofia, 2002), pp. 392–402
127. S.M. Stefanov, A Lagrangian dual method for solving variational inequalities. Math. Inequalities Appl. **5**(3), 597–608 (2002)
128. S.M. Stefanov, Method for solving a convex integer programming problem. Int. J. Math. Math. Sci. **2003**(44), 2829–2834 (2003)
129. S.M. Stefanov, Convex quadratic minimization subject to a linear constraint and box constraints. Appl. Math. Res. Express AMRX, Oxford University Press **2004**(1), 17–42 (2004)
130. S.M. Stefanov, Polynomial algorithms for projecting a point onto a region defined by a linear constraint and box constraints in $\mathbb{R}^n$. J. Appl. Math. **2004**(5), 409–431 (2004)
131. S.M. Stefanov, Convex separable minimization problems with a linear constraint and bounded variables. Int. J. Math. Math. Sci. **2005**(9), 1339–1363 (2005)
132. S.M. Stefanov, An efficient method for minimizing a convex separable logarithmic function subject to a convex inequality constraint or linear equality constraint. J. Appl. Math. Decis. Sci. (now Adv. Decis. Sci.) **2006**(1), 1–19 (2006), Article ID 89307
133. S.M. Stefanov, Minimization of a convex linear-fractional separable function subject to a convex inequality constraint or linear equality constraint and bounds on the variables. Appl. Math. Res. Express AMRX, Oxford University Press **2006**(4), 1–24 (2006), Article ID 36581
134. S.M. Stefanov, Minimizing a convex separable exponential function subject to linear equality constraint and bounded variables. J. Interdiscip. Math. **9**(1), 207–226 (2006)
135. S.M. Stefanov, On the solution of variational inequality problems by using cutting plane methods. Math. Inequalities Appl. **10**(2), 427–436 (2007)
136. S.M. Stefanov, Applications of convex separable unconstrained nonsmooth optimization to numerical approximation with respect to ℓ_1 and ℓ_∞ norms. J. Interdiscip. Math. **10**(6), 767–788 (2007)
137. S.M. Stefanov, Convex separable minimization with box constraints. Proc. Appl. Math. Mech. PAMM, Wiley-Interscience **7**(1), 2060045–2060046 (Special Issue: *Sixth International Congress on Industrial and Applied Mathematics ICIAM-07 and GAMM Annual Meeting, July 2007, Zurich, Switzerland*) (2008)
138. S.M. Stefanov, Convex separable minimization with bounded variables, in *Proceedings of 8th World Congress on Structural and Multidisciplinary Optimization WCSMO-8 (June 2009, Lisbon, Portugal)* (2009), pp. 1– 10
139. S.M. Stefanov, Minimization of a strictly convex separable function subject to convex separable inequality constraint and bounds on the variables. J. Interdiscip. Math. **12**(5), 647–673 (2009)
140. S.M. Stefanov, Solution of some convex separable resource allocation and production planning problems with bounds on the variables. J. Interdiscip. Math. **13**(5), 541–569 (2010)
141. S.M. Stefanov, Valid inequalities, cutting planes and integrality of the knapsack polytope. J. Interdiscip. Math. **14**(4), 389–406 (2011)
142. S.M. Stefanov, Well-posedness and primal-dual analysis of some convex separable optimization problems. Adv. Oper. Res. **2013**, 1–10 (2013), Article ID 279030
143. S.M. Stefanov, On the application of iterative methods of nondifferentiable optimization to some problems of approximation theory. Math. Probl. Eng. **2014**, 1–10 (2014), Article ID 165701
144. S.M. Stefanov, On the solution of multidimensional convex separable continuous knapsack problem with bounded variables. Eur. J. Oper. Res. **247**(2), 366–369 (2015)
145. S.M. Stefanov, Strictly convex separable optimization with linear equality constraints and bounded variables. J. Stat. Manag. Syst. **21**(2), 261–272 (2018)
146. S.M. Stefanov, On the solution of quadratic programming problem with a feasible region defined as a Minkowski sum of a compact set and finitely generated convex closed cone. J. Inf. Optim. Sci. **39**(6), 1223–1230 (2018)
147. S.M. Stefanov, Applications of iterative nondifferentiable optimization to some engineering approximation problems, in *EngOpt2018 Proceedings of the 6th International Conference on Engineering Optimization* (Springer, 2019), pp. 80–92

148. S.M. Stefanov, On the Cauchy-Schwarz inequality approach for solving a quadratic optimization problem. J. Inf. Optim. Sci. **40**(4), 973–981 (2019)
149. S.M. Stefanov, Characterization of the optimal solution of the convex generalized nonlinear transportation problem. J. Interdiscip. Math. **22**(5), 745–756 (2019)
150. S.M. Stefanov, Theorems of the alternative revisited. J. Interdiscip. Math. **22**(5), 787–796 (2019)
151. S.M. Stefanov, Characterization of the optimal solution of the convex separable continuous knapsack problem and related problems. J. Inf. Optim. Sci. **42**(1), 1–16 (2021)
152. S.M. Stefanov, On the numerical solution of separable stochastic inventory control problems. J. Inf. Optim. Sci. **42**(3), 533–561 (2021)
153. S.M. Stefanov, Numerical solution of some systems of nonlinear algebraic equations. J. Interdiscip. Math. **24**(6), 1545–1564 (2021)
154. S.M. Stefanov, Numerical solution of systems of nonlinear equations defined by convex functions. J. Interdiscip. Math. published online
155. H.A. Taha, *Operations Research*, 10th edn. (Pearson, London, 2017)
156. T.V. Tchemisova, On the constructive solution of convex programming problems in separable form. J. Math. Sci. **120**(1), 1016–1031 (2004)
157. L.S. Thakur, Error analysis for convex separable programs: the piecewise linear approximation and the bounds on the optimal objective value. SIAM J. Appl. Math. **34**(4), 704–714 (1978)
158. M.C. Trevisan, Résolution numérique de problèmes d'optimisation à une seule contraine linéaire. Application au problème de Thomas-Fermi. Thèse de 3ième cycle, Université des Sciences et Techniques du Languedoc, Montpellier (1980)
159. A.N. Tychonov, Stability of inverse problems. Dokl. Akad. Nauk SSSR **39**(5), 195–198 (1943) (in Russian)
160. A.N. Tychonov, Solution of ill-posed problems and the regularization method. Dokl. Akad. Nauk SSSR **151**(3), 501–504 (1963) (in Russian)
161. A.N. Tychonov, On the regularization of ill-posed problems. Dokl. Akad. Nauk SSSR **153**(1), 49–52 (1963) (in Russian)
162. A.N. Tychonov, On the regularization methods for optimal control problems. Dokl. Akad. Nauk SSSR **162**(4), 763–765 (1965) (in Russian)
163. A.N. Tychonov, Ill-posed problems of optimal planning and stable methods for solving them. Dokl. Akad. Nauk SSSR **164**(3), 507–510 (1965) (in Russian)
164. A.N. Tychonov, Ill-posed optimal planning problems. USSR Comput. Math. Math. Phys. **6**(1), 81–89 (1966) (in Russian)
165. A.N. Tychonov, On the stability of the functional optimization problems. USSR Comput. Math. Math. Phys. **6**(4), 631–634 (1966) (in Russian)
166. A.N. Tikhonov, V.Y. Arsenin, *Solutions of Ill-Posed Problems* (Wiley, New York, 1977)
167. A.N. Tychonov, V.Y. Arsenin, *Methods for the Solution of Ill-Posed Problems* (Nauka, Moscow, 1986) (in Russian)
168. S. Urasiev, Adaptive stochastic quasigradient procedures, in *Numerical Techniques for Stochastic Optimization*. ed. by Yu. Ermoliev, R.J.-B. Wets (Springer, Berlin, 1988), pp. 373–384
169. S.A. Vavasis, Local minima for indefinite quadratic knapsack problems. Math. Program. **54**(2), 127–153 (1992)
170. G.A. Watson, On two methods for discrete ℓ_p approximation. Computing **18**(3), 263–266 (1977)
171. C. Wilkinson, S.K. Gupta, Allocating promotional effort to competing activities: a dynamic programming approach, in *IFORS Conference, Venice* (1969), pp. 419–432
172. J.M. Wolfe, On the convergence of an algorithm for discrete ℓ_p approximation. Numer. Math. **32**(4), 439–459 (1979)
173. P. Wolfe, Methods of nonlinear programming, in *Recent Advances in Mathematical Programming*. ed. by R.L. Graves, P. Wolfe (McGraw-Hill, New York, 1963), pp. 67–86
174. P. Wolfe, Algorithm for a least-distance programming problem. Math. Program. Study **1**, 190–205 (1974)

175. G. Xue, Y. Ye, An efficient algorithm for minimizing a sum of Euclidean norms with applications. SIAM J. Optim. **7**(4), 1017–1036 (1997)
176. E.K. Yang, J.W. Tolle, A class of methods for solving large convex quadratic programs subject to box constraints. Technical report, Department of Operations Research, University of North Carolina, Chapel Hill (1988)
177. V.A. Yemelichev, M.M. Kovalev, M.K. Kravtsov, *Polytopes, Graphs and Optimisation* (Cambridge University Press, Cambridge, 1984)
178. K. Yosida, *Functional Analysis*, 6th edn. (Springer, Berlin, 1980)
179. Yu.P. Zaichenko, *Operations Research* (Slovo, Kiev, 2003) (in Russian)
180. P.H. Zipkin, Error bounds for aggregated convex transportation problems. Working Paper, Graduate School of Business, Columbia University, New York (1978)
181. P.H. Zipkin, Simple ranking methods for allocation of one resource. Manag. Sci. **26**(1), 34–43 (1980)

Notation

$=$	is equal to, equals
$\neq$	not equal to
$\equiv$	identical with, identically equal to
$\stackrel{\text{def}}{=}$	equal by definition
$:=$	the left-hand side is notation of the right-hand side
$=:$	the right-hand side is notation of the left-hand side
$\leq$	less than or equal to
$\geq$	greater than or equal to
$\exists$	existential quantifier, "there is/are", "there exist(s)"
$\forall$	universal quantifier, "for every" or "for all"
$\blacksquare$, Q.E.D.	end of proof, Quod Erat Demonstrandum
∞	infinity
$\mathbf{R}$	the real numbers
$\overline{\mathbf{R}} = \mathbf{R} \cup \{+\infty\}$	the extended real numbers
$\mathbf{Z}$	the integers
0 $(\mathbf{0})$	zero (zero vector of appropriate dimension)
$[x]$	the largest integer less than or equal to x
$[a, b]$	closed line segment
$[a, b)$, $(a, b]$	closed-open and open-closed line segments
(a, b)	open line segment
$x \in X$	x is a member of X, x belongs to X
$x \notin X$	x is not a member of X, x does not belong to X

© The Editor(s) (if applicable) and The Author(s), under exclusive license
to Springer Nature Switzerland AG 2021

S. M. Stefanov, *Separable Optimization*, Springer Optimization and Its Applications 177,
https://doi.org/10.1007/978-3-030-78401-0

$\mathbf{x} = (x_1, \ldots, x_n)$	a 1 by n vector of variables
$\mathbf{x} \geq \mathbf{y}$	$x_j \geq y_j$ for all $j = 1, \ldots, n$. In particular, $\mathbf{x} \geq \mathbf{0}$ means $x_j \geq 0,\ j = 1, \ldots, n$
$\mathbf{R}^n$	the n-dimensional real space
$\mathbf{R}^n_+$	the nonnegative orthant $\{\mathbf{x} \in \mathbf{R}^n : \mathbf{x} \geq \mathbf{0}\}$ of $\mathbf{R}^n$
$\mathbf{R}^n_{++}$	the positive orthant $\{\mathbf{x} \in \mathbf{R}^n : \mathbf{x} > \mathbf{0}\}$ of $\mathbf{R}^n$
$\mathbf{Z}^n$	the space of n-dimensional vectors whose components are integers
A	$(\in \mathbf{R}^{m \times n})$ an $m \times n$ matrix
I	the unit matrix of appropriate dimension (the identity mapping, identity operator)
$\mathcal{M}_n$	the set of all $n \times n$ square matrices over a field F
$X \subseteq Y\ (X \subset Y)$	X is a subset (proper subset) of Y, set X is contained (properly contained) in set Y
$\|X\|$	the number of elements in X, cardinality of X
$\cap X_i$	intersection of all X_i
$\cup X_i$	union of all X_i
$\emptyset$	the empty set
$X \setminus Y$	difference of sets X and Y, $X \setminus Y = \{x : x \in X, x \notin Y\}$
$X \times Y$	Cartesian product of the sets X and Y, set of all (x, y) such that $x \in X$, $y \in Y$
∂X	boundary of X
$\overline{X}$, cl X	closure of X
int X	interior of X
ri X	relative interior of X
$\Pi_X(\mathbf{y})$	projection mapping, projection of $\mathbf{y}$ onto the set X
$f : X \to Y$	function mapping
$\lim_{x \to a} f(x) = b$	b is the limit of $f(x)$ as x approaches a
$\lim_{x \to -a} f(x)$	limit of $f(x)$ as x approaches a from the left
$t \to +0,\ t \downarrow 0$	t approaches 0 from the right
$f(n) = \mathcal{O}(n)$	$\dfrac{f(n)}{n}$ is bounded as $n \to \infty$
$f(n) = o(n)$	$\dfrac{f(n)}{n}$ approaches 0
$\mathbf{f}'(\mathbf{x}),\ \nabla f(\mathbf{x})$	the gradient of $f(\mathbf{x})$
$\frac{\partial f}{\partial \mathbf{x}},\ \mathbf{f}'_{\mathbf{x}}$	partial derivative of f with respect to $\mathbf{x}$

$\widehat{\mathbf{f}}(\mathbf{x})$	subgradient of f at $\mathbf{x}$
$\partial f(\mathbf{x})$	subgradient set, subdifferential of f at $\mathbf{x}$
$f'(\mathbf{x}_0; \mathbf{d})$, $\frac{\partial f}{\partial \mathbf{d}}(\mathbf{x}_0)$	directional derivative of f at $\mathbf{x}_0$ along $\mathbf{d}$
$F'(\mathbf{x})$	the Jacobian of $F : \mathbb{R}^m \to \mathbb{R}^n$ at $\mathbf{x} \in \mathbb{R}^n$
$\nabla F(\mathbf{x}) = (F'(\mathbf{x}))^T$	the transposed Jacobian of $F : \mathbb{R}^m \to \mathbb{R}^n$ at $\mathbf{x} \in \mathbb{R}^n$, where $F(\mathbf{x}) = (F_1(\mathbf{x}), \dots, F_m(\mathbf{x}))$ and $\nabla F(\mathbf{x}) = \begin{pmatrix} \nabla F_1(\mathbf{x})^T \\ \vdots \\ \nabla F_m(\mathbf{x})^T \end{pmatrix}_{m \times n}$
e^x	the exponential function
$C^k(X)$	the set of all functions defined in X which have continuous partial derivatives of order up to and including k
dom f	effective domain of f
epi f	epigraph of f
$\{\mathbf{x}_k\}$	sequence with terms $\mathbf{x}_k$
$\langle \mathbf{x}, \mathbf{y} \rangle = \sum_{i=1}^{n} x_i y_i$	canonical inner (scalar, dot) product of $\mathbf{x}$ and $\mathbf{y}$
ℓ_p, $1 \le p \le \infty$	the totality of all sequences $\{a_k\}_{k=0}^{\infty}$ such that $\left(\sum_{k=0}^{\infty} \lvert a_k \rvert^p \right)^{\frac{1}{p}} < \infty$
$\lVert . \rVert_{\ell_1}$	the 1-norm (ℓ_1-norm). For $\mathbf{x} \in \mathbb{R}^n$, $\lVert \mathbf{x} \rVert_{\ell_1} = \sum_{i=1}^{n} \lvert x_i \rvert$
$\lVert . \rVert$, $\lVert . \rVert_{\ell_2}$	the usual Euclidean norm (ℓ_2-norm). For $\mathbf{x} \in \mathbb{R}^n$, $\lVert \mathbf{x} \rVert_{\ell_2} = \langle \mathbf{x}, \mathbf{x} \rangle^{\frac{1}{2}} = \left(\sum_{i=1}^{n} x_i^2 \right)^{\frac{1}{2}}$
$\lVert . \rVert_{\ell_\infty}$	the ∞-norm (ℓ_∞-norm). For $\mathbf{x} \in \mathbb{R}^n$, $\lVert \mathbf{x} \rVert_{\ell_\infty} = \max_{i=1,\dots,n} \lvert x_i \rvert$. (Note that $\lVert \mathbf{x} \rVert_{\ell_\infty} \le \lVert \mathbf{x} \rVert_{\ell_2} \le \lVert \mathbf{x} \rVert_{\ell_1}$ for any vector $\mathbf{x}$)
$B(\mathbf{x}_0; r)$	open ball of radius r around the point $\mathbf{x}_0$ in $\mathbb{R}^n$
$\overline{B}(\mathbf{x}_0; r)$	closed ball of radius r around the point $\mathbf{x}_0$ in $\mathbb{R}^n$

$\mathbf{x}_k \to \mathbf{y}$	the sequence $\{\mathbf{x}_k\}$ converges (in norm) to $\mathbf{y}$
min	minimum
max	maximum
inf	infimum, greatest lower bound
sup	supremum, least upper bound
lim inf, $\underline{\lim}$	limit inferior
lim sup, $\overline{\lim}$	limit superior
$\min_{\mathbf{x}\in X} f(\mathbf{x})$, $\min\{f(\mathbf{x}) : \mathbf{x} \in X\}$, $f(\mathbf{x}) \to \min_{\mathbf{x}\in X}$	minimum of $f(\mathbf{x})$ when $\mathbf{x}$ is in the set X
$\arg\min_{\mathbf{x}\in X} f(\mathbf{x})$, $\arg\min\{f(\mathbf{x}) : \mathbf{x} \in X\}$	the set of optimal solutions to the above problem
$dist\ (\mathbf{x}, \mathbf{y})$	the distance between the points $\mathbf{x}$ and $\mathbf{y}$
$\delta(\mathbf{x}, X) = \min_{\mathbf{z}\in X} \lVert\mathbf{x} - \mathbf{z}\rVert$	the distance from $\mathbf{x}$ to the set X
$L(\mathbf{x}, \mathbf{u}, \mathbf{v})$	the Lagrangian function $f(\mathbf{x}) + \sum_{i=1}^{m} u_i g_i(\mathbf{x}) + \sum_{j=1}^{p} v_j h_j(\mathbf{x})$ associated with the problem: min $f(\mathbf{x})$ subject to $g_i(\mathbf{x}) \le 0, i = 1, \ldots, m,$ $h_j(\mathbf{x}) = 0, j = 1, \ldots, p.$ The Lagrange multipliers u_i, $i = 1, \ldots, m$ are nonnegative.
$\mathbf{E}_\omega f(x, \omega)$	mathematical expectation of f with respect to ω
$\mathbf{E}_\omega(\xi/x_0, \ldots, x_k)$	conditional mathematical expectation of ξ with respect to ω subject to $x_0, \ldots, x_k$.

List of Statements

© The Editor(s) (if applicable) and The Author(s), under exclusive license to Springer Nature Switzerland AG 2021

S. M. Stefanov, *Separable Optimization*, Springer Optimization and Its Applications 177,
https://doi.org/10.1007/978-3-030-78401-0